S0-CCX-887

DESIGN of WATER-RESOURCE SYSTEMS

DESIGN of WATER-RESOURCE SYSTEMS

New Techniques for Relating Economic Objectives, Engineering Analysis, and Governmental Planning

—— ARTHUR MAASS

—— MAYNARD M. HUFSCHMIDT

—— ROBERT DORFMAN

—— HAROLD A. THOMAS, JR.

—— STEPHEN A. MARGLIN

—— GORDON MASKEW FAIR

HARVARD UNIVERSITY PRESS · CAMBRIDGE · MASSACHUSETTS · 1966

Second printing

Library of Congress Catalog Card Number 62–8181

Printed in the United States of America

Principal Authors

ARTHUR MAASS
MAYNARD M. HUFSCHMIDT
ROBERT DORFMAN
HAROLD A. THOMAS, JR.
STEPHEN A. MARGLIN
GORDON MASKEW FAIR

Other Authors

BLAIR T. BOWER
WILLIAM W. REEDY
DEWARD F. MANZER
MICHAEL P. BARNETT
MYRON B FIERING
PETER WATERMEYER

Contents

CONTENTS

Part II · METHODS AND TECHNIQUES

CONTENTS

Tables

TABLES

TABLES

Graphs

GRAPHS

GRAPHS

GRAPHS

DESIGN of WATER-RESOURCE SYSTEMS

1 Introduction

ARTHUR MAASS

THE purpose of this book is to improve the methodology of designing water-resource systems. In this respect it differs basically from the many volumes on water resources published in the last decade, which have been concerned mainly with governmental organization for development of resources, the economics of project evaluation, and the collection and evaluation of basic data.[1] Systematic research on the methodology of system design, such as underlies this book, has been neglected in spite of the marked growth during the last quarter-century of public interest in

[1] Books and reports concerned largely with organization include: Arthur Maass, *Muddy waters: the Army Engineers and the nation's rivers* (Harvard University Press, Cambridge, 1951); Charles McKinley, *Uncle Sam in the Pacific Northwest* (University of California Press, Berkeley, 1952); Gordon Clapp, *The TVA: an approach to the development of a region* (University of Chicago Press, Chicago, 1955); Henry Hart, *The dark Missouri* (University of Wisconsin Press, Madison, 1957); Roscoe Martin and associates, *River basin administration and the Delaware* (Syracuse University Press, Syracuse, 1960); U. S. Commission on Organization of the Executive Branch of the Government (First Hoover Commission), *Report on Department of the Interior* and *Task force report on natural resources* (1949); U. S. Missouri Basin Survey Commission, *Missouri: land and water* (1953); U. S. Commission on Organization of the Executive Branch of the Government (Second Hoover Commission), *Water resources and power* and *Task force report on water resources and power* (1955); U. S. Presidential Advisory Committee on Water Resources Policy, *Water resources policy* (1955).

Books concerned largely with economics of project evaluation include: Otto Eckstein, *Water-resource development, the economics of project evaluation* (Harvard University Press, Cambridge, 1958); Roland McKean, *Efficiency in government through systems analysis, with emphasis on water resources development* (John Wiley and Sons, New York, 1958); John Krutilla and Otto Eckstein, *Multiple purpose river development: studies in applied economic analysis* (Johns Hopkins University Press, Baltimore, 1958); Jack Hirshleifer, James C. De Haven and Jerome W. Milliman, *Water supply: economics, technology, and policy* (University of Chicago Press, Chicago, 1960).

Books concerned largely with basic data and water technology include: Walter Langbein and William Hoyt, *Water facts for the nation's future* (Ronald Press, New York, 1959); Edward Ackerman and George Löf, *Technology in American water development* (Johns Hopkins University Press, Baltimore, 1959).

In addition, the following are noteworthy: Report of the U. S. President's Water Resources Policy Commission, vol. 1, *A water policy for the American people* (U. S. Govt. Printing Office, Washington, 1950), which deals broadly with many aspects of water development; and Luna Leopold and Thomas Maddock, Jr., *The flood control controversy* (Ronald Press, New York, 1954), which alone of the books referenced in this note deals with system design, but largely in connection with upstream programs for flood control.

multiunit, multipurpose river developments. Indeed, few comprehensive statements on methods of design have appeared since the now-classic reports of Arthur Morgan and his associates of the Miami (Ohio) Conservancy District on their work during the second decade of this century.[2] Some important additions to their writings were made in the 1930's, to be sure, when the U. S. Government became interested in multipurpose river development,[3] and certain refinements did follow when high-speed digital computers became available to engineers and economists,[4] but on the whole considerably less attention has been paid to system design since 1935 than to many other aspects of water-resource development. This book reports the results of research on system design conducted over several years by the Harvard Water Program of the Graduate School of Public Administration.

THE FOUR STEPS OF DESIGN

The methodology of system design, as defined in this book, involves four related steps: identifying the objectives of design; translating these objectives into design criteria; using the criteria to devise plans for the development of specific water-resource systems that fulfill the criteria in the highest degree; and evaluating the consequences of the plans that have been developed.

IDENTIFYING AND EVALUATING OBJECTIVES

As we shall see, a choice of objectives is open to the community from among a variety of combinations of economic efficiency and development, income redistribution by region or economic class, and aesthetic and other essentially noneconomic values. The first step in design is to determine which of these alternatives reflects the consensus of the community concerned with river development. A broad framework for this step is given in Chapter 15, *System Design and the Political Process: A General Statement.* Many words have been written on the objectives that should prevail in given regions or countries; we take no part in such discussion,

[2] Miami Conservancy District, *Technical reports*, pts. 1–10 (1917–36).

[3] See Joseph Sivera Ransmeier, *The Tennessee Valley Authority: a case study in the economics of multiple purpose stream planning* (Vanderbilt University Press, Nashville, 1942), and references cited therein.

[4] For example, Humphrey Morrice and William Allan, "Planning for the ultimate hydraulic development of the Nile Valley," *Proc. Inst. Civil Engrs. 14*, 101 (1959).

because it has been one of the principal aims of the Harvard Water Program to improve the methodology of system design in such a way that it will meet any reasonable economic objectives within any reasonable institutional constraints.

For example, let us consider federal irrigation projects in the southwest United States. The policy-makers, to be defined in Chapter 15, may select as the objective for the projects the most efficient food production for the nation as a whole — economic efficiency; or the attainment of a given increase in income for the American Indians in that region — regional-class income redistribution; or the greatest increase in income for the Indians provided the project breaks even on benefits and costs measured in efficiency terms — maximum regional-class income redistribution subject to a constraint relating to economic efficiency; or efficient food production for the nation as a whole provided the Indians in the area receive a certain income supplement. We are not concerned with which of the objectives is chosen. Our approach is likewise indifferent as to whether the scarce water resources of the southwest United States should be put to use for irrigation or for urban and industrial water supply; for this decision, too, depends on the objective. If the objective is economic efficiency in national terms, a larger amount of water is likely to be allocated to urban and industrial water uses than if the objective is to stabilize the income of farmers within the region.

Once the objective has been determined, our methodology leads to the selection of that combination of structures, levels of development for different water uses, and operating procedures that will best achieve the objective. Also, the methodology demonstrates the cost to any one objective of substituting another — for instance, the cost in terms of efficient food production for the nation as a whole of providing a given increase in income to the Indians; or the cost in terms of efficiency of honoring an institutional constraint, such as an interstate compact which divides the waters of a stream between neighboring states. The information obtained in these ways can be used by the policy-maker to evaluate the consequences of his objectives, which is the important fourth step in our methodology.

It should be clear from this discussion that the design techniques developed in this book are applicable not to the United States alone, but to other parts of the world as well. Our most limiting assumptions are those of the last chapter, in which it is presumed that river development takes place under a constitutional democratic system of government.

FROM OBJECTIVES TO DESIGN CRITERIA

The second step in our methodology, that of translating objectives into design criteria, demands essentially greater precision in the statement of the objectives. As many terms as possible are expressed as benefits and costs, the values of which may reflect market values or vary from them, depending on the objective. Although it is easier to define benefit and cost measures for the objective of economic efficiency than for the other objectives, we undertake in Chapter 2, *Objectives of Water-Resource Development: A General Statement*, to demonstrate how a variety of objectives can be translated into useful design criteria.

An objective contains, explicitly or implicitly, the policy-maker's attitudes toward fundamental matters such as the relative values to be placed on outputs received in the present and in the future, which is expressed finally in the interest or discount rate; the necessity for a constraint on budget or expenditure and how it should vary, if at all, over successive time periods; and the degree of certainty desired in meeting planned outputs. The incorporation of discount rates and budgetary constraints into design criteria is discussed in general in Chapter 2 and in greater detail in Chapter 4, *Economic Factors Affecting System Design;* uncertainty is treated in Chapter 3, *Basic Economic and Technologic Concepts: A General Statement.*

FROM DESIGN CRITERIA TO SYSTEM DESIGN

The third step in our methodology of system design translates the design criteria into actual designs for the development of water-resource systems through combined economic and engineering analysis and procedures for identifying optimal designs.

Economics and engineering. One of the principal aims of the Harvard Water Program has been to improve the methodology of system design by joining engineering and economics more effectively than has been done in the past. For simple problems in production economics, the economist is accustomed to receiving from the engineer a production function that summarizes the technologic data for all system designs that are neither impracticable nor inefficient. The economist then derives a net-benefit function as a device for evaluating system designs in terms of the objective. This he applies to the production function to select that point, representing a unique design, which confers the highest value on the benefit function.

The complexities of the production function for water-resource development, however, defy so neat a division of labor between engineering and economics. This is explained in Chapter 3 on basic concepts and illustrated in Chapter 5, *Application of Basic Concepts: Graphic Techniques*. Since the production function is stochastic rather than determinate (that is, its precise values are unpredictable because of the irregularity and unpredictability of streamflows), optimization requires early introduction into the analysis of economic attitudes on uncertainty and of economic measures of failure to achieve planned outputs. The production function is further characterized by complementary outputs and by dynamic rather than static conditions over the period of analysis. For instance, water used to produce hydroelectric energy can be used again for irrigation, and the construction of system units can be staggered to account for increasing demands for system outputs within the period of analysis. Such factors, too, necessitate the early introduction of economic considerations into concepts of production.

Present procedures, called conventional techniques in Chapter 8, *Conventional Methods of Analysis*, fail to consummate an effective union of engineering and economic analysis. In the first place, conventional design techniques make use of physical constraints the economic consequences of which are not explicit. This is particularly true in connection with uncertainty, as illustrated by failures to meet planned levels of development, or target outputs as we shall call them. A typical physical constraint for irrigation is of the following form: in any one year, the monthly deficits should not exceed 50 per cent of the annual target output, and in a study period of 50 years all deficits should not exceed one and one-half times the total annual target output. Although it is recognized that economic considerations are usually taken into account implicitly when physical constraints such as this one are proposed and adopted, such considerations would be reflected more adequately if physical constraints were replaced by continuous economic loss functions that relate economic losses in dollar terms to the degree or intensity of deficits in physical terms. Furthermore, there has been a marked tendency to convert physical constraints into rigid professional engineering standards that soon lose all contact with the economic realities that once entered into their acceptance. This is especially true of physical constraints in the design of flood-protection systems. Accordingly, the concept and use of loss functions appear throughout this book; they are explained most fully in Chapter 7, *A Simplified River-Basin System for Testing Methods and Techniques of Analysis*.

What has been said about constraints that derive from uncertainty is true also about institutional and legal constraints. Conventional techniques have led to the practice of honoring these factors without examining the costs of doing so in terms of the appropriate design criterion.

A second evidence of the failure of conventional techniques fully to unite economic and engineering analysis is the frequent practice of designing systems as if there were no limits on expenditures, that is, no budgetary constraints, and as if the entire system were to be constructed in a single time period rather than over several periods. Such assumptions are realistic in limited circumstances only. The consequences, in terms of system design, of using them more broadly are discussed briefly in Chapter 2 and more fully in Chapter 4.

Finally, conventional techniques seldom state the objective function clearly, nor do they translate it into a useful design criterion. For many water-resource activities in this country, for example, the design criterion (as set forth in the Flood Control Act of 1936) is that "the benefits to whomsoever they may accrue should exceed the costs." This is so imprecise that it encourages engineers to design river systems for maximum physical outputs and confine economic considerations solely to cost minimization.

The optimal design. An optimal system design can be defined as that combination of system units, outputs, and operating procedures which fulfills the objective better than any other. In terms of this definition, one of the principal aims of the Harvard Water Program has been to improve the methodology of system design so as to identify the optimal design or, if this is not practicable, to evaluate readily a sufficiently large number of possible designs to justify assurance that the best of these approximates the theoretical optimum. Conventional design techniques cannot identify the optimal design directly; nor can they, as a rule, provide a comparison of a large number of alternative designs, along with a systematic selection of the samples of designs to be compared.

For identifying optimal designs, we have experimented with recently developed, powerful mathematical techniques that produce optimal decisions for relatively complex objective functions. Our results are summarized in Chapter 6, *Methods and Techniques of Analysis of the Multiunit, Multipurpose Water-Resource System: A General Statement*, and the models are developed in detail in Chapter 13, *Mathematical Models: The Multistructure Approach*, and Chapter 14, *Mathematical Models: A Stochastic Sequential Approach.* For examining a large number of possible designs, we have experimented with simulation of the behavior of a water-resource system on high-speed digital computers. To select from among the multitude of possible designs those to be run on the computers, we have tested different combinations of statistical sampling techniques. These results, too, are summarized in Chapter 6. The several procedures themselves are presented in Chapter 9, *Analysis by Simulation: Programming Techniques for a High-Speed Digital Computer*, Chapter 10, *Analysis by Simulation: Examination of Response Surface*, and Chapter 11, *Operating Procedures: Their Role in the Design of Water-Resource Systems by Simulation Analyses.* The final strategy we recommend combines both the mathematical model and simulation techniques.

WORDS OF CAUTION

Lest we be charged with claiming too much, we should insert a few words of caution at the very beginning of this volume. These issue, we hope, from wisdom rather than from cowardice; for as the ancients have warned us, both have been known to inspire caution.[5]

The water-resource systems with which we have experimented are confined largely to water flowing in river channels and to structures that alter the times and places of availability of this water for different uses. Thus, we have not dealt explicitly with groundwater nor with nonstructural alternatives to storage, such as flood-plain zoning. However, the system elements we have chosen lie at the heart of the multiunit, multipurpose problem. If we can improve design methods for them, these methods can then be elaborated to account for the important aspects that have been by-passed.

We have successfully simulated the behavior of a typical, although simplified, water-resource system on a high-speed computer; however, we see the need for improvement in our techniques, especially for greater flexibility in choosing the operating procedures, which are the instructions to the computer on when and how to store and release water.

To select the samples of possible alternative designs to be compared by computer simulation, we have applied several sampling techniques

[5] "... how cautious are the wise!," Homer's *Odyssey*, bk. XIII, l. 375, Pope's translation. "The coward calls himself cautious," Syrus' *Maxims*.

in unique combination and have identified an approximately optimal design with relatively few studies. But we cannot say with assurance that we have found the best combination of techniques or that our combination is applicable to systems that differ significantly from those we have examined.

Through new means for examining a large number of combinations of system units, target outputs, and operating procedures, we have learned much about the relative importance of these several factors in producing a high value of net benefits; but we realize that much more remains to be learned. The range of system designs over which a typical operating procedure is approximately optimal should be investigated, and similar studies should be made of typical combinations of outputs and of structures. Also, many systems with variant hydrologies, topographies, and combinations of outputs should be analyzed in order to identify model types that will serve as starting points for the analysis of any water-resource system.

The potential of our optimizing mathematical models is both great and exciting, because they operate internally so as to reject all nonoptimal solutions. At the same time, we must acknowledge that at this stage of development the models are embryonic and that substantial simplification is required in their use.

The science of production economics is not sufficiently advanced to deal fully with complex stochastic functions such as those created by the variable nature of streamflows. Hence the task of uniting engineering and economics in system design is a formidable one. We have made certain contributions to this area of knowledge, we believe, but we hasten to point to the need for considerably more research. The historical record of streamflows, for example, may be a very limited representation of the possible configuration of such flows. To adjust for this fact in computer simulation of river-basin systems, we have generated, from the statistical characteristics of the historical record, synthetic flows for long consecutive periods. This technique is described in Chapter 12, *Mathematical Synthesis of Streamflow Sequences for the Analysis of River Basins by Simulation.* We have divided the synthetic flows into replicate records and used these, along with the historical record, to test alternative designs. In this way we have taken account, although theoretically not full account, of the stochastic component of streamflow. To the same end, in the mathematical models we have substituted for actual streamflows probability distributions of flow. But how to evaluate the results achieved by these methods — how to relate attitudes on uncertainty to the replicate records in the case of simulation, and to the probability distributions of outputs in the case of mathematical models — requires further research.

We have undertaken to translate into a design criterion any reasonable objective. We should repeat, however, an earlier warning: that it is considerably easier to do this for economic efficiency than for other objectives.

Finally, there are cautions that result from our having avoided almost entirely a large area of research by assuming that, in addition to the hydrology, most of the physical and economic data necessary to apply our methods of system design are available. Practical men may have reservations about the utility of our methodology because required data are not available at this time, or because the problems of organizing existing data in the forms needed are immense. But the capacity to use data should serve as an incentive for the devising of techniques to collect, organize, and interpret them more effectively. In other words, existing data-collecting programs can be modified and new programs created to supply the kinds of information needed, once the need has been demonstrated.

Along much the same line, our design methods are based on the assumption that we have adequate techniques to measure benefits and costs in accordance with the standards of the objective function; but apart from certain generalizations in Chapters 2 and 4, which provide a theoretical basis for cost and benefit evaluation, we have not gone into these techniques. Government agencies annually spend large sums of money on perfecting them, and, as we have noted, economists have written extensively on the subject.[6] We have felt that we should make our contribution elsewhere. However, it should be pointed out that our design techniques assume the availability of a type of benefit data to which relatively little attention has been given in the past, namely, losses incurred from failures to meet planned or target outputs. Again the fact that such data can be put to good use should result in greater attention to their collection, organization, and interpretation.

[6] See reference 1; also, California Department of Water Resources, *Selected bibliography of resource development* (1959), sections on "Project evaluation" and "Economic considerations in water development."

PRESENTATION OF RESULTS

This book is intended for engineers, economists, and administrators interested in the development of water resources and in similar public activities and systems of production. The research that lies behind it has been conducted cooperatively by engineers, economists, and political scientists. For these reasons, however, the book has not been easy to write. How much should be said about the fundamentals of welfare and production economics, of hydrology and hydraulics, and of constitutional democratic government in order that readers who are engineers, economists, and administrator-politicians may understand equally well the basic doctrines of the fields of learning that we seek to unite for the purpose of system design? Also, how technical should we be in presenting results of research that have special significance for one of these fields?

We have answered these questions in part by introducing two levels of discussion: a series of generalized chapters for readers in all three fields of learning who are somewhat familiar with problems of system design, and a series of specialized chapters for those whose competence lies in one field in particular. The generalized chapters have the phrase *A General Statement* at the end of the title. By perusing these four chapters in succession, a reader should be able to acquire a comprehensive view of the conception, conduct, results, and deficiencies of our research. The specialized chapters that follow the generalized ones record results that have justified more detailed or particularistic treatment.

Composing a book in this way has inherent weaknesses, some of which we have not been able to avoid. The style, for instance, is not uniform, although a fairly high level of uniformity has been achieved in concepts, terminology, and methods of presenting figures, tables, and equations. Furthermore, there is inevitably some repetition between the generalized and the specialized chapters. We hope that the reader will not be too sorely tried by these defects.

HISTORY OF THE HARVARD WATER PROGRAM

There have been three stages in the history of the Harvard Water Program: one year (1955–56) of exploration to determine whether or not a study should be undertaken in this field, and, if so, what type of study it should be; three years (1956–57 to 1958–59) of combined training and research, during which senior employees of federal and state water-resource agencies came to Harvard to assist in the research and at the same time to prepare themselves for positions of greater responsibility in the public service; and one final year (1959–60) of research and writing.

The participants in the program are identified in Table 1.1 by academic year. Attributing authorship in a cooperative effort of this sort is obviously difficult. The men who actually wrote the several chapters are listed as their authors in the Table of Contents and on the chapter heads but behind each chapter lies a great deal of research, and in most cases several research papers, by other participants. The names of chapter authors are indicated in bold type in Table 1.1. Harvard faculty members whose names appear in boldface contributed to the writing or review of all chapters of the book; along with Stephen A. Marglin, they are its chief authors. A number of participants, although not authors, made a special contribution to this published version of the Harvard research; their names are indicated in italics.

The Harvard Water Program has been financed by generous grants from the Rockefeller Foundation to the Harvard Graduate School of Public Administration, and has enjoyed professional support from several public water-resource agencies, notably from the U. S. Bureau of Reclamation and the U. S. Army Corps of Engineers. Dr. Robert P. Burden of the foundation, Messrs. N. B. Bennett, Jr., and Eugene W. Weber of the bureau and corps respectively, and successive Deans Edward S. Mason and Don K. Price of the graduate school have been model sponsors.

Table 1.1. Participants in Harvard Water Program.[a]

Harvard faculty — regular participants[b]	Government agencies — regular participants[c]	Harvard University — regular participants	Consultants — participants over extended periods	Assistants
1955–56:				
ARTHUR MAASS Professor of Government and Director, Harvard Water Program **MAYNARD M. HUFSCHMIDT** Research Associate, Graduate School of Public Administration				
1956–57:				
ARTHUR MAASS **MAYNARD M. HUFSCHMIDT** **ROBERT DORFMAN** Professor of Economics *Otto Eckstein* Associate Professor of Economics **GORDON MASKEW FAIR** Professor of Engineering **HAROLD A. THOMAS, JR.** Professor of Engineering	**BLAIR T. BOWER** Calif. Dept. of Water Resources Paul W. Eastman U. S. Public Health Service Charles O. Eshelman U. S. Army Corps of Engineers Ellis L. Hatt U. S. Soil Conservation Service Frankie L. Johnson U. S. Bureau of Reclamation Charles E. Knox U. S. Geological Survey Edgar W. Landenberger U. S. Army Corps of Engineers Laurence G. Leach U. S. Army Corps of Engineers Charlie M. Moore U. S. Soil Conservation Service **WILLIAM W. REEDY** U. S. Bureau of Reclamation	Nicholas C. Matalas Graduate School of Arts and Sciences	*Peter O. Steiner* Professor of Economics, Univ. of Wisconsin	Ellen Rome Ehrlich
1957–58:				
ARTHUR MAASS **MAYNARD M. HUFSCHMIDT** **ROBERT DORFMAN** (fall term only) *Otto Eckstein* **GORDON MASKEW FAIR** (fall term only) John R. Meyer Professor of Economics **HAROLD A. THOMAS, JR.**	Dante A. Caponera U.N. Food and Agriculture Organization William H. Davis U. S. Public Health Service Charles O. Eshelman Clyde W. Graham U. S. Soil Conservation Service James S. King U. S. Army Corps of Engineers *Joseph G. Polifka* U. S. Soil Conservation Service Kenneth E. Ristau U. S. Army Corps of Engineers	*Hugh Blair-Smith* Harvard College **STEPHEN A. MARGLIN** Harvard College Donald R. Spuehler Graduate School of Public Administration	**BLAIR T. BOWER** Delaware River Basin Advisory Committee Morris E. Garnsey Professor of Economics, Univ. of Colorado	Cleo Staciva Tomei

	Jabbar K. Sherwani Pakistan Planning Board Robert E. Whiting Calif. Dept. of Water Resources John M. Wilkinson U. S. Bureau of Reclamation			
1958–59 **ARTHUR MAASS** **MAYNARD M. HUFSCHMIDT** **ROBERT DORFMAN** *Otto Eckstein* **GORDON MASKEW FAIR** **HAROLD A. THOMAS, JR.**	Howard R. Bare U. S. Army Corps of Engineers Sheldon G. Boone U. S. Soil Conservation Service Edward Hill U.S. Army Corps of Engineers Howard J. Mullaney U. S. Army Corps of Engineers Alan S. Payne U. S. Soil Conservation Service Ralph M. Peterson U. S. Forest Service William B. Shaw Calif. Dept. of Water Resources Robert F. Wilson U. S. Bureau of Reclamation	*David W. Bernstein* Harvard College George T. Berry Graduate School of Arts and Sciences *Hugh Blair-Smith* *Louis M. Falkson* Harvard College *Donald E. Farrar* Graduate School of Arts and Sciences **MYRON B. FIERING** Graduate School of Arts and Sciences Robert S. Gemmell Graduate School of Arts and Sciences **DEWARD F. MANZER** Harvard College **STEPHEN A. MARGLIN**	**MICHAEL P. BARNETT** Visiting Associate Professor in the Dept. of Physics, MIT **BLAIR T. BOWER** **WILLIAM W. REEDY** U. S. Bureau of Reclamation Allyn St. C. Richardson Assistant Professor of Civil Engineering, Tufts Univ. *Loren A. Shoemaker* U. S. Army Corps of Engineers	Cleo Staciva Tomei Irene Siebens Willis Carolyn Sullivan Cohen
1959–60: **ARTHUR MAASS** **MAYNARD M. HUFSCHMIDT** **ROBERT DORFMAN** **GORDON MASKEW FAIR** **HAROLD A. THOMAS, JR.**		*David W. Bernstein* Law School George T. Bryant Graduate School of Arts and Sciences *Louis M. Falkson* Graduate School of Arts and Sciences **MYRON B. FIERING** **DEWARD F. MANZER** **STEPHEN A. MARGLIN** Henry Fellow Cambridge University **PETER WATERMEYER** Graduate School of Arts and Sciences	**MICHAEL P. BARNETT** **BLAIR T. BOWER** Meramec Basin Research Project, Washington Univ. **WILLIAM W. REEDY**	Irene Siebens Willis Carolyn Sullivan Cohen Naomi Hanser McCann Marian Stanwood Adams Janet Pearson Pinkerton Vivian Barbara Wheeler

[a] Boldface indicates chapter authors; italics indicate special contribution to this published volume. Where position or affiliation is not given, see prior listing in same column.
[b] Present position or status given in each case.
[c] Either assigned, or on leave, to Harvard for educational purposes.

Part I OBJECTIVES AND CONCEPTS

2 Objectives of Water-Resource Development: A General Statement

STEPHEN A. MARGLIN

THE prime objective of public water-resource development is often stated as the maximization of national welfare. That this is a goal to be desired, few would question; that it cannot be translated directly into operational criteria for system design, few would deny. Translation would require not only agreement on a definition for the deceptively simple phrase "national welfare" but also some assurance that the defined concept is measurable.

One possibility is to define national welfare as national income. The objective of system design then becomes maximization of the contribution of the system to national income. This definition is measurable, but it has implications for the meaning of national welfare that make us unwilling to accept it as a complete expression of the broad objective. Identifying national welfare with the size of national income not only excludes noneconomic dimensions of welfare but also implies either that society is totally indifferent as to the recipient of the income generated by river-development systems, or that a desirable distribution of gains will be made by measures unrelated to the manner in which the system is designed.

Social indifference to the distribution of income generated by the system suggests that the marginal social significance of income is the same regardless of who receives it; a dollar of extra income to Jones is no more desirable socially than a dollar to Smith, despite differences in initial incomes, employment opportunities, and the like. Such an attitude connotes a satisfaction with the status quo that few would share. The alternative implication — that, with maximization of national income as the sole design criterion, a socially optimal redistribution of system-generated income will take place nonetheless — calls for resort to means such as taxation and subsidies that redistribute income directly from one group of individuals to another. This alternative, in other words, ties us to specific means for effecting a desired redistribution of income that the community may not accept. The community's preferences should not be dismissed as irrational; indeed, its choice of means may be one component of its concept of national welfare. The community may, for example, wish to use a desired redistribution of income as a criterion for the design of water-resource systems, even at the expense of potential increases in national income.

In view of the three-dimensional nature of national welfare — the size of the economic pie, its division, and the method of slicing — we believe it unwise to attempt to define a single index for this broad objective; instead we shall develop alternative objectives for the most important ways in which water-resource development can contribute to national welfare. These are broadly two: efficiency, which expresses the objective of maximization of the size of the economic pie in a more sophisticated manner than does maximization of national income, and income redistribution, which expresses the objective of achieving a desired slicing of the economic pie by a knife that suits community values.

The objective of economic efficiency will be discussed shortly, but first a few words on certain problems of methodology and definition that arise in this chapter.

A design criterion selects from among alternative designs the one that performs "best" in terms of given objectives. It is theoretically possible to employ the design criterion to select the best design without explicit reference to the relative rankings of alternative designs; however, behind the implementation of any design criterion always lies a complete ordering of alternatives in terms of attainment of the objectives. The device for ordering the entire set of alternative designs, which we have called a *ranking function*, plays a key role in the search for the optimal design in the strategy presented in Part II of this book.[1] There will be, of course, a different ranking function for each objective.

We may require that a design criterion choose the system design that performs best in terms of one objective, subject to a specified level of performance in terms of another, which we shall call a *constraint*.[2] In

[1] Theoretically, marginal conditions can be employed to identify a subset of designs in which the optimal design must lie. Whereas these conditions implicitly are based upon a ranking function, the function itself need be identified only if more than one design satisfies the marginal conditions. For a discussion of marginal conditions, see pp. 31–34 and Chapter 3. An exceptional strategy not making explicit use of the ranking function right from the outset rests on direct implementation of the marginal conditions. It is explained in Chapter 5.

[2] Maximization of an objective function may also be subject to institutional constraints (for example, at least X acre feet of water to California and Y acre feet of water to Arizona)

this case one ranking function is used to select from among the alternatives those satisfying the constraint, and another ranking function to select from them the one that performs best in terms of the first objective. It is important to be clear on this point. It is not sufficient to develop a design that satisfies a constraint; for instance, to provide a specified level of income for the farm families in a given region by means of irrigation. Almost always, more than one design will satisfy the constraint, and the planner must have a further criterion for selecting among these. In other words, he needs a ranking function that is to be maximized, which can be called an *objective function.* A design criterion always includes an objective function, which may or may not be subject to constraints. In translating verbal objectives into design criteria, the planner must choose with care and discrimination the objectives to be placed in the objective function and those to be treated as constraints.

Two requirements must be fulfilled if the ranking function is to be effective. First, it should provide a well-defined ordering of alternatives; if it says that design A is preferable to design B, for example, it should not say at the same time that B is better than A. Also, it should provide a transitive ordering of alternatives; if it says that A is preferable to B and B is preferable to C, for instance, it should not say that C is preferable to A. To ensure fulfillment of these requirements, we shall define the ranking functions as *scalar functions;* for scalar orderings have the property that they are well-defined and transitive. Therefore, to ensure that a scalar function $U(A)$ is a valid ranking function, we need only verify that $U(A^i) > U(A^j)$ if and only if A^i performs better than A^j in terms of the given objective for all A^i and A^j.

Traditionally, the scalar measure of performance of a design with respect to a given objective provided by the ranking function is called the *net benefits of design.* Benefits and costs, it should be emphasized, have only instrumental significance; we can speak properly of net benefits only as applying to given objectives, such as "efficiency net benefits" or "redistribution net benefits." A design criterion of maximizing net benefits in the abstract is meaningless.[3] We risk laboring this point be-

---

or to budgetary constraints (for instance, no more than $2,000,000 to be spent on construction of water-resource developments). Such constraints also may be thought of as embodying social objectives.

[3] For example, the standard set forth in the Flood Control Act of 1936 (49 Stat. 1570) for evaluating water-development proposals, that "the benefits to whomsoever they may accrue are in excess of the estimated costs" gives no guidance as to the type of benefits. The Corps of Engineers has generally (but not invariably) interpreted this as meaning efficiency benefits.

cause two alternative, and in our view inferior, concepts of the role of benefits and costs are widely held. The first is that benefits and costs measure respectively all the good and bad consequences of water-resource development. This notion is the logical counterpart of an immediate objective of maximization of national welfare, and it is no more usable than the latter. The second concept of benefits and costs is that they should measure only a single, most-favored objective. Other objectives are admitted but at a different level of significance, as factors that may, in certain circumstances, justify "sacrificing net benefits" — meaning, of course, net benefits with regard to the favored objective; but this important qualification is often omitted, and the bond between benefits and costs and the objective is then lost. It is the efficiency objective that is identified typically with benefits and costs, usually because efficiency can be measured more precisely and reliably than other objectives. Although we do not dispute that this is so, we cannot accept it as adequate reason for not translating verbal concepts of other objectives into design criteria.

THE EFFICIENCY OBJECTIVE

We have referred to economic efficiency as a concept for expressing the size of the economic pie that is superior to the concept of national income. The difficulty with national income is that it is too closely tied to market values. Intuitively we sense, for instance, that if a water-resource development made great amounts of electric energy available for residential consumers at low cost, it would add more to the size of the economic pie than another development scheme which made smaller amounts available at higher cost, even if the national-income value of the latter scheme measured at market prices were to be higher than the former. Such considerations have led economists to speak of the size of the economic pie in terms of a more fundamental principle, namely, economic efficiency.

In terms of water-resource development, the traditional definition of economic efficiency identifies a proposed design as economically efficient if no alternative design would make any member of the community better off without making some other member worse off.[4] Unfortunately, this definition does not provide a decisive criterion of choice among designs, because often there will be more than one design that satisfies it. For example, design A^1, yielding large amounts of irrigation water, might

[4] Economists will recognize this as the definition of Pareto optimality.

make farmers better off than any alternative design, whereas design A^2, providing a deep navigation channel, might make shippers better off than all alternatives. Each design is therefore efficient. To choose between them, we can expand the traditional definition of efficiency by introducing the notion of compensating side payments. We then say that design A^1 is relatively more efficient than design A^2 if those affected by A^1 are willing to pay those affected by A^2 a sum sufficient to persuade them to agree to the construction of A^1.

One alternative is always to do nothing; this we call the zero design. It is convenient to measure willingness to pay for "positive" designs in terms of the zero design, that is, in terms of willingness to pay for a specific design rather than have no project at all. If $W(A^1)$ and $W(A^2)$ denote the aggregate net willingness of people affected by designs A^1 and A^2 respectively to pay rather than have no project, and if $W(A^1) > W(A^2)$, it can be shown that the people affected by design A^1 would be willing (under certain simplifying assumptions to be considered presently) to pay the people affected by design A^2 an amount sufficient to make them willing to agree to the building of A^1 instead of A^2. And conversely, it can be shown that if those affected by A^1 are willing to pay those affected by A^2 an amount sufficient to go along with A^1, then $W(A^1) > W(A^2)$. Therefore, for the efficiency objective, $W(A)$ (where A is a variable representing alternative designs $A^1, \ldots, A^j, \ldots$) meets the requirements for a scalar ranking function listed above.

The aggregate net willingness to pay, $W(A)$, of everyone affected by a particular design, A, is of course a confusing concept; perhaps it can be made clearer if those affected by the design are divided into two categories, those who are made better off and those who are made worse off. To simplify the exposition, let us temporarily make the additional assumption that system outputs are distributed free of charge. Let $E(A)$ denote the willingness of the people who benefit from outputs of design A to pay for the design rather than have no design at all, and let $C(A)$ denote the cost of A measured in terms of the money value of the goods foregone to construct and operate A. If we group those who pay for the project under the label "taxpayers," $C(A)$ represents the amount that taxpayers (who are worse off if the project is built, because they have to pay for it) would be willing to pay not to have the project built.[5] It

[5] For convenience we assume that the shift in resources from other uses to project construction is accomplished by taxation; those who are made worse off can thus be designated as taxpayers. Other means, such as a budget deficit, could of course be used to make the shift of resources.

follows that $W(A)$ is equal to the willingness to pay of beneficiaries, $E(A)$, minus the amount taxpayers would be willing to pay not to have the project, $C(A)$. Thus the efficiency ranking function may be rewritten as

$$W(A) = E(A) - C(A). \tag{2.1}$$

This ranking function will meet the requirements that it provide a well-defined and transitive ordering of alternatives only if the amount the beneficiaries of one design are willing to *accept* as compensation to do without their project is equal to the amount they are willing to *offer* as compensation to the beneficiaries of other designs to persuade them to do without their projects. For example, if a group of beneficiaries are willing to accept $100 to forego a project, they must be willing to offer exactly $100 to persuade others to go along with their project. This condition will be met if we assume for consumers' goods (such as domestic water supply) that the marginal utilities of incomes of the beneficiaries are constant over the relevant ranges of income,[6] and for producers' goods (such as industrial water supply) that the beneficiaries are motivated in their economic decisions solely by the goal of maximization of profits and are not hindered by lack of finance capital. These assumptions will be discussed later.

EFFICIENCY BENEFITS AND COSTS

We have seen that the terms "benefits" and "costs" have meaning only in relation to a well-defined objective. For our definition of efficiency, then, gross benefits $E(A)$ are the willingness of beneficiaries to pay for design A; gross costs $C(A)$, the money value of goods foregone by individuals throughout the economy in order to construct and operate design A; and net benefits $E(A) - C(A)$, the difference between willingness to pay and the value of goods displaced.

To be used in a ranking function, willingness to pay and value of goods displaced must be measurable. Although, as stated in Chapter 1, we are not concerned with the technical problems of benefit and cost measure-

[6] The role of the assumption of constancy of the marginal utility of income in the compensation criterion of efficiency and in the measurement of efficiency benefits is critical; the discussion of the assumption itself has a long history in economic theory.

We are simplifying matters by assuming a cardinal rather than an ordinal utility in postulating a constant marginal utility of income. A more precise statement would be in terms of parallel indifference curves. See Kenneth E. Boulding, *Economic analysis* (Harper and Brothers, New York, ed. 3, 1955), pp. 816 ff., and J. R. Hicks, *Value and capital* (Clarendon Press, Oxford, ed. 2, 1955), chap. 2.

ment, we should at least ensure that these concepts are defined in terms that can be measured. To this end, we temporarily make the following assumptions: (1) that no individual derives economic gain or loss from the provision of goods for another's consumption (that is, we assume that there are no external effects in the consumption of water-resource outputs); (2) that prices throughout the economy equal marginal costs; (3) that the scale of water-resource development is not so large that it will affect prices in general in the economy, the only exception being the immediate products of water-resource systems; although we continue to assume that these products are given to their recipients free of charge, we do not assume that the recipients' willingness to pay for them is constant.

These assumptions describe a situation in which it is proper to employ the economists' competitive model; with them our attention may be limited to system outputs and inputs in measuring benefits and costs, and we may ignore widely diffused changes induced throughout the economy by the system. In terms more familiar to economists, these competitive assumptions permit substituting a partial-equilibrium analysis of immediate system outputs and inputs for the general-equilibrium analysis of all changes throughout the economy induced by water-resource development that our definition of efficiency logically requires.[7] In language in which engineers may find themselves more at home, the competitive assumptions permit making a "cut" between the inputs and outputs of the water-resource system and the rest of the economy and analyzing the system as a "free body." The significance of these assumptions will be examined in a later section.

Identification of a system design with its immediate outputs and inputs requires specifying these in more detail. By immediate outputs we mean the primary products of water-resource systems rather than the final goods produced by use of system outputs — for example, electric energy or irrigation water rather than the aluminum or rice produced by use of the energy or irrigation water. A system output has three dimensions: its annual quantity, the time pattern in which it is available, and the certainty of obtaining the output. We shall deal here with outputs only in terms of annual aggregates, leaving the time pattern and uncertainty aspects for treatment in other chapters.

[7] A general-equilibrium analysis can be identified with the "with and without" principle. See, for example, Otto Eckstein, *Water-resource development, the economics of project evaluation* (Harvard University Press, Cambridge, 1958), pp. 37–38 and 51–52. A partial-equilibrium analysis is consistent with this principle only when the competitive assumptions hold.

The concept of inputs is more flexible. System inputs can be thought of as the physical structures (the dams, power plants, and irrigation canals) that define system design, and throughout Part II of this book this is the viewpoint adopted. However, in the present chapter it is often more illuminating to deal with inputs as the primary resources, namely, the various kinds of land, labor, and capital, by means of which a water-resource system is built and operated. Thus in this and succeeding chapters of Part I there will be occasion to use both concepts.

We shall denote outputs by the variables $y_1, \ldots, y_m$ or the more compact vector notation $\mathbf{y} = (y_1, \ldots, y_m)$ and inputs by the symbols $x_1, \ldots, x_n$ or the vector $\mathbf{x} = (x_1, \ldots, x_n)$. The ranking function, Eq. (2.1), becomes

$$W(A) = E(\mathbf{y}) - C(\mathbf{x}), \tag{2.2}$$

outputs and inputs being related to each other by a production function

$$f(\mathbf{x}, \mathbf{y}) = 0. \tag{2.3}$$

The meaning of the production function will be explored in Chapter 3. For the present, the following definition will suffice: Equation (2.3) excludes from consideration all combinations of outputs and inputs (that is, all designs) save those by which it is impossible to produce more of one output without either producing less of some other output or requiring more of some input. In other words, the level of any output in the production function, Eq. (2.3), and hence in the ranking function, Eq. (2.2), is the maximum compatible with preassigned levels of the other $m - 1$ outputs and the n inputs. Similarly, the level of any input in Eq. (2.3), and thus in Eq. (2.2), is the minimum level with which, in conjunction with preassigned levels of the other $n - 1$ inputs, the preassigned levels of m outputs can be attained.

In these terms the value of goods displaced, given our competitive-model assumption, is equal to the money cost $C(\mathbf{x})$ of the inputs that will produce the outputs. Willingness to pay of positive beneficiaries, $E(\mathbf{y})$, is a bit more difficult to identify; for each purpose it is equal graphically to the area under the aggregate demand curve, representing how much the beneficiaries of a system output would purchase at successive prices if the output were sold in a market. (The assumptions of constant marginal utilities of income and of profit maximization are necessary for this definition.) For output level y^1 of irrigation water, for example,

$E(y^1)$ is the shaded area under the aggregate demand curve DD in Fig. 2.1.

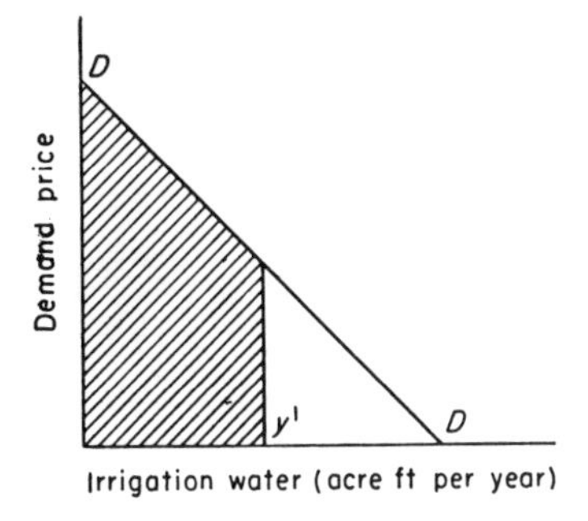

Fig. 2.1. Aggregate demand curve for irrigation water. Shaded area represents willingness to pay for a specified quantity (y^1) of irrigation water.

If aggregate demand is represented analytically as a function $y(p)$ of the price p, willingness to pay will equal the integral of the inverse $D(y)$ of this function, that is, the integral of the "demand-price" function representing the prices at which successive quantities of irrigation water could be sold in a competitive market. Formally,

$$E(y) = \int_0^y D(\eta)d\eta,$$

where η represents the dummy variable of integration.

Since one cannot "test the market" to learn the quantities of outputs that beneficiaries would demand at different prices, demand curves in general must be imputed rather than measured directly. For a producers' good such as irrigation water, demand curves can be synthesized by studies of production profitability — in other words, studies of the quantities of output that the range of productive techniques available to the producer lead him to purchase at various prices in order to maximize his profits. For consumers' goods it may be necessary to rely on historical data and on information from other regions because of the impossibility of measuring subjective utility functions — the functions which have the same relationship to consumer demands that production functions have to producer demands.

Our understanding of efficiency benefits and costs can be improved by relating them to market values of outputs and inputs. For costs, the two are identical. In the competitive model, efficiency costs, or the value of goods displaced by a water-resource system, are equivalent to the market value of the resources employed in the system. For benefits, there is an important difference. If the aggregate demand curve for irrigation water of

Fig. 2.1 is reproduced as in Fig. 2.2, the market value of the quantity y^1 is the rectangle of area $p_{y^1}y^1$ formed by the broken lines and the two axes.

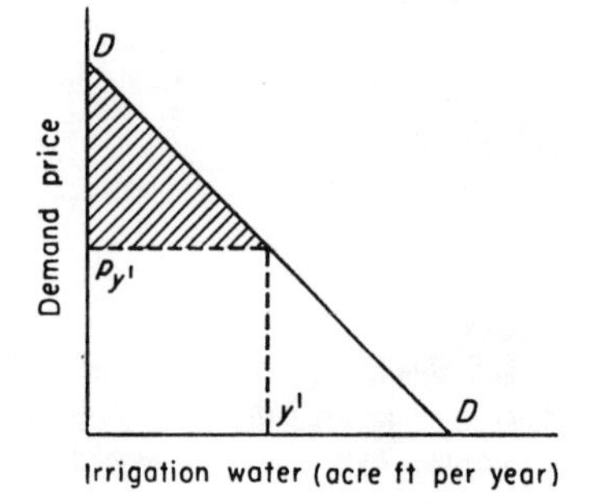

Fig. 2.2. Aggregate demand curve for irrigation water. Shaded area shows consumers' surplus for a specified quantity (y^1) of irrigation water.

As we have seen, the willingness of beneficiaries to pay for the output rather than do without is this rectangle plus the shaded triangle above it. It consists of market value plus what economists sometimes call "consumers' surplus." Market value will coincide with willingness to pay only when the demand curve is horizontal.[8]

However, it is not reasonable to assume that demand curves for outputs of water-resource systems will always be horizontal. For irrigation, for example, a horizontal demand curve implies that the potential levels of system output represent so small an addition to the total supply of water available that they cannot make a dent in the productivity of irrigation and hence in the price per acre foot that the beneficiaries are prepared to offer. In actuality, a new irrigation system is usually if not the only

[8] This distinction between willingness to pay and market value sheds light on the proper manner of interpreting the following quotation from the Report of the U. S. President's Water Resources Policy Commission, vol. 1, *A water policy for the American people* (U. S. Govt. Printing Office, Washington, 1950), pp. 60–61 (italics in original):

> There are many types of public objectives and gains which defy analysis in commercial terms. For example, market price valuation can give quite misleading results when applied to public measures designed to stop resources deterioration and to strengthen the resources base of the country. Market prices reflect scarcity values; thus, if half our forests were burned, the remainder would be worth more money than the existing forests because of the very sharp rise in prices of forest products that would take place.
>
> A commercial property has *dollar* value according to scarcity; a Nation's resources have *security* value if they are so plentiful as to be cheap. Thus, a successful irrigation or drainage program might lead to a great reduction in the dollar value of some particular crop; the more that is planted and protected, the less it is worth; and *the stronger is the Nation.* This basic conflict between private values based on scarcity and public values based on plenty lies at the very heart of the conservation problem.

The conclusions are correct but the reasoning confusing. The *market value* of forests might be increased by burning half of them (technically, the "demand for forests" might be inelastic), but the *willingness to pay* for forests in the aggregate never could be increased in this manner.

source, at least a substantial addition to the available supply, of irrigation water. Thus successive increases in output, whether they lead to more intensive or more extensive irrigation, generally result in significantly smaller increases in crop production; in consequence, beneficiaries will offer successively smaller prices as the level of system irrigation output increases. Note that the nonconstancy of the demand price for irrigation water is independent of our assumption of constant prices throughout the economy. The fact that the quantity of rice grown with the aid of system irrigation is considered to be too small to affect the price of rice will not ensure the equality of the market value of, and the willingness to pay for, the irrigation water itself.

Some outputs of certain river-basin developments, however, are small in relation to existing sources of the good available to beneficiaries. A small public-power project in the Columbia River Basin, for example, may represent but a minor addition to the pooled energy of the existing northwest power system. As a general empirical proposition, however, the coincidence of market value and willingness to pay cannot be upheld.

Turning to the national-income measure of efficiency benefits, a fairly precise definition of the social accountant's measure of national income is the market value of all final goods and services produced in the economy plus the value of any change in the stock of producers' plants and inventories. Thus for outputs that, like electric energy consumed in the home, are final goods, the national-income measure of efficiency suffers from the defect of a market-value measure, namely, failure to take account of consumers' surplus. For outputs that are producers' goods (electric energy for manufacturing uses, for example), the social accountant's concept of national income and our efficiency definition of benefits are identical, although this may not be obvious at first glance. Whereas our definition in terms of producers' goods is related to the system outputs themselves (electric energy, for example) rather than to the final goods, the social accountant does not include intermediate producers' goods in his evaluation of national income. To evaluate the energy used to fabricate final goods as well as the final goods themselves would be double counting. Thus the contribution to national income of the productive outputs of a river-basin system is the market value of the final goods less the costs of all factors of production other than those provided by the system itself. In turn, this is the maximum that producers, seeking maximum profits, will pay for water rather than do without, or, simply, the immediate recipients' willingness to pay, which is our definition of efficiency benefits.

To conclude, net efficiency benefits measure the net contribution of a water-resource system to national income — or the difference between national income generated by system outputs on the one hand and national income displaced by system construction and operation on the other — except for the discrepancy occasioned by the surplus enjoyed by consumers on system outputs that are final goods.

THE MULTIPURPOSE, MULTIPERIOD RANKING FUNCTION

The efficiency ranking function reduces in effect to a balancing of individual preferences (backed by willingness to pay) for the final goods of a water-resource system against individual preferences for the goods displaced by the system. The examples that we have used so far to illustrate the efficiency ranking function have been for single-purpose, single-time-period systems. We now extend our results to multipurpose and multiperiod systems.

Evaluation of the efficiency costs of a single-period, multipurpose system does not differ significantly from that of a single-purpose system. To be sure, the inputs may be composed of a larger number of variables, but conceptually this makes for no difficulty; efficiency costs are still evaluated by the money cost of the system, $C(\mathbf{x})$. Evaluation of efficiency benefits also poses no new problems so long as willingness of the beneficiaries to pay for each output is independent of the quantities of other outputs provided by the system. The efficiency benefits of each output y_j remain the willingness to pay for that output, and aggregate efficiency benefits are the sum of the efficiency benefits of each output, or

$$E(\mathbf{y}) = \sum_{j=1}^{m} \int_0^{y_j} D_j(\eta_j)d\eta_j,$$

where $\mathbf{y}$ represents a vector of outputs $y_1, \ldots, y_m$. If the assumption that willingness to pay for each output is independent of the levels of other outputs is no longer valid (as would be the case if farmers purchased from the same project both irrigation water and energy to pump it to their farms), the measurement of efficiency benefits becomes more complicated and requires the derivation and manipulation of joint demand functions.[9]

[9] In general, if the willingness to pay for outputs y_j and y_h depends on the levels at which both are provided, we must deal with demand-price functions of the form $D_j(y_j, y_h)$ and $D_h(y_j, y_h)$, representing respectively the prices p_{y_j} and p_{y_h} for which the quantities y_j and y_h could be sold in the market place. (These functions are the inverses of the joint demand functions $y_j(p_{y_j}, p_{y_h})$ and $y_h(p_{y_j}, p_{y_h})$, which represent the quantities of each output as func-

The multiple-period aspect of water-resource development adds significantly to the complexity of system design. Decisions must be reached today that may have consequences over 50 or 100 years. However, because design decisions must be made today, we cannot deal with the willingness of individuals to pay for irrigation water in 1975, but only with the equivalent willingness of our contemporaries to pay for this future output. Thus we cannot merely sum the benefits and costs of all periods; instead we need an interest or discount rate to determine the present value of future benefits and costs.

The first problem posed by the multiperiod life of water-resource systems is, therefore, the choice of an interest rate for system design. One alternative, derived from the competitive model, is to use the market rate of interest. We shall adopt this approach for the present, but shall return to question it later in the chapter.

Changes in willingness to pay over the economic life of a system present other complications in multiperiod analysis. For example, the demand for system energy in 1985 will not be the same as in 1965. But secular changes in demand take place comparatively slowly. This allows us to define demand periods (of unspecified length but assumed to be of the order of a decade) within which annual demands are assumed to be constant but between which annual demands vary. Finally, we assume that shifts in aggregate demands between demand periods are independent of the levels of outputs during previous periods.

An additional problem posed by multiperiod systems concerns the scheduling of construction. Here we shall answer it by assuming, as an

---

tions of the prices of both.) Total willingness to pay for the two outputs becomes the sum of line integrals

$$\int_{(0,0)}^{(y_j,y_h)} D_j(\eta_j, \eta_h)d\eta_j + \int_{(0,0)}^{(y_j,y_h)} D_h(\eta_j, \eta_h)d\eta_h$$

independent of the path along which one integrates. By choosing the convenient path from the origin along the y_j-axis to $(y_j, 0)$, thence parallel to the y_h-axis to (y_j, y_h), this expression can be written as the following sum of "ordinary" integrals:

$$\int_0^{y_j} D_j(\eta_j, 0)d\eta_j + \int_0^{y_h} D_h(y_j, \eta_h)d\eta_h,$$

which is more susceptible to intuitive interpretation. This says that the total efficiency benefits from, or the total willingness to pay for, the economically interdependent outputs y_j and y_h is the sum of the willingness to pay for y_j on the assumption that none of the *h*th output is provided and the willingness to pay for y_h on the assumption that y_j units of the *j*th output are provided. Graphically, efficiency benefits of the two outputs are the sum of the area under the demand curve for the *j*th output (truncated at the level of output y_j) drawn on the assumption that none of the *h*th output is provided, and the area under the demand curve for the *h*th output (truncated at y_h) drawn on the assumption that the level of the *j*th output is y_j.

approximation of the fact that construction outlays are usually bunched heavily in the early years of a system's life, that all construction takes place in year zero. However, in Chapter 4 we shall show that when system demands are increasing with time, it may be desirable to postpone some construction until later.

The efficiency ranking function and design criterion for multipurpose, multiperiod river-basin systems can now be formulated. To begin with, efficiency benefits in year t are:

$$E_t(\mathbf{y}_t) = \sum_{j=1}^{m} \int_0^{y_{tj}} D_{tj}(\eta_{tj}) d\eta_{tj}, \tag{2.4}$$

where $D_{tj}(y_{tj})$ represents the demand for the jth output in year t. $E_t(\mathbf{y}_t)$ is measured in terms of demands for outputs in year t because in the future, just as in the present, willingness to pay is measured by the area under the demand curve. The efficiency net benefit or ranking function for a multiple-purpose water-resource system with an economic life of T years can therefore be expressed as:

$$\sum_{t=1}^{T} \frac{E_t(\mathbf{y}_t) - M_t(\mathbf{x})}{(1 + r)^t} - K(\mathbf{x}), \tag{2.5}$$

where $M_t(\mathbf{x})$ denotes operation, maintenance, and replacement costs in year t (which we assume to be constant within each demand period), $K(\mathbf{x})$ denotes construction cost, and r denotes interest rate.

By virtue of our division of the economic life of systems into demand periods over which demands and operation, maintenance, and replacement costs (hereafter referred to as OMR costs) are constant, we can replace Exp. (2.5) by a ranking function that has fewer variables. We shall use the ranking function

$$\sum_{q=1}^{Q} \theta_q[E_q(\mathbf{y}_q) - M_q(\mathbf{x})] - K(\mathbf{x}), \tag{2.6}$$

where θ_q is the discount factor applicable to the demand period q, given by

$$\theta_q = \frac{1 - (1 + r)^{-T/Q}}{r(1 + r)^{(q-1)T/Q}};$$

$q = 1, \ldots, Q$ indicates the number of a particular demand period; Q denotes the number of demand periods in the economic life of the system, T/Q therefore being the length of each demand period (Q is assumed to be a proper divisor of T); $E_q(\mathbf{y}_q)$ denotes the annual efficiency benefits in period q as a function of the outputs in period q, that is,

$$E_q(\mathbf{y}_q) = \sum_{j=1}^{m} \int_0^{y_{qj}} D_{qj}(\eta_{qj})d\eta_{qj}, \tag{2.7}$$

where $D_{qj}(y_{qj})$ represents the demand for the jth output in period q; the term $M_q(\mathbf{x})$ denotes the constant annual OMR costs in period q, replacing the term $M_t(\mathbf{x})$ in Exp. (2.5).

The efficiency design criterion is, therefore, the maximization of the ranking function Exp. (2.6) subject to the production function,

$$f(\mathbf{x}, \mathbf{y}_q) = 0. \qquad q = 1, \ldots, Q \tag{2.8}$$

We have avoided an important complication introduced by the extension of the ranking function to systems with economic lives of many years, namely, the uncertainty of future willingness to pay. The uncertainty inherent in projections of demand and the technical problems of making forecasts over lengths of time as great as the economic lives of most systems are generally recognized. Consequently we shall not discuss them here other than to point out that, since benefits in distant years when discounted to their present values are smaller dollar for dollar than benefits in near-by years, difficulties and errors in projections become less important as they become more pronounced for distant years.

MARGINAL CONDITIONS OF EFFICIENCY MAXIMIZATION

The marginal conditions of maximization of the efficiency objective function are obtained mathematically by setting the first derivatives of the Lagrangian form of Exp. (2.6) and Eq. (2.8) equal to zero.[10] We can state these conditions in the form

$$\sum_{q=1}^{Q} \theta_q D_{qj}(y_{qj}) \frac{\partial y_{qj}}{\partial x_i} = \sum_{q=1}^{Q} \theta_q \frac{\partial M_q}{\partial x_i} + \frac{\partial K}{\partial x_i}. \qquad \begin{matrix} i = 1, \ldots, n \\ j = 1, \ldots, m \end{matrix} \tag{2.9}$$

[10] In the event that two or more designs satisfy the marginal conditions, the design maximizing the value of the ranking function must be chosen by direct comparison of the ranking-function values for all these designs. (Of course these designs must satisfy second-order conditions as well, distinguishing designs conferring a maximum value on the ranking function from those conferring a minimum or stationary value.)

These marginal conditions are applicable only to inputs and outputs at positive (or "interior") levels. For inputs and outputs at zero (or "boundary") values, inequalities replace the equalities appearing in this section. Zero levels of outputs and inputs will always satisfy the appropriate marginal inequalities in the presence of indivisibilities, that is, when any level of development of an input or output, no matter how small, entails a positive "fixed" cost. Thus, in the presence of indivisibilities, if a positive-level design satisfies the marginal conditions of this section, its rank relative to a design with zero levels of inputs or outputs must be found by direct comparison of the values of the ranking function for the two designs.

This formidable-looking expression[11] can be simply interpreted, element by element, as follows: θ_q is the discount factor defined above. The $D_{qj}(y_{qj})$ term represents the demand price in period q, that is, the price each beneficiary would pay for the last unit of output in period q rather than go without — in short, the marginal willingness to pay. The $\partial y_{qj}/\partial x_i$ term represents the marginal productivity in period q of the ith input (input is now defined as physical structure) when devoted to the jth output, in other words, the extra y_{qj} obtained from increasing the size of the ith structure by one unit. The product of these three terms measures the indirect present willingness to pay for an extra unit of the ith input by the recipients of the jth output in each year of the qth demand period. Equal for all outputs, this expression is simply the marginal benefit from the ith input in each year of the qth period. The left-hand side of Eq. (2.9) is thus the present value of the annual marginal efficiency benefit, and the right-hand side is the sum of the present value of annual marginal OMR costs of the ith input, $\partial M_q/\partial x_i$, and the marginal capital cost, $\partial K/\partial x_i$. In simpler notation, Eq. (2.9) then becomes, by virtue of the equality of the left-hand side for all outputs,

$$\sum_{q=1}^{Q} \theta_q \mathrm{MVP}_q(x_i) = \sum_{q=1}^{Q} \theta_q \mathrm{M}M_q(x_i) + \mathrm{M}K(x_i) \qquad i = 1, \ldots, n \tag{2.10}$$

where $\mathrm{MVP}_q(x_i)$ indicates the marginal annual benefit ("marginal value product") in period q from an extra unit of input x_i (the partial derivative $D_{qj}(y_{qj})\partial y_{qj}/\partial x_i$), and the terms $\mathrm{M}M_q(x_i)$ and $\mathrm{M}K(x_i)$ similarly represent marginal annual OMR and construction costs.

Equation (2.10) says that the present value of the marginal benefit derived from an extra unit of each input must equal the present value of its marginal cost. This mathematically and intuitively obvious condition states that sizes of structures (and consequently levels of outputs) cannot be optimal, in efficiency terms, if the present value of the extra willingness to pay exceeds the present value of the extra goods displaced by further increasing the size of the system.

We can rewrite Eq. (2.10) as

$$\frac{\sum_{q=1}^{Q} \theta_q[\mathrm{MVP}_q(x_i) - \mathrm{M}M_q(x_i)]}{\mathrm{M}K(x_i)} = 1. \qquad i = 1, \ldots, n \tag{2.11}$$

[11] Actually, Eq. (2.9) is not the original form of the marginal conditions; some of the conditions obtained by setting the derivatives of the Lagrangian form of Eq. (2.4) and Exp. (2.6) equal to zero have been employed to eliminate the Lagrangian multipliers from other marginal conditions, thus simplifying the discussion.

This says that for a design to be optimal the ratio of the present value of the marginal efficiency benefit less the marginal OMR cost to the marginal construction cost, or more simply the marginal benefit-cost ratio, must be unity. Stated in this form, the marginal conditions will be recognized by water-resource-planners as identical with those in the "Green Book." [12]

Further examination yields an interpretation of marginal conditions more akin to the economist's notions of the significance of efficiency maximization. This interpretation can be presented in the simpler context of a system with a single demand period, that is, a system for which demands and OMR costs are constant over its economic life. Accordingly, the marginal conditions of the equality of marginal benefit and cost, Eq. (2.9), become

$$\theta_1 D_{1j}(y_{1j}) \frac{\partial y_{1j}}{\partial x_i} = \theta_1 \frac{\partial M_1}{\partial x_i} + \frac{\partial K}{\partial x_i}. \qquad \begin{matrix} i = 1, \ldots, n \\ j = 1, \ldots, m \end{matrix} \qquad (2.12)$$

Dividing by $\theta_1 \partial y_{1j}/\partial x_i$, we obtain

$$D_{1j}(y_{1j}) = \frac{\partial M_1}{\partial x_i} \frac{\partial x_i}{\partial y_{1j}} + \frac{1}{\theta_1} \frac{\partial K}{\partial x_i} \frac{\partial x_i}{\partial y_{1j}}. \qquad \begin{matrix} i = 1, \ldots, n \\ j = 1, \ldots, m \end{matrix} \qquad (2.13)$$

Since the inverse $\partial x_i/\partial y_{1j}$ of the marginal productivity $\partial y_{1j}/\partial x_i$ measures the increase in the size of the ith unit of the system required to provide an increase of one unit annually of the jth output, and since the marginal OMR and capital-cost terms $\partial M/\partial x_i$ and $\partial K/\partial x_i$ represent the extra annual and initial costs per extra unit of the ith input, the products $(\partial M_1/\partial x_i)(\partial x_i/\partial y_{1j})$ and $(\partial K/\partial x_i)(\partial x_i/\partial y_{1j})$ represent the marginal OMR and construction outlay entailed by an increase of one unit annually of the jth *output*, which we denote by $MM_1(y_{1j})$ and $MK_1(y_{1j})$ respectively. As before, the annual demand price $D_{1j}(y_{1j})$ of the jth output is a measure of the marginal willingness of each beneficiary to pay for output y_{1j}, denoted by $ME_1(y_{1j})$. Thus Eq. (2.13) expresses the marginal benefit-cost relationship in terms of outputs rather than inputs:[13]

[12] *Proposed practices for economic analysis of river basin projects*, a report to the Inter-Agency Committee on Water Resources by its Subcommittee on Evaluation Standards (May 1958).

[13] There is only one complication. Marginal benefits and OMR costs are expressed on an annual basis, whereas marginal construction outlay embodies the different dimension of a once-and-for-all expenditure. It is legitimate, however, in view of the assumed constancy of demands and OMR costs, to regard the inverse of the present-value factor, namely $1/\theta_1$, as an "amortization factor" or annual "interest and depreciation charge" per dollar of original construction outlay. Thus $MK_1(y_{1j})/\theta_1$ can be thought of as the *annual* (amortized) marginal construction cost of the jth output.

$$\mathrm{M}E_1(y_{1j}) = \mathrm{M}M_1(y_{1j}) + \frac{\mathrm{M}K(y_{1j})}{\theta_1}. \qquad j = 1, \ldots, m \qquad (2.14)$$

In Eq. (2.14) the marginal willingness to pay for the jth output is the rate at which each immediate recipient of outputs is willing to substitute money for the jth output in each year, and the sum of the marginal annual OMR and construction costs represents the marginal rate at which money can be "transformed" into the jth output by construction and operation of a water-resource system. If we express the marginal rates of substitution and transformation by $\mathrm{MRS}_{\$y_{1j}}$ and $\mathrm{MRT}_{\$y_{1j}}$ respectively, Eq. (2.14) becomes

$$\mathrm{MRS}_{\$y_{1j}} = \mathrm{MRT}_{\$y_{1j}}. \qquad j = 1, \ldots, m \qquad (2.15)$$

This states that, when efficiency is the sole consideration, system design is not optimal if the rate at which the recipients are willing to substitute dollars for outputs in each year differs from the rate at which dollars can be transformed into outputs, including the dollars required for interest and depreciation of the system.

MAXIMIZING EFFICIENCY IN THE PRESENCE OF A BUDGETARY CONSTRAINT

The efficiency design criterion as it has been stated above is concerned solely with maximization of efficiency unconstrained by considerations of other objectives, limited budgets, or institutional factors. Let us now examine its form in the presence of a budgetary constraint.

A budgetary constraint is simply another objective of public policy, a manifestation of a desire to limit the scale of public development of water resources. Although it can take many forms, including a constraint on construction cost alone or together with OMR outlays, we shall for the present use the constraint on construction cost alone. In Chapter 4 we shall discuss the alternative forms of the constraint in more detail.

Let us assume an over-all water-resource construction budget of $\overline{K}$ dollars to be allocated among S systems and designate the subbudget allocated for the sth system as $\overline{K}_s$ dollars. The efficiency design criterion for the sth system then becomes the choice of the most efficient design among all designs with costs less than or equal to $\overline{K}_s$ dollars; or what is the same, maximization of

$$\sum_{q=1}^{Q} \theta_q[E_{sq}(\mathbf{y}_{sq}) - M_{sq}(\mathbf{x}_s)] - K_s(\mathbf{x}_s), \qquad (2.16)$$

subject to the subbudget constraint

$$K_s(\mathbf{x}_s) \leq \overline{K}_s \tag{2.17}$$

as well as to the production function for system s.

Differentiating the Lagrangian form of Eqs. (2.16) and (2.17) yields, in lieu of Eq. (2.9), the following marginal condition for maximization:

$$\sum_{q=1}^{Q} \theta_q D_{sqj}(y_{sqj}) \frac{\partial y_{sqj}}{\partial x_{si}} = \sum_{q=1}^{Q} \theta_q \frac{\partial M_{sq}}{\partial x_{si}} + (1 + \lambda_s) \frac{\partial K_s}{\partial x_{si}} \qquad \begin{matrix} i = 1, \ldots, n \\ j = 1, \ldots, m \end{matrix} \tag{2.18}$$

and, upon simplification,

$$\frac{\sum_{q=1}^{Q} \theta_q [\text{MVP}_{sq}(x_{si}) - \text{M}M_{sq}(x_{si})]}{\text{M}K_s(x_{si})} = 1 + \lambda_s \qquad i = 1, \ldots, n \tag{2.19}$$

in place of Eq. (2.11). Equation (2.19) indicates that the marginal benefit–cost ratio on each system structure will exceed unity by λ_s in the presence of an effective budgetary constraint. If, for example, λ_s were to equal 0.5, no system increments should be undertaken for which the benefit–cost ratio would be less than or equal to 1.5. This is to be expected, as the limitation on outlays makes it impossible to design all system units to the scale at which the present value of the marginal annual benefit minus OMR costs no longer exceeds the marginal construction outlay. Since it becomes necessary to hold expenditure on some units below the level at which the marginal benefit–cost ratio equals unity, it is apparent that, for the design attaining the maximum efficiency permitted by the constraint, the marginal benefit-cost ratios on all system units must be equal. For as long as the margins are not equalized, funds can be transferred from units with lower marginal benefit–cost ratios to those with higher ratios, thus leading to a new design with higher net benefits.

Of course, the greater the scale of system units, the more nearly does the value of λ_s, the "cutoff benefit–cost ratio," approach unity. Why not therefore increase the subbudget for system s enough to drive the marginal benefit–cost ratio all the way down to unity and thereby reap all the benefits that the system potentially offers? The answer lies in the fact that to do so we would have to take funds away from some other system; it clearly would not pay to employ an extra dollar in the construction of system s to reap a reward of $1.01 if this dollar must come from a system u where it could earn $1.50.

This impels us to consider the problem of determining the appropriate subbudget for each system. Unless the reason for the budgetary constraint goes beyond limiting the over-all size of the public water-resource development program, the goal of allocating a given budget among the component systems is to maximize efficiency net benefits for the program as a whole. Consider for the moment the entire program as a single system consisting of S subsystems. Expressing the *present value* of gross benefits and OMR costs of subsystem s (s now being employed as a running subscript over the S subsystems in the program) as functions $E^{(s)}(\overline{K}_s)$ and $M^{(s)}(\overline{K}_s)$ of the subbudget allocated for outlay on subsystem s, $\overline{K}_s$, the criterion for efficient subdivision of the over-all budget, becomes maximization of

$$\sum_{s=1}^{S} [E^{(s)}(\overline{K}_s) - M^{(s)}(\overline{K}_s) - \overline{K}_s],$$

subject to the constraint that the sum of the subbudgets for the subsystems must not exceed the entire program budget, $\overline{K}$, or

$$\sum_{s=1}^{S} \overline{K}_s \leq \overline{K}.$$

If the constraint is effective, the marginal conditions for efficient allocation of the budget are

$$\text{M}E^{(s)}(\overline{K}_s) - \text{M}M^{(s)}(\overline{K}_s) = 1 + \lambda. \qquad s = 1, \ldots, S$$

According to this equation, the present value of the marginal efficiency benefit less the marginal OMR cost contributed by the last dollar of construction outlay must be the same, $1 + \lambda$, for all subsystems. In other words, the marginal cutoff benefit–cost ratio, $1 + \lambda_s$, must not only be identical for all segments in each subsystem, it must also be the same for all subsystems.

The problem of implementing budgetary constraints has been greatly simplified in this section by assuming that all construction takes place in the year zero. In Chapter 4 the complications entailed in relaxing this assumption will be considered.

EXAMINING THE ASSUMPTIONS

In our exposition of the efficiency ranking function and design criterion, we made several assumptions that, with the exception of assumptions regarding uncertainty, we shall examine in this section.

Constant marginal utility of income and unfettered profit maximization. The assumptions of constant marginal utility of income for consumers' goods and unfettered profit maximization for producers' goods were essential not only to the expression of efficiency in terms of willingness to pay but also to the equating of a beneficiary's willingness to pay for the output he receives with the area under his demand curve.

Constancy of marginal utility of income for an individual means, in general, that each of his successive dollars of income provides the same satisfaction. Applied to his full range of income, this would mean that the last hundred dollars of income, spent for example on chrome trim for a car, must give the same satisfaction as the first hundred dollars, spent for food, clothing, and shelter. This is clearly a dubious assumption. The "income" received from system outputs that are consumers' goods (for example, electric energy and water consumed in the home), however — measured in terms of the willingness of individuals to pay for the quantities of outputs they receive[14]— is likely to be sufficiently small in relation to an individual's total income to justify the simplifying assumption of constant marginal utility of income over the range of increase in "income" from system outputs that are consumers' goods.

In assuming unfettered profit maximization for producers' goods, we are asserting that beneficiaries of outputs such as irrigation water and industrial energy, whether large corporations or small enterprises, are motivated solely by the goal of maximizing the difference between revenues received from the sale of goods produced and outlays for intermediate goods purchased (including the imputed value of labor and other factors supplied by the producer himself). Further, we assume that limitations of finance capital do not restrict the scale of producers' operations — that to producers lacking the capital to finance a profit-maximizing scale of operations, adequate credit facilities will be made available.

Each of these assumptions, quite obviously, entails some simplification of reality. Large enterprises are motivated by considerations such as market power as well as by maximization of profits; and for small, personal enterprises such as family farms it may be impossible to isolate a profit-maximization motive from the general pursuit of utility by the individual or family.[15] Also, in the world in which enterprise operates,

[14] Willingness to pay measures the incomes of individuals from system outputs provided they pay nothing. If they are required to pay something, the amount of repayment must be subtracted from their willingness to pay to calculate their net gain.

[15] Why should it be necessary or desirable to separate a family's pursuit of profit from its pursuit of utility? Why not simply rely on the assumption of the constancy of the marginal

shortages of finance capital do exist. Nevertheless, for our purposes these complications can be waived, and the assumption of profit maximization can be accepted as an approximation of a capitalist economy.

Beneficiary repayment. We have ignored beneficiary repayment in our derivation of the efficiency design criterion, because potential system revenues play no part in the ranking of designs in terms of efficiency. Revenues can be looked upon simply as transfers of income from direct beneficiaries to the nation at large, that is, from the "positive beneficiaries" to the "taxpayers" of pages 21–22. To demonstrate the irrelevance of revenues, let us return to our derivation of the ranking function. There the relevant comparison, given the assumptions of constant marginal utility of income and unfettered profit maximization, was between the willingness of the beneficiaries of alternative designs A^1 and A^2 to compensate one another, including the willingness of taxpayers to compensate positive beneficiaries in order to keep their sacrifice at a minimum. Since revenues work symmetrically on positive beneficiaries and taxpayers, they do not affect over-all willingness to compensate. To the extent that revenues decrease the willingness of positive beneficiaries to compensate other beneficiaries on behalf of a design, they improve the relative disposition of taxpayers toward the design by lightening their tax burden.

Expressed in mathematical terms, the argument is as follows. Let $E(A^1)$ and $E(A^2)$ denote, as before, the total willingness of beneficiaries to pay for the two designs. Then if $R(A^1)$ and $R(A^2)$ denote the revenues that beneficiaries would actually be compelled to pay for the designs, $E(A^1) - R(A^1)$ represents the net willingness of beneficiaries of design A^1 to compensate those of design A^2, and similarly for $E(A^2) - R(A^2)$. If we denote, as before, the value of goods displaced by the system, or costs, by $C(A^1)$ and $C(A^2)$, the net sacrifices of taxpayers for the two designs are $C(A^1) - R(A^1)$ and $C(A^2) - R(A^2)$. If the net sacrifice is assumed to be smaller for A^1, then taxpayers will be willing to add the difference $[C(A^2) - R(A^2)] - [C(A^1) - R(A^1)]$ to the compensation offered the beneficiaries of A^2 to go along with A^1. Accordingly, design A^1 is more efficient than A^2 if the total compensation offered on behalf of A^2 is greater than that offered on behalf of A^1, or

utility of income? The answer is that the range of variation occasioned in producers' incomes by a water-resource development is very often too large to make the constancy of the marginal utility of income a valid approximation to reality. Contrast the change in income of a farmer from an irrigation project, for example, with the change in income of a residential household from the provision of system energy.

$$E(A^1) - R(A^1) + [C(A^2) - R(A^2)] - [C(A^1) - R(A^1)] > E(A^2) - R(A^2),$$

which reduces to the form

$$E(A^1) - C(A^1) > E(A^2) - C(A^2).$$

The argument is symmetric if taxpayers' net sacrifice is smaller for A^2, so it follows that revenues are immaterial in paired comparisons of alternative designs, and that designs $A^0, \ldots, A^j, \ldots$ can still be ranked by Eq. (2.1),

$$W(A) = E(A) - C(A).$$

Although potential revenues do not affect system design, design does affect revenues, and it is instructive to examine this relationship. For simplicity in exposition let us assume a single-demand-period, single-purpose irrigation project, such as that outlined in Fig. 2.3. Here curves

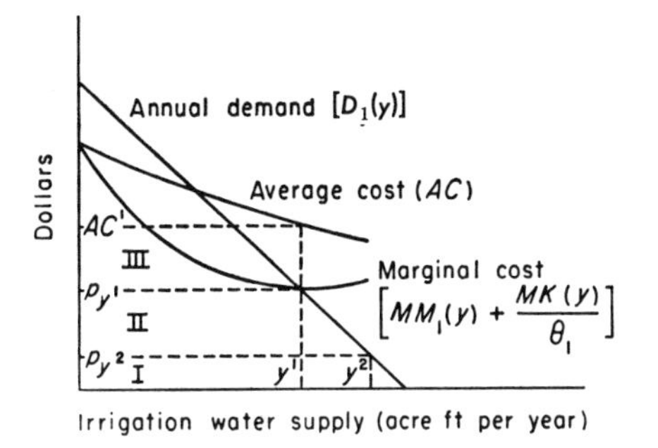

Fig. 2.3. Cost and revenue functions for a single-purpose irrigation project.

representing annual demand, marginal cost, and average cost of irrigation water are labeled $D_1(y)$, $MM_1(y) + MK(y)/\theta_1$, and AC respectively. It is evident from the local marginal condition of Eq. (2.14) that y^1 acre ft of irrigation water per year represents the optimal scale of development, if efficiency is the sole objective of design. However, the annual revenues yielded by the y^1 output depend on which of several options is adopted by the authority in charge of distribution of irrigation water. It can use, for example, the market mechanism to distribute the water, charging the demand price p_{y^1} to all farmers, at which price farmers will want no more and no less than y^1 units. In this case the revenue is simply the market value $p_{y^1}y^1$, given by the sum of the areas of rectangles I and II. If the distribution authority wants to give farmers more help, however, it can charge a lower price, say p_{y^2}. For this policy, system revenues equal $p_{y^2}y^1$, or, graphically, the area of rectangle I. At price p_{y^2}, however, farmers would want to purchase quantity y^2. Thus a policy of charging p_{y^2}, or any price less than the market clearing price p_{y^1}, would necessarily entail rationing the y^1 acre ft of water among the farmers. On the other hand, should the distribution authority wish to recover a greater portion of the gains of farmers than the market value of y^1, it must employ one or more forms of price discrimination: higher prices in accordance with the higher willingness to pay of some groups or individuals, successively lower prices in accordance with the number of units of output purchased, or two-part pricing schemes that include an "entry fee" for the privilege of purchasing output at a given price, say p_{y^1}, per unit.[16]

There is a further consideration. In our example, the average cost of y^1 acre ft of water, AC^1, exceeds the demand price. The total cost, $(AC^1)(y^1)$, or the sum of rectangles I, II, and III, thus exceeds the revenue obtained by charging the market clearing price p_{y^1}, namely, the market value of y^1. Therefore, if cost recovery is an objective of public policy, it is incumbent upon the distribution authority to employ one or more forms of price discrimination.

We should note here that the distribution authority is not so free to choose among distribution policies as we have implied. Our discussion of revenues has, in fact, been purely theoretical, taking no account of the administrative problems inherent in maintaining efficiency in the presence of deviations from a policy of charging the market clearing price.

The competitive assumptions. In effect, the efficiency ranking function reduces to balancing the preferences of individuals, backed by willingness to pay, for the final goods of a water-resource system against their preferences for the goods displaced by the system. The competitive assumptions allow us to consider preferences solely in the market place — to rely, therefore, solely upon market demand curves for the calculation of benefits, upon market prices for the evaluation of costs, and upon the market interest rate for converting the stream of benefits and costs to their equivalent present value.

This principle of "consumer sovereignty," as reliance upon market-exhibited preferences is termed, is the distinguishing characteristic of the efficiency ranking function in a competitive model. In the sections that follow, we shall relax the competitive assumption in order to render

[16] An example of a two-part pricing scheme is a capacity charge coupled with an energy charge in the sale of electric power.

the efficiency design criterion more applicable to the environment in which planning decisions must be made. The effect, as we shall see, is to deemphasize market-exhibited willingness to pay, as well as market costs, in order to take account of certain aspects of costs and willingness to pay not exhibited in the market place. In other words, in place of the competitive framework in which market prices are the final and ultimate revelations of individual preferences, we shall substitute a broader framework that takes into account willingness to pay and costs not embodied in market demand curves and market prices; in short, we shall substitute a broader notion of "individual sovereignty" for the narrower concept of consumer sovereignty.

External effects. We shall use the term "external effect" to describe a situation in which one individual derives gain or loss from the provision of goods for someone else.[17] Many kinds of external effects are encountered in the development of water resources, and here we examine a few that are typical. These are external effects involving (1) time, (2) collective goods, (3) altruistic and misanthropic considerations, and (4) jurisdictional relationships.

The first of these deals with the relationship of identical physical outputs of a water-resource system at different times of the year. When provision of an output in one period of the year inevitably implies its provision in another period, there are external effects between the individuals receiving the outputs in the different periods. Of course, this would not be true if outputs could be stored, but some outputs, such as recreation, obviously cannot be. How do these external effects come about? Suppose, for example, that I wish to use the recreational facilities — for camping, boating, fishing — afforded by a particular water-resource development only in summer, and that you wish to use them only in winter. In this case provision of "recreation output" for me means recreation for you as well, since the facilities I use in summer are available for your use in winter. Only the specific operating costs of winter recreation detract from your economic gain from the provision of summer recreation for me.

Not only is such an output nonstorable; no more can be produced in one period by cutting down on production in another period. This is in contrast with other joint products for which the level of production in one period can be increased by reducing the level of production in another

[17] While this is a somewhat broader definition of external effects than is customary, we feel it is more useful for our purposes.

period, or vice versa. This point can be expressed technically by comparing the production-function relationship between summer and winter recreation with the usual production-function relationship between outputs as presented in Chapter 3. As an example, let us compare the relationship between a single input x (the capacity of a single reservoir, for instance) and, alternatively, irrigation divided into two seasons y_{1s} and y_{1w} and recreation likewise divided into two seasons y_{2s} and y_{2w}.

The "variable-proportion" irrigation production function takes the form

$$f(y_{1s}y_{1w}, x) = 0 \tag{2.20}$$

and can be graphically represented as in Fig. 2.4. There isoquants

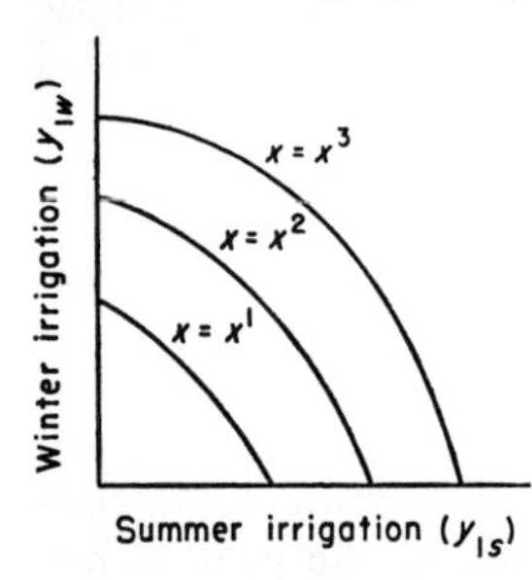

Fig. 2.4. Irrigation production function showing variable proportions of summer and winter irrigation obtainable from reservoirs of given sizes ($x = x^1$, x^2, and x^3).

labeled x^1, x^2, and x^3 show the boundaries of the sets of feasible combinations of summer and winter irrigation for the given level of x; and, by definition, only the points on the boundary loci themselves lie on the production function. For input level x^3, for example, points below locus $x = x^3$ are inefficient and points above it are unattainable. The possibility of substitution between summer and winter irrigation is verified by the absence of vertical or horizontal stretches in the isoquants, more irrigation being made possible in summer by reducing the level of irrigation in winter and storing the unused water, and vice versa. The negative of the slope of the isoquants, furthermore, indicates the marginal rate at which outputs can be substituted for one another while the level of input is held constant.

The production-function relationship between summer and winter recreation and the system input,[18] on the other hand, takes the more complicated "fixed-proportion" form

$$\begin{aligned} f^{(s)}(y_{2s}, x) &= y_{2s} - y(x) = 0 \\ f^{(w)}(y_{2w}, x) &= y_{2w} - y(x) = 0 \end{aligned} \tag{2.21}$$

[18] Expressing recreation output as a function of reservoir capacity alone is patently unrealistic. It is done for expositional simplicity only.

as shown graphically in Fig. 2.5. Once again the isoquants are seen to be the boundaries of the sets of feasible combinations of outputs for

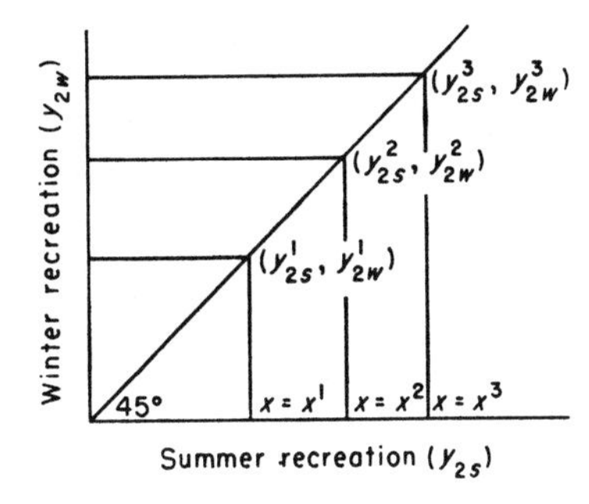

Fig. 2.5. Recreation production function showing fixed proportions of summer and winter recreation obtainable from reservoirs of given sizes ($x = x^1$, x^2, and x^3).

each given level of input, but this time only points such as $(y^1{}_{2s}, y^1{}_{2w})$, $(y^2{}_{2s}, y^2{}_{2w})$, and $(y^3{}_{2s}, y^3{}_{2w})$, on the 45-degree line lie on the production function. The slope of the isoquants therefore changes abruptly in each case from infinity to zero at the 45-degree line, these right angles indicating the impossibility of increasing recreation in one season by reducing it in the other, as is possible for irrigation.

The consequences of this external effect show themselves in the marginal conditions of efficiency maximization rather than in the efficiency ranking function. If the demands for recreation in the two seasons, denoted now by y_s and y_w, are independent of one another and constant in each season over the life of the system, then the ranking function for a single-purpose recreation system is simply the familiar difference between the sum of the efficiency benefits of the two outputs and costs,

$$\theta_1 \left[\int_0^{y_s} D_s(\eta_s)d\eta_s + \int_0^{y_w} D_w(\eta_w)d\eta_w - M(\mathbf{x}) \right] - K(\mathbf{x}), \qquad (2.22)$$

in which $D_s(y_s)$ and $D_w(y_w)$ denote the demands for summer and winter recreation respectively. In Exp. (2.22) we assume for simplicity that there are no separable OMR costs associated with recreation output in each season, and the variable $\mathbf{x}$ can once again be thought of as a vector of inputs rather than a single input.

The marginal conditions of maximization of the ranking function (2.22), subject to the production function Eq. (2.21), however, become

$$\theta[D_s(y_s) + D_w(y_w)] \frac{\partial y}{\partial x_i} = \theta \frac{\partial M}{\partial x_i} + \frac{\partial K}{\partial x_i}. \qquad i = 1, \ldots, n \qquad (2.23)$$

Stated in words, we compare the present value of the sum of demand prices for the two outputs with the present value of the joint marginal cost instead of comparing, as in Eq. (2.12), the demand price of each output with the present value of its marginal cost.

Since our two system outputs are joined like Siamese twins in the present case, the one being produced as an automatic and costless counterpart of production of the other, it is evident that the relevant marginal comparison is between the willingness to pay and the combined cost of summer and winter recreation, as expressed in Eq. (2.23), rather than the cost of each separately. In the terms of efficiency maximization familiar to economists, when dealing with joint outputs we replace Eq. (2.15) relating the marginal rate of substitution and the marginal rate of transformation by the following:

$$\mathrm{MRS}_{\$y_s} + \mathrm{MRS}_{\$y_w} = \mathrm{MRT}_{\$y(\mathbf{x})}, \tag{2.24}$$

where $y(\mathbf{x})$ denotes recreation output in each season.

The "time" external effect is not limited to recreation. It obtains also when, for example, electric energy is produced by a run-of-the-river power plant, or water transpoıt is provided by a navigable channel. Today's kilowatt hour or transport ton mile is either used or wasted; a reduction of output today, therefore, does not make possible an increase in output tomorrow. In contrast, there are the outputs in one season (such as irrigation or municipal water supply, and energy produced by power plants operated in conjunction with storage reservoirs) that can be substituted for the physically identical outputs in another season.

The second class of external effects is associated with production of collective goods. "Normal" goods have the property that, given a fixed amount, more units for ourselves mean fewer for others and vice versa. However, there are categories of goods for which this is not true. National defense is the leading example. Each one of us "consumes" the same amount of national defense, and although each person added to the population also consumes national defense, his consumption does not necessarily make less available for the rest of us. Goods of this kind are called "collective" or "public" goods; in contrast, so-called "normal" goods are called "personal" or "private" goods.[19]

Flood control through reservoir storage is a striking example of a col-

[19] Although others have addressed themselves to the problem of collective goods, the name that comes to mind in any discussion of the subject is that of Paul Samuelson. See his articles "The pure theory of public expenditure"; "Diagrammatic exposition of public expenditure"; and "Aspects of public expenditure theories" in *Rev. Econ. and Stat. 36*, 387 (1954); *37*, 350 (1955); and *40*, 332 (1958) respectively.

lective good associated with water-resource systems. Within the protected area, all benefit from this type of flood control; provision of protection for one automatically protects all others in the area too. In other words, one individual gains from provision of flood control for another, which is the criterion of an external effect.

Collective external effects, like time external effects, do not alter the basic definition of efficiency benefits as willingness to pay for outputs, but they do change the relationship of willingness to pay to individual demand curves. Willingness to pay is still defined as the area under the aggregate demand curve, but the way in which this curve for collective goods is derived from individual demand curves is different from the way it is derived for personal goods.

To illustrate this difference, let us return to the determination of the aggregate demand curve for a personal good such as irrigation water in a single demand period. If we assume that there are only two beneficiaries, with individual demand curves such as lines D_1D_1 and D_2D_2 in Fig. 2.6,

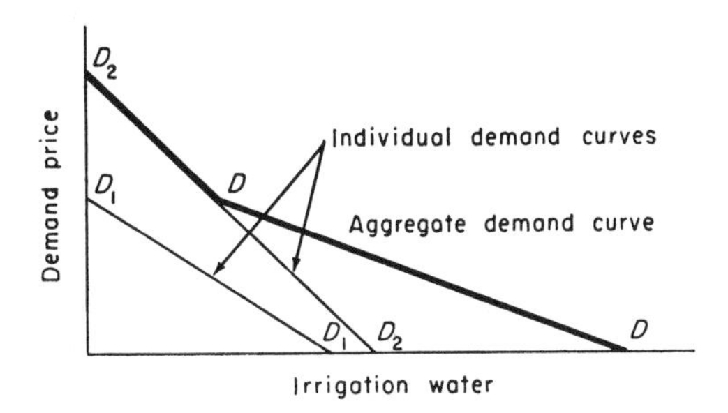

Fig. 2.6. Individual and aggregate demand curves for a personal good, such as irrigation water.

the aggregate demand schedule for irrigation water is the *horizontal* sum *DD* of the individual demands, because each beneficiary derives gain only from the irrigation water he consumes. The resulting heavy line D_2DD, therefore, is the locus of sums of the quantities that each beneficiary demands at each price.

In Fig. 2.7, which shows the derivation of the aggregate demand curve for flood control, the same D_1D_1 and D_2D_2 now represent the respective individual demands for flood control. Since both beneficiaries consume flood protection simultaneously, the aggregate demand curve for flood control is the *vertical* sum DDD_2 of the individual prices for each quantity, not the horizontal sum of the individual demands at each price. With this definition of the demand-price function $D_{qj}(y_{qj})$ for collective goods,[20] efficiency benefits continue to be expressed as the discounted sum of the integrals of aggregate demands for outputs. This presents the problem of measuring the aggregate demand curve for a collective good. Apart

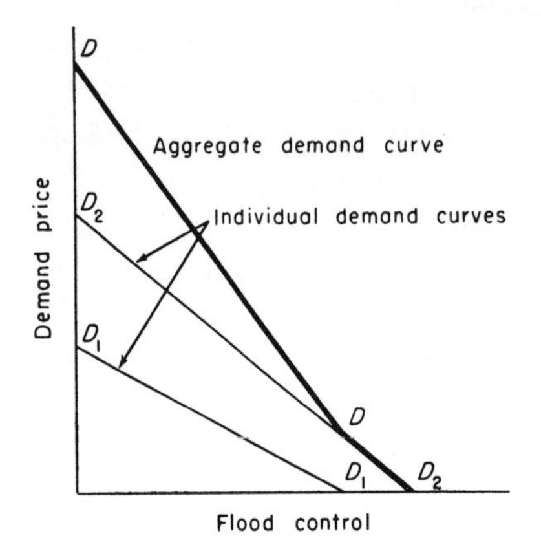

Fig. 2.7. Individual and aggregate demand curves for a collective good, such as flood control.

from uncertainty, the problems of measuring demand for personal goods are empirical rather than conceptual; we can imagine testing the market for irrigation water, for example, to determine the quantities farmers would demand at successive prices. However, because of the nonmarketable nature of flood control, it is not possible to visualize, even conceptually, the aggregate demand curve for flood control as a schedule of the various quantities that would be demanded in a market at successive prices. For this reason actual measurement of willingness to pay for flood control is ordinarily framed in terms of direct evaluation of the damages that successive degrees of flood protection prevent, on the assumption that this reflects the individual willingness to pay for protection.

Since the ranking function and the production function for personal goods are not formally altered by collective goods, the marginal conditions

[20] Analytically, the demand-price function $D_{qj}(y_{qj})$ for a collective good is defined as the sum of the inverses $p_{qj}{}^{s}(y_{qj})$ of individual demand functions, $y_{qj}{}^{s}(p_{qj})$. For S individuals, therefore, the demand-price function is

$$D_{qj}(y_{qj}) = \sum_{s=1}^{S} p_{qj}{}^{s}(y_{qj}).$$

This is to be contrasted with the analytic definition of the demand-price function for private goods. $D_{qj}(y_{qj})$ is itself the inverse of the sum of individual demand curves, that is, the inverse of

$$\sum_{s=1}^{S} y_{qj}{}^{s}(p_{qj}).$$

of efficiency maximization also remain formally unchanged. As for personal goods, the marginal condition for flood control, denoted by y_{fc} for the time being, is

$$\theta_1 D_{fc}(y_{fc}) \frac{\partial y_{fc}}{\partial x_i} = \theta_1 \frac{\partial M_1}{\partial x_i} + \frac{\partial K}{\partial x_i}, \qquad i = 1, \ldots, n$$

provided that the demand for flood control is constant over the life of the system. However, the significance of the marginal condition is altered part and parcel with the changed meaning of the demand-price function for a collective good; whereas for a private good the marginal annual efficiency benefit $D_{fc}(y_{fc})$ represents the marginal willingness of each recipient of the output to pay, it represents for a collective good the sum of the marginal willingness of all individuals to pay for the output. Accordingly, the economist's statement of the marginal condition for personal goods, Eq. (2.15), is replaced by the following statement for a collective good such as flood control:

$$\Sigma \, \text{MRS}_{\$y_{fc}} = \text{MRT}_{\$y_{fc}}, \tag{2.25}$$

where the summation extends over the marginal rates of substitution of money for flood control of all the individuals obtaining flood protection.[21]

The third type of external effects are psychic interrelationships between one individual's consumption and another's satisfaction. Through jealousy or inconvenience my satisfaction may be decreased when your consumption is increased; altruism or civic interest may have the opposite effect. Moreover, these external effects may be collective as well: you and I simultaneously may derive pleasure or displeasure from a third person's consumption.

The most important altruistic external effects are associated with the desire to provide for future generations. Each of us probably wishes to see future generations have higher standards of living than would be the case if the future were left to "take care of itself." No one of us, however, is willing to sacrifice his own consumption to this end, for the psychic reward to each from such sacrifice is not so great as the satisfaction obtainable from present consumption. Nevertheless, each would like to

[21] Compare the modifications in the marginal conditions introduced because of time external effects, Eq. (2.24), and collective external effects, Eq. (2.25). In both instances the single marginal rate of substitution is replaced by a sum of individual marginal rates of substitution. This similarity in the modification required by the two types of external effects suggests a close relation between them.

see others sacrifice to this end. We may all agree, therefore, that the community as a whole should undertake more investment than each would undertake individually. In this case, we can be sure that everyone will be compelled to cooperate, each of us viewing his own sacrifice as a small price paid to get others to make the investments he wants. In other words, because of these altruistic external effects a political distillation of individual time preferences for consumption over investment may well be different from a market distillation of these preferences, and may be preferred by each to the market distillation. The obverse of this is that a socially determined interest rate reflecting a different emphasis on the future relative to the present may be different from the competitive-market interest rate. This concept of the interest rate is discussed at greater length in Chapter 4.

The fourth and final category of external effects can be termed jurisdictional. If, for example, the Upper and Lower Colorado River systems were considered to be separate developments, increases in energy produced on the lower river by releases of water from storage for energy generation on the upper river[22] would be labeled an external effect in the benefit–cost analysis of the upper system. Optimization of development plans would not then include coordination of upstream and downstream design. Similarly, if a river development were planned primarily in terms of a single output, other outputs provided as by-products would be external effects in relation to the primary purpose. The obvious remedy is to break down the jurisdictional barriers that separate one part of a river system from another (or one river system from another) and to plan river developments for all relevant outputs rather than in terms of one or more primary outputs.[23]

All four of these categories of external effects share the characteristic that provision of a good for one individual leads to gains or losses for others. The first two can be taken into account within the framework of economic efficiency by comparatively modest changes of the definition of willingness to pay, the third by a change in the interest rate, and the fourth as just described.

[22] Downstream production is increased by upstream storage insofar as the latter makes the water available downstream in a more useful time pattern, that is, insofar as it "firms up" the downstream flow.

[23] A more detailed treatment of this kind of external effect is given in John Krutilla and Otto Eckstein, *Multiple purpose river development: studies in applied economic analysis* (Johns Hopkins University Press, Baltimore, 1958), chap. 3.

Equality of prices and marginal costs throughout the economy. Economic efficiency throughout the economy requires that prices reflect marginal costs, just as efficient design requires that system outputs be provided at levels for which prices reflect marginal costs. If prices in general fail to do so, others in addition to the immediate recipients of outputs and taxpayers may enjoy gains or suffer losses from water-resource developments which they would pay to secure or to avoid. While in this case the efficiency criterion is still based upon willingness to compensate, the term must be redefined to include the willingness of *indirect* beneficiaries (both positive and negative) to compensate, if alternative courses of action are to be evaluated properly.

Consider, for example, potential irrigation development with immediate benefits to a group of sugar-beet farmers. Let us assume that the harvest is sold to processers who, by virtue of their monopoly or oligopoly market position, sell refined sugar at a price above marginal cost and thereby enjoy an abnormal profit. In this case, although the relative efficiency of a potential design is still measured by the difference between costs and willingness of positive beneficiaries to pay, willingness to pay for irrigation water should include both that of the farmers and of the refiners, the willingness to pay of the refiners being their extra profits.

To take another case, if prices of resources employed in the construction of river-development systems are greater or less than their marginal opportunity costs,[24] money costs overstate or understate the sacrifices that individuals throughout the economy must make to provide the system and hence their willingness to pay to forego its construction. The money costs will continue to measure the sacrifices of taxpayers, but if, for example, prices exceed opportunity costs, the part of this sacrifice in excess of the value of goods displaced represents merely a transfer payment from taxpayers to the owners of the resources. In the efficiency calculus such transfers are properly ignored; only the value of goods displaced is to be included in measuring the cost of resources required for the construction of water-resource systems. This is so because transfer payments do not enter into the aggregate willingness to pay to forego the system, the sacrifice of the taxpayer being exactly balanced by the gain of the resource-owner.[25] Similarly, if opportunity

[24] For simplicity the comparatively minor OMR costs are neglected in the discussion that follows. Our results hold as well for these costs as they do for construction cost.

[25] Costs could be defined so as to include these transfers; then we would be obliged to count them as benefits to the resource-owners as well. On the whole, however, it seems easier to ignore transfer payments than to perform the double-entry exercise of counting them as both costs and benefits.

costs exceed prices, individuals other than taxpayers throughout the economy sacrifice consumption with each dollar of construction outlay; these sacrifices must also be counted into the willingness to pay to forego construction of the system. When opportunity costs are used in lieu of money prices of resources in efficiency-cost measurement, we shall call them the "shadow prices" of resources.

As an example of such shadow prices, let us consider first the evaluation of efficiency costs in the presence of involuntary unemployment. Here the market price of labor (the money wage rate) remains positive. However, the marginal opportunity cost of unemployed labor is zero;[26] no production is displaced by using formerly unemployed labor in constructing the system, and hence the "value of goods displaced" is zero. Conceptually, the required modification in the ranking function is quite simple: substitute a zero shadow wage rate for the positive money wage rate in evaluating the efficiency cost of labor which, were it not for the construction of the system, would be involuntarily unemployed. In actual operations, however, differences in the nature of involuntary unemployment give rise to differences in the manner of modifying the ranking function.

Basically, we can distinguish four broad types of unemployment: (1) cyclical unemployment of national dimensions, as in periods of business recession; (2) long-run "Keynesian" unemployment, characteristic of severe depression or "secular stagnation"; (3) local, spot unemployment, often chronic, such as characterizes depressed areas in a mature mixed-enterprise economy like that of the United States; and (4) structural unemployment, often disguised and perhaps more appropriately called "underemployment," characterizing underdeveloped economies.

The four types of unemployment do not affect in the same way the efficiency evaluation of construction costs occurring at different times, once the expositional assumption that all construction costs occur at once is dropped. Cyclical unemployment is characteristically short-lived in present-day "managed" economies, and thus zero shadow wage rates should be used to evaluate only the labor costs occurring in the early years of system construction, that is, within the expected length of the depression or recession. On the other hand, the remaining three categories

[26] Technically, leisure is a "good" produced by unemployment, and its value is the marginal disutility of employment. When unemployment is involuntary, the marginal disutility of employment can be considered negligible.

of unemployment tend to be long-lived, and in these cases future, as well as present, outlays for labor can properly be valued at zero shadow wage rates.[27]

Divergences also exist in the efficiency evaluation of labor costs under different types of unemployment because of qualitative variations in labor. Structural unemployment, for example, by definition manifests itself more in one category of labor skill than in another. Clearly, only the labor drawn from the unemployed categories can properly be evaluated at a zero shadow wage rate. Also, in spot unemployment technical and managerial labor must often be brought into depressed areas from areas of full employment. Their money costs should then not be discounted. And, with regard to cyclical unemployment, opportunity costs of all types of labor will fall to zero in only the severest of recessions. More mild downturns may not affect the opportunity cost of managerial and technical labor; only that of semiskilled and unskilled labor may be affected.

Rarely will all labor of any kind employed in system construction have zero opportunity cost, regardless of the type of unemployment. Some labor will inevitably be drawn from production of other goods. The average marginal opportunity cost of each category employed in system construction accordingly will be positive, not zero. Its appropriate shadow wage rate is the marginal opportunity cost of the force actually drawn from alternative employments multiplied by the percentage which this force forms of the total labor employed in this category in system construction.[28]

Two other examples of institutionally imposed shortages of resources that cause discrepancies between prices and marginal opportunity costs of individual resources are shortages of foreign exchange and budgetary constraints. When government places limitations on imports of capital goods in order to conserve foreign exchange, domestic prices, calculated

[27] Just as unemployment affects costs, so unemployment among beneficiaries affects benefits. A farmer's willingness to pay for irrigation water, for example, is measured by his income over and above what he would earn without irrigation. This is obviously greater if he is totally unemployed in the absence of the irrigation project. Unlike modifications on the cost side, this modification requires no specific change in the measurement of benefits, for it will be fully reflected in the willingness of the beneficiaries to pay.

[28] Insofar as jobs vacated are filled by otherwise unemployed individuals, the effect with respect to efficiency is the same as if the unemployed labor were employed directly in construction of the system; the average marginal opportunity cost must be adjusted downward accordingly.

at official rates of exchange, on imported capital goods used in water-resource development[29] will underestimate marginal opportunity costs. The import limitation makes it impossible to bring in sufficient capital equipment to drive the domestic value of the extra goods produced per unit of imported capital goods throughout the domestic economy down to official prices of the foreign capital goods calculated at the official rate of exchange. The (shadow) domestic prices of imported factors are correctly computed on the basis of a shadow exchange rate reflecting the domestic currency value (in terms of goods produced) throughout the economy per unit of foreign currency embodied in imported capital goods.[30]

Budgetary constraints give rise to institutionally imposed shortages of resources that lead to opportunity costs in excess of prices. This is not the interpretation originally placed on a limitation on construction expenditures, but it can be shown that our earlier discussion amounts to this. The marginal condition of efficiency maximization in the presence of a budgetary constraint, Eq. (2.19), tells us that no system units should be constructed beyond the scale at which the marginal benefit–cost ratio equals $1 + \lambda$. Funds allocated for development beyond this point would be taken away from other units or systems, thus leading to a contraction of outputs from other units or systems whose marginal net present value per dollar of construction outlay equals $1 + \lambda$. This means that the marginal opportunity cost or shadow price of construction outlay is $1 + \lambda$ dollars per dollar (although the nominal price of outlay is simply one dollar per dollar), because the loss of present value of outputs from other units and systems is the "value of goods displaced," previously defined as the opportunity cost.

This interpretation of the cutoff benefit–cost ratio as a shadow price of construction outlay provides the key to the allocation–design problem that arises because of the practical impossibility of simultaneously determining designs of all systems falling under the budgetary constraints. If the appropriate value of the shadow price were known, each system design could be determined independently, simply by maximizing the

[29] The term "capital goods" here refers both to capital goods used in construction (earthmovers, for example) and capital goods forming part of the system (such as generators).

[30] This modification in evaluation of efficiency costs has a counterpart in benefit-counting: in computing the willingness to pay for outputs of producers' goods as the residual profits after payments (explicit or implicit) to other factors of production, the payments for imported capital goods should be evaluated on the basis of their shadow prices. This procedure parallels that suggested in footnote 27, except that the adjustment is not automatic.

ranking function, Exp. (2.16), with each dollar of construction outlay evaluated at $1 + \lambda$ dollars. The factor limiting this use of the shadow price is that knowledge of the value of λ presupposes knowledge of the optimal designs for all systems. However, in Chapter 4 we present methods of overcoming this limitation through procedures of independent design of each system and determination of λ by successive approximations.

There are two important differences between deviations of opportunity costs from money costs arising from budgetary constraints and deviations caused by unemployment and foreign-exchange shortages. First, the deviation due to budgetary constraints is internal to water-resource development — that is, it is caused by a difference between the value of goods displaced and their money costs within the development program. By contrast, deviations between opportunity costs and prices of labor and imported capital goods are external — determined by differences between the goods displaced and their money costs outside the water-resource development. Second, deviations caused by budgetary constraints apply to all construction outlays equally, whereas deviations caused by unemployment and shortages of foreign exchange apply only to outlays for labor and for imported capital goods, respectively.

This second distinction has important consequences for system design. Use of shadow prices higher than money prices in evaluating resources employed in construction always reduces the scale of efficient development. (The effect of a shadow price lower than the money price naturally is the reverse.) Although this scale effect is the only result of a symmetric change in the operative prices of all resources (such as is imposed by a budgetary constraint), changes in the operative prices of individual resources (such as labor or certain capital equipment) lead, in addition, to a substitution effect. By virtue of least-cost considerations, a rise in the operative price of an individual resource leads to the use of less of this resource in construction, both absolutely and in relation to other resources. The substitution effect of a decrease in the price of a resource operates in the opposite direction. Thus evaluation of labor costs at shadow wage rates in unemployment situations, for example, means substitution of labor for capital as well as bigger systems. In comparison with full-employment designs or unemployment designs based on money wage rates, therefore, the unemployment designs based on shadow wage rates will tend to be more labor-intensive in addition to being larger. The effects of using a shadow price that is higher than the money price of foreign capital goods calculated at the official rate of exchange is, of course, the reverse: in the direction of smaller systems and less utilization of foreign machines.

Empirical estimation of the appropriate values of resource shadow prices, when these differ from nominal money prices, is a formidable task. Comparable to that of estimating demands, it is less imposing only because single parameters, rather than functions, are in question and because construction costs, unlike demand functions, need not in our model be projected 20, 50, or even 100 years ahead. In one aspect, however, determination of shadow prices may be even more difficult. The values of resource shadow prices may depend on system designs themselves, as do budgetary shadow prices, and consequently may have to be determined by an iterative process of successive designs and approximations to the shadow prices.

Because of this basic similarity in the determination of correct shadow prices and designs, the discussion of the iterative implementation of budgetary constraints in Chapter 4 is applicable as well to the determination of shadow prices of individual resources when the water-resource program is large enough in its own right to be a determinant of the marginal opportunity costs of the resources.

When we move from consideration of isolated discrepancies of prices from marginal costs such as those considered above to pervasive and large deviations throughout the economy, the simple techniques that we have described are no longer adequate to identify the potential gains and losses of each design throughout the economy.[31] We cannot then expect to measure even all the "tangible" gains and losses. But we can reasonably take the view that significant economy-wide deviations of prices from marginal costs are prima facie evidence of the lack of importance to society of economic efficiency in general, at least in terms of the criterion of willingness to compensate; for if efficiency were a major consideration, society presumably would already have taken steps, through the political process, to correct the wholesale discrepancies. And if efficiency were not an important social objective, it would follow that it should not be of much concern in the design of river systems.

[31] We can, however, be sure that in the presence of deviations from the marginal efficiency conditions (that is, price equals marginal cost) in other sectors of the economy, implementation of the unmodified marginal conditions of the competitive model in the water-resource sector, Eq. (2.11), will not lead to the optimum. For discussion of this problem in a more general context, see R. G. Lipsey and Kelvin Lancaster, "The general theory of the second best," *Rev. Econ. Stud. 24*, No. 11, 1956–57. Our treatment above of isolated discrepancies of prices from marginal costs is compatible with the conclusions of "second-best" reasoning.

There are known situations in which prices differ from marginal costs, and efficiency, construed more broadly than in terms of our criterion of willingness to compensate, is a public-policy objective. We shall examine later one such situation, namely, that faced by underdeveloped nations where economic growth is a primary concern. We shall see that in spite of many differences, efficiency in the context of economic growth continues to resemble, in important respects, efficiency based on willingness to compensate.

The constancy of prices throughout the economy. Whereas we have assumed right along that the prices of immediate outputs would vary, we have assumed up to now that other prices throughout the economy remain constant regardless of the scale of the water-resource development. This is as important as the assumption of equality of prices and marginal costs in permitting reliance on a partial-equilibrium analysis when we use the willingness-to-pay definition of efficiency for system design. Without this assumption we would have to follow the entire sequence of changes induced by river developments in a general-equilibrium framework. As is true for deviations of prices from marginal costs, our ability to trace the gains and losses caused by water-resource development in terms of the willingness-to-compensate criterion depends upon the extent to which prices change: just as only isolated deviations of prices from marginal costs are tractable in this definition of efficiency, so are only isolated price changes tractable in terms of willingness to compensate. More precisely, we define isolated price changes as changes only in the prices of final goods produced with system outputs (to be called "system final goods" for short) and the resources used in the construction of systems (to be called "construction resources"). Widespread changes, on the other hand, are said to occur when water-resource development triggers a chain reaction of price changes throughout the economy.

Price changes in system final goods or construction resources can be further classified according to whether they come from the construction of the water-resource program as a whole or of only an individual system. For the program as a whole, price changes, from the point of view of design of a component system, present a problem similar to discrepancies of prices from marginal costs and can be treated in simple cases by the use of shadow prices. Suppose, for example, that the extra demand for earth-moving machines created by construction of a water-resource program as a whole (but not the extra demand created by construction of any one system) is great enough to increase the price per hour of use of earth-moving machines. Use of the *ex-ante* price (the price which would obtain in the absence of the water-resource program) then underestimates the actual cost of construction of a component system, even though that system by itself is not large enough to affect the demand for earth-moving machines. The correction is to substitute an estimate of the *ex-post* price of earth-moving machines in calculating system costs, and to revise the estimate iteratively, in the manner outlined in Chapter 4 for the implementation of budgetary constraints.

Price changes in system final goods associated solely with the development program as a whole require only slightly different treatment than price changes in construction resources. Let us suppose, for example, that the proposed over-all river-development program would provide irrigation water to grow rice in quantities sufficient to depress the price of rice. The *ex-post* demand for irrigation water to grow rice would then decrease in comparison with the *ex-ante* demand based on the *ex-ante* price of rice. From the point of view of a component system that itself provides too little rice irrigation to affect the price of rice, the relevant demand for irrigation water is then this *ex-post* demand. The process of identifying *ex-post* demand is analogous to the iterative process for identifying the *ex-post* price of a resource; the only difference is that here a shift in the entire demand-price function for irrigation water from the system is in question, not merely a shift in a single price.

Compensation for both input and output price changes becomes more complicated when the individual system is itself large enough to affect prices of system final goods or construction resources. If, to continue our rice example, the rice grown with system irrigation water itself represents an addition to existing annual supplies sufficient to depress the price of rice, the *ex-post* demand for irrigation rather than the *ex-ante* demand is again relevant. But unlike the example discussed before, the system itself is responsible for significant changes in the quantity and price of rice in the economy; consequently the gains and losses of consumers and producers of rice associated with the quantity and price change must also be incorporated into the net-benefit function of the system. These gains and losses are easily identified in terms of the aggregate demand-price function for rice, depicted as the line *DD* in Fig. 2.8. The annual production of rice without the system is here presumed to be r^1, which fetches a market price of p_{r^1}, and we assume that the system provides enough irrigation water to increase aggregate rice production to r^2, which reduces the market price to p_{r^2}.

The consumers of rice then gain on two counts. First, they save $p_{r^1} - p_{r^2}$ per unit on the r^1 units already purchased before the system output was added to the total supply. This is shown in Fig. 2.8 by the crosshatched

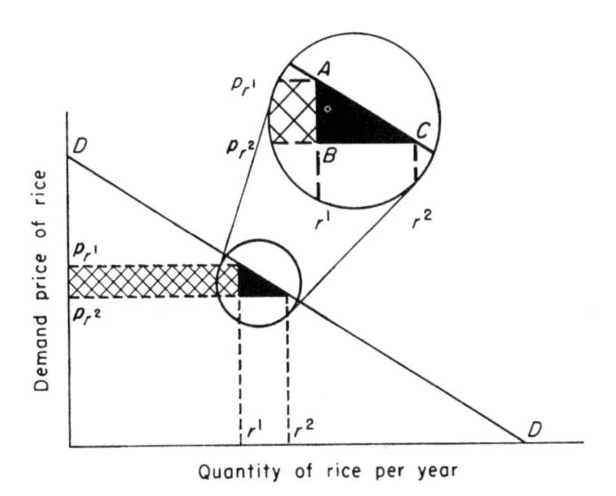

Fig. 2.8. Assumed demand-price function for rice. Solid triangular area shows gain in net benefits — consumers' surplus less producers' losses.

rectangle. Next, they gain on the $r^2 - r^1$ additional units provided by the system to the extent that their willingness to pay for these units exceeds the price p_{r^2} they actually pay. This consumers' surplus is shown in Fig. 2.8 by the solid triangle *ABC* (provided, as assumed, that consumers' marginal utilities of income remain constant).

On the producers' side, the loss is confined to the decrease in revenue, $(p_{r^1} - p_{r^2})r^1$, suffered by producers of the r^1 presystem units.[32] This also is the crosshatched area in the figure and cancels out the corresponding consumers' gain. There is no "producers' surplus" on the $r^2 - r^1$ system units to be accounted for, because producers' gains are fully reflected in their willingness to pay for irrigation water.

The net gain of consumers over the loss of producers, which is the appropriate term for inclusion in the willingness-to-pay side of the net-benefit function, is thus the consumers'-surplus triangle *ABC*. Analytically, this difference between consumers' willingness to pay and market value is

$$\int_{r^1}^{r^2} D(\eta)d\eta - D_r(r^2)(r^2 - r^1),$$

[32] In a competitive situation this loss would be manifested by a reduction in the rents of lands employed in the production of rice.

where $D_r(r)$ represents the aggregate demand-price function for rice.

Changes in the prices of construction resources caused by the size of component systems are treated analogously. Resources for which prices change are evaluated at *ex-post* rather than *ex-ante* prices, and the producers' surplus gained by suppliers of resources used in system construction is subtracted from system costs (or, alternatively, added to system benefits).

As is true when deviations of prices from marginal costs are widespread, the gains and losses of all individuals cannot be measured practically when system construction triggers a chain reaction of large price changes throughout the economy. This would be the case, for example, when the change in the price of rice affects the demand for, and hence the price of, wheat, which in turn reacts upon the price of corn, which in its own turn changes the price of hogs, and so on. Where water-resource development is a major factor in the economy as a whole, we must turn in such cases from the concept of efficiency based on willingness to compensate to modified efficiency concepts adapted to the specific case. As an example, we shall consider later the formulation of an efficiency-like objective and its translation into a design criterion in the context of underdeveloped nations for whom economic growth is a primary concern.

As a preliminary to this, however, let us consider the general implications for efficiency of growth.

EFFICIENCY IN THE CONTEXT OF ECONOMIC GROWTH

GENERAL

To increase output tomorrow, we must undertake investment today. For the economy to grow, therefore, we must sacrifice some current consumption for future consumption. A concern for growth on the part of the government is not in itself cause for abandonment of our previous criterion of efficiency based on willingness to compensate. For the interest rate serves precisely the purpose of balancing future against present consumption by placing future willingness to pay and costs on a par with their present counterparts. In the competitive model, indeed, there is no need whatsoever for governmental concern with the pattern of growth. Use of the competitive-market interest rate in system design leads to attainment of the efficient pattern of outputs over time, that is, the optimal rate of growth,[33] in terms of preferences for consumption over time evidenced by individuals in the market place. In short, determination of the pattern of growth can be "left to the market" in the competitive model.

However, even if the other competitive assumptions hold, the external effects of investment make it necessary, as we have seen, to reject the competitive-model solution of the problem of choice of interest rate and to substitute a political distillation of individual preferences for the market distillation. The necessity for a political decision on the choice of interest rate means that it is impossible for government to avoid the question of the pattern of economic growth. And it is precisely the degree of concern which the community in its political capacity places upon the future relative to the present (that is, upon the rate of growth) that determines the extent of the divergence between the political and the market distillation of individuals' preferences for consumption over time. The greater the concern for the future, the lower is the politically determined interest rate applicable to public investment. The effects upon system design thus are significant, for the lower the interest rate, the more capital-intensive and the more extensive — the more durable and larger — are system designs planned in terms of pure efficiency.[34]

In economies such as that of the United States which approximately fulfill the assumptions that prices equal marginal costs and are independent of the scale of water-resource development throughout the economy, the choice of interest rate may well be the only way in which concern for a pattern of economic growth weighted toward the future need be reflected in system design.

EFFICIENCY IN THE CONTEXT OF GROWTH IN UNDERDEVELOPED NATIONS

Efficiency, when complicated by the need for economic growth in underdeveloped nations, presents a different problem and is perhaps the

[33] Note that we say "optimal," not "maximal," rate of growth. Maximization of the rate of growth would dictate maintenance of consumption at subsistence levels (in order to free the greatest possible quantity of resources for investment), hardly a policy in keeping with the wishes of individuals living today.

[34] If the discount rate is viewed as a price ratio of present to future consumption, the greater system durability and size induced by decreases in the discount rate are substitution and scale effects.

best illustration of violation of the assumptions of equality of prices and marginal costs and constancy of prices. On the one hand, market imperfections of all sorts — from immobility of underemployed labor in rural areas to insufficiency of demand for industrial goods to support a competitive-market organization — are endemic; on the other hand, water-resource systems often form integral parts of development programs having as their goal such large changes in the size and composition of gross national product that prices throughout the economy shift radically as an integral part of the process of economic growth.

In such underdeveloped economies we believe that the criterion of efficiency can be stated more usefully in terms of the contribution of water-resource systems to over-all development goals of the economy than in "willingness-to-compensate" efficiency terms. We presuppose goals for future aggregate production in the various sectors of the economy (for instance, industrial and agricultural) further classified into target-output levels, set in physical terms, for the various commodities (such as aluminum and rice). These targets can be presumed set in terms of annual goals for successive "five-year plans" based on rough projections of the technology and resources available to the economy at each stage in the growth process.

The problem of translating this notion of efficiency into a usable ranking function is as follows: given targets of, say, increases of aluminum and rice output from 100,000 tons and 5,000,000 bushels respectively per year to 500,000 tons (400 percent) and 7,500,000 bushels (50 percent), how do we determine whether a system makes the greater contribution to the over-all program if it is designed to supply much electricity for electroprocessing of aluminum and little water for irrigation of rice fields, or if it is designed to provide little electric energy and correspondingly more irrigation water? Do we look only at proposed percentage increases and accordingly establish a priority for electric energy, or do we consider absolute magnitudes of resources required to fill the respective quotas and perhaps establish a priority for irrigation? Neither of these procedures would be sensible, for neither takes into account the *relative* desirability of using the system under consideration rather than other means to produce rice and aluminum.

The notion of relative desirability is essentially one of relative values — although, to be sure, values in terms of the over-all development plan rather than money values. How are these relative values computed? If, for example, in the development plan for the economy, five extra units of rice could be produced by sacrificing two units of aluminum at the margin, then five units of rice are worth two units of aluminum in terms of the plan. For operational purposes these relative values, or marginal rates of transformation throughout the economy, are expressed in terms of plan shadow prices: if five units of rice are worth two units of aluminum in terms of the plan, then the relative shadow prices of rice and aluminum are two and five respectively per unit. Similarly, resources are priced in terms of the marginal rates of transformation between the resources and the final products.

With shadow prices based on such marginal physical rates of transformation replacing money prices, the efficiency ranking function remains formally Exp. (2.6). Benefits represent a "shadow willingness to pay" for system outputs, that is, the value of the immediate outputs of water-resource development in terms of the extra goods produced by virtue of their provision, the extra goods being valued in terms of their plan-determined shadow prices. Costs represent the value of goods displaced by the system, also valued in terms of their plan-determined shadow prices.

The practical problem of translating into shadow prices the physical marginal rates of transformation between "target commodities" (such as aluminum and rice, for which production goals are set), between resources, and between resources and target commodities is obviously a formidable one. However, high-powered mathematical techniques such as input-output analysis and linear and nonlinear programming provide an operational framework for simplifying and organizing physical data relative to target production levels, the availability of resources, and the projected technology — and for drawing from these data the shadow prices consistent with the over-all development plan.

Just as system design affects the shadow price of outlay in the presence of a budgetary constraint, so does it also affect the shadow prices generated by the over-all development plan. Water-resource systems must logically form part of the technology with which target production levels are planned and shadow prices are calculated, since they form part of the technology with which plan goals will actually be achieved. The repercussions of design on shadow prices cannot be dismissed as negligible if, as we are assuming, water-resource development forms an integral and important part of the over-all development program.

The method of dealing with repercussions of design on plan shadow prices is analogous to the case of budgetary constraints discussed in

Chapter 4. If proposed designs differ radically from the gross estimates of designs on which the development targets are based, the estimates and production targets (and consequently the shadow prices) must be revised and new designs formulated on the basis of the new plan. This iterative process of interaction between estimates of the capabilities of the economy and production goals, on the one hand, and detailed system designs, on the other, must continue until the detailed designs coincide with the rough concept of the designs used in estimating the capabilities of the economy and production goals.

Marginal rates of transformation among resources and target commodities thus replace the willingness of individuals to compensate as the external criteria for determining prices used in the design of water-resource systems. Does this mean that efficiency in the context of growth of an underdeveloped economy ceases to be related to individual preferences? Not necessarily. In democratic nations, planning goals will be set in terms of the "individual sovereignty" in the sense discussed earlier; individual preferences will count, in that planning authorities will attempt to provide in the long run the goods and services that individuals want. The essential difference between efficiency in the context of development and efficiency based on willingness to compensate is that, in the first instance, a political distillation of the preferences of individuals replaces the market distillation revealed in market demand curves and money costs.

THE INCOME-REDISTRIBUTION OBJECTIVE

Our goal in this section is to translate a general governmental policy of using water-resource developments to redistribute income into a criterion for the design of specific systems. As a preliminary, our study of efficiency will be used to examine more systematically the reasons cited at the start of the chapter for a government's pursuit of a redistribution goal through water-resource development. That is, we propose to examine the shortcomings of the efficiency objective that keep us from basing design upon it alone.

There are only two situations in which efficiency might be the only economic concern. First, if each dollar of net income — real or money — generated by water-resource development is assumed to be of equal social significance regardless of who receives it, maximization of efficiency — which amounts to maximization of real national income — would then be a sensible objective. And second, if, rejecting this position, we assume that the government will take necessary steps to obtain optimal distribution patterns by redistributing income from the original beneficiaries to those whom the government feels to be more deserving, then once again the original distribution of income generated by system development would be of little concern, and in drawing up designs we could concentrate on the size of income generated, that is, on efficiency.

The first position, we agreed, represents a value judgment few would share: that an extra dollar to an oil millionaire is as desirable socially as a dollar to a struggling tenant farmer. The second situation is more challenging, for it seems to permit the best of both worlds — efficiency and redistribution. But there may be administrative difficulties in achieving a given redistribution; the compensation criterion of efficiency is purely hypothetical. More important, the community may not find it wholly satisfactory to achieve a given redistribution by simply transferring cash from one individual to another, even if this is administratively possible; as pointed out earlier, the size of the economic pie and its division may not be the only factors of concern to the community — the method of slicing the pie may also be relevant.

In the following section we shall describe in more detail the administrative and institutional barriers to a combination of the most efficient system design and the optimal distribution patterns for the benefits of river development.

INCOME REDISTRIBUTION WITH EFFICIENCY

Suppose that it is public policy to redistribute income in favor of a group of impoverished Indians barely getting along on their subsistence farms in the Western Desert. One way of achieving this goal might be to build a project on the *X* River to provide irrigation water for them. To maximize their gain, the government might even give the water to them or sell it to them at a nominal price. But suppose that, although feasible, irrigation is relatively inefficient in the *X* River Basin; that is, efficient design of the *X* River project calls not for irrigation water, but instead for electric energy for industrial and urban consumers. According to the definition of efficiency, the economic gains from the efficient electric-energy design must be great enough, relative to the inefficient irrigation design, to permit a side payment to the Indians that would give them more additional income than they would enjoy from the irrigation project. This, in a nutshell, is the redistribution-with-efficiency argument. It, and the barriers to it, can advantageously be represented graphically.

For simplicity, let us assume that the X River project is costless, so that the gains of beneficiaries from the system (their willingness to pay) represent net as well as gross system benefits. Let us suppose further that annual demands for outputs remain constant over time, so that annual benefits also remain constant. Annual benefits to energy consumers and Indians are represented on the horizontal and vertical axes, respectively, of Fig. 2.9; it follows that the point B (30, 0), the horizontal-axis intercept

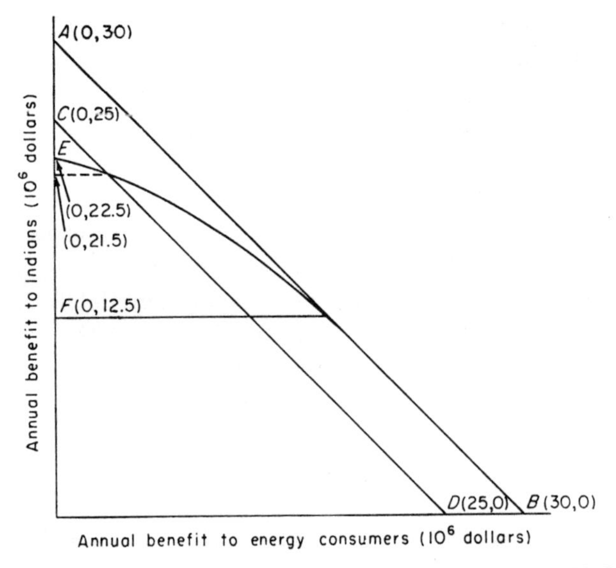

Fig. 2.9. Map of benefits to Indians and electric-energy consumers, X River project.

of line AB, defines the annual gain of $30,000,000 that would be enjoyed by energy consumers from the efficient design of the X River project under the supposition that they receive all the gains. Similarly, the point C (0, 25), the vertical-axis intercept of CD, represents the annual gain of $25,000,000 that Indians would enjoy if the irrigation-oriented design were constructed and they received irrigation water free of charge.

The lines CD and AB form the loci of all the divisions of the gains from each design between the two groups that are theoretically attainable by means of side payments. Since the locus AB lies completely to the northeast of CD, the energy project represents the superior design no matter what the division (in terms of the percentage of the total gain that accrues to each group). So runs the argument for redistribution with efficiency.

The difficulty lies in the means by which the side payments must be effected. To move to any other point on AB from point B (30, 0), the government could not rely upon individual energy consumers voluntarily to make the necessary side payments directly to individual Indians, nor upon an association of energy consumers to make voluntary payments to an association of Indians. Rather the government would have to step in, exploiting the willingness of energy consumers to pay and using the proceeds to finance subsidies to the Indians. This leads to conflicts between efficiency and redistribution on both the collection and the disbursement sides.

The collection difficulties were previously labeled as administrative. To exploit willingness to pay is to employ one or another of the forms of price discrimination listed previously. Their use creates administrative difficulties, for, as noted in the discussion of revenues, any deviation from a policy of charging the market clearing price leads to such problems. The nature of these difficulties stems from the marginal conditions of efficiency maximization themselves: even the simplest form, Eq. (2.15), states that the amount of money each consumer of a system output would be willing to pay for an extra unit must be the same. The logic behind this condition is straightforward: for a design to be efficient, outputs must go to those who are willing to pay the most for them; but as long as the marginal willingness of individuals to pay differs, units of output can be transferred from individuals with lower marginal willingness to pay to individuals with higher willingness to pay, thus increasing the over-all willingness of beneficiaries to pay.

Why does price discrimination lead to this difficulty? In theory, price discrimination can be imposed in such a way that each individual will be willing to pay the same amount for an extra unit, but this would require such precise knowledge of individual demand curves and so extensive a collection apparatus that it cannot, in fact, be accomplished. To be practicable, price discrimination must be applied to groups rather than to single individuals, and such classification schemes inevitably group individuals with different demand schedules. Under price discrimination it becomes operationally impossible both to place outputs in the hands of individuals who will pay the most for them and to exploit fully the willingness of individuals to pay. Price discrimination is thus compatible neither with efficiency maximization nor with complete redistribution of the gains of the original beneficiaries to another group.

In terms of Fig. 2.9 the locus *EB*, which represents the possible distribution combinations for the electric-energy design (taking into account the problems of implementing price discrimination), replaces the locus *AB*. As indicated by the intercept of *EB* on the vertical axis, the maximum amount of gain transferable from energy consumers to Indians is $22,500,000. Indeed, if the government wishes to redistribute any amount to the Indians in excess of $21,500,000 annually (the level of benefits to Indians at which the electric-energy-design locus *EB*, as modified, and the irrigation-design locus *CD* intersect), then the irrigation project is preferable to the energy project.

More important than conflicts between efficiency and redistribution on the collection side are barriers on the disbursement side — institutional and political bars to payments of subsidies to the group to whom the government wishes to redistribute income. The word "subsidy" itself is enough to set many otherwise rational individuals off into diatribes against the "weakening of moral fiber." With reference again to Fig. 2.9, when such barriers are taken into account, the distribution possibilities of the efficient energy design might become the locus *FB*, which shows that no more than $12,500,000 annually could be transferred to the Indians in the form of subsidies. Under this assumption distribution with efficiency becomes impossible for redistribution goals in excess of $12,500,000; to redistribute any amount of income in excess of this amount, the irrigation project becomes the only feasible means — up to $25,000,000.

The redistribution problems discussed above are not peculiar to the division of gains of water-resource systems; they are equally applicable when income of any sort is to be redistributed. Of all methods of distribution, only lump-sum taxes and subsidies do not violate efficiency conditions somewhere in an economy fulfilling the assumptions of the competitive model.[35] And, although direct subsidies are vigorously opposed, lump-sum taxes are even more impractical politically.

Thus income must be redistributed through such means as development

[35] For example, a sales tax in a competitive model raises the price to consumers above marginal cost, even though the net receipts of the seller just equal marginal cost.

of water-resource projects, with accompanying inefficiencies, if desired income-redistribution goals are to be achieved without violating institutional and political arrangements valued by the community. Economists especially must avoid the temptation to dismiss as "irrational" institutional and political barriers to full realization of the economic potentialities of efficiency plus the desired distribution of income, which necessitates sacrifice of a bit of one to achieve some of the other. The means by which a desired distribution of income is achieved may be of great importance to the community. Stripped of its emotional content, much remains in the "moral-fiber" argument. Many, perhaps most, people would be willing to give up a bit of efficiency to see the living standards of Indians of the *X* River Basin improved by their own labors rather than by the dole. Once more we realize that the community may not be indifferent to the choice of knife with which the economic pie is sliced.

INCOME REDISTRIBUTION THROUGH SYSTEM DESIGN

The redistribution ranking function. In this section we shall consider the problem of defining a function that ranks alternative designs in terms of their contribution to income redistribution; in the subsequent sections, we shall outline three alternative methods of using the ranking function to determine the system design, that is, three alternative methods of translating a verbal objective of improving income distribution by means of water-resource development into a criterion for design of specific systems.

Let us assume to start that the objective is to raise the income of the *X* River Indians as a group without any concern for the distribution of income within the group. With this assumption, construction of the redistribution ranking function is relatively easy, at least if our ranking is confined to performance within periods during which the community is indifferent to the timing of redistribution, which we shall normally take to be one year in length.

We shall define the criterion of performance of a design in terms of the net income generated to the *X* River Indians; for any one year design A^1 will be considered superior to design A^2 in terms of redistribution if the amount of annual net income generated by A^1 to the group of Indians as a whole exceeds that generated to the group by A^2. The net income generated to the *X* River Indians by a design is equal to the gross increase in income derived by them from its outputs less the actual charges levied upon them for these outputs. If annual net income from design A is denoted by $I(A)$, gross income by $G(A)$, and system revenues by $R(A)$, the criterion for superiority of A^1 over A^2 for a given year is

$$I(A^1) > I(A^2), \quad \text{or} \quad G(A^1) - R(A^1) > G(A^2) - R(A^2).$$

Thus the function

$$I(A) = G(A) - R(A) \tag{2.26}$$

fulfills our requirements of a ranking function with respect to redistribution of income for a single year. We define this expression as the redistribution benefits of design A.

To obtain the measurable content of Eq. (2.26), we again assume that prices equal marginal costs throughout the economy and are unaffected by water-resource development, that there are no external effects, and that the marginal utilities of incomes of consumers are constant and producers are unfettered in a single-minded pursuit of profits.

For a single-purpose irrigation system on the X River the gross income of the X River Indians from irrigation water, as a producers' good, is the additional income from the sale of extra crops grown by virtue of irrigation, less any additional factor payments (actual or imputed) complementary to irrigation. If the Indians, in their capacity as agricultural producers, are concerned only with profits, they will be willing to pay any amount up to the entire extra income they receive from irrigation rather than go without it. Thus, if irrigation water goes to those Indians who are willing to pay the most for it, then

$$G(y) = E(y) = \int_0^y D(\eta)d\eta, \tag{2.27}$$

where $D(y)$ represents the aggregate demand schedule of the Indians for irrigation water, denoted by y. If irrigation water were a consumers' good, willingness to pay would represent the equivalent value in money of the increase in satisfaction enjoyed by Indians from the provision of y units; the increase in income measured by willingness to pay would represent a gain in real income but not in money income. This increase would continue to be measured by Eq. (2.27) as long as the marginal utility of income is assumed to be constant.

If the Indians are charged a constant price $\bar{p}_y$ for each acre foot of water they receive, system revenues are $\bar{p}_y y$. The annual net increase in the income of the Indians, or the redistribution benefits, then becomes $E(y) - \bar{p}_y y$.

Figure 2.10 shows the net income of the Indians from an arbitrarily

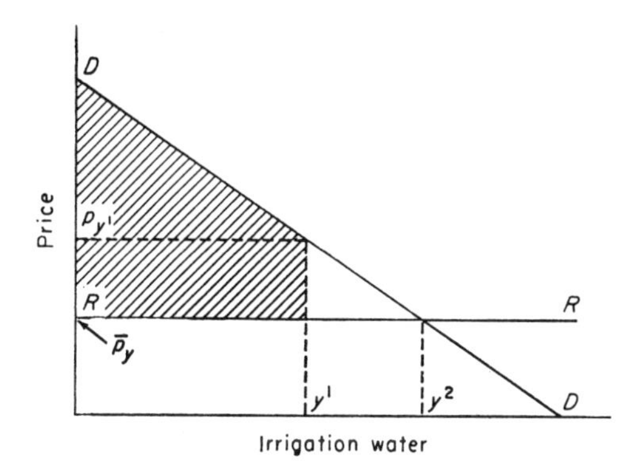

Fig. 2.10. Aggregate demand schedule (DD) of Indians for irrigation water. Shaded area represents net increase in income of Indians if quantity y^1 is provided and price $\bar{p}_y$ is charged for irrigation water.

chosen level of irrigation, y^1 acre ft, for the single-purpose case. The line *DD* represents their aggregate demand schedule for irrigation water and the line *RR* the assumed public policy to charge the price $\bar{p}_y$ per acre foot of water. Since the willingness to pay of the Indians for y^1 acre ft of water is the area under the demand curve truncated at y^1, that is, the shaded area plus the rectangle beneath it, and since the actual charge is this rectangle itself, the net increase in the income of the Indians from y^1 acre ft of water is the shaded area between the demand curve *DD* and the average revenue curve *RR*.

The multiple-purpose case presents little additional complication in measuring redistribution benefits, at least from the formal point of view. As long as the demands of the Indians for system outputs $y_1, \ldots, y_m$ are independent,

$$I(y) = \sum_{j=1}^{m} \left[\int_0^{y_j} D_j(\eta_j) d\eta_j - \bar{p}_{y_j} y_j \right]. \tag{2.28}$$

In words, annual redistribution benefits become the sum of the willingness to pay for all outputs less the actual payment for these outputs.

Two problems have been avoided throughout this discussion of redistribution ranking function. The first is that of determining the prices actually charged for outputs. In contrast to the efficiency objective, it is clear from the redistribution ranking function that actual revenues play an important role in determining the performance of a design in terms of redistribution. Thus the prices actually charged play a crucial role in determining the optimal design. We shall at first simply assume without explanation that prices are set in advance by policy-makers; in the course of the discussion we shall seek to justify this assumption.

Second, we have dealt only with the performance of a design within a single year. Here too the redistribution ranking function differs from the efficiency function. Under the efficiency objective intertemporal comparisons were taken into account by dealing with the equivalent willingness to pay today of beneficiaries for outputs provided over the economic life of a development scheme and with the equivalent cost to taxpayers today of future costs; that is, the criterion was framed in terms of present values based on an interest rate reflecting the preferences in consumption over time of those who reap the gains and pay the costs. A parallel procedure is not appropriate for redistribution, however. For as it is the community as a whole that undertakes to redistribute income to a special group of individuals, it is the community's attitude toward redistribution at different times that is relevant, rather than the preferences of the group benefiting from the community's redistribution policy. The group preferences count only insofar as the community wishes to heed them. The significance for design of this complication will also be explored in the course of the discussion.

It is more difficult to translate the objective of income redistribution into a design criterion than it is for efficiency. When efficiency is the sole objective, the goal is to choose the design with the greatest efficiency net benefits. On the other hand, maximization of redistribution, that is, maximization of Eq. (2.28) in each year, is never a sensible sole objective, for this would logically lead to the absurdity of employing the entire national income to redistribute income to the X River Indians by means of water-resource development. Instead, income redistribution is always circumscribed by the community's notion of how much redistribution it wants, how much efficiency it is willing to sacrifice in order to obtain redistribution, or both. In the sections that follow, we shall present three alternative methods of incorporating such notions about income redistribution, as refined by government policy-makers, into criteria for the design of river-basin systems.

Method I — redistribution constraints. Let us suppose that the redistribution goal formulated by policy-makers is raising the level of income of the X River Indians as a group by at least $20,000,000. At the start, for simplicity, we assume that the life of the system and the government's interest in the Indians terminate simultaneously at the end of one year. For further simplicity, we shall assume once again that a single-purpose irrigation system furnishes the particular means for accomplishing the given goal.

If we continue to denote the amount of irrigation water by y, this goal imposes the constraint upon system design that there be sufficient irrigation water to ensure that $I(y)$, the net increase in income of the Indians, is at least $20,000,000. System design is thus subjected to the constraint

$$\int_0^y D(\eta)d\eta - \bar{p}_y y \geq 20{,}000{,}000, \tag{2.29}$$

where $\bar{p}_y$ represents the preassigned price of irrigation water to Indians. With the redistribution goal framed in this way, the design goal becomes the meeting of this constraint at the smallest sacrifice in income for the community as a whole. By the very simplified nature of this problem, the goal reduces to finding the least-cost design that provides sufficient irrigation water for one year to satisfy requirement (2.29), or, mathematically, to minimize $K(\mathbf{x})$ subject to the redistribution constraint (2.29) as well as the production function Eq. (2.3).

The design yielding the required output in terms of this redistribution goal will, in general, differ from the design providing the most efficient output. It is likely, indeed, that costs will exceed the willingness of the Indians to pay for all levels of output, so that efficiency dictates no output whatsoever. Furthermore, except when the redistribution constraint can be met simply in the course of efficient development, any increase in income of the Indians beyond $20,000,000 adds needlessly to cost.[36] Thus, barring the fortuitous achievement of the desired redistribution in the course of efficiency maximization, we can be sure that, for the design fulfilling the redistribution goal, the equality rather than the inequality will hold in Exp. (2.29).

Given a policy of charging $\bar{p}_y$ per acre foot of irrigation water, Fig. 2.10 demonstrates that there is no point in generating output beyond y^2 acre ft, where the demand and revenue functions intersect; the gross

[36] Should the redistribution constraint be ineffective, cost minimization is no longer a sensible objective. As we shall see, maximization of net efficiency benefits subject to the redistribution constraint is a more general design criterion, reducing to cost minimization as a special case. See footnote 39.

income of the Indians per acre foot of water in excess of y^2 is less than the price $\bar{p}_y$, and therefore quantities in excess of this level detract from rather than add to their net income. Thus, if the constraint cannot be met by a level of output less than or equal to y^2, it cannot be met at all.

Furthermore, if the constraint can be met by a level of output less than y^2, for example y^1, the discrepancy between marginal willingness to pay, p_y^1, and the actual price, $\bar{p}_y$, will necessitate rationing in the distribution of irrigation water. This deviation from charging the market clearing price, like the deviation in the opposite direction entailed in price discrimination, implies sacrificing the iron-clad guarantee associated with distribution of outputs by the market, namely, that irrigation water will go to the users willing to pay the most for it. To the extent that water is nót rationed in accordance with willingness to pay, it will not be consumed most profitably; in consequence, the net increase in the income of Indians as a group indicated by aggregate willingness to pay would not be realized.[37]

But might we not employ the pricing policy itself to ensure that each unit of output goes to the beneficiary who is in a position to employ it most profitably, simply by choosing a price for which there is only one level of output which fulfills the constraint — namely, the level that clears the market at the price chosen? In other words, should we not change the design decision from the choice of the *output* generating \$20,000,000 income to Indians at the given price $\bar{p}_y$ to the choice of the *price* at which the beneficiaries are willing to take just enough irrigation water to yield them a net income of \$20,000,000? Graphically, the design decision would become choice of a price such that the area of the consumers' surplus triangle formed by the demand curve, the revenue curve, and the vertical axis just equals \$20,000,000.

If the market is allowed to clear, that is, if output is provided in sufficient quantity to drive the marginal willingness of the beneficiaries to pay down to the price, then the beneficiaries derive no net income from the marginal unit of output. On the other hand, the marginal unit of

[37] Insofar as income differences within the group of Indians are a public-policy consideration, most profitable employment of water within the group is properly not an overriding consideration of design. For example, laws and regulations, such as the "160-acre law," which limit the amount of irrigation water supplied to any person to the quantity deemed necessary to irrigate 160 acres, cease to appear arbitrary when viewed in terms of a redistribution objective instead of an efficiency objective. Even though such laws might preclude maximization of the net income of the group of redistribution beneficiaries as a whole, they can be justified as a means of implementing a more sophisticated redistribution objective concerned with the allocation of income within the group as well as redistribution to the group.

output entails a cost that taxpayers must bear. Therefore, this unit is provided not only to no avail for the redistribution goal, but also — barring a fortuitous fulfillment of the redistribution goal consistent with efficiency — to the positive detriment of taxpayers who must forego the goods or services which the resources required to provide the marginal unit of output could otherwise produce. Here, in a nutshell, is the argument against choosing a price that allows the redistribution beneficiaries' market for output to clear.

How then is pricing policy determined? Clearly, the lower the price of the system output, the greater is the net income of the redistribution beneficiaries per unit of output; thus the lower the price, the smaller is the level of output required to meet the redistribution constraint, and, consequently, the smaller is the gross cost to taxpayers of accomplishing the desired redistribution. The logical consequence of this reasoning is to *give* the system output to the redistribution beneficiaries, and, indeed, viewed solely in terms of meeting the constraint of Exp. (2.29), this is the best procedure. However, against this desideratum must be balanced the political and institutional factors that originally determined the use of water-resource systems to meet the redistribution goal. The same objections to outright cash subsidies may make it necessary to charge something for the outputs that beneficiaries receive.

Such political and institutional considerations may lead to nothing more than a token price independent of system costs. On the other hand, the pricing policy may require total repayment of the entire cost of the system, or repayment of some part of system costs — for example, the OMR costs but not the construction cost. In these cases the marginal revenue function would be much more complicated than the constant price $\bar{p}_y$ shown in Fig. 2.10, which is independent both of level of output and levels and costs of inputs. We shall not, however, strive for realism in this regard, the simple function $\bar{p}_y y$ serving to remind us of the role of revenues as effectively as more complicated functions.

To summarize: We take pricing policy as an established matter at the level of system design, determined at a higher echelon of the planning process by a balancing of institutional considerations leading to a comparatively high price for system output, against redistribution considerations leading to a zero price. We accept the consequence of taking price as given; namely, the necessity for rationing output among beneficiaries. The comparatively small discrepancies between the actual group income and aggregate willingness to pay that arise from rationing will be ignored.

Extension of the illustrative single-purpose example above to the multiple-purpose case involves only one new consideration of importance. This can be stated in the form of a question: If we can, for example, raise the real incomes of the Indians to the specified level by provision of electric energy, flood control, and a recreational lake as well as by irrigation water, on what grounds do we decide to fulfill the redistribution constraint by designing for more of one output than another? "On the basis of least-cost considerations, as before," is the correct answer; but what does "least cost" mean in the presence of a choice among alternative combinations of outputs, each of which is capable of doing the redistribution job? Least cost does *not* mean simple minimization of the costs of constructing and operating the system. The efficiency benefits of a combination of outputs represent gross gains of individuals throughout the economy from the combination,[38] and in choosing a design to meet the redistribution constraint, it is as desirable to maintain efficiency gains at a high level as it is to hold costs down; efficiency benefits and costs are simply two sides of the same coin. Least cost in the presence of alternatives therefore means the minimization of efficiency costs minus efficiency benefits.[39]

The goal of minimization of efficiency costs minus efficiency benefits subject to fulfillment of the redistribution constraint is algebraically equivalent to maximization of efficiency benefits minus efficiency costs subject to the same restriction. Thus the design criterion can be interpreted as the more intuitively appealing objective of choosing, from the set of all designs meeting the redistribution constraint, that design which maximizes the efficiency objective function. Analytically, the design criterion for a multiple-purpose system providing outputs $y_1, \ldots, y_m$ is to maximize $E(\mathbf{y}) - K(\mathbf{x})$ by choosing from among all designs for which

$$E(\mathbf{y}) - \sum_{j=1}^{m} \bar{p}_{y_j} y_j \geq 20{,}000{,}000 \qquad (2.30)$$

and subject to the production function $f(\mathbf{x}, \mathbf{y}) = 0$. Here $E(\mathbf{y})$, as before,

[38] Given our assumptions of constancy of prices and equality of prices to marginal costs throughout the economy, in this instance the "individuals throughout the economy" who gain from the system are the Indians and the taxpayers. The taxpayers' share in the gross efficiency benefits is the gross revenue collected from the Indians.

[39] The need to supplant the objective of money-cost minimization with the more general objective of efficiency-cost minimization stems from the greater number of degrees of freedom introduced in the multiple-purpose case. We can ignore efficiency benefits in the single-purpose case because, with only one output at our disposal, there is no way to avoid losses in efficiency benefits.

denotes aggregate efficiency benefits, that is, the aggregate willingness of the Indians to pay for outputs.

The multiple-year case adds little complexity. If a year is the length of time within which the community is indifferent to the time pattern of the income of the Indians, a set of constraint levels $\bar{I}_1, \ldots, \bar{I}_T$ can be substituted for the single constraint in Exp. (2.30). In theory, these levels of specified annual increases in income can vary over time according to any pattern, but we shall assume that the specified level of annual redistribution benefits is constant over the group of years within the demand period as defined on p. 30. The restrictions upon design imposed by income redistribution are embodied in a set of constraints $\bar{I}_1, \ldots, \bar{I}_Q$ representing the annual increases in income to be achieved within each demand period. With this assumption the multiple-period design criterion becomes maximization of the efficiency objective function

$$\sum_{q=1}^{Q} \theta_q[E_q(\mathbf{y}_q) - M_q(\mathbf{x})] - K(\mathbf{x}), \tag{2.31}$$

subject to

$$E_1(\mathbf{y}_1) - \sum_{j=1}^{m} \bar{p}_{y_{1j}} y_{1j} \geq \bar{I}_1$$

$$\vdots$$

$$E_Q(\mathbf{y}_Q) - \sum_{j=1}^{m} \bar{p}_{y_{Qj}} y_{Qj} \geq \bar{I}_Q \tag{2.32}$$

and to the production function of Eq. (2.8). Note that in Exp. (2.32) the $\bar{P}_{y_{qi}}$ term can change with time. The price charged for irrigation water in 1975 need not be the same as the price charged in 1965.

By setting the derivatives of the Lagrangian form of this constrained maximization problem equal to zero and substituting, the following marginal conditions of optimization result:[40]

$$\sum_{q=1}^{Q} \theta_q \left[D_{qj}(y_{qj}) \frac{\partial y_{qj}}{\partial x_i} - \frac{\partial M_q}{\partial x_i} \right] - \frac{\partial K}{\partial x_i}$$

$$= \sum_{q=j}^{Q} \lambda_q [D_{qj}(y_{qj}) - \bar{p}_{x_{qj}}] \frac{\partial y_{qj}}{\partial x_i} \qquad \begin{array}{l} i = 1, \ldots, n \\ j = 1, \ldots, m \end{array} \tag{2.33}$$

[40] The qualifications set forth in footnote 10 as to second-order conditions, replacement of equalities by inequalities at the boundary or zero levels of inputs and outputs, and the consequent need for direct comparison of the values of the ranking function apply here as well.

The left-hand side of this equation is familiar as the marginal net efficiency benefit from an increase of one unit in the size of the ith structure:

$$\sum_{q=1}^{Q} \theta_q[\mathrm{MVP}_q(x_i) - \mathrm{M}M_q(x_i)] - \mathrm{M}K(x_i).$$

On the right-hand side we have a sum of λ_q's, numbers the significance of which will be explained shortly, each multiplying the difference between the gross income derivable by the Indians in each year of period q from a unit increase in the quantity of the ith input, $D_{qj}(y_{qj})\partial y_{qj}/\partial x_i$, and the marginal charge they will actually be called upon to pay in period q, $(\bar{p}_{y_{qj}}\partial y_{qj}/\partial x_i)$. The right-hand side is thus simply the sum of multiplicative products of the λ_q's and the marginal net increases in annual income of the Indians in period q from a unit increase in the size of the ith structure, denoted by $\mathrm{M}I_q(x_i)$. In simpler notation Eq. (2.33) then becomes

$$\sum_{q=1}^{Q} \theta_q[\mathrm{MVP}_q(x_i) - \mathrm{M}M_q(x_i)] - \mathrm{M}K(x_i) = \sum_{q=j}^{Q} \lambda_q \mathrm{M}I_q(x_i). \qquad i = 1, \ldots, n \qquad (2.34)$$

Equation (2.34) says that, for the most efficient design fulfilling the redistribution constraints, the present value of the marginal net efficiency benefit of each structure must equal a weighted sum of the marginal net increases in the income of the redistribution beneficiaries over the life of the system.

It is precisely these weights — the λ_q's — that attract our attention. To the mathematician these numbers are familiar as Lagrangian multipliers, variables introduced to facilitate maximization or minimization of a constrained objective function, the values of which are determined as a by-product of optimization. For the economist, however, these auxiliary variables play an important substantive role: each Lagrangian multiplier can be viewed as the marginal "cost" of honoring the applicable constraint, "cost" being measured in terms of the loss of value of the objective function.[41] This interpretation of Lagrangian multipliers really is not new: the cutoff benefit–cost ratios met in the context of budgetary constraints (representing the efficiency benefits per dollar sacrificed at the margin of system development in order to honor the

[41] More rigorously, each Lagrangian multiplier represents the total derivative of the maximum value of the objective function with respect to a relaxation in the constraint to which it applies.

budgetary constraint) were nothing more nor less than Lagrangian multipliers. In the present instance the λ_q's represent the marginal opportunity costs, or shadow prices, of redistribution in terms of efficiency.

This interpretation is best explained by an example. Consider a system with a one-year economic life. Equation (2.34) then reduces to the simpler form

$$\text{MVP}(x_i) - \text{M}K(x_i) = \lambda \text{M}I(x_i),$$

from which

$$\lambda = \frac{\text{MVP}(x_i) - \text{M}K(x_i)}{\text{M}I(x_i)}.$$

The value of λ is seen to be the ratio of the (negative)[42] marginal efficiency gain, $\text{MVP}(x_i) - \text{M}K(x_i)$, to the marginal net redistribution gain, $\text{M}I(x_i)$, or, in other words, the efficiency loss per dollar of net income provided for redistribution beneficiaries at the margin.

Extension of this interpretation to a system with an economic life of Q periods requires little modification. In the multiple-period case, λ_q is the loss of efficiency, in terms of the (negative) present value of marginal net efficiency benefits, incurred by the last dollar of net income provided annually for redistribution beneficiaries in period q.

As a limiting case, the marginal efficiency benefit will be zero if the redistribution constraints are met simply in the course of efficient development. A non-binding constraint implies a zero opportunity cost and vice versa.

Interpreted as opportunity costs, the λ_q's are of great importance in making up an inherent deficiency of the expression of the redistribution objective as a set of constraints upon system design, that is, as annual net increases in the income of the Indians to be attained whatever the cost in terms of efficiency. In view of the complex process lying behind the formulation of objectives (discussed in some detail in Chapter 15), constraints ordinarily cannot be set with full confidence on the first attempt. Rarely does a constraint such as "The incomes of Indians in the X River Basin are to be increased by \$20,000,000 annually" mean literally that this must be accomplished no matter what the costs in efficiency are. If the last \$5,000,000 per year entails an inordinate cost in terms of efficiency, public-policy-makers might well settle for an annual increase

[42] How can we be sure that marginal efficiency will represent a negative rather than a positive gain? Clearly, if $\text{MVP}(x_i) - \text{M}K(x_i)$ were positive, further development would be justified simply on efficiency grounds regardless of redistribution. Thus, at the margin this expression, marginal efficiency, cannot be positive.

of $15,000,000 by means of water-resource development and search for other means to make up some or all of the remainder. On the other hand, if the incomes of the Indians could be increased $5,000,000 per annum above the constraint level, or $25,000,000 in all, at comparatively little extra cost in efficiency, policy-makers might think the extra redistribution gain worthwhile.

In short, a redistribution objective formulated in terms of constraints appears properly to be implemented by means of an interaction between constraint levels and design — an interaction turning on the values of the λ_q's, the marginal opportunity costs of redistribution in terms of efficiency. Rather than being the end of the planning process, design on the basis of initial constraints may be virtually the beginning. On the basis of the shadow prices of redistribution generated by the initial constraint levels, policy-makers must decide whether more or less redistribution should be obtained through water-resource development. Some or all constraints may be revised, not necessarily in the same direction. If changes are made, the system must be redesigned on the basis of the new set of constraints. The interaction between design and constraints continues until policy-makers are satisfied that, in terms of the requisite sacrifices of efficiency, the levels of increase from river development in the annual income of the redistribution beneficiaries are satisfactory.

This interaction clearly makes design in terms of a redistribution objective a complex affair. However, there is one short cut. Inasmuch as early designs are likely to require revision, a strategy of successive refinements can be adopted in a form somewhat as follows: (1) first designs, based as they must be on constraint levels set more or less arbitrarily, are determined in rough-and-ready form, primarily to learn what the magnitudes of the shadow prices of redistribution are at various times; (2) constraint levels are then reformulated, taking into account the opportunity costs found in step (1); and (3) new designs are determined for the new constraint levels, this time in greater detail and with greater precision. Steps (2) and (3) can be repeated, with ever-greater attention to detail and precision, until policy-makers are satisfied that, given the opportunity costs in terms of efficiency, the levels of increase in the incomes of beneficiaries from the system are nearly optimal.

Method II — redistribution combined with efficiency in the objective function. The second method works from the outset in terms of designated opportunity costs instead of designated increases in the level of income of redistribution beneficiaries. Policy-makers might state the objective in terms of their willingness to sacrifice efficiency for redistribution. In this case, instead of basing designs on maximization of efficiency compatible with preassigned levels of income for the Indians, planners would formulate designs to maximize a weighted sum of redistribution and efficiency, the weights representing the preassigned values of the opportunity costs. From the process of determining the optimal design, the levels of income to be provided for the redistribution beneficiaries would be obtained as by-products.

When the opportunity cost is preassigned, it is more conveniently employed with positive sign; that is, μ_q (the negative of the marginal opportunity cost, λ_q) replaces λ_q and represents the efficiency value per dollar of redistribution benefits. The μ_q's are means of rendering redistribution benefits comparable with efficiency benefits in a composite objective function. Given values of $\mu_1{}^0, \ldots, \mu_Q{}^0$, the objective function becomes a weighted sum of redistribution and efficiency net benefits,

$$\sum_{q=1}^{Q} \mu_q{}^0 \left[E_q(\mathbf{y}_q) - \sum_{j=1}^{m} \bar{p}_{y_{qj}} y_{qj} \right] + \sum_{q=1}^{Q} \theta_q [E_q(\mathbf{y}_q) - M_q(\mathbf{x})] - K(\mathbf{x}), \qquad (2.35)$$

the weights being the efficiency values of redistribution benefits. Maximization of Exp. (2.35) is of course constrained by the production function, Eq. (2.8).

Marginal conditions are formally identical with the marginal conditions of Eq. (2.33) for maximization of efficiency constrained by redistribution. There is the substantive difference, however, that in the present case the opportunity costs are determined prior to, rather than as a by-product of, maximization. The formal identity and the substantive difference arise from the mathematical relation between the efficiency objective function (2.31) subject to redistribution constraints (2.32) and the composite objective function (2.35): function (2.35) is the Lagrangian form of function (2.31) with the Lagrangian multipliers (the λ_q's or μ_q's) determined in advance.

The composite-objective-function form of method II clarifies the observation of footnote 42: if the efficiency values of redistribution are all zero, then the composite objective function reduces to efficiency maximization. For this reason an objective of efficiency maximization alone can be regarded as a special case of a composite efficiency and income-redistribution objective in which the community is neutral to the distribution of the income generated by river development, the neutrality of the community being reflected by the zero values of the μ_q's.

The composite objective function also permits comparison of the manner in which redistribution and efficiency benefits occurring in different years are placed on a common footing. The intertemporal question, it will be recalled, was raised during the discussion of the redistribution ranking function, and it was perhaps a bit surprising that design criteria for systems with multiple-period economic lives were introduced without mention of a redistribution interest rate. The interest rate has been present all along, however, as is indicated by the common dimensionality of the θ's — the efficiency interest factors — and the μ_q^0's. In fact, in the process of defining the relative values of the μ_q^0's for different periods, policy-makers are defining a redistribution interest rate.[43]

The simplest way to appreciate this is to assume that demand periods are each one year in length. Then, provided that the ratio of prearranged efficiency values in every pair of years is constant, this ratio itself defines a constant redistribution interest rate, r_I. The interest rate is simply the solution of $\mu_{t+1}^0/\mu_t^0 = 1/(1 + r_I)$ where $t = 1, \ldots, T$. In other words, $r_I = (\mu_t^0 - \mu_{t+1}^0)/\mu_{t+1}^0$. Of course, we still require a weight for placing the "present value" of redistribution benefits on a common basis with efficiency benefits; this is provided by μ_1^0, the efficiency value of redistribution benefits in year one. The composite objective function (2.35) then becomes

$$\mu_1^0 \sum_{t=1}^{T} \frac{E_t(\mathbf{y}_t) - \sum_{j=1}^{m} \bar{p}_{y_{tj}} y_{tj}}{(1 + r_I)^t} + \sum_{t=1}^{T} \frac{E_t(\mathbf{y}_t) - M_t(\mathbf{x})}{(1 + r)^t} - K(\mathbf{x}).$$

The redistribution interest rate is only formally analogous to the efficiency interest rate. For example, whereas the efficiency rate is invariably nonnegative, the redistribution rate might well be negative, indicating that the efficiency value to policy-makers of redistribution benefits increases with time. This and other substantive differences stem from the fundamental difference in the determination of the two rates discussed earlier in this chapter.

Insofar as policy-makers cannot set efficiency values initially with complete confidence, the objective function (2.35) would be implemented in

[43] Whereas the redistribution interest rate r_I makes an explicit appearance for the first time in connection with the present method of implementing the redistribution objective, it would be incorrect to say that it exists for the first time in connection with this method. In method I the opportunity costs implicitly define a redistribution interest rate: $r_I = (\lambda_t - \lambda_{t+1})/\lambda_{t+1}$.

a manner similar to the way in which the objective of efficiency maximization constrained by redistribution was implemented. Policy-makers would examine the redistribution consequences of a preliminary design that maximizes the objective function for the initial efficiency values $\mu_1^0, \ldots, \mu_Q^0$. Then the efficiency values for years in which the increases in income to redistribution beneficiaries are felt by policy-makers to be unsatisfactory would be revised, and new designs would be formulated by maximizing the objective functions for the new set of efficiency values. The interaction between efficiency values and design, in this case revolving about the levels of redistribution benefits provided, would presumably continue until efficiency values and redistribution benefits were felt to be in satisfactory relationship to each other.

Method III — maximization of redistribution subject to an efficiency constraint. The two previous methods of expressing the redistribution objective function are predicated on the assumption that policy-makers can state their attitudes toward redistribution in terms of target levels of increases in income for the redistribution beneficiaries, the $\bar{I}_q$'s of Exp. (2.32), or in terms of rates at which they are content to trade efficiency for redistribution, the μ_q's of Exp. (2.35). A third method expresses the redistribution objective in terms of the amount of efficiency net benefits, viewed as incremental present value of national income, that policy-makers are willing to sacrifice for income redistribution. For our example, the objective might be stated in the following way: we shall provide *all* the income for Indians that can be generated by the X River project without inducing a decrease in the present value of national income.

Such a formulation still requires a definition of weights to place redistribution benefits occurring in different years on a commensurate level; there is no reason to suppose that policy-makers will deem redistribution benefits 20 years hence of equal value with redistribution benefits tomorrow. These redistribution weights serve only the role of an interest rate, r_I. Unlike the weights in Exp. (2.35), they do not also put redistribution benefits on a par with efficiency benefits. Assuming constancy over time of the community's time preference for redistribution benefits, we denote the relative weight for period q by $\theta_q{}^I$, the $\theta_q{}^I$s being related to the redistribution interest rate r_I just as the θ_q's are related to the efficiency interest rate r. With this notation the design criterion implementing the present objective is maximization of the redistribution objective function

$$\sum_{q=1}^{Q} \theta_q{}^I \left[E_q(\mathbf{y}_q) - \sum_{j=1}^{m} \bar{p}_{y_{qj}} y_{qj} \right]. \tag{2.36}$$

subject to the constraint on net efficiency benefits

$$\sum_{q=1}^{Q} \theta_q[E_q(\mathbf{y}_q) - M_q(\mathbf{x})] - K(\mathbf{x}) \geq 0 \tag{2.37}$$

and subject to the production function.

The constraint level in Exp. (2.37) need not be zero. Policy-makers may, for example, be willing to tolerate a $100,000,000 loss in the present value of national income, in which case the constraint level becomes −100,000,000. Or policy-makers may insist that the system generate a minimum net addition to the present value of national income of $100,000,000, in which case the constraint level becomes +100,000,000.

Once again, the marginal conditions of optimization are Eq. (2.33); and the composite objective function (2.35) is the Lagrangian form of the redistribution objective function (2.36) subject to the efficiency constraint (2.37). There is, however, a difference in detail. As we have seen, two factors enter into the determination of opportunity costs: the value of redistribution benefits at different times relative to one another, and the present value of redistribution benefits as a whole relative to efficiency benefits. In method I, both of these factors are generated by the process of optimization in the form of single numbers, the λ_q's, combining both factors, that is, representing the loss of present value of efficiency benefits incurred per dollar of redistribution benefits in each period q. In method II both the value of redistribution benefits at different times relative to one another and the value of redistribution relative to efficiency are preassigned in the form of efficiency values, the μ_q's. The present method lies midway between I and II: the value of redistribution benefits in period q relative to its value in period 1, $\theta_q{}^I$, is preassigned, while the value of redistribution in period 1 relative to efficiency, denoted by λ^*, is determined by the optimization process. The product of these two factors, $\theta_q{}^I\lambda^*$, the value of redistribution in period q relative to the present value of efficiency, is the opportunity cost generated by method I, λ_q. Its negative, $-\theta_q{}^I\lambda^*$, is the efficiency value, μ_q of method II.

Once more, ultimate design is likely to be the product of interaction between the consequences of preliminary designs and the attitudes of policy-makers. The levels of income provided for redistribution beneficiaries and the marginal rate of substitution of redistribution for efficiency resulting from design on the basis of the initial efficiency constraints serve to guide policy-makers in revising constraints.

Comparison of the three methods of implementing the redistribution ob-

jective. Schematically, the three formulations of the redistribution objective can be compared as follows:

	I	II	III
1. Policy-makers set *decision parameters:*	Level of redistribution to be attained	Efficiency values of redistribution	Level of efficiency to be attained and relative redistribution values
2. Planners design to maximize this *objective function:*	Efficiency	Weighted sum of redistribution and efficiency	Redistribution
3. Subject to *constraints:*	Redistribution	None	Efficiency
4. Results of design revealed in terms of these *consequence parameters:*	Opportunity costs	Levels of redistribution attained	Opportunity costs and levels of redistribution attained

It might appear that in method III two items of information are generated as consequence parameters, while the previous two generate only one each. This is true solely because we chose to ignore the second item generated by each of the previous formulations, which is the level of net efficiency benefits provided by the design, assuming implicitly that only the *marginal* rates of substitution of efficiency for redistribution are of interest to policy-makers. The absolute level of efficiency benefits is information that is available free of extra planning cost in the event that it is of interest to policy-makers.

However, once policy-makers become interested in the total efficiency benefits of a redistribution-oriented design, it is likely that they will be equally if not more interested in the loss of efficiency benefits than in the absolute efficiency of the design alone. And indeed they ought to be; policy-makers ought not embark on redistribution through system design without knowing what they are giving up in terms of efficiency, totally as well as marginally. To discover the magnitude of the loss, planners must, in addition to designing the system to meet the redistribution objective, design the system to maximize the efficiency objective function as if they were unrestricted by redistribution considerations.

Clearly, the choice among the three formulations of the redistribution objective depends on the manner in which planners initially prefer to frame value judgments between redistribution and efficiency. But policy-makers need not make a final commitment to a single form of the objective, even in the design of a single system. For example, they may not be able to set appropriate efficiency values of the objective as required in method II, and hence may initially set target levels for increases in redistribution beneficiaries' income and use method I. In considering the concrete opportunity costs of the design based on the initial redistribution-constraint levels, however, policy-makers may find it easier to define appropriate rates of substitution than appropriate revisions of the redistribution constraints. Determination of the revised design would consequently be based on method II. In a similar manner, method III might be used in combination with I or II, or indeed all three forms might be used alternatively, as one or another value judgment becomes easiest for policy-makers to formulate.

Redistribution to a region. Up to this point we have discussed income redistribution to a selected group of individuals, the hypothetical *X* River Indians. The methods presented are equally applicable to other kinds of redistribution, including redistribution of income to all the individuals of a particular region. The only methodological difference between the types of objective lies in the makeup of redistribution benefits. Since for redistribution to a group we wished to include only the increases in income actually accruing to the group, only the demands and payments of individuals within the group were included in the functions $E_q(\mathbf{y}_q)$ and $\sum_{q=1}^{m} \bar{p}_{y_{qi}} y_{qj}$. On the other hand, if the objective is redistribution to all inhabitants of a region, the increased incomes of all individuals and firms within the region should be comprehended in the redistribution-benefit terms.

Redistribution within a group or region. We have stated at several points that our redistribution ranking function draws no distinction between incomes accruing to different individuals within the beneficiary group; the same applies to redistribution to a region. By and large, however, when redistribution is a public-policy objective, the distribution within the group or region of the income created by a water-resource development will also be an important concern of policy-makers. For example, the goal probably would not be merely to raise the mean income of the *X* River Indians by $2000 annually without reference to the variance of income within the group, because an increase of $2000 to a family with a presystem income of $5000 would hardly be as desirable as the same increase to a family with a presystem income of $1000.

Dealing adequately with redistribution within the group or region naturally introduces new complexities. Additional ranking functions dealing with the variance of income must be introduced, and either brought to bear as constraints or incorporated directly into the objective function. The considerations explored earlier would still be relevant; for example, measurement of efficiency costs of achieving our redistribution goal would still be of interest.

Simplifying assumptions. The remaining simplifying assumptions are the familiar ones of absence of external effects and of deviation of prices from marginal costs, and independence of prices throughout the economy from the scale of system development. Treatment of these largely parallels their treatment in regard to efficiency benefits. Indeed, there is no difference in treatment of external effects in the two cases, and the chief qualification of our earlier discussion is in the direction of simplification: deviations of prices from marginal costs and changes in prices enter the redistribution picture only insofar as they affect individuals of the group or region to which it is desired to redistribute income.

Consider, for example, a redistribution-oriented system providing irrigation water to grow sugar beets, in the processing of which a monopoly profit is earned by virtue of a discrepancy between the price and the marginal cost of sugar. The profit on the processing of each ton of sugar beets grown with system irrigation water certainly represents an increase in income for the processers. But the increase counts as a redistribution benefit only if the processers are members of the group or inhabitants of the region to which it is desired to redistribute income. Similarly, the increase in the real incomes of consumers caused by a change in the price of rice resulting from a large increase in rice production on system-irrigated lands would count as redistribution benefits only for consumers in the group or region.[44] With this qualification our earlier discussion of external effects in relation to evaluation of efficiency benefits becomes equally applicable to evaluation of redistribution benefits. Of course, widespread departures from these assumptions will, as under the effi-

[44] To include the families of farmers consuming rice grown on the farm as "consumers" in the sense used in the text would be double counting, since they do not actually purchase the rice. Their increase in income is fully reflected in their willingness to pay for the water with which the rice is grown.

ciency framework, render our benefit-measurement framework nonoperational.[45]

Other objectives. Up to this point we have described how ranking functions involving the objectives of income redistribution and economic efficiency can be constructed, and how national economic growth can be handled within the context of efficiency. Still other objectives are possible. For example, regional economic development can be, and often is, considered a goal in itself. It has been suggested, for example, that resource-development programs in the western United States be used to develop the economy so as to achieve balance with other regions of the country, *for the sake of balance itself.* Such regional-development objectives can be handled in exactly the same manner as the objective of economic growth in underdeveloped nations; the word "regional" is merely inserted before the phrase "economy as a whole" in our discussion of efficiency in the context of growth in underdeveloped nations.

More important, we have dealt only with the economic objective of increasing national income and improving its distribution. Our failure to deal with non-economic components of national welfare such as preservation and enhancement of scenic and cultural values does not indicate a desire to denigrate these objectives; rather, it stems from a desire to simplify the discussion. The conflict between efficiency and redistribution serves to illustrate the general nature of the conflict among objectives, whatever they might be.

CONCLUSION

In Chapter 1 we observed that the efficiency objective is by far the most tractable of all development objectives, and now we can see why. It is not so much the increased difficulties of benefit measurement that render the redistribution objective less tractable; it is rather the problems of specifying the constraints on redistribution or setting their efficiency values that makes implementing a redistribution objective such an involved procedure. However, it cannot be otherwise; added complexity is the price we pay for dropping the "neutral" position toward income distribution expressed in the efficiency objective.

[45] In Chapter 4 we deal with further complications. For example, we discuss the impact on regional income-redistribution benefits of the multiplier expenditures generated by the system, that is, of the chain reaction of expenditure starting with the income created by the system. Of course, only income created within the region is relevant. This complication does not occur with a group-redistribution objective because the qualification that only income created within the group counts essentially reduces the value of the multiplier to unity.

These additional complexities can lead one to question the very idea of implementing a redistribution objective in this way. The argument is that, realistically, such formal criteria cannot be set. Instead, objectives other than efficiency should be brought to bear on design only on the *ad hoc* basis of whether the marginal opportunity cost of redistribution in terms of efficiency seems "reasonable" to system-designers. Policy-makers would have their say only when they reviewed the design plans. If then they were dissatisfied with the proposed combination of efficiency and other objectives, the designers would be told to try again, perhaps incorporating specific modifications proposed by the policy-makers.

The gulf between this view and the position taken in this chapter is not wide in all respects. We recognize, for example, the need for interaction between constraints, values (opportunity costs), and design plans in view of the difficulty of specifying such parameters at the outset. The basic difference between the two views is that the one ignores what we consider to be a significant distinction between the setting of policy objectives and their implementation. As will be explained in Chapter 15, mere veto power by policy-makers over designs prepared in the absence of specific criteria neither permits consideration of a sufficient number of alternatives to ensure achievement of the optimal combination of efficiency and nonefficiency objectives nor, indeed, does it permit refinement of nonefficiency objectives in an atmosphere in which national, rather than special, interests dominate.

3 Basic Economic and Technologic Concepts: A General Statement

ROBERT DORFMAN

THE fundamental viewpoint of this volume is that the design of a water-resource project depends on an intimate interplay of economic and engineering considerations. Questions of what can be done and how much it will cost are largely in the province of the engineer; questions of what is worth doing and how worthwhile it is are mostly within the competence of the economist. But neither discipline can contribute its share to a design without the collaboration of the other. The economist cannot ascertain what is worth doing without the cost estimates and the estimates of physical output that the engineer must provide. The engineer cannot arrive at his estimates without calling upon the economist for the prices with which to convert physical quantities into economic values and for judgments as to the kinds of physical output that are likely to be most desirable.

To achieve the necessary close collaboration, specialists in each discipline must have a rudimentary acquaintance with the point of view and vocabulary of the other. The purpose of this chapter is to present those rudiments. Most of the space, however, will be devoted to the economic rather than to the engineering rudiments, for several reasons. First, the essential ideas that economics has to contribute can be conveyed in much briefer compass than can the highly elaborated technology of civil engineering; accordingly, it appears easier for engineers to learn the relevant economic doctrines than for economists even to approach the techniques of engineering design. Second, the common ground where economists and engineers must meet appears to be mostly on the economist's side of the borderline. The economist does not concern himself with the engineer's conclusions about the dimensions of structures or the design of turbines; but the engineer, even in his private thinking, cannot avoid the problems of costing and weighing of alternative uses of resources.

In this chapter, therefore, we shall start by recapitulating the "economics of production" since, from the economist's point of view, a water-resource project is simply a special kind of productive organization. Then we shall deal with some of the technical concepts that distinguish water-resource projects from other kinds of productive enterprise. Finally we shall treat some of the special considerations that arise from the irregularity and unpredictability of streamflow and water availability. The conceptual framework arising from these three discussions will underlie most of the analysis in the later chapters of the book although, as in Chapter 2, the concepts will rarely be useful in the pure, uncomplicated form in which they are first presented.

THE CONCEPT OF THE PRODUCTION FUNCTION

Production is defined in economics as any activity intended to convert resources of given forms and locations into other resources of forms and locations deemed more useful for purposes of further production or consumption. In this definition the word "location" should be understood in a four-dimensional sense, incorporating time along with the three spatial coordinates. This fourth dimension is particularly important in water-resource projects which have as one of their prime purposes improvement of the time pattern of water availability.

The resources employed by water-development projects (their inputs) are naturally available water, land, construction materials and manpower, and a variety of machinery and equipment. The resources produced (the outputs) are water at the times and places required for irrigation, industrial and municipal water supply, hydroelectric power, navigation, waste-water transport, low-water regulation, vacant spaces behind dams for flood control, and a variety of recreational and aesthetic facilities.[1]

The economist is in the habit of expressing the relationship between the inputs to a system and its outputs by means of a "production function." It will be helpful in introducing this concept to distinguish between "short-run" and "long-run" production functions. In the short-run production function the durable plant and equipment are assumed to be in place and unalterable, so that the relationship required connects outputs with readily variable inputs. The long-run production function applies

[1] The temperature, purity, mineral content, and other characteristics of water comprehended in the phrase "water quality" are important for most uses of water. Although we shall not mention water quality explicitly in this discussion, it should be kept in mind that when we write "water," we mean "water of requisite quality."

to situations where the physical plant can be altered, that is, where the problem includes the design as well as the operation of a water-resource project. The resources required to build or alter the plant are then counted among the variable inputs, and the plant is not regarded as an established fact conditioning the relationship between inputs and outputs. The short-run production function, simpler although less relevant to design problems, will be discussed first.

SHORT-RUN PRODUCTION FUNCTION

The short-run production function applies to a stream with specific structures (perhaps none at all) in place. In the course of a year, or a month, a variety of outputs can be produced by the combination of stream and structures, depending on how the system is operated — a matter that is referred to as the operating procedure. The operation of the system for a period of time also requires certain inputs which come under the heading of operation, maintenance, and replacement (OMR) expenses. These quantities too may depend on the operating procedure, although to a lesser extent than the quantities of the outputs. Stated in mathematical terms, a system that employs n distinguishable inputs to produce m distinguishable outputs can be described by an input vector $\mathbf{x} = (x_1, \ldots, x_n)$, where x_i is the quantity, let us say per year, of the ith input, and an output vector $\mathbf{y} = (y_1, \ldots, y_m)$, where y_j is the quantity of the jth output in the same interval of time. If specific numerical values are assigned to $\mathbf{x}$ and $\mathbf{y}$, it may or may not be technologically possible to produce the output vector $\mathbf{y}$ with the input vector $\mathbf{x}$ and the given system of stream and structures. If it is possible, we say that the given pair of vectors $(\mathbf{x}, \mathbf{y})$ belongs to the technologically feasible region; otherwise it does not. In other words, the technologically feasible region is the region of all physically attainable combinations of inputs and outputs.

Of all the vector pairs in the technologically feasible region, our interest centers on certain ones to be called "efficient" in a very specialized sense not to be confused with the meaning of efficiency in Chapter 2. Suppose that a certain vector pair $(\mathbf{x}^1, \mathbf{y}^1)$ is in the region, and also another pair $(\mathbf{x}^1, \mathbf{y}^2)$. Note that these two pairs have the same input vector, $\mathbf{x}^1$, but different output vectors. Suppose also that every component of $\mathbf{y}^2$ is at least as great as the corresponding component of $\mathbf{y}^1$ but that they are not all identical. Then it is technologically possible to employ the input quantities $\mathbf{x}^1$ to obtain either the outputs $\mathbf{y}^1$ or the outputs $\mathbf{y}^2$, and the second choice is clearly better (assuming that all the outputs are desirable) since $\mathbf{y}^2$ provides at least as much of every output as does $\mathbf{y}^1$, and more of some. We then say that the input-output pair $(\mathbf{x}^1, \mathbf{y}^1)$ is wasteful or "inefficient." On the contrary, a particular pair $(\mathbf{x}^0, \mathbf{y}^0)$ is said to be "efficient" if the technologically feasible region contains no different pair that either produces the outputs $\mathbf{y}^0$ with inputs that do not exceed $\mathbf{x}^0$ in any component, or utilizes the same inputs $\mathbf{x}^0$ to produce outputs that do not fall short of $\mathbf{y}^0$ in any component.

It should be emphasized that efficient points, as just defined, are not necessarily desirable or even sensible ones. For example, opening all the gates and going away is an efficient point since its input vector is $\mathbf{x} = (0, \ldots, 0)$ and it is impossible to improve on its outputs, however small they may be, without exceeding its input vector in at least one component.

We can now define the short-run production function: it is simply the locus of all efficient points, that is, of all technologically feasible combinations of inputs and outputs such that the output vector cannot be unambiguously exceeded without increasing some input, and the input vector cannot be unambiguously reduced without decreasing some output. All points not on the production function are clearly either infeasible or wasteful. Alternatively, if a point is feasible but not on the production function, there must be some feasible point that is an improvement on it, as long as all the outputs are desirable and costs are attached to all the inputs. On the other hand, as noted before, some of the points on the production function are extremely undesirable by any reasonable standards and, indeed, there will always be some points on the production function that are less desirable than some points off it. In practical problems, it is often advisable to adopt a design or operating procedure that is "inefficient" or "off the production function" in the strict technical sense of those words.

Somewhat gratuitously, economists assume (and so shall we) that the production-function locus can be described by an equation like

$$f(x_1, \ldots, x_n, y_1, \ldots, y_m) = 0,$$

and that this function is continuous and has at least two partial derivatives in each of its variables. Nearly all economic discussion of production begins with a function like this one, without the elaborate introduction that we have given. We hope that by spending words we have spared later confusion about the meaning of this kind of formula.

Although we wish to postpone criticism and evaluation until we have introduced the long-run production function, which is the important one for design purposes, we may as well acknowledge right now that this tidy concept does not apply very well to water-resource projects. A leading source of difficulty is the irregularity and unpredictability of natural-water flow. At the beginning of a year any input vector **x** can be considered, but it is impossible to predict which output vector **y** will result from it. That depends on the amount and timing of the natural inflow. In technical terminology the relationship between **x** and **y** is stochastic rather than determinate; that is, to each value of **x** there corresponds at best a probability distribution of **y** rather than a definite determinate value. The most we can do is predict the probability distribution of the outputs resulting from a given system, given inputs, and a given operating procedure if we know enough about the hydrology.

While this is a much more sophisticated and less satisfactory kind of relationship than a determinate production function, we must be content with it. We now reformulate our concept of a production function in the light of this complication. Consider a particular vector of controllable inputs,[2] **x**, and suppose that the probability distribution of natural-water availability is known or at least can be estimated adequately. Then we can, in principle at least, calculate the joint probability distribution of the outputs, $y_1, \ldots, y_m$. To incorporate these probabilities in a production function, suppose that we are concerned with only the first- and second-order moments of the probability distribution. Let $\bar{y}_j$ be the expected value of y_j, let σ_j be its standard deviation, and let ρ_{jk} be the correlation coefficient of y_j and y_k. Then we can write the output-moment vector as

$$\mathbf{M} = (\bar{y}_1, \ldots, \bar{y}_m, \sigma_1, \ldots, \sigma_m, \rho_{1,2}, \ldots, \rho_{m-1,m}).$$

The reformulated definition of the feasible region is now evident. The output-moment vector **M** and its components are clearly functions of the vector of decidable inputs **x** and the operating procedure. That is, the input vector and the operating procedure together determine the probability distribution of outputs and the moments of that distribution. Therefore we can define the feasible region as those vector pairs (**x**, **M**) that are physically realizable.

A definition of the efficient points of this region and of the production

[2] Here and in later discussion we shall not include naturally available water among the inputs, because its quantity is not subject to decision or control.

function presents more difficulties. Other things being equal, a moment vector $\mathbf{M}$ is more desirable the larger the expected values in its first m components and the smaller the standard deviations in its second m components. The desirability of the output correlation coefficients is more ambiguous; this and other sources of ambiguity will be examined more closely when we come to our detailed discussion of uncertainty. At any rate, we can say that an output-moment vector $\mathbf{M}^1$ is unambiguously better than some other vector $\mathbf{M}^2$ if, when compared element by element, all the differences are either zero or in the desirable direction. For that matter, extending this kind of comparison to pairs like $(\mathbf{x}^1, \mathbf{M}^1)$ and $(\mathbf{x}^2, \mathbf{M}^2)$, we can say that a pair is efficient if it is feasible and there is no other feasible pair that is unambiguously better on the basis of a component-by-component comparison. The production function is then the locus of efficient pairs $(\mathbf{x}, \mathbf{M})$.

LONG-RUN PRODUCTION FUNCTION

The long-run production function is an elaboration of this same theme. To formulate it, we consider a specified stream with perhaps some structures in place. The relevant inputs are the resources needed to construct additional installations and modify existing ones as necessary, together with the resources required to operate and maintain the old and new structures. The outputs are the same as in the short-run case, except that it is no longer reasonable to relate inputs to outputs for a single month or year. We must think of much longer spans of time. A simplifying device that we shall follow in the hope of rendering our explanation comprehensible is to pretend that all construction is accomplished in a brief time after the initiation of the project, say one year, and that thereafter throughout the life of the project, say T years, each year is just like every other. Then the output vector $\mathbf{y}$ is the vector of outputs in a typical year. The input vector $\mathbf{x}$ is composed of two parts. The more important part is the resources consumed during the construction phase; the remaining inputs are the resources expended for operation, maintenance, and replacement during a typical year of operation. This, indeed, is the conceptual framework employed in the familiar benefit–cost analysis, which will be presented more fully below.

Adopting these conventions, we can formulate input-output pairs $(\mathbf{x}, \mathbf{y})$ just as in the short-run case, and when $\mathbf{x}$ and $\mathbf{y}$ are given, we can determine whether or not they constitute a physically attainable combination of original-resource investment, annual OMR expenditure, and annual output. The technologically feasible region is again those pairs $(\mathbf{x}, \mathbf{y})$ that are physically realizable. A feasible pair is efficient if no other feasible pair is unambiguously superior to it on the basis of a component-by-component comparison. The production function is the locus of all efficient pairs.

Even more than in the short-run case, we must take account of the fact of hydrologic uncertainty. The means is clearly the same: instead of relating actual outputs to inputs in the production function, we relate a vector of first and second moments of the probability distribution of outputs to the vector of inputs. Although it is unreasonable to regard the relationship of $\mathbf{y}$ to $\mathbf{x}$ as being determinate, it does seem reasonable so to regard the relationship of $\mathbf{M}$ to $\mathbf{x}$, in the long run as well as in the short.

There is another difficulty in the long-run production function as we have just formulated it. We really cannot pretend that all investment in a project is accomplished in the first year (or two, or three) of its life and that thereafter one year is just like another. In actuality, river-basin developments grow slowly — the TVA system is still expanding — and outputs change substantially during the five decades or more of useful life. There are a number of procedures for formulating production functions that take account of the dynamic character of water-resource projects. The most prevalent one derives from the idea that physically identical articles available at different times are different commodities. By a "commodity" an economist means a collection of articles that are perfect substitutes for one another. In less technical words, if consumers do not care which of two articles they receive, the two are units of the same commodity; otherwise not. Now a farmer is not indifferent to whether an acre foot of water is made available to him in January or in June. Similarly 1000 cubic yards of concrete consumed in 1960 is not interchangeable with an identical quantity available in 1961. In short, the date of availability is one of the distinguishing specifications of a commodity.

With this concept in mind we can conceive of a vector of dated inputs and outputs like $(\mathbf{x}^1, \mathbf{y}^1, \ldots, \mathbf{x}^t, \mathbf{y}^t, \ldots, \mathbf{x}^T, \mathbf{y}^T)$, where $\mathbf{x}^t$ denotes a vector of inputs in the tth period and $\mathbf{y}^t$ denotes a vector of outputs in the tth period, where $t = 1, \ldots, T$. This enlarged vector can be employed in the same way as the simpler vectors introduced previously: all those vectors $(\mathbf{x}^1, \mathbf{y}^1, \ldots, \mathbf{x}^T, \mathbf{y}^T)$ that are physically realizable constitute the technologically feasible region. A feasible vector is efficient if there is no distinct feasible vector that is unambiguously superior to it on the basis of a component-by-component comparison. The production function is the locus of all efficient vectors.

In this formulation, clearly, the dams and other structures are not counted among the inputs — they are called "intermediate products"; the inputs are only the resources consumed in constructing, operating, and maintaining the project. The outputs in each period are the usual ones. To take account of uncertainty, moment vectors $\mathbf{M}^t$ can be substituted for the output vectors $\mathbf{y}^t, t = 1, \ldots, T$, if it can be assumed that the probability distributions of outputs in the several periods are statistically independent. If this assumption cannot be made, then one huge output-moment vector is required.

The technique of using dated inputs and outputs, while logically sound, is unsatisfactory for several reasons. Its chief shortcoming conceptually is its inherent rigidity: it implies that the complete plan of operation and development of the project throughout its life is to be determined from the beginning and that the most desirable plan is to be selected, once and for all, at the initiation of the project. Actually, of course, much more flexibility is available and desirable. At the beginning of a project one need plan in detail only for the first few periods, with provision for subsequent revision and development as events mature. For instance, instead of using a 50-yr vector of dated inputs and outputs, it is preferable to think in terms of, say, the first five years and include among the outputs of the abbreviated vector the physical structures that will be in place at the end of the fifth year and that will facilitate operation of the system in the years beyond the more limited horizon. The concepts of the feasible region, efficient vectors, and production function then follow immediately.

In summary, a production function expresses the technologic limits on the outputs attainable from given inputs. The outputs may be regarded as functions of the inputs or vice versa, whichever may be convenient. From this discussion of the production function, three things should now be evident. First, the concept is a very adaptable one; the arguments of the function may be chosen with great freedom as circumstances dictate. Second, from a conceptual point of view, the production function is by no means a straightforward, elementary idea, for it incorporates functional relationships of great intricacy. In expounding the production function, economists prefer to deal with cases with only a single output, one or two inputs, and no uncertainty. Such cases (we shall examine one example below) provide useful insight into the properties of production functions and into the nature of production itself, but they exist only in textbooks. Third, from an empirical, quantitative point of view, it is out of the question actually to determine a production function except in the most highly simplified circumstances.

These last two remarks may appear to constitute a devastating critique of the concept, but in reality they only place unfortunate limitations on a fundamental and useful idea. The countervailing considerations are at least two. From a conceptual point of view, the production function is an indispensable tool of thought because it alone incorporates the connection between inputs and outputs and provides a framework for keeping this connection always in mind. The production function reminds us that we cannot make bricks without straw or store water without a reservoir, and that in general it takes a larger reservoir to store more water, although the exact connection between water in storage and capacity of reservoir cannot be defined. From an empirical point of view, even though a production function in its full comprehensiveness can almost never be determined, many of its essential properties (in particular, its partial derivatives at points of interest) can frequently be ascertained. We shall take advantage of both of these virtues many times in the remainder of this volume.

A SIMPLE EXAMPLE

At this stage it will be helpful to consider a concrete example.[3] Let us turn to the problem of designing a single reservoir for a single purpose, the provision of irrigation water. This system has only a single output in each year of its life. By the following reasoning, we can also think of it as having only a single input. The amount of irrigation water furnished in each year depends only on the capacity of the reservoir and on uncontrollable hydrologic circumstances. Thus we can say that the output, or at least its probability distribution, depends only on the capacity of the reservoir, which in turn (the topography of the region being given) depends principally on the height of the dam. For simplicity we shall suppose that the capacity of the reservoir depends only on the height of the dam. We shall also assume that engineers can ascertain the best

[3] In this example, as well as in later examples in this chapter, we make no pretense of realism. In particular, we assume that dams and power plants are the only structures that need be considered in a water-resource system, whereas in fact a variety of canals, conduits, transmission lines, and other facilities are integral parts of any such system.

combination of resources for constructing a dam of any specified height. Thus the height of the dam is the only variable about which a decision has to be made. Given the height of the dam, the resource inputs are determined by engineering considerations and the usable-water output

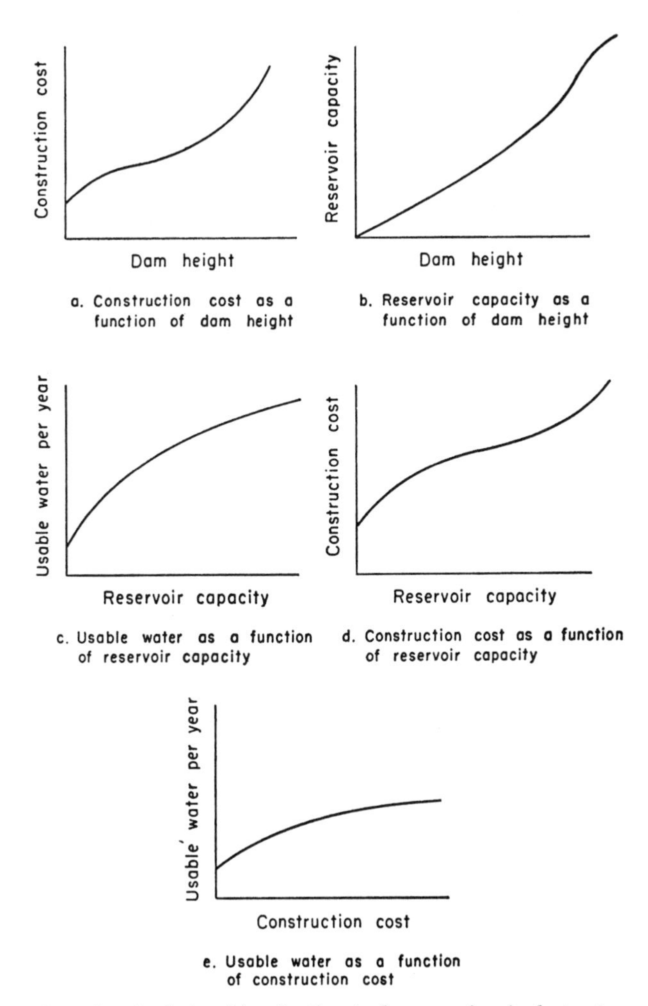

Fig. 3.1. Functional relationships for the single-reservoir, single-purpose example.

is determined by a combination of physiographic and hydrologic data.

The situation can be depicted by a series of simple graphs, as in Fig. 3.1, where parts *a*, *b*, and *c* express the basic data of the problem. The term "usable water" in *c* means the number of acre feet made available at the time required for irrigation. Its relation to reservoir capacity, sketched in *c*, is a complicated one depending on the time profiles of water inflow and water use, and will be discussed in more detail in the section on hydrologic concepts. Here we note that as the figure is drawn, successive increases in reservoir capacity yield smaller and smaller additions to the quantity of usable water. These diminishing marginal physical returns are typical of water-storage reservoirs. Because the runoff of most streams occurs in irregular spurts and varies from year to year, a larger reservoir is not filled as frequently or utilized as fully as a smaller one. Even a reservoir approaching infinite size cannot provide an annual usable output greater than the average annual runoff.

Part *d* is derived from *a* and *b* simply by reading off the dam height corresponding to each reservoir capacity and the construction cost corresponding to that dam height. Part *e*, derived similarly from *c* and *d*, is the production function, the relationship between inputs (summarized in construction cost) and the corresponding maximum attainable outputs.

In this simple case, where we have ignored hydrologic uncertainty, the production function is derivable from empirically ascertainable data. As we remarked above, this happy situation is not typical. Nevertheless, it illustrates the way in which a production function arises from the circumstances of the case in all instances, however complicated.

THE NET-BENEFIT FUNCTION AND OPTIMALITY CONDITIONS

NET-BENEFIT FUNCTION

The production function is only one ingredient of the economic theory of production. It summarizes the technologic data of a problem and incorporates all the possible plans of design and operation that are neither impracticable nor wasteful. In order to go further and determine which of the plans contained in the production-function locus is to be adopted, we must turn to the purposes that the project is intended to serve. From a consideration of these purposes we deduce a criterion, which we shall call interchangeably a "net-benefit function" or an "objective function," that will enable us to compare two or more plans and ascertain which of them is most desirable.

To this end we shall assume that the results of any plan are fully incorporated in its input-output vector $(\mathbf{x}, \mathbf{y})$. That is, we shall assume that if we know the quantities of all inputs called for by a plan (and their dates, if relevant) and have similar information about the outputs, then we have all the information about the plan itself necessary to evaluate its desirability, the precise manner in which the inputs are converted into outputs being irrelevant for this purpose. We have already invoked this assumption and indulged in a limited amount of desirability comparison in constructing the production functions. There we compared two plans with input-output vectors $(\mathbf{x}^1, \mathbf{y}^1)$ and $(\mathbf{x}^2, \mathbf{y}^2)$ and asserted that if on a component-by-component comparison each component of $(\mathbf{x}^1, \mathbf{y}^1)$ differed from the corresponding component of $(\mathbf{x}^2, \mathbf{y}^2)$ in the favorable direction if at all, then the plan giving rise to $(\mathbf{x}^1, \mathbf{y}^1)$ was unambiguously superior to the plan giving rise to $(\mathbf{x}^2, \mathbf{y}^2)$. Technically, such a comparison is called a partial ordering. If all the discrepancies are in the same direction, favorable or unfavorable, we can say which of two plans is superior. But if some of the component-by-component comparisons go in one direction and some in the other, then that method of comparison is inconclusive. Now we need something more: a complete ordering that will never be inconclusive, although it may in some instances assert that two or more plans are indifferent. It will be complete in the sense that it will never assert that two plans are incomparable.

This complete ordering is achieved by constructing a scalar-valued function of the input-output vectors, say $u(\mathbf{x}, \mathbf{y})$, in such a way that higher values of $u(\mathbf{x}, \mathbf{y})$ will indicate more desirable plans. (In this notation u is intended to suggest utility.) This function u is a quantitative expression of the extent to which the plan fulfills the various objectives of the project; it is a summary of the values of the inputs and outputs. To be more specific, suppose that the ith input to a project has an average value of p_i' per unit and the jth output an average value of p_j per unit. Then a plausible utility function, or objective function, or net-benefit function, would be $u(\mathbf{x}, \mathbf{y}) = \Sigma p_j y_j - \Sigma p_i' x_i$. In this formula the unit values p_i' and p_j may be either constants or functions of the vector $(\mathbf{x}, \mathbf{y})$ as circumstances indicate. We shall make extensive use of this form of net-benefit function although, as we shall see shortly, it has some serious shortcomings. For the rest of this chapter net-benefit functions will be assumed to be of this form, except when explicitly noted otherwise.

The truly enormous difficulties involved in deciding on the objectives to be served by a water-resource project and in devising an objective function that will measure their attainment have been discussed at length in Chapter 2 and will not be reiterated here. We saw that the consequences of a water-resource project, all of which should be reflected in the objective function, include not only the production and consumption of marketable goods and services but also redistribution of income and much more. In consequence, estimation of the unit values p_i' and p_j, difficult as it often is, frequently presents the least of the perplexities that have to be resolved in constructing an objective function.

Over and above the conceptual problems of formulating objectives and measuring their attainment, there is an additional technical difficulty to be mentioned: that resulting from uncertainty due to hydrologic, economic, and other factors. Generally speaking, in a short-run plan the inputs can be ascertained in advance, while the outputs are likely to be influenced by unpredictable hydrologic factors. In a long-run plan not even the inputs can be foreseen reliably, since many repairs, alterations, and deferred construction expenses depend on unpredictable contingencies. Consequently the net benefits of a project constitute a random variable, and it is impossible, even in principle, to determine what the net benefits of a proposed plan will turn out to be. However, it is possible in principle (although usually not in practice) to determine the probability distribution of net benefits. Since the actual net benefits $u(\mathbf{x}, \mathbf{y})$ yielded by a plan are not ascertainable, we often employ as our criterion the expected value of net benefits,[4] $Eu(\mathbf{x}, \mathbf{y})$. In adopting this criterion, we are ignoring all characteristics of the probability distribution except for its first moment; in particular we are ignoring the amount of uncertainty surrounding our prediction of $u(\mathbf{x}, \mathbf{y})$. The justifiability of this simplification will be discussed more fully in the section on uncertainty.

It is probably wise at this stage to point out that the entire foregoing discussion applies equally to long-range and to short-range plans. The pertinence to short-range plans is obvious; indeed, it may appear from the notation that the discussion applies only to short-range plans. But in reality there is nothing that prevents inputs required at different dates from appearing as components of $\mathbf{x}$, or outputs occurring at different dates from appearing as components of $\mathbf{y}$. With the understanding that $\mathbf{x}$ and $\mathbf{y}$ may include goods pertaining to different dates, all the foregoing expressions can be interpreted so as to apply indifferently to single-period plans or to plans of any length.

[4] The symbol E stands for expected value, a concept that will be explained more fully in a later section of this chapter. If p_i' and p_j are constants, $Eu(\mathbf{x}, \mathbf{y}) = u(E\mathbf{x}, E\mathbf{y})$; otherwise not.

OPTIMALITY CONDITIONS

To proceed with the economic analysis, let us assume that all difficulties have been surmounted or evaded and that we have at hand a satisfactory single-valued net-benefit function $u(\mathbf{x}, \mathbf{y})$ that expresses the desirability of the plan that yields the input-output vector $(\mathbf{x}, \mathbf{y})$. We can then bring the production function and the net-benefit function together and say that we seek that member of the production-function locus that confers the highest possible value on the net-benefit function. Algebraically stated, we seek to maximize $u(\mathbf{x}, \mathbf{y})$ subject to the constraint that the plan be on the production function, or $f(\mathbf{x}, \mathbf{y}) = 0$. If we assume that both of these functions are differentiable, we have a straightforward problem in the calculus. To solve it, we set up the Lagrangian function

$$L(\mathbf{x}, \mathbf{y}) = u(\mathbf{x}, \mathbf{y}) + \lambda f(\mathbf{x}, \mathbf{y}),$$

where λ is an undetermined Lagrangian multiplier.[5] A necessary condition for a point $(\mathbf{x}, \mathbf{y})$ to maximize $u(\mathbf{x}, \mathbf{y})$ subject to our constraint is that a value of λ can be found for which all the first partial derivatives of $L(\mathbf{x}, \mathbf{y})$ vanish at the point $(\mathbf{x}, \mathbf{y})$. These conditions amount to:

$$\frac{\partial u}{\partial x_i} = -\lambda \frac{\partial f}{\partial x_i} \qquad (i = 1, \ldots, n)$$

and

$$\frac{\partial u}{\partial y_j} = -\lambda \frac{\partial f}{\partial y_j} \qquad (j = 1, \ldots, m)$$

for some value of λ. From these we deduce by division:

$$\frac{\partial u}{\partial x_i} \Big/ \frac{\partial u}{\partial y_j} = \frac{\partial f}{\partial x_i} \Big/ \frac{\partial f}{\partial y_j},$$

$$\frac{\partial u}{\partial x_i} \Big/ \frac{\partial u}{\partial x_h} = \frac{\partial f}{\partial x_i} \Big/ \frac{\partial f}{\partial x_h},$$

and

$$\frac{\partial u}{\partial y_j} \Big/ \frac{\partial u}{\partial y_k} = \frac{\partial f}{\partial y_j} \Big/ \frac{\partial f}{\partial y_k},$$

where i and $h = 1, \ldots, n$ and j and $k = 1, \ldots, m$, provided that the derivatives in the denominators do not vanish. Furthermore, from the production-function constraint, $f(\mathbf{x}, \mathbf{y}) = 0$, we have

[5] For an economic interpretation of the Lagrangian multiplier λ, see Chapter 2, p. 76.

$$\frac{\partial f}{\partial x_i}\Big/\frac{\partial f}{\partial y_j} = -\frac{\partial y_j}{\partial x_i},$$

$$\frac{\partial f}{\partial x_i}\Big/\frac{\partial f}{\partial x_h} = -\frac{\partial x_h}{\partial x_i},$$

and
$$\frac{\partial f}{\partial y_j}\Big/\frac{\partial f}{\partial y_k} = -\frac{\partial y_k}{\partial y_j}.$$

Therefore we can write the necessary conditions for an input-output vector that maximizes the net-benefit function as

$$\frac{\partial u}{\partial x_i}\Big/\frac{\partial u}{\partial y_j} = -\frac{\partial y_j}{\partial x_i},$$

$$\frac{\partial u}{\partial x_i}\Big/\frac{\partial u}{\partial x_h} = -\frac{\partial x_h}{\partial x_i},$$

and
$$\frac{\partial u}{\partial y_j}\Big/\frac{\partial u}{\partial y_k} = -\frac{\partial y_k}{\partial y_j}.$$

Our next task is to interpret these rather imposing equations. The last three are the important ones, and we shall interpret them term by term.

The first term $\partial u/\partial x_i$ is the rate of change of the net-benefit function with respect to the ith input. It is the marginal cost of the ith input, and we shall give it the more suggestive symbol MC_i. If p_i' the unit price of the ith input is a constant with respect to $(\mathbf{x}, \mathbf{y})$, $\text{MC}_i = p_i'$. Otherwise the relationship is more complicated.

Similarly, $\partial u/\partial y_j$ is the marginal benefit of the jth output, MB_j for short. If p_j is a constant, $\text{MB}_j = p_j$; otherwise not. That takes care of all the expressions to the left of the equal signs, since $\partial u/\partial x_h$ and $\partial u/\partial y_k$ are analogous to $\partial u/\partial x_i$ and $\partial u/\partial y_j$, respectively.

On the right sides of the equations we find the partial derivatives of the components of $(\mathbf{x}, \mathbf{y})$ with respect to one another. These have clear economic meanings. Thus $\partial y_j/\partial x_i$ is the rate at which the jth output can be increased (or must be decreased) for each unit increase (or decrease) in the ith input. It is the marginal productivity of the ith input when devoted to the jth output. We shall denote it by MP_{ij}. Correlatively, $\partial x_h/\partial x_i$ is the rate at which the hth input can be reduced for each unit increase in the ith input, with all outputs and all other inputs unchanged. It is also the rate at which the hth input must be increased for each unit decrease in the ith input, if all outputs and all other inputs are to be maintained at constant levels on the production function. Since it is called the marginal rate of substitution of the hth input for the ith input, we shall denote it by MRS_{hi}. Finally, $\partial y_k/\partial y_j$ is the rate at which the kth output must be reduced per unit increase in the jth output, all inputs and all other outputs remaining fixed. The interpretation for decreases in the jth output is analogous. This derivative is called the marginal rate of transformation of output k for output j, and is symbolized by MRT_{kj}. With these concepts the necessary conditions for a plan that maximizes net benefits can be written:

$$MC_i/MB_j = MP_{ij} \tag{3.1}$$

$$MC_i/MC_h = MRS_{hi} \tag{3.2}$$

$$MB_j/MB_k = MRT_{kj}. \tag{3.3}$$

In spite of their elaborate derivation these equations are simply common sense expressed in algebra. Consider the last one, for example. Suppose $MRT_{kj} = q$. Then a one-unit decrease in output j, which would reduce net benefits by p_j, would permit a q-unit increase in output k, worth qp_k. If $p_j < qp_k$, then output j should be decreased and output k increased correspondingly. On the other hand, if $p_j > qp_k$, net benefits could be increased by increasing output j at the expense of output k. If either of these inequalities held, therefore, net benefits could be increased and the plan could not be optimal. The optimality condition is therefore $p_j = qp_k$. The common sense underlying the other conditions is similar.

All these results have been deduced by straightforward application of the theory of maximization of the differential calculus. They are therefore subject to all the limitations of that theory. There are three such limitations particularly that should be kept in mind. First, the conditions that we have applied are necessary but not sufficient conditions for a maximum. With certain exceptions, to be mentioned below, every maximum satisfies them, but there may be some values of $(\mathbf{x}, \mathbf{y})$ that satisfy them without maximizing $u(\mathbf{x}, \mathbf{y})$. To determine whether a vector $(\mathbf{x}, \mathbf{y})$ that satisfies the necessary conditions for a maximum is indeed a maximizing vector, we may either inspect the purported solution informally to see whether small variations in all components of $(\mathbf{x}, \mathbf{y})$ decrease the value of the objective function or, more formally, apply the sufficient conditions for a maximum, which depend on the second partial derivatives.[6]

Second, even when the sufficient conditions for a maximum are met,

[6] These conditions are too technical to be developed here. A good standard reference is Theodore Chaundy, *The differential calculus* (Clarendon Press, Oxford, 1935).

the methods of the calculus determine only local maxima, that is, they guarantee only that if a vector satisfies the conditions, there is no other permissible vector within a small (strictly speaking, infinitesimal) region around it that confers a larger value on the objective function. If there are several solutions to the maximizing conditions, the maximum maximorum must be determined by enumeration.

And finally, if the permissible range of variation of the components of $(\mathbf{x}, \mathbf{y})$ is delimited, it is unfortunately possible for a maximum to occur at a point that does not even meet the "necessary" conditions for a maximum. This paradox arises because the necessary conditions of the differential calculus are valid only when the variables can vary freely. But in most economic applications the ranges of the disposable variables are bounded; in particular, many of them are inherently nonnegative and they often have upper bounds as well. A maximum can occur where one or more of the variables takes on an extreme value, even though the conventional maximizing conditions are not satisfied. Thus the theorems we have deduced must be understood in the light of all these qualifications.

SOME EXAMPLES

Now it is time for some examples intended to illuminate these somewhat abstract equations. All the examples will deal with long-run problems — that is, problems in which a water-resource system is to be designed, built, and operated for some period of time. A more explicit notation than the very general one used heretofore will be helpful in these examples. Hence we shall denote a plan by a triple vector such as $(\mathbf{x}^0, \mathbf{x}', \mathbf{y})$ instead of by the double vectors used previously. In this notation $\mathbf{x}^0$ will represent the resource quantities used in original construction, $\mathbf{x}'$ will symbolize the inputs consumed in a typical year of operation, and $\mathbf{y}$ will signify the levels of outputs in a typical year. Since the outputs particularly are not exactly predictable, we shall regard $\mathbf{y}$ as meaning the average or expected values of the outputs in a typical year, but this refined question of interpretation will be irrelevant to our reasoning for the present.

In these examples we shall assume that the planned economic life of each project is T years, that the rate of discount used to render comparable inputs and outputs occurring in different years is r percent per year, and that the unit prices of the components of $\mathbf{x}^0$, $\mathbf{x}'$, and $\mathbf{y}$, respectively, are p_i^0, p_i', and p_j. The length of life of the project, the rate of interest, and the unit prices will all be regarded as given constants. Then the cost of construction is $K = \Sigma p_i^0 x_i^0$, and the present value of the outputs is

$$\Sigma p_j y_j \left(\frac{1}{1+r} + \frac{1}{(1+r)^2} + \ldots + \frac{1}{(1+r)^T} \right) = \frac{1-(1+r)^{-T}}{r} \Sigma p_j y_j.$$

Similarly, the present value of the inputs is

$$\frac{1-(1+r)^{-T}}{r} \Sigma p_i' x_i'.$$

We shall sometimes use the symbol v to stand for the present-value factor occurring in the last two formulas and shall use the notations $B = \Sigma p_j y_j$ and $C = \Sigma p_i' x_i'$ for annual gross benefits and annual costs, respectively. The net benefit from the plan $(\mathbf{x}^0, \mathbf{x}', \mathbf{y})$ then becomes

$$u(\mathbf{x}^0, \mathbf{x}', \mathbf{y}) = vB - vC - K = v\Sigma p_j y_j - v\Sigma p_i' x_i' - \Sigma p_i^0 x_i^0. \qquad (3.4)$$

Single reservoir, single purpose. Let us begin with the example of the previous section: a single reservoir designed to supply irrigation water. The input vector $\mathbf{x}^0$ is a compilation of the quantities of all the goods and services, including labor, required to construct a dam of specified size and design, along with the conduits, gaging stations, access roads, and all other apparatus and installations required for efficient operation. Just as before, we assume that for any given reservoir capacity the most economical design of dam and appurtenances is known. Therefore we can regard the cost of this most economical design as summarizing and measuring the entire resource cost of a reservoir of the given capacity and can treat the input vector, $\mathbf{x}^0$, as if it had only a single component, the cost of resources. Since we shall regard OMR costs as negligible, we shall omit $\mathbf{x}'$ from the current analysis. The only component of $\mathbf{y}$ is the quantity of water released in a typical year at the times required for irrigation. Equation (3.1) with $i = j = 1$ is the only applicable optimizing condition, the other two equations being irrelevant. Since the input is measured in terms of cost, $p_i^0 = 1$ by definition. Then $\text{MC}_1 = 1$, $\text{MB}_1 = vp_1$ (both derived by taking partial derivatives of the net-benefit function), and the sole optimizing condition is $\text{MP}_{11} = 1/vp_1$. This asserts that a necessary condition for the reservoir to be of optimal size is that an additional dollar spent on construction will purchase an increase in reservoir size just sufficient to permit an increase of $1/vp_1$ units of irrigation water a year. It is easy to see the intuitive reasonableness of this result. Since each unit of irrigation water is worth p_1, $1/vp_1$ units are worth $1/v$. This, then, is the increase in the value of output in each year of operation. The present value of $1/v$ a year for T years is $v(1/v) = 1$, which is the cost of procuring that increase. Thus slight changes in the size of the reservoir will not increase the value of the net-benefit function, which is surely a necessary condition for optimal size.

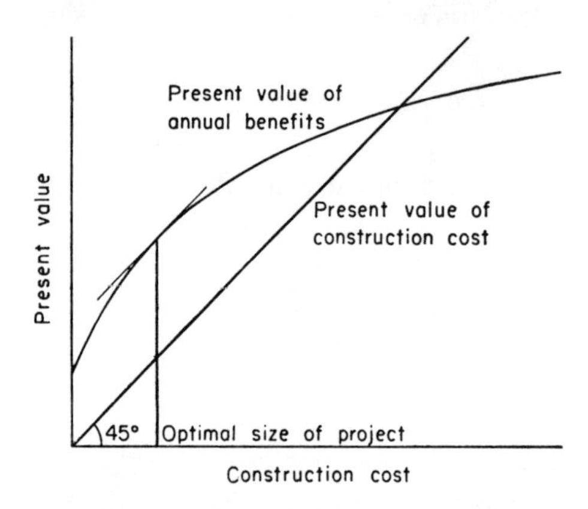

Fig. 3.2. Relationship of benefit and cost curves for the single-reservoir, single-purpose example.

This same solution is frequently presented graphically, as in Fig. 3.2, which is derived from part *e* of Fig. 3.1. The horizontal axis is the same. The vertical axis, instead of being scaled in physical quantities of usable water per year, is scaled in terms of the present value of those quantities, computed at the assigned unit price and discount factor. In addition, the graph includes the present value of construction cost, which (assuming that all construction is undertaken at the initiation of the project) is simply the construction cost itself and is therefore represented by the 45-degree line. The net benefits corresponding to each scale of construction are represented on the graph by the excess of the height of the annual-benefits present-value curve over the height of the construction-cost line. The optimal scale is the one for which the curve is as far as possible above the line; this clearly is where the slope of the curve equals the slope of the line, or 45 degrees. The slope of the present-value-of-annual-benefits curve is simply the present value of the marginal productivity, or $vp_1\text{MP}_{11}$. Hence the optimizing condition found in the graphic analysis is $vp_1\text{MP}_{11} = 1$, which is the same as the condition determined algebraically.

But in this instance we cannot reasonably assume that p_1, the unit price of irrigation water, is a constant. The greater the quantity of irrigation water provided, the farther some of it will have to be transported and, in general, the lower will be the productivity of some of it in agriculture. Hence to dispose of a larger quantity of irrigation water, the price will have to be lowered and, indeed, some of it will have to be devoted to economically less productive uses.[7] We must therefore regard p_1 as a decreasing function of y.

Let us now deduce the optimizing condition for this case. We can write $p_1 = D(y)$. It is still true that $p_1^0 = 1$ and $MC_1 = 1$. But, on differentiating $u(x_1^0, y_1)$, we find

$$MB_1 = v[D(y_1) + y_1 D'(y_1)]. \tag{3.5}$$

The expression $D(y_1) + y_1 D'(y_1)$ that appears in this formula occurs so frequently in economic analysis that it is worth remarking on. It asserts, speaking heuristically, that if y_1 units of output are being produced each year, a one-unit increase in annual output will have two effects. In the first place, the additional unit of output will add approximately $D(y_1)$ units to the annual benefit. In the second place, the additional unit of output will cause the value placed on each unit of output to change by approximately $D'(y_1)$. The sum of these two effects is the expression in question.

Inserting this result in Eq. (3.1), we obtain the optimizing condition

$$v[D(y_1) + y_1 D'(y_1)] = 1/MP_{11}, \tag{3.6}$$

which asserts that the present value of the change in annual benefits must be equated to the cost of increasing the size of the reservoir enough to permit a unit increase in annual output. Readers familiar with economic literature will see at once that this equation indeed yields the size of project corresponding to maximum net benefits, as defined, but it does not lead to an economically optimal size of project or quantity of output. Instead it corresponds to the behavior of a monopolist.

The reason is most clearly seen by considering Fig. 3.3. Since x_1^0 and y_1 are functionally related by the production function (Fig. 3.1 *e*), we are free to use y_1 as our independent variable in place of x^0, and we have done so in drawing this figure. The function $D(y_1)$ is really the demand

[7] At this point we ignore some complications such as the possibility of charging different prices to different users, although they are important in practice. When such discriminating price scales are used, it is necessary to regard the output vector as containing several components: y_1, the quantity of water sold at price p_1; y_2, the quantity sold at p_2; and so forth. We prefer to stay within the single-component framework for the present.

curve for irrigation water, the functional relationship between the quantity of irrigation water and the price at which purchasers are just willing to take that quantity.

Now consider any particular annual output y_1. To it there corresponds a value $D(y_1)$. Each unit of output will therefore confer a benefit of $D(y_1)$ each year, the present value for all T years being $vD(y_1)$. The curve $vD(y_1)$ in Fig. 3.3 shows how the present value of benefits per unit of

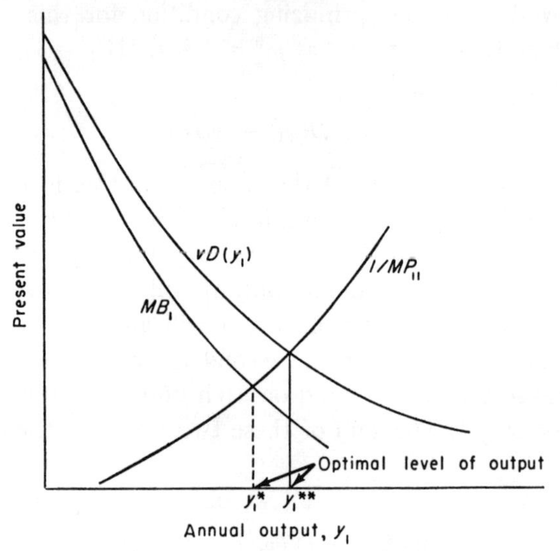

Fig. 3.3. Demand and marginal-cost curves for the variable-value case of the single-reservoir, single-purpose example.

water falls as the number of units provided increases. The curve MB_1, derived from the demand curve by Eq. (3.5), and the curve $1/MP_{11}$, the reciprocal of the marginal productivity of construction expenditure in producing irrigation water, are also plotted on the figure. These two curves cross at the value of y_1 which satisfies the optimality condition, Eq. (3.6), say y_1^*.

At this level of output an additional unit of irrigation water per year would have a present value of $vD(y_1^*)$ but would cost only $1/MP_{11}$ to obtain. From an over-all economic point of view, therefore, it would be worthwhile to increase the scale of the project even though it would not increase the value of net benefits according to the net-benefit formula we have adopted. This shows that our net-benefit formula is misleading; it recommends projects of economically inadequate size when the unit value of output falls in response to increases in quantity.

In Chapter 2 we developed a net-benefit formula that is free of this defect. Consider the following equation:

$$u(x_1^0, y_1) = v \int_0^{y_1} D(\eta)d\eta - p_1^0 x_1^0.$$

With this formula $MB_1 = vD(y_1) = vp_1$ and $MC_1 = p_1^0$, and the optimizing condition is $p_1^0 = vp_1 MP_{11}$. This corresponds to the solution y_1^{**} on Fig. 3.3, which is the economically appropriate one.

The difference between this formulation and the previous one is that earlier we assumed that each unit of output had the same value, namely p_1, whereas now each unit is valued according to its marginal contribution to the objective function. In other words, the ηth unit is valued at $D(\eta)$.

This elementary single-input, single-output example has taught us a lesson of rather broad significance which we can express in general form. If the output prices are functions of the output quantities, the net-benefit formula of Eq. (3.4) will not lead to an economically optimal design, but to an inadequate scale of development. If the marginal value of each output depends only on the quantity of that output, so that the relationship can be written $p_j = D_j(y_j)$, then the following net-benefit function adapted from Chapter 2 will give correct results:

$$u(\mathbf{x}^0, \mathbf{x}', \mathbf{y}) = v\Sigma \int_0^{y_j} D_j(\eta_j)d\eta_j - v\Sigma p_i' x_i' - \Sigma p_i^0 x_i^0.$$

However, it is quite likely that the values appropriate to some outputs are influenced by the levels of other outputs. For example, the unit value of power provided for rural electrification may well be influenced by the quantity of irrigation water provided. Furthermore, the unit values of the inputs may be influenced by the quantities of the inputs, which would have consequences similar to those just considered. Net-benefit functions appropriate to these conditions can be constructed, but they are quite complicated (see footnote 9 in Chapter 2). Since we know already what the optimizing conditions are, it seems wiser for present purposes to drop the approach of maximizing some explicit function and, instead, to present forthwith the economically correct optimizing conditions. Necessary conditions for a plan $(\mathbf{x}, \mathbf{y})$ to be economically optimal are:

$$p_i'/p_j = MP_{ij},$$

$$p_i'/p_h' = MRS_{hi},$$

and

$$p_j/p_k = MRT_{kj},$$

where p_i', p_j, and the rest are the marginal unit values for the inputs and outputs at the planned levels $(\mathbf{x}, \mathbf{y})$.

This whole discussion results from the fact that the profit-maximizing monopolist does not, while the resource-development agency should, take account of consumers' and suppliers' surpluses when evaluating the results of a plan of construction and operation. The relationship between demand curves for outputs and objective functions was explored in Chapter 2; the relationship between supply curves for inputs and objective functions is strictly analogous. Here our concern is with the use of objective functions, not with their formulation, and as long as we are aware of the pitfalls, we shall have little to say about this aspect of the problem.

Single reservoir, dual purpose. Let us turn now to a slightly more elaborate example: the planning of a single reservoir with two useful outputs, say irrigation water and water supply for a municipality. In order to keep the argument as simple as possible, we shall use the net-benefit function of Eq. (3.4) and assume that the price of the input and the prices of the outputs are constants. In this case there is still only a single input quantity, x_1^0, but there are two output quantities, y_1 and y_2. We continue to assume either that OMR costs are negligible or that changes in the plan do not influence them significantly. Differentiating Eq. (3.4), we find:

$$\mathrm{MB}_1 = vp_1,\ \mathrm{MB}_2 = vp_2,\ \mathrm{MC}_1 = 1,$$

where again we take construction cost to be the measure of input. There are now three optimizing conditions:

$$1/vp_1' = \mathrm{MP}_{11}, \tag{3.7}$$

$$1/vp_2 = \mathrm{MP}_{12}, \tag{3.8}$$

and

$$p_1/p_2 = \mathrm{MRT}_{21}. \tag{3.9}$$

A necessary condition for a plan (x_1^0, y_1, y_2) to be optimal is that its three arguments satisfy these three equations. The meaning of these equations can be appreciated most readily by constructing a graphic solution to the problem. Since there are now three variables to be considered, the problem can be represented by a construction similar to the contour-line representation of a relief map, called an isoquant diagram. Geometrically, an isoquant diagram is the projection of a three-dimensional production function into the plane of two of its variables. In Fig. 3.4 we have drawn five isoquants, labeled $x_1^0 = h_1, \ldots, h_5$. The curve $x_1^0 = h_1$ depicts all combinations of y_1 and y_2 such that the point (h_1, y_1, y_2) is on the production function, that is, it represents all combinations of outputs that can be produced efficiently with the input equal to h_1. The other four isoquants are constructed similarly. Note that there is one isoquant for each possible level of the input, in this case an infinite number, although only a few are sketched.

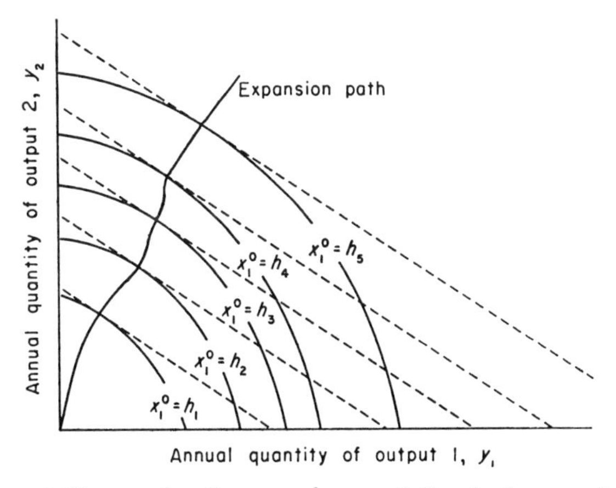

Fig. 3.4. Isoquant diagram for the normal case of the single-reservoir, dual-purpose example, showing increasing marginal rates of transformation.

The shape of the isoquants is worth noting. The slope of an isoquant at any point is the marginal rate of transformation of the outputs at that point. As we have drawn them, for low values of y_1, unit increases in y_1 with x_1^0 constant can be obtained at the cost of only small decreases in y_2, but as the level of y_1 increases, the sacrifice in y_2 necessary to obtain further unit increases also grows. This characteristic, of increasing marginal rates of transformation, is regarded by economists as the normal or typical case in production problems. But it is surely not the invariable case. For example, suppose that the two outputs had been municipal water supply and hydroelectric energy, and that adequate reregulating storage capacity was available in the municipal water system to smooth out releases for power. Each curve in the diagram would then correspond to the release of a certain amount of water, each gallon of which could be used for power production, water supply, or both. Therefore, for any given total release of water the maximum output of energy would be the same for all quantities of municipal water supply, and vice versa. Each isoquant would then be a horizontal straight line up to the technically maximum production of the output plotted horizontally, followed by a vertical line down to the horizontal axis.

It is also conceivable that the isoquants could be downward sloping, but convex rather than concave. This possibility is depicted in Fig. 3.5.

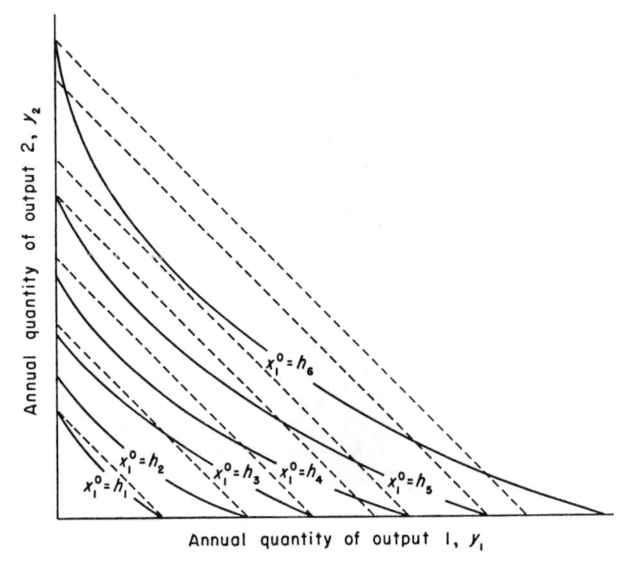

Fig. 3.5. Isoquant diagram for the convex case of the single-reservoir, dual-purpose example, showing decreasing marginal rates of transformation.

The various possible shapes of the isoquant diagrams reflect the technical and timing relationships among the different uses of water which economists classify as "competitive" or "complementary." Two uses of a resource are said to be competitive if an increase in the quantity of the resource devoted to one use decreases the quantity available for the other, other factors remaining the same. Otherwise they are said to be complementary.

Consider, for example, a single reservoir used both to generate hydroelectric energy and to supply an irrigated area. If all water passed through the turbines can be used to irrigate crops, these two uses are complementary. More likely, however, it will be necessary at times — in the winter months, for example — to use more water for energy generation than is needed simultaneously for irrigation. Water used at those times for energy generation will be lost to irrigation, assuming capacity in the reservoir to store it, and to this extent these uses of water will be competitive. Similarly, at times it may be necessary to supply more water to the irrigated district than is needed to meet the firm-energy demands of the system. Any energy generated by this water will then have to be sold at a low dump-energy price, and the irrigation use will reduce the quantity of water available for firm-energy commitments. If the power plant is located downstream from the irrigated district, the relationships are somewhat altered. Then all water used consumptively for irrigation is competitive with energy generation, but the return flow is complementary to the extent that its timing coincides with energy demand.

One of the most important types of mixed complementarity-competitiveness arises when a reservoir is used for both water storage and flood control. Water in the reservoir at the onset of the flood season may have to be spilled in order to vacate storage capacity for flood control, even though that water could be put to good use during and after the flood season if the anticipated floods do not materialize. To that extent, flood control competes for storage capacity with other uses that require water to be stored for timely release. On the other hand, storage capacity created for flood control can be used during the nonflood seasons to improve the time pattern of releases and, to that extent, flood control is complementary with other uses of water storage. In water use, therefore, although complementarity is prevalent, it is typically admixed with an element of competitiveness.

The dotted diagonal lines in Figs. 3.4 and 3.5 are known as price lines. Along each of them the gross value of output is a constant; thus each of them is the graph of the equation $p_1y_1 + p_2y_2 =$ constant. Now in Fig. 3.4 consider any isoquant, starting from the left and moving gradually to the right. It starts by climbing up the ladder of price lines until it reaches the price line to which it is tangent, then it descends. Since all points on a single isoquant have the same resource input, the point at which the isoquant is tangent to a price line gives the highest value of output attainable with that input, that is, the optimal combination of outputs for that input. These points are the ones that satisfy Eq. (3.9). The optimal design will be one of these points — the one corresponding to the highest excess of the present value of gross benefits over the resource cost. The expansion path shown on the figure is the locus of all these points of tangency.

In Fig. 3.5, however, as we trace any isoquant from left to right, we see that it starts by moving down the scale of price lines, rather than up, and that the price line to which it is tangent is the lowest one that it intersects in the course of its path. In this case the tangency points correspond to minimum values of output attainable with efficient use of given inputs, rather than to maximum values. The maximum values are given by the corner solutions, at which only one of the two outputs is produced. This situation illustrates two of the limitations on the formal solution by means of the differential calculus: the possibility that the solution of the necessary conditions will yield a minimizing vector rather than a maximizing one, and the possibility that the maximizing solution will occur at a corner where the necessary conditions are not fulfilled.

This is as far as these diagrams will take us. They are sufficient to determine the optimal values of y_1 and y_2 as functions of x_1^0. To determine the optimal values of x_1^0 for this example, we must turn again to Fig. 3.2. In our current interpretation of the figure, the horizontal axis still represents construction cost, x_1^0, while on the vertical axis we plot the present value of the gross benefits yielded by the values of y_1 and y_2 optimal for that value of construction cost. The optimal value of construction cost x_1^0 is the one for which the gross-benefit curve is as far as possible above the construction-cost line or, alternatively, where the gross-benefit curve has the same slope as the construction-cost line. By arguments that are already familiar, it will be seen that this point satisfies Eqs. (3.7) and (3.8).

Dual reservoir, single purpose. Our final example is much like the second. Let us consider the design of a project consisting of two reservoirs on two branches of a stream for the single purpose of providing irrigation water. The input-output vector is now (x_1^0, x_2^0, y_1) with unit values 1, 1, p_1 for the three components. The optimizing conditions, Eqs. (3.1), (3.2), and (3.3), become

$$1/vp_1 = \text{MP}_{11}, \tag{3.10}$$

$$1/vp_1 = \text{MP}_{21}, \tag{3.11}$$

and

$$1 = \text{MRS}_{21}.$$

This case also can be represented by an isoquant diagram, as in Fig. 3.6, where each isoquant (three of the infinite number are shown) portrays all combinations of expenditures for the two reservoirs that can be drawn upon to obtain a constant aggregate output of irrigation water without waste. This figure is essentially the same as Fig. 3.4, except that now as we move from left to right along an isoquant, we see that, at first, small increases in the size of reservoir 1 can be substituted for large decreases in the size of reservoir 2 and that the marginal rates of substitution decrease as the size of reservoir 1 increases. Two considerations account for this shape. First, for each reservoir more output per unit of

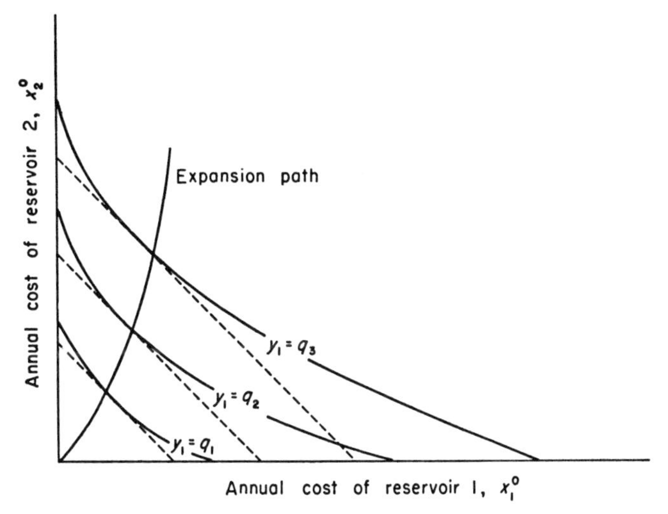

Fig. 3.6. Isoquant diagram for the dual-reservoir, single-purpose example.

storage capacity is obtained at the lower scales of development for the reasons we encountered in discussing the single-reservoir, single-purpose example. With two reservoirs it is possible to use the more productive lower capacities of each to obtain the total output, instead of being forced to use the less productive higher capacities of a single reservoir. Second, with two reservoirs it is possible to take advantage of the diversity of streamflows at the two sites to obtain a larger total output from the combination than from a single reservoir. It follows that the early stages of development of either reservoir will be more productive than the later stages of development of the other.

The price lines now depict pairs of reservoir sizes that have constant aggregate cost; by virtue of the fact that we have chosen to measure reservoir size by cost (this was purely voluntary; we might equally validly have taken height of dam or capacity of reservoir as our measure of reservoir size), they all have slopes of -45 degrees. The point of interest on each isoquant is the point at which the given quantity of output (q_1 or q_2 or q_3) can be obtained for minimal total cost, or the point of tangency of the isoquant with a price line. Just as in the previous case, this figure determines optimal values of x_1^0 and x_2^0 as functions of y_1. To find the optimal value of y_1, we require a figure similar to Fig. 3.2. To each value of y_1 there correspond, as we have noted, optimal values of x_1^0 and x_2^0 and, corresponding to them, a minimal cost at which the output y_1 per year can be obtained. This relationship is portrayed by the total-cost curve in Fig. 3.7. The optimal level of output is the one for

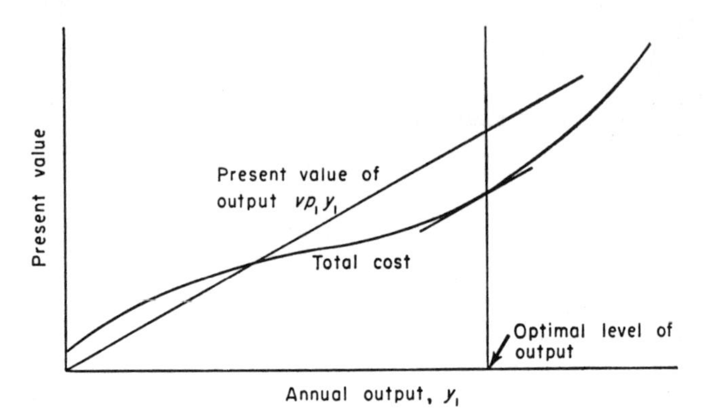

Fig. 3.7. Total-cost curve and optimal level of output for the dual-reservoir, single-purpose example.

which the gap between the present-value-of-output line and the total-cost curve is as great as possible or, alternatively, where the slope of the cost curve equals the slope of the output line.

Two important functions derived from the total-cost curve are shown in Fig. 3.8. The marginal-cost curve is the slope or derivative of the total-cost curve, that is, the rate of change of cost with respect to output, or the cost of an additional unit of output. Since there is likely to be some minimal size of installation required to obtain any usable output, the marginal-cost curve starts out rather high but diminishes at first, as increasing use is made of the capacity of the irreducible minimal structures. Marginal costs continue to decrease beyond this point because constant increments in the height of the dam are rewarded, up to a point, by increasing increments in the capacity and yield of the reservoir. But eventually a stage is reached where (because of physiographic problems, increasingly complete utilization of natural inflow, and the like) further increases in the capacity of the reservoir become decreasingly productive; in the end, when all available water is being utilized, further increases are not productive at all.

The average-cost curve is self-explanatory. It is horizontal when it crosses the marginal-cost curve (this is an easy theorem in the calculus and also intuitively obvious) and the optimal plan is not, in general, the one corresponding to minimum average cost.

The marginal-benefit curve is the slope of the present-value-of-output line, in this case a constant equal to vp_1. It measures the contribution to gross benefits of an additional unit of output. The optimal design is the

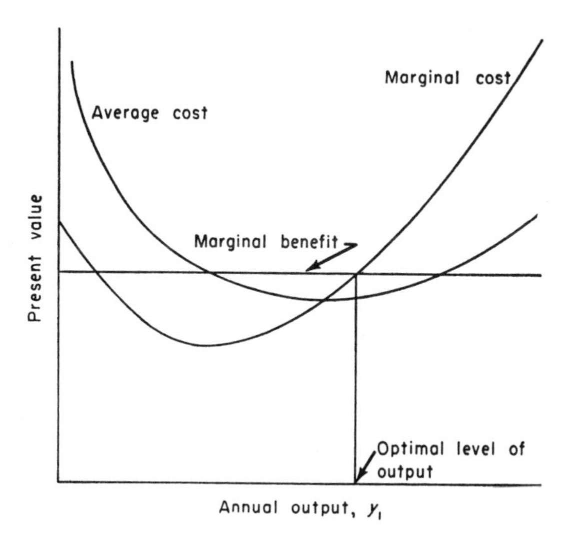

Fig. 3.8. Marginal- and average-cost curves for the dual-reservoir, single-purpose example.

one for which marginal cost equals marginal benefit, as shown on both Fig. 3.7 and Fig. 3.8.

When there is only a single output, the problem can be conceived most simply in terms of these marginal cost and benefit curves. Since net benefits are merely total benefits minus total cost, differentiation yields those net benefits which are maximized when marginal benefit equals marginal cost. This statement is the same as that made in Eqs. (3.10) and (3.11), since the phrase "marginal cost" in the present connection means the cost of a unit increase in output, or p_1^0/MP_{11} or p_2^0/MP_{21}, whereas in earlier passages this phrase designated the cost of a unit increase in some input. Where there are several outputs, the earlier formulation and phraseology are more appropriate than the one utilizing cost and benefit curves as functions of output.

THE PRODUCTION FUNCTION: HYDROLOGIC ASPECTS

Since most of the economic inputs absorbed by a water-resource project are consumed in constructing the installations, the production function can be split conveniently into two parts: the relationship between the resources consumed and the resultant structures, and the relationship between the structures and the useful outputs. About the first of these parts we shall have little to say; it lies entirely in the realm of engineering. The second part, however, is so basic to problems of system design and operation that we shall have to discuss some of its fundamentals.

The reader will have noticed that in discussing inputs to water-resource systems we have not, up till now, mentioned water itself. This is because the amount of natural-water inflow absorbed by a system is not a variable about which any decision can be made or whose quantity can be varied at will.[8] The inflow of water, at least to the installations farthest upstream, is one of the data of the problem. We shall use the term "inflow hydrology" to denote the rate of inflow to a system regarded as a function of time.

The rate of inflow at any particular time is a random variable, and a very complicated one. Even if we assume (and the assumption is very dubious) that rainfall and other climatic conditions in successive months are statistically independent, we cannot realistically so regard the rates of flow in streams. One of a number of reasons is that an abnormal proportion of the rain falling after a dry spell will be retained on the surface of the ground and in the soil itself. Furthermore, some of the rainfall in any month will not reach hydraulic works as inflow until after the month has expired. Finally, insofar as inflow results primarily from snowmelt, all months of a given season will be influenced by a common set of causes. For all these reasons and a few others, the probability distribution of inflow to a system in any month depends upon the inflows in immediately preceding months. A rigorous treatment of water inflow, with due attention to its probabilistic aspects, is therefore very complicated and has never been achieved fully. We shall not pursue these complexities further, since they will be treated more extensively in Chapter 12.

The logical partner of inflow hydrology is outflow hydrology, or the

[8] However, a limited degree of control can be exercised by management of watersheds and snowpack, and by weather modification. See Edward Ackerman and George Löf, *Technology in American water development* (Johns Hopkins University Press, Baltimore, 1959), chaps. 8 and 14.

characteristics of drafts from a system regarded as a function (a random function, of course) of time. The planners of a water-resource system have substantial control over its outflow hydrology. They influence its general magnitude by designing the system for specified levels of demand (the y of the previous sections), and they can influence its seasonal pattern by choosing appropriate types of demand. Water for municipal supply, for example, is required throughout the year, although normally with a distinct seasonal peak in the summer; water for industrial use is likely to be even more regularly demanded throughout the year, although the exact pattern depends on the nature of the industry; water for hydroelectric power is also required throughout the year, with some seasonal variation and with important peaks at certain times of day and certain days of the week; water for irrigation has a marked seasonal demand the pattern of which depends on the requirements of the crops cultivated. The total outflow hydrology of a system is a kind of sum of the patterns of demand for water for these various uses. Straightforward addition of the separate demands will not give the correct total requirement at any time, however, because of their complementarity. A substantial proportion of the water released for irrigation reenters the stream as return flow useful for other purposes farther downstream. Water required for any purpose can be passed through one or more power plants on its way, provided that the loss in head does not interfere with delivery of the water at adequate pressure or elevation. The total outflow hydrology can be determined by dividing demands into noncomplementary segments and adding them. Thus the outflow hydrology, like the inflow hydrology, is a complicated matter.

It is useful to conceive of the purpose of a storage dam and reservoir as being one of regulation, the conversion of a natural inflow hydrology into a desired outflow hydrology. The quantity of water a given reservoir can release per year in accordance with a given time pattern of outflows is called the yield of the reservoir. The relationship between the storage capacity of a reservoir and its yield, called the storage–yield function, is clearly one of the most important ingredients of its production function. This relationship depends in a very complicated way on the inflow and outflow hydrologies. At one extreme, if the seasonal patterns of natural inflow and of useful outflow are the same and if the region is humid enough not to require overyear storage, full use of available water can be obtained with no reservoir capacity other than that needed to produce adequate head for moving the water and generating energy. At the other extreme, if inflow and outflow are completely out of phase, the useful output of a system is limited to its storage capacity. Intermediate cases are, naturally, the prevalent ones.

An instructive, and standard, method for analyzing the relationships among inflow hydrology, outflow hydrology, and storage capacity was developed by W. Rippl; since adequate descriptions do not seem to be readily available, we shall discuss it in detail. Rippl's method, or the mass diagram, is used to determine the required storage capacity of a reservoir when the inflow and outflow hydrologies are both given functions of time. It is also necessary to assume that these patterns repeat periodically once every T months, time being measured in months for convenience. Since the Rippl method employs deterministic inflows and outflows, it can provide only a first approximation to the solution of its problem. Nevertheless, it provides illuminating insights into the factors at work.

Since the inflows and outflows are assumed to be periodic, it is sufficient to design the reservoir to meet requirements for a single typical period of length T. Thus we can choose any moment as the origin of our time axis and consider a time variable t over the range 0 to T. We let $S(t)$ denote the total cumulated inflow from the origin to time t (adjusted for evaporation and seepage if desired), $D(t)$ the total cumulated draft to meet demands, and $C(t)$ the quantity of water in storage at time t. The obvious basic relationship connecting these quantities is

$$C(t) + S(t + x) - S(t) - [D(t + x) - D(t)] \geq C(t + x) \geq 0 \qquad (3.12)$$

for all times t, $0 \leq t \leq T$, and all x, $0 \leq x \leq T - t$. This says simply that the contents of the reservoir on any date plus the inflow in the next x months less the useful draft in those same months must leave at least as much water as there will be in the reservoir x months later. The inequalities are necessary because some water may be spilled between t and $t + x$.

The argument will be simplified if we consider two cases separately, although it is not logically necessary to do so. In case I the inflow in the entire period is just adequate to meet demands, that is, $S(T) = D(T)$; in case II there is some surplus water, or $S(T) > D(T)$. If the water supply over a cycle is insufficient, the problem is insolvable.

In case I we first determine the initial storage $C(0)$ by setting $t = 0$ in Exp. (3.12) and noting that $S(0) = D(0) = 0$. Thus

$$C(0) + S(x) - D(x) \geq 0$$

for all x, $0 \leq x \leq T$. The smallest initial storage that meets this condition is

$$C(0) = \max_x [D(x) - S(x)].$$

If we assign this value to $C(0)$ and note that under the assumption of case I no water can be spilled, the amount of water in storage at any time t is

$$C(t) = C(0) + S(t) - D(t).$$

The required capacity of the reservoir, say C^*, is the largest value of $C(t)$ for any value of t, or

$$C^* = C(0) + \max_t [S(t) - D(t)].$$

Since $\max [D(x) - S(x)] = -\min [S(x) - D(x)]$, and since the symbol used for the running subscript is immaterial, we can write this solution for case I:

$$C^* = \max [S(t) - D(t)] - \min [S(t) - D(t)] = \text{range of } [S(t) - D(t)].$$

To start the solution of the more general case, we note that there must be some date, to be called t_0, at which the reservoir can be drawn down completely. That is, there must be some date such that

$$S(t_0 + x) - S(t_0) \geq D(t_0 + x) - D(t_0) \tag{3.13}$$

for all x in the range $0 \leq x \leq T$. This assertion is proved by construction. Let t_0 be the earliest date for which

$$D(t_0) - S(t_0) = \max_t [D(t) - S(t)].$$

For $0 \leq x \leq T - t_0$ we have

$$D(t_0) - S(t_0) \geq D(t_0 + x) - S(T_0 + x)$$

by construction, and Exp. (3.13) follows immediately. If $T - t_0 \leq x \leq T$, we invoke the cyclic assumption to obtain

$$D(t_0 + x) - D(T) - [S(t_0 + x) - S(T)]$$
$$= D(t_0 + x - T) - S(t_0 + x - T).$$

Since the right-hand side cannot exceed $D(t_0) - S(t_0)$, we have

$$D(t_0 + x) - D(T) - [S(t_0 + x) - S(T)] \leq D(t_0) - S(t_0),$$

$$D(t_0 + x) - D(t_0) \leq S(t_0 + x) - S(t_0) - [S(T) - D(T)]$$
$$\leq S(t_0 + x) - S(t_0),$$

which completes the proof. This argument amounts to showing that if there were no date t_0, there would be some water that is never used.

This date, t_0, is clearly the end of the drawdown phase of the cycle.[9] Since the date at which we break into the cycle is completely arbitrary, we shall choose the end of the drawdown phase. Thus we can assume, without loss of generality, $t_0 = 0$, $S(t) \geq D(t)$, all $0 \leq t \leq T$, and $C(0) = C(T) = 0$. The quantity in storage at any intermediate date, $C(t)$, must satisfy three conditions. First, it must be nonnegative. Second, it must be sufficient, together with later inflow, to meet all later requirements, that is,

$$C(t) + S(t + x) - S(t) \geq D(t + x) - D(t), \qquad (3.14)$$

all $0 \leq t \leq T$ and all $0 \leq x \leq T - t$. Finally, it must be possible to attain that storage beginning with the storage available at each previous date, that is,

$$C(t - y) + S(t) - S(t - y) - [D(t) - D(t - y)] \geq C(t), \qquad (3.15)$$

all $0 \leq t \leq T$ and all $0 \leq y \leq t$.

We now construct a storage function that not only meets these conditions but is the smallest storage function that does, moment by moment, for each value of t. Let

$$C(t) = \max_x [D(t + x) - D(t) - S(t + x) + S(t)].$$

Then $C(t)$ has the asserted properties. It is nonnegative since the maximand is zero for $x = 0$. It satisfies Exp. (3.14) by construction and is the smallest number that does. We must finally confirm that it satisfies Exp. (3.15). If $C(t) = 0$, (3.15) reduces to (3.14) and is satisfied. If $C(t) > 0$, let t^* be the smallest number in the range $t \leq t^* \leq T$ such that

$$C(t) + S(t^*) - S(t) = D(t^*) - D(t), \qquad (3.16)$$

that is, such that $C(t)$ is just sufficient to meet the demand at t^*. By our construction,

$$C(t - y) + S(t^*) - S(t - y) \geq D(t^*) - D(t - y). \qquad (3.17)$$

If we subtract (3.17) from (3.16),

$$C(t) - C(t - y) + S(t - y) - S(t) \leq D(t - y) - D(t),$$

or

$$C(t) \leq C(t - y) + S(t) - S(t - y) - [D(t) - D(t - y)],$$

[9] There may be several drawdown periods in a single cycle. If so, we define t_0 as the end of the first of them.

which is Exp. (3.15). Thus this function, $C(t)$, meets all the conditions and is the smallest value of storage that does so. Finally, the required storage capacity is $C^* = \max C(t)$. This solution can be written in somewhat simpler terms. For $t' \geq t$, let $D(t, t')$ denote the draft between t and t' and $S(t, t')$ denote the inflow. Then

$$C^* = \max_{t} C(t) = \max_{t,t'} [D(t, t') - S(t, t')],$$

for $0 \leq t \leq t' \leq T$. This completes the solution.

Although the proof is sophisticated, the underlying idea is simple: the minimum storage at any date that permits future requirements to be met is the smallest number $C(t)$ satisfying

$$C(t) \geq S(t') - [S(t) + D(t, t')] \tag{3.18}$$

for all dates t' subsequent to t. If $S(t') - S(t) \geq D(t') - D(t)$ for all t' later than t, then $C(t) = 0$. Otherwise $C(t)$ must equal the greatest excess of cumulated demand, $D(t') - D(t)$ or $D(t, t')$, over cumulated inflow, $S(t') - S(t)$.

In actuality, the solution is best carried out graphically by a method that exploits this idea. Plot $S(t)$ and $D(t)$ over a period T units long,

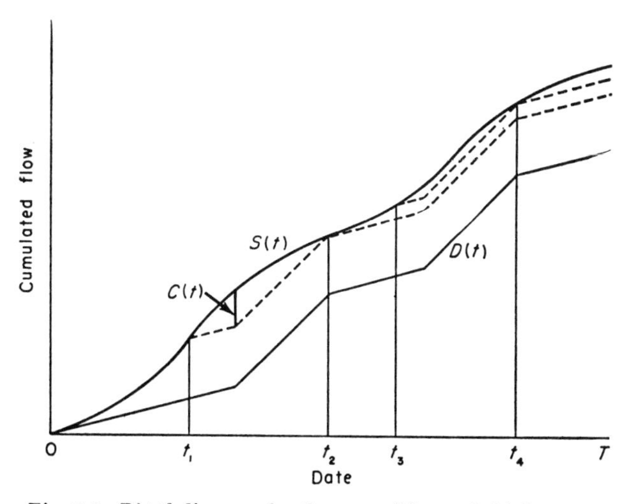

Fig. 3.9. Rippl diagram for the case with zero initial storage.

beginning at any date inasmuch as the end of the drawdown period is as yet unknown. Two slightly different cases can arise. In the first case, illustrated in Fig. 3.9, the $S(t)$ curve never drops below the $D(t)$ curve, indicating that no water is required on the initial date. In the second case, illustrated in Fig. 3.10, the $S(t)$ curve does fall below the $D(t)$ curve, showing that unless there is some water in storage on the initial date the cumulated demands cannot be met.

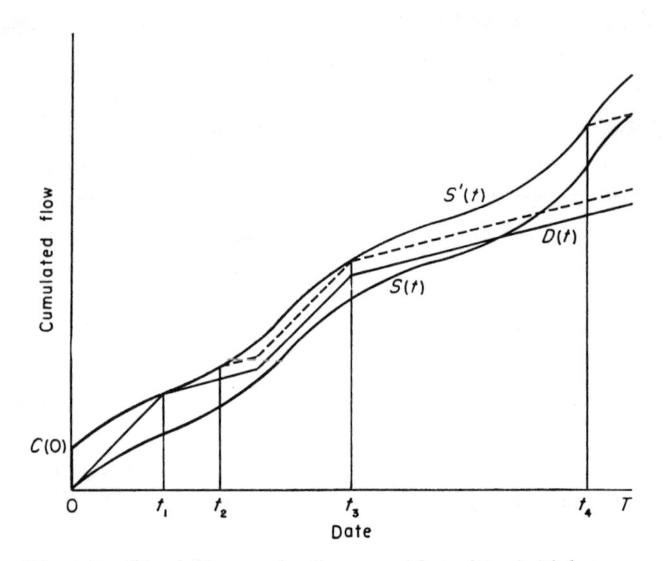

Fig. 3.10. Rippl diagram for the case with positive initial storage.

The first case is slightly simpler than the second. To determine the amount of storage required at each date in the period, move along the $S(t)$ curve beginning at the left, and through each point imagine drawing a curve parallel to the $D(t)$ curve extending to the right (but not to the left) of the point selected on the $S(t)$ curve. The dotted curves in the figure are examples of such imagined curves. For any date t' subsequent to the date at which the imagined curve originates, the height of the imagined curve is $S(t) + D(t, t')$. Hence, from Exp. (3.18), if the imagined curve always lies below the cumulated-inflow curve, no water is required in storage at date t. If any storage is ever required, as we move along $S(t)$ we shall come to some date, say t_1, for which the imagined curve just touches $S(t)$ at some subsequent date. Let t_2 be the latest date for which the imagined curve emanating from $S(t_1)$ touches $S(t)$. Then for any date between t_1 and t_2 the excess of $S(t)$ over the imagined curve gives $C(t)$, the storage required at that date.

This determines storage requirements for all dates up to t_2. To find storage requirements for later dates, the process is repeated. That is, beginning at t_2 we imagine drawing curves parallel to $D(t)$, extending to the right from $S(t)$, until we come to some date t_3 whose imagined curve touches $S(t)$ at some later date t_4. Then the interval from t_3 to t_4 is an interval during which the reservoir should have positive contents in amount equal to the excess of $S(t)$ over the imagined curve beginning at $S(t_3)$. This process is repeated until the concluding date T is reached. The required reservoir size is the greatest excess of $S(t)$ over the applicable imagined curve. In practice, the critical imagined curves should be drawn in as in the figure.

If positive initial storage is required, the graphic solution begins by determining its amount. The cumulated-inflow curve is raised rigidly to a new position where it just touches the $D(t)$ curve but never falls below it (Fig. 3.10). This raised inflow curve is denoted by $S'(t)$. The amount by which the inflow curve must be raised is the required initial storage, $C(0)$. If t_1 is the last date at which $S'(t)$ touches $D(t)$, the excess of $S'(t)$ over $D(t)$ is the required storage at all dates up to t_1. In addition, to assure that adequate water is in the reservoir at the end of the cycle, the terminal point of $D(t)$ is raised by $C(0)$ units, as shown in the figure. The amount of storage required at all dates between t_1 and T is determined as in the case of no initial storage, except that $S'(t)$ is used in place of $S(t)$, and the $D(t)$ curve modified by raising its terminal point is used in place of the unmodified one. When this procedure is applied to Fig. 3.10, it is found that there should be water in the reservoir from t_2 to t_3, and that at t_4 the reservoir should begin to fill so as to have the required initial contents at the start of the next cycle.

In this way the Rippl diagram can be used to determine the amount of water required in storage at each date of the cycle and, thereby, the necessary capacity of the reservoir. But the underlying assumption, that the history of a reservoir can be regarded as a sequence of identical periods, is a gross simplification. When we dispense with it, two problems arise that the Rippl method in its pure form cannot contend with. The simpler problem arises in the case of streams for which the annual flow is invariably greater than the annual demand for water. No overyear storage is required for such streams, and therefore it is legitimate to consider each year in isolation from its neighbors, although it is not true that the storage capacity found to be adequate for any single year will be adequate for all the others. The more difficult problem arises in analyzing streams for which the year-to-year variation in inflow is so great that the flow is frequently less than the annual demand. For such streams overyear storage must be provided so that surplus water in the wet years can be conserved for use in the dry.

The classic work on both of these problems is that of Allen Hazen.[10] Hazen's method for dealing with the random variability of flows in streams where overyear storage is not needed is essentially to construct a Rippl diagram for each year of record and from it to compute the storage that would have been required in that year to meet the assumed drafts. This computation results in a frequency distribution of estimated storage requirements, one estimate for each year of record. From this frequency distribution a probability distribution of storage requirements can be estimated,[11] and the size of reservoir that would be adequate in any given proportion of years can be deduced from this probability distribution. The estimation of requirements for overyear storage in more arid regions is a much more difficult problem, involving the theory of stochastic processes, a smattering of which will be sketched in the next section. In practice, engineers are pragmatists and design according to empirical formulas that experience has shown to be reasonably safe.[12] One frequent expedient is to construct a Rippl diagram for some critical period of record, four or five years long. Although we do not deal with overyear storage here, we must mention that in dry regions the storage of water in years of ample rainfall for use in periods of drought is often far more important than the storage of wet-season water for use in the dry season of the same year.

The Rippl method falters also when a system incorporates several reservoirs either in series or in parallel. In such cases the concept of a given inflow hydrology still applies to the works farthest upstream on each branch or tributary. But the inflows to all other reservoirs are no longer data but consequences, at least in part, of releases farther upstream. And the outflow patterns from the various reservoirs can no longer be derived directly from the patterns of demand for water, since the man-

[10] Allen Hazen, "Storage to be provided in impounding reservoirs for municipal water supply," *Trans. Am. Soc. Civil Engrs. 77*, 1539 (1914).

[11] Hazen assumed that this probability distribution was approximately normal and made some other simplifying assumptions, but these are not central to his approach. Since streamflow records for most streams are too short to permit reliable estimation of the probability distributions of their storage requirements, Hazen developed "normal storage curves" based on data for some dozen streams. These curves, he asserted, could be applied to most streams in the United States to obtain acceptable estimates of storage requirements as functions of the ratio of daily draft to mean annual flow and of the desired degree of assurance that the storage will be adequate.

[12] For discussion of such formulas, see Hazen, reference 10, and Charles E. Sudler, "Storage required for the regulation of stream flow," *Trans. Am. Soc. Civil Engrs. 91*, 622 (1927).

agers of the system have substantial scope for deciding which reservoir should provide how much water for which purpose and when. The problem of ascertaining adequate and efficient designs and operating procedures then becomes so complicated that trial-and-error methods must be used. These methods operate by considering various designs and operating procedures and, for each in turn, simulating the operation of the system by hand calculations or, nowadays, with the help of computing machines. Such methods are the subject of Chapters 8 to 10.

If we ignore these difficulties for the moment, it is interesting to note that the Rippl method can be used to estimate the marginal productivities that played such an important role in our analysis of design and investment criteria. To be used for this purpose, the method is applied a number of times, once for each level of a particular output, its normal outflow hydrology being associated with that output. The rate of change of reservoir capacity with respect to changes in the level of the output leads at once to the required marginal productivities. If there are several reservoirs, a base set of structures and operating procedures is assumed at the start. Then, by increasing the output slightly and holding fixed the drafts on all reservoirs but one, estimates are made of the increase in capacity of that reservoir required by the increased output and of the associated cost of construction. This calculation can be performed for each structure in turn. Again, estimates of the marginal productivities and costs are found directly.

The storage function $C(t)$ encountered in developing the Rippl method illustrates one of the main concepts of water-system operation. This $C(t)$ is a "rule curve," a curve giving the desired contents of a reservoir as a function of time, as illustrated in Fig. 3.11. In practice, the engineers in charge of a system spill water when the contents of a reservoir exceed the quantity specified by the rule curve at a given date and hope for rain when it falls below. When the contents fall substantially below the rule curve, they may even take steps to conserve water, such as discouraging or forbidding use of water for sprinkling lawns, washing cars, or other low-priority purposes. During extreme shortages the voltages and cycles of hydroelectric power may be reduced at noncritical times of day. The use of rule curves and other aspects of operating procedure are dealt with more fully in Chapter 11.

One purely technologic relationship that will be of importance in several of the later chapters is the connection between the volume of water

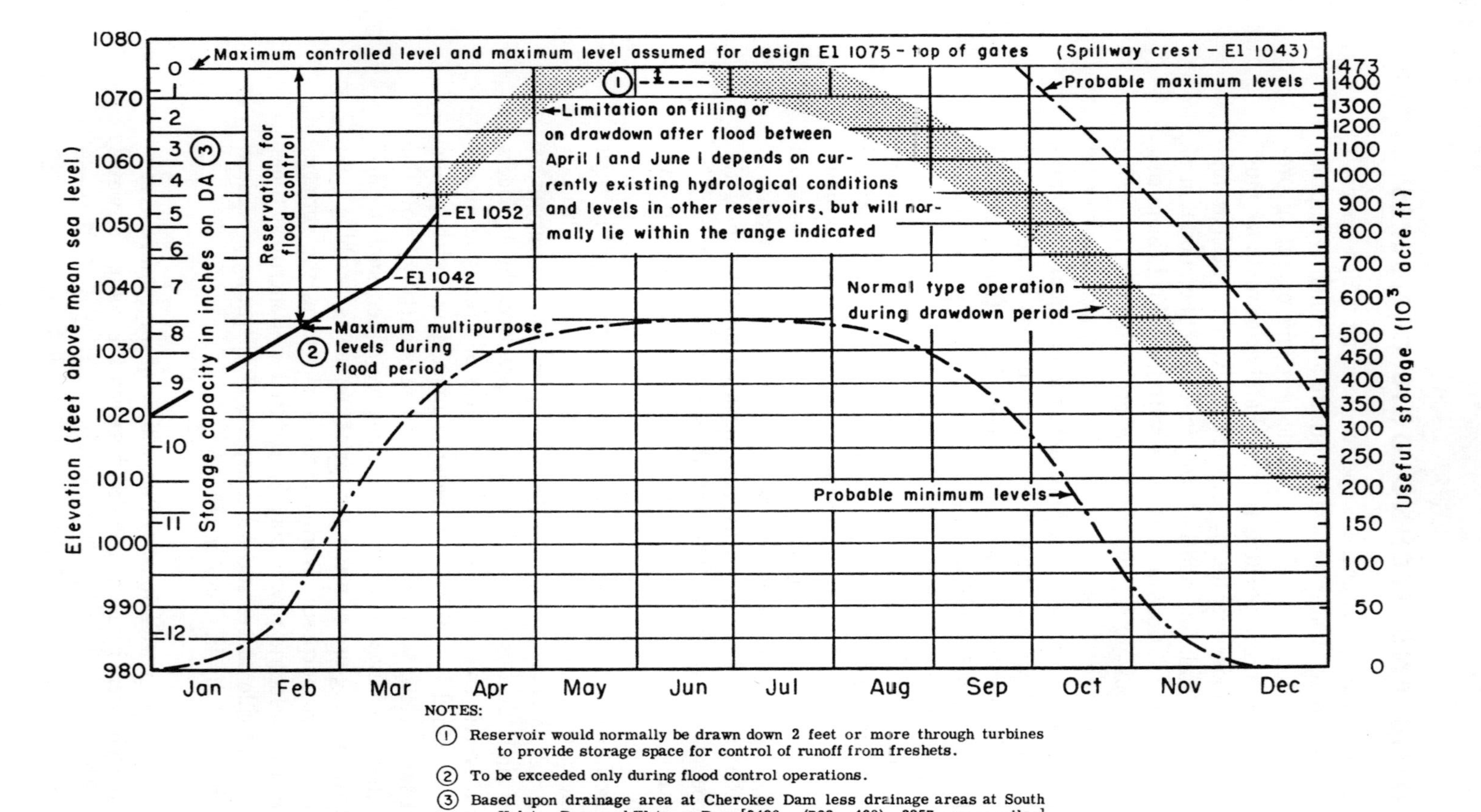

Fig. 3.11. "Guide curve" or "rule curve" for operation of the Cherokee Dam and Reservoir. Operating guides similar to this chart have been prepared for all TVA multiple-purpose reservoirs, and are based upon the annual cycle of runoff revealed in a study of the more than 100 years of hydrologic records. The example illustrates in graphic form the seasonal operation of a tributary multiple-purpose reservoir. The water level is held low during the flood season, except in flood-control operations; is filled in late spring; and is then drawn down in late summer and fall. The shape of the curve is largely dependent upon the primary objective — flood regulation — with such storage space reserved at all times of the year as is required to achieve that purpose.

passed through a hydroelectric plant and the amount of energy produced. This relationship follows from the basic conversion formulas for units of power:

$$1 \text{ hp} = 0.746 \text{ kw} = 550 \text{ ft lb/sec.}$$

To transform these power conversion factors into conversion factors for energy, we multiply through by a time unit to obtain:

$$1 \text{ hp hr} = 0.746 \text{ kw hr} = 550 \times 3600 \text{ ft lb} = 1.98 \times 10^6 \text{ ft lb.}$$

We are interested only in the conversion between kilowatt hours and foot pounds, or:

$$1 \text{ ft lb} = \frac{1}{2.655 \times 10^6} \text{ kw hr.}$$

Now, one acre foot of water weighs 2.718×10^6 lb. So Q acre ft of water falling through h ft release $2.718 \times 10^6\, Qh$ ft lb of energy, equivalent to

$$\frac{2.718 \times 10^6}{2.655 \times 10^6} Qh \text{ kw hr} = 1.024\, Qh \text{ kw hr.}$$

Finally, the energy-flow relationship for a power plant of $100e$ percent efficiency is

$$E = 1.024\, Qhe \text{ kw hr.}$$

UNCERTAINTY

Uncertainty enters into economic decisions whenever the consequences of a decision cannot be foretold with confidence, which is to say almost always. Both in practical affairs and in theorizing about them, it is frequently expedient to ignore uncertainty — to act as if consequences could be predicted accurately — in order to simplify problems. By and large this is the course we have followed up to this point. But in design and operating decisions the results of which are influenced by chance or unknown factors such as the whims of the weather, this simplification is clearly untenable. Therefore we must analyze decision-making in circumstances where uncertainty is taken explicitly into account, and for this purpose we turn to decision theory, a new discipline being developed largely by statisticians.

OBJECTIVES AND CONCEPTS

DECISION THEORY

A decision problem can be visualized as follows. A choice is to be made among a number of alternative courses of action. The practical consequences of adopting any particular course depend not only on the choice made but also on other data, called with purposeful vagueness "the state of nature," which cannot be known at the time the choice is made. Therefore the decision must be based on three types of consideration: (1) some judgment as to the likelihood of nature being in each of its possible states, (2) predictions as to the consequences of each course of action assuming that nature is in each of its possible states, and (3) evaluations of the desirability of each possible outcome of the situation.

To illustrate these concepts, let us suppose that a reservoir used for both irrigation and flood protection is full at the beginning of the flood season, and that only one type of flood occurs in the region. (Of course, at the beginning of the flood season one cannot predict whether or not a flood will occur.) The allowable decisions are to spill one-third of the water in the reservoir, spill two-thirds of it, or spill all of it. Table 3.1

Table 3.1. Physical consequences of decisions in text example: reservoir used for both irrigation and flood protection.

	Flood		No flood	
Decision	Harvest (10^3 bu)	Flood damage (10^3 dollars)	Harvest (10^3 bu)	Flood damage (10^3 dollars)
Spill one-third	950	250	1000	0
Spill two-thirds	600	100	650	0
Spill all	200	0	200	0

shows the physical consequences of each permissible decision in each of the two possible states of nature, flood or no flood. To compress the two dimensions of these consequences into one to facilitate comparisons, we assume that the crop is worth $0.40 a bushel and thus derive the monetary consequences of each decision-state contingency,[13] shown in Table 3.2.

[13] It should be noted that the method employed in this illustration for adding the benefits from flood control to the benefits from other purposes such as irrigation is different from that used elsewhere in this book. Here we subtract flood damages incurred from irrigation benefits, whereas elsewhere flood damages prevented (that is, flood damages that would occur if the reservoir were not used for flood protection minus actual flood damages) are added to the benefits from other sources. Thus, where our figures for benefits under the assumption that a

Table 3.2. Monetary consequences of decisions in text example.

Decision	Returns (10^3 dollars)	
	Flood	No flood
Spill one-third	130	400
Spill two-thirds	140	260
Spill all	80	80

From Table 3.2 it is seen that the monetary consequences of spilling all the water are inferior to those of spilling two-thirds, whether or not a flood occurs. Such a decision is said to be "dominated" or "inadmissible," and an obvious principle of decision theory is that all dominated decisions can be discarded forthwith. They are analogous to the inefficient points we encountered in discussing production theory.

The choice between the other two decisions is much more difficult. If a flood is in the offing, two-thirds of the water should be spilled, otherwise only one-third. But whether or not a flood impends is precisely what cannot be known at the time the decision has to be made.

The maximin-returns principle. One plausible decision would be to release two-thirds of the contents of the reservoir, on the grounds that this act would assure a return of at least $140,000 whatever happens. The underlying principle invoked here is that each alternative should be evaluated by the minimal return that it guarantees, and the one with the highest guarantee should be adopted. This principle is known as the "maximin-returns principle" and is one of the most frequently advocated bases for decisions under uncertainty.

It suffers from two serious defects. In the first place, it is exceedingly conservative and pays no attention to the potentialities of a decision in any circumstances except the worst possible ones for that decision. In the present illustration, the maximin-returns decision would not be very appealing if floods are rare, for then its consequence would be to accept returns of $260,000 in many years when $400,000 are attainable in order to avoid losses of $10,000 during a few years.

The other defect is that this principle can lead to illogical behavior. This can be seen by considering a modification of the problem. Suppose

flood would occur are 130, 140, and 80, depending on the decision on spill, the level of benefits under the other method of measurement would be 380, 390, and 330. This difference in method of computation does not affect the ranking of alternatives or the difference in the monetary consequences of any two decisions.

that if no flood occurs, insect damage amounting to $135,000 will be sustained no matter how much water is released. Since the release decision will have no effect on the insect damage, one feels that this additional datum should not influence the release decision. But it does influence decisions based on the maximin-returns principle. Table 3.3 shows the

Table 3.3. Monetary consequences of decisions in text example, allowing for insect damage.

Decision	Returns (10^3 dollars)	
	Flood	No flood
Spill one-third	130	265
Spill two-thirds	140	125
Spill all	80	−55

monetary returns of the three permissible decisions allowing for the predaciousness of the insects. The maximin-returns principle now instructs us to release only one-third of the water, thus securing a guaranteed return of $130,000 and avoiding the possibility of only $125,000. If the insect damage had been estimated at $125,000, the maximin-returns policy would have been to release two-thirds. Thus it appears that decisions based on the maximin-returns principle are influenced by irrelevant considerations.[14]

The minimax-risk principle. To avoid this objection, most decision-theorists recommend an alternative decision criterion based on the idea that a decision should take the fullest possible advantage of the potentialities of a situation. Where uncertainty prevails, these potentialities are unknown, but we can analyze them in the following manner. Releasing one-third of the water will secure the maximum possible return if no flood impends but will result in a return $10,000 below the maximum possible in the alternative state of nature. Thus that decision exposes

[14] We did not really need to drag the insects into the story. The same paradoxical behavior would result if a flood-relief agency stood ready to pay an indemnity of $135,000 to the members of the water district in case of flood, irrespective of the release policy followed. This alternative supposition perhaps makes clearer the reasons for the paradox. Either modification of the problem makes the consequences of a flood less serious as compared with the consequences of no flood, and therefore reduces the need to guard against them. Even though the paradox is explicable, it is damaging; for the fact remains that neither the insect damage nor the indemnity alters the differential consequences of the three decisions in either possible state of nature. Therefore it seems illogical for them to alter the relative desirability of the three decisions.

the water district to the risk of a loss of $10,000. On the other hand, releasing two-thirds of the water will secure the maximum attainable return if it is a flood year at the risk of sacrificing potential returns of $140,000 ($400,000 minus $260,000) if it is a no-flood year. In a sense, then, releasing two-thirds is the riskier alternative since it has the potentiality of falling farther short of the actual returns permitted by the situation.

To generalize these ideas, define the risk of any combination of a decision and a state of nature by the excess of the maximum return attainable in that state of nature over the return that actually results from the given decision in that state of nature.[15] Thus the risks in an uncertain situation can be computed from a table of returns by subtracting each return from the highest figure in its column. The risks for our example are given in Table 3.4. It seems reasonable to choose the alternative

Table 3.4. Risk table for text example.

	Risk (10^3 dollars)	
Decision	Flood	No flood
Spill one-third	10	0
Spill two-thirds	0	140
Spill all	60	320

for which the maximum possible risk, defined in this way, is as small as possible. This criterion is called the "minimax-risk principle." In this case it recommends releasing one-third of the water. Since a risk table computed from Table 3.3 will be identical with the one computed from Table 3.2, the minimax-risk principle is free of the distorting effect of irrelevant circumstances.

The minimax-risk principle is a bit more sophisticated and less intuitive than the maximin-returns principle, so perhaps one more word should be written in its behalf. It is based on the notion that the absolute magnitudes of the returns in the two states of nature are useful only to determine how earnestly we should hope that one or the other will prevail. But they are irrelevant to any practical decision, which should depend only on the comparative consequences of the various alternative decisions. The subtraction involved in computing risks has the effect of

[15] Later on we shall use the word "risk" in a different sense, but until further notice it will have the meaning just given.

canceling out the absolute levels attributable to the various states of nature and placing them all on the same footing, so that the consequences of each decision as compared with the maximum attainable in the circumstances are no longer obscured. In our example the minimax-risk decision assures a return within $10,000 of the maximum possible, whatever happens. To be sure, this return may be only $130,000 in an adverse state of nature, but this low result is the cost of a flood, not to be attributed to the decision.

Whether or not we have made the minimax-risk principle seem plausible, we now have to confess that it too suffers from a logical defect. We have just seen that, according to this principle, releasing one-third of the water is preferable to releasing two-thirds. Now suppose that another alternative becomes available. Rather than try to conjecture as to its nature, let us just call it Decision *X* and suppose that it yields a return of $300,000 if a flood occurs and $50,000 if not. Adding this alternative to the data of Table 3.2, we obtain Table 3.5. The evaluation of the returns

Table 3.5. Returns and risks of four alternatives in text example.

	Returns (10^3 dollars)		Risks (10^3 dollars)	
Decision	Flood	No flood	Flood	No flood
Spill one-third	130	400	170	0
Spill two-thirds	140	260	160	140
Spill all	80	80	220	320
Decision *X*	300	50	0	350

to the three original alternatives is unaltered, but the values of the risks are radically different. In fact, the minimax-risk criterion now recommends releasing two-thirds (maximum risk = $160,000) instead of releasing one-third (maximum risk = $170,000). The new alternative has reversed our choice between the old two, in spite of the fact that it is inferior to both of them in terms of the criterion. (The maximum risk for Decision *X* is $350,000, greater than for any other alternative.) This, surely, is illogical behavior.

We have demonstrated that both the maximin-returns and the minimax-risk principles can lead to illogical behavior and consequently must violate some fundamental axiom of rational choice. Rigorous scrutiny from the standpoint of formal logic indeed shows this to be the case. These are by no means the only principles that can be devised for decision-making under uncertainty, although they are probably the two most popular ones. For example, an optimistic decision-maker might advocate a maximax principle, selecting the alternative whose maximum return is as great as possible. But, just as in our two previous cases, skeptically minded critics have found a logical defect in every proposal thus far advanced for making decisions without forming a judgment of the relative likelihoods of the various states of nature. Furthermore, there seems to be no logical basis for preferring any of these principles over any other. It seems fair to conclude that no satisfactory basis for decision can be found that does not invoke judgments concerning the likelihoods of the various states of nature. We turn now to the question of how such judgments can be incorporated into the decision process.

Probabilistic approach. Judgments about the likelihoods of various states of nature are expressed by assigning probabilities to them, making use of whatever statistical evidence and other information is available at the time the decision is to be made. These probabilities are therefore relative to the state of information on the decision date. To define the probability of any state of nature, we conceive of all instances with the given state of information and define the probability as the proportion of those instances in which nature turns out to be in the given state.[16]

The probabilities of the various states of nature can be applied to each decision to arrive at a probability distribution of the consequences of that decision. For example, if the probabilities of flood and no-flood are judged to be 0.4 and 0.6, the probability distribution of the consequences of releasing one-third of the water would be $130,000 with the probability 0.4 and $400,000 with the probability 0.6. In general, the probability to be assigned to any outcome of a particular decision is the sum of the probabilities of all the states of nature that give rise to that outcome when the particular decision is taken.

[16] The very concept of the probability of a state of nature raises philosophical difficulties that are at present controversial. One difficulty is that at any given time nature is in some particular state, unknown to be sure, and there is no question of probability about it. As Einstein said, "God does not play at dice." Another line of objection is that the concept is not operational. It does not give rules for deciding which data are relevant and therefore to be included in the "state of information," and except in artificially contrived cases there will never occur two instances in which the state of information is identical. And even if a number of instances of sufficiently identical information have occurred, past experience is only a rough indicator of the proportion of future times in which various states of nature will actually occur. The approach taken in the text is just a rough-and-ready cut through this morass. In the context of our example, the two possible states of nature are that a flood is impending and that it is not. The probabilities in question are simply the proportions of the time that one would expect to be right or wrong if one predicted that there would be a flood on the basis of available evidence and standard methods of prediction.

The choice among decisions, then, amounts to the choice among the probability distributions of outcomes associated with them. Of course, it is more difficult to choose among a number of probability distributions of outcomes than among a number of definite outcomes, but it is by no means a sophisticated or rare accomplishment. In fact, every businessman, government administrator, and housewife performs it many times a day. A common and important example is the decision as to whether to insure a risk (automobile collision, flood, warehouse fire, or the like) and if so, how fully. The purchase of insurance converts a situation in which there is a small probability of a great loss into one with a certainty of a small loss (the premium) plus a small probability of a middling loss (the uninsured or uninsurable part of the risk).

The considerations involved in choosing among probability distributions can be illustrated by analyzing Table 3.2, assuming particular probabilities for the two states of nature. The augmented data, assuming probabilities of 0.4 and 0.6 for flood and no-flood respectively, are given in Table 3.6. The reader may have either of two reactions to this table.

Table 3.6. Probability distributions of the outcomes of three decisions in text example.

Decision	Return (10^3 dollars)		Weighted average return (10^3 dollars)
	Flood	No flood	
Spill one-third	130	400	292
Spill two-thirds	140	260	212
Spill all	80	80	80
Probability	0.4	0.6	

One is that the additional data have solved everything. If one-third of the water is released, an average return of $292,000 will result; if two-thirds, only $212,000; therefore one-third should be released. We shall discuss this conclusion at length below.

The other possible reaction is that the added data have solved nothing. Releasing one-third of the water still offers the enticement of a $400,000 return offset by the risk of returning only $130,000; releasing two-thirds still guarantees a larger minimum at the cost of reducing the maximum. How is one to evaluate these contingencies, even given the probability estimates?

The answer is almost immediately apparent. A listing of outcomes and their associated probabilities is not the only way to describe a probability distribution, and for the purpose of choosing among distributions it is not the best way. For this reason, among others, probability distributions are often described by their moments instead of by the outcome probabilities themselves.

The first moment of a probability distribution, often called the expected value or mathematical expectation (two misnomers), is defined by the formula $\mu = \sum_i p_i x_i$, where x_i is the value of the ith possible outcome and p_i is the probability of its occurrence. The second moment, called also the variance or the square of the standard deviation, is defined by $\mu_2 = \sigma^2 = \sum_i p_i(x_i - \mu)^2$ and the nth moment is given by $\mu_n = \sum_i p_i(x_i - \mu)^n$.

It is often said that the first moment, μ, is the most important characteristic of a probability distribution and therefore a very useful summary of the array of probabilities that comprise it. This assertion is justified as follows. Suppose that an insurance company or fund undertakes to assume the risks resulting from an uncertain situation by making a guaranteed payment of g dollars to the people exposed to the risk and keeping for itself the actual outcome of the situation, x_i dollars. Then the central-limit theorem, which is perhaps the most fundamental result of mathematical probability, asserts that in a long series of such contracts (1) if $g < \mu$ the insurance company is almost certain to receive more than it pays out, (2) if $g > \mu$ the insurance company is almost certain to pay out more than it receives and, indeed, to go bankrupt, (3) if $g = \mu$ it is impossible to predict whether the insurance company will pay out more or less than it receives. Thus in a sense μ, the mathematical expectation, is a fair certainty equivalent to the proceeds of the risky situation; it is the only amount which in a long series of such undertakings is not certain to average out at either more or less than the actual result.

While this reasoning has some appeal, there is legitimate question about its applicability to actual situations where the number of replications is small (often only one) and where no guarantee fund is actually established. The expected value does measure the "central tendency" of the probability distribution in the sense we have described, but there are other useful measures of central tendency too, for example, the mode (the outcome with a probability at least as great as that of any other) and the median (roughly, the outcome which has a 0.5 probability of being exceeded).

The second most important characteristic of a probability distribution is the degree of uncertainty, that is, the spread of the possible outcomes around their expected value. This is measured by the second moment. The first two moments together are often considered to contain most of the essential information about a probability distribution, and we shall discuss below some proposals for choosing among probability distributions that make use of only these two moments. Some alternative measures of the degree of uncertainty are the average deviation and the interquartile range, discussed in standard statistics texts.

The third moment measures the asymmetry of the probability distribution, and the fourth and higher moments measure other more esoteric characteristics. All the moments together constitute as complete a specification of the probability distribution as the list of probabilities itself.[17] In addition, as we have just seen, they have considerable intuitive appeal as summary indicators of the main characteristics of the distribution. Table 3.7 gives statistics computed from the first few moments of the

Table 3.7. Statistics of outcome distributions in text example, computed from the first few moments.

Decision	Expected value μ	Variance σ^2	Standard deviation σ	μ_3/σ^3	μ_4/σ^4
Spill one-third	292	17,500	132	−0.41	1.17
Spill two-thirds	212	3,460	59	−0.41	1.17
Spill all	80	0	0	—	—

three possible decisions in our example. Because there are only two possible states of nature, only the first two moments differ in this instance.

Although a complete description of a probability distribution requires all its moments, much of the information and most of the intuitive significance is contained in the first two. For this reason most practicable proposals for choice among probability distributions fall into two broad

[17] If the uncertain situation has k possible outcomes, then it is a matter of elementary algebra to see that if the first $k - 1$ moments are known or, for that matter, any $k - 1$ moments, then the probabilities of the k outcomes can be computed. If the probability distribution is continuous, so that an infinite number of outcomes is possible, then it is a very difficult theorem, but true, that knowledge of all the moments is equivalent to knowledge of the probability distribution itself. It is for this reason that we are justified in saying that an uncertain situation can be described fully by the moments of the probability distribution of its outcomes. These are not the only means used for describing probability distributions. For some purposes, for example, the generating function or the characteristic function (the Laplace and Fourier transforms of the probability distribution, respectively) are more useful than either the distribution itself or its moments.

classes: those that employ only the first moment (that is, choose the distribution with the highest possible expected value) and those that employ only the first two. We now consider the first, and simpler, of these proposals.

Expected-value approach. In the first place, it is clear that in a comparison between two decisions the one that gives rise to the higher expected value is not invariably to be preferred. Expected value, after all, is only a mathematical artifact; there is nothing compelling about it. The most familiar example of situations in which it is considered reasonable to choose an alternative that does not have the highest possible expected value is provided by insurance. An individual contemplating insurance of some risk, say collision damage to his automobile, faces two alternatives. If he decides not to insure, his expected disposable income for a year will be his expected income from other sources minus the expected value of collision damage to his car. If he insures, his expected disposable income will be his expected income from other sources minus the insurance premium. Since insurance premiums are invariably greater than the expected value of the losses insured against, in order to cover the expenses and profits of insurance companies, the expected disposable income will be smaller if insurance is taken out than if it is not. Nevertheless, people do insure; they knowingly choose the alternative with smaller expected value. From this we see that it is quite reasonable to give up some expected value in order to reduce uncertainty, that is, in order to reduce the standard deviation of the distribution of outcomes.

There are certain circumstances, however, in which the expected value alone is an adequate basis for choice among alternatives. Most trivially, this is the case when all the alternatives have approximately the same spread or standard deviation of outcomes. A more important case that leads to the same result is where the decision-maker is in the position of an insurance company. We shall call this the case of "actuarial risk." Actuarial risk arises when the decision to be made is one of a large number of similar, and independent, decisions and when major importance attaches to the average or over-all result of all the decisions.[18]

In the water-resource field actuarial risk can arise in at least two ways. First, an agency such as the Soil Conservation Service may construct a large number of relatively small projects. It would like to choose and

[18] Note the asymmetry between the rational behavior of the insurer and the insured. The insured reduces the expected value of his net benefits when he takes out insurance, and may be behaving rationally when he does so, but the rational policy for the insurer, who has the advantage of pooling numerous risks, is to maximize his expected net benefits.

design them so that each year the aggregate net benefits arising from its program, per dollar expended, are as large as possible. We shall now show that such an agency should design each project so as to obtain the maximum expected net benefits per dollar expended, without regard to any other characteristic of the probability distribution of net benefits. Let n be the number of projects undertaken by the agency, x_i the actual net benefits of the ith project, and y_i the cost of the ith project. We assume that the benefits x_i are independent random variables but that the costs y_i are known.[19] The agency is supposed to be concerned with its aggregate benefits per dollar, or $A = \Sigma x_i/\Sigma y_i$. The expected value and variance of A are $E(A) = \Sigma E x_i/\Sigma y_i$, and $\sigma_A{}^2 = \Sigma \sigma_i{}^2/(\Sigma y_i)^2$, where $\sigma_i{}^2$ is the variance of the net benefits of the ith project. Now the variance of A is small if the number of projects is large, as can be seen most readily by writing $\bar{\sigma}^2$ for the average variance, averaged over all projects, and $\bar{y}$ for the average cost of a project. Replacing the sums in the formula for $\sigma_A{}^2$ by these symbols, we have $\sigma_A{}^2 = n\bar{\sigma}^2/(n\bar{y})^2 = (1/n)(\bar{\sigma}^2/\bar{y}^2)$, which clearly decreases as n increases provided only that $\bar{\sigma}$ does not grow in proportion to n. Since, therefore, the variance of A is very small when n is large, we can be assured [20] that the actual realized value of A will be close to its expected value. But the expected value of A depends only on the expected values of the net benefits resulting from the individual projects. This argument is simply an application of the law of averages.

Actuarial risks can also arise in conjunction with a single project if the net benefits in the individual years of its life are independent random variables, and if we are more concerned with the average net benefits of the project over its lifetime than with the net benefits in particular years. The argument is similar to the one just presented.

The third circumstance in which the expected value suffices as a basis for choice among probability distributions or the decisions that give rise to them is suggested by decision theory. Clearly it is easy to choose among probability distributions if the expected value is an adequate measure of desirability, but otherwise it is substantially more difficult. Decision theory responds to this fact by asserting that if the various outcomes in the returns table are measured properly, the expected value will be an adequate measure of desirability and the problem of choosing among

[19] This last assumption is not necessary, but simplifies the algebra considerably.

[20] The assurance is provided by "Tchebycheff's inequality," which asserts that for any probability distribution that has a finite standard deviation σ, the probability that an observation will fall q or more units away from the mean is at most σ^2/q^2. See, for example, H. Cramer, *The elements of probability theory* (John Wiley, New York, 1955), pp. 81–82.

alternatives can be solved. Let us consider the rationale of this position.

In Tables 3.1 and 3.2 the outcomes of the various possible combinations of decision and state of nature are expressed in terms of physical and monetary results. The underlying idea of decision theory, however, is that we are not really interested in these results per se, but rather in the utilities that they provide. Thus we construct a table of returns, called a "payoff table," in which the entries are not physical or monetary quantities but the utilities corresponding to them — the easy affluence of a bountiful year, the pain and disruption of a heavy flood, or the sober prosperity of an average crop. Estimating these utilities is naturally a bit of a problem, but the technique is as follows.

To set up a scale, we choose two results arbitrarily, say $100,000 and $300,000, and define a utility of zero as that yielded by a return of $100,000 and a utility of 100 as that resulting from a return of $300,000. Thus we can write $u(100) = 0$, $u(300) = 100$. Now we have to determine the utilities of all other possible returns consistently with these two values. By way of example, let us consider a return of $260,000. We can present the members of the water district with a series of options as in Table 3.8.

Table 3.8. Options for estimating the utility of a return of $260,000.

Option	Returns (10^3 dollars)	
0	260	260
All other	100	300
	Probabilities of returns of —	
	100×10^3 dollars	300×10^3 dollars
1	1.0	0.0
2	0.9	0.1
3	0.8	0.2
.	.	.
.	.	.
.	.	.
9	0.2	0.8
10	0.1	0.9
11	0.0	1.0

In this table, option 0 is a certainty of $260,000, the result the utility of which we wish to evaluate. The other options involve uncertainty. They offer returns of $100,000 and $300,000, whose utilities we have already established by definition, with the varying probabilities listed in the lower part of the table. Clearly option 1, which amounts to \$100,000 for certain, is inferior to option 0, and option 11, which offers \$300,000 for certain, is preferable to option 0. It is highly plausible to assume that somewhere between options 1 and 11 we can find one that is indifferent to option 0, although we may have to interpolate some additional options if, for example, option 0 is preferred to option 9 but option 10 is preferred to option 0.

Let us suppose that it turns out that option 10 is indifferent to option 0. Then a return of \$260,000 is the "certainty equivalent" of a 0.1–0.9 chance of returns of \$100,000 and \$300,000, respectively. Now since option 10 and option 0 are indifferent, their utilities are the same. Up to this point these utilities, although equal, have been undefined; but let us now define the utility of an uncertain event like option 10 to be the expected value of the utility that will result from adopting it. Making the calculation, we find

$$u(\text{option } 10) = 0.1\, u(100) + 0.9\, u(300) = 0.1(0) + 0.9(100) = 90, \qquad (3.19)$$

which is also the utility of the certain return of \$260,000. Thus $u(260) = 90$.

The calculation of the utility of a yield outside the range of the two arbitrarily established values is slightly different. We illustrate it by calculating the utility of \$400,000 in Table 3.9. The main difference be-

Table 3.9. Options for estimating the utility of a return of \$400,000.

Option	Returns (10^3 dollars)	
0	300	300
All other	100	400
	Probabilities of returns of —	
	100×10^3 dollars	400×10^3 dollars
1	1.00	0.00
2	0.90	0.10
3	0.80	0.20
.	.	.
.	.	.
.	.	.
9	0.20	0.80
10	0.13	0.87
11	0.10	0.90
12	0.00	1.00

tween this table and Table 3.8 is that we now take a yield of $300,000 as our certain outcome and a yield of $400,000, whose utility we wish to ascertain, as one of the outcomes of the uncertain options. Suppose that option 10 in this table is indifferent to option 0. Then, by the same definitions and reasoning:

$$u(\text{option } 10) = u(300) = 0.13u(100) + 0.87u(400). \qquad (3.20)$$

Since $u(100)$ and $u(300)$ are known, we can solve for $u(400)$ to obtain $u(400) = 100/0.87 = 115$.

As these calculations indicate, once utilities have been assigned to two outcomes, the utilities of all other outcomes can be derived from them by some introspection followed by a simple calculation. The advantage of this procedure, as von Neumann and Morgenstern have shown,[21] is that under some very plausible axioms if utilities are computed in this manner, the utility of any uncertain situation equals the expected value of the utilities of the possible outcomes. We shall make use of this fact in a moment, but first note that although a bit shocking — it says that any two uncertain situations with the same mathematical expectation of utility are indifferent, even though the range of possible outcomes in one situation is much greater than the range in the other — this conclusion should not be surprising, since it is just an extension of the definition of our utility scale.

To apply these concepts to the problem presented by Tables 3.1 and 3.2, we assume that we have ascertained the utilities corresponding to the physical and monetary consequences in those tables, and so deduce Table 3.10. The final column is now the basis of an unambiguous choice. Since the utility of any decision is the expected value of the utilities to which it can give rise, we see that the best decision is to release one-third of the water in the reservoir. This is decision theory's solution to the problem of planning under uncertainty. Its essence lies in converting from physical units to utility units. Its virtue is that after this conversion a choice can be made among a number of alternatives simply by comparing the mathematical expectations of the utilities of their outcomes.

It must be recognized that decision theory has provided the most thoroughgoing and profound analysis thus far achieved of the problem

[21] J. von Neumann and Oskar Morgenstern, *Theory of games and economic behavior* (Princeton University Press, Princeton, ed. 2, 1947, or ed. 3, 1953); Robert Dorfman, Paul A. Samuelson, and Robert M. Solow, *Linear programming and economic analysis* (McGraw-Hill, New York, 1958).

Table 3.10. Utility payoff table for text example of reservoir used for both irrigation and flood protection.

Decision	Utility in case of — Flood	No flood	Expected value of utility
Spill one-third	30	115	81
Spill two-thirds	37	90	68.8
Spill all	−23[a]	−23[a]	−23[a]
Probability	0.4	0.6	

[a] These negative utilities do not mean that a harvest of 200,000 bushels or a return of $80,000 has a "disutility" of 23 units, but simply that the utility of this outcome is 23 units below that of the outcome that was assigned arbitrarily a utility of zero.

of reaching decisions under uncertainty, and that one must give serious consideration to the results of research in this area. For this reason we have presented its main tenets fully and fairly. Nevertheless, we feel that at the present stage of development its conclusions cannot be applied to decisions in the water-resource field. For this there are several reasons.

First, decision theory does not so much solve the difficulties as shift their locale. It makes deciding among alternatives easy at the cost of requiring us to determine a scale of utilities in a particular way. To appreciate the severity of this requirement, try to visualize a water-district meeting at which government experts try to ascertain the sense of the meeting with respect to a long list of options like those in Table 3.8 and 3.9. In effect, this approach replaces the problem of choosing among the actual alternatives with a number of artificial-choice problems, each of which is simpler than the actual one (which is to the good) but also more artificial (which is to the bad).

The second objection to the decision-theory approach pertains more to principle. It assumes that there is no cost of uncertainty per se. This is implicit in the postulate that all options with the same expected utility are indifferent regardless of the spread of possible results. To see where the cost of uncertainty has its impact, consider Table 3.6 again. If two-thirds of the water is released, the members of the district can count on returns of at least $140,000 whatever happens and can plan their affairs with the assurance that funds to this extent will be forthcoming. But if only one-third of the water is released, the members are assured of only $130,000. Under no circumstances can they undertake obligations costing $292,000, still less $400,000. Their arrangements must provide for the possibility of the least favorable outcome. This consideration indicates that any income expected with assurance is more valuable, because it can be utilized more fully, than that same income resulting from an uncertain situation.

Another way to see the same point is to turn to Table 3.8 or 3.9 and suppose that one of the conditions of the choice is that the chooser will not be informed for several months whether the right- or the left-hand column is to be applicable. This delay is of no consequence if he chooses option 0, but in every other option the utility of the more favorable result is diminished by the fact that no firm plans for employing it can be made. The utility of any physical or monetary option depends not only on the possible outcomes and their probabilities, but also on the promptness with which the result of the choice will be known. The procedure given for evaluating the utilities of options makes no allowance for this fact. Thus in Eq. (3.19) the symbol $u(300)$ stands for the utility of $300,000 received as the favorable outcome of an uncertain situation, whereas in Eq. (3.20) the same symbol denotes the utility of $300,000 for certain. It is illegitimate to assume, as we have done, that these two utilities are the same.

Since a substantial delay in learning the results of uncertain situations is inherent in the uncertainties encountered in water-resource planning, the technique of utility scaling will tend to yield unduly high estimates of certainty equivalents by ignoring the cost of uncertainty. This is regrettable because, as we have said, where the techniques of decision theory apply, uncertain alternatives can be ranked simply by computing the expected values of their consequences, measured in utility units; if decision theory cannot be applied, a judgment must somehow be made as to whether an alternative with higher expected value but more uncertainty is or is not preferable to one with lower expected value and lower uncertainty. There seem to be no well-established rules for making such judgments. Nevertheless, since such judgments must be made, we turn to the discussion of a number of expedients that seem to be useful for the purpose.

MORE TRADITIONAL APPROACHES

There are three time-honored approaches to the problem of uncertainty that occur repeatedly in the literature of economics although, as we shall see, all are too vague to constitute usable solutions to the problem.

Certainty equivalents. The simplest, and vaguest, is the concept of a "certainty equivalent." The idea is that to every uncertain situation there corresponds some riskless one that is indifferent to it. For example, consider the situation resulting from releasing one-third of the water in the reservoir, as in Table 3.6. The members of the water district presumably would prefer that situation to a guaranteed yield of $130,000 but, at the other extreme, would prefer a guaranteed yield of $400,000 to the uncertain situation. Between these two extremes there must be some guaranteed yield, perhaps $250,000, which would content them just as well as the risky situation. This is its certainty equivalent. Where a choice is to be made among a number of uncertain situations, the one with the highest certainty equivalent should be selected.

This concept is clearly close to the one we used in deriving the von Neumann utility function but is free of its defect of introducing an artificial intermediate choice. However, it really begs the question at issue, since the problem of ascertaining certainty equivalents is simply the problem of evaluating uncertain situations in thin disguise, and what we are seeking is a rational basis for solving either of these problems.

Gambler's indifference map. A somewhat more elaborate version of the uncertainty equivalent is the "gambler's indifference map." Suppose, for simplicity, that all that matters in an uncertain situation is the expected value and the standard deviation of its probability distribution. Then the essence of any uncertain situation can be represented by a point on a graph with axes that measure expected value and standard deviation. Each of the crosses in Fig. 3.12 represents an uncertain situation in this manner. This form of diagram derives its name from the curved lines that have been drawn in, each of which connects a family

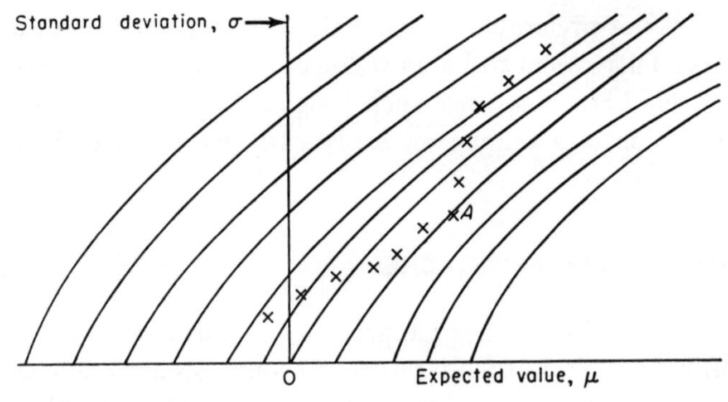

Fig. 3.12. Representation of an uncertainty situation by a gambler's indifference map.

of expected-value–standard-deviation pairs that are indifferent to one another; as we move along any of these lines, the increase in expected value is just sufficient to compensate for the increase in standard deviation or riskiness. The shape of these lines is purely hypothetical; all one can feel confident of *a priori* is that they slope upward to the right. If the crosses represent a number of alternatives from which one is to be selected, the one marked A should be chosen, since it lies on a more desirable indifference curve than any of the others. The certainty equivalent of any uncertain situation can be read from a gambler's indifference map by tracing back the indifference curve on which its expected-value–standard-deviation point lies until the curve reaches the axis of abscissas. Thus a gambler's indifference map is a device for determining certainty equivalents. It has, however, more predictive content than the certainty-equivalent concept itself, because once the indifference map for an individual or group has been determined, the certainty equivalent of any new risky situation can be ascertained by plotting its first two moments on the map.

Unfortunately, it is generally impracticable to establish gambler's indifference maps, and, besides, the concept contains a fundamental logical inconsistency which is disclosed by the following example. Suppose that P white balls and Q black balls are in an urn. One ball is to be drawn by a blindfolded blonde. If it is white you get x dollars, if it is black you get nothing. What value of x should you select if the choice is up to you? Obviously you should choose x as large as permissible. But suppose you decide to be scientific about it and make the choice by consulting your indifference map. The situation is depicted in Fig. 3.13. In drawing this

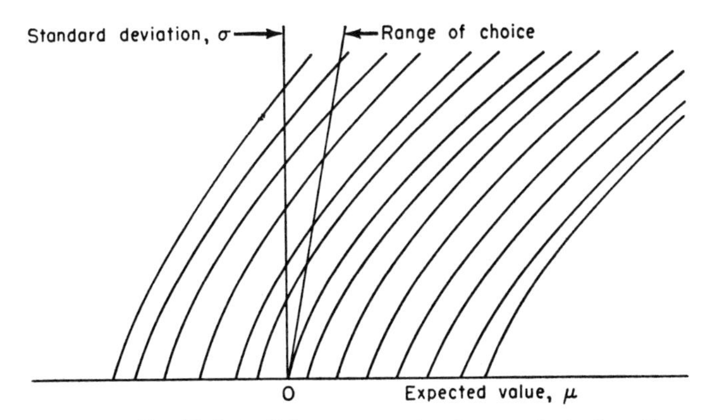

Fig. 3.13. Gambler's indifference map for the paradoxical problem.

diagram, we assume that the slopes of the indifference curves are everywhere positive. The line representing the range of choice is computed from the formulas for the expected value and standard deviation of this chance event as functions of x, which are:

$$\mu = \text{expected value} = x[P/(P + Q)]$$

and $$\sigma = \text{standard deviation} = x\sqrt{PQ/(P + Q)^2}$$

or $$\sigma = \mu\sqrt{Q/P}.$$

By choosing a large enough ratio of black balls to white, we can make the range-of-choice line as steep as desired. In particular it can be made steeper than any of the indifference curves, as drawn. But when this is done, the point on the line that touches the most preferable indifference curve attainable corresponds to $x = 0$, which we already know to be the least desirable point on the line. Hence the use of indifference curves has led to a nonsensical recommendation. This shows that, more generally speaking, an analysis based on the principle of the gambler's indifference map can lead to the adoption of an inadmissible alternative.

Risk-discounting. The third traditional expedient for dealing with uncertainty in economics is "risk-discounting." In this approach the certainty equivalent to a risky venture is computed by multiplying the expected value of the outcome by a factor between zero and one which is proportionately smaller as the risk is higher. A typical risk-discount factor is $1/(1 + c\sigma)$, where σ denotes the standard deviation of outcomes and c is a behavioral constant which we shall consider in a moment. By this discount formula, if the expected net benefits of a plan are denoted by μ, its risk-discounted net benefits, or certainty equivalent, is $\mu/(1 + c\sigma)$. Differentiating this formula shows readily that $c/(1 + c\sigma)$ is the percentage increase in expected net benefits necessary to compensate for a one-unit increase in the standard deviation of the outcome distribution, so that c expresses the additional enticements required to compensate for additional risks.

The risk-discounting approach has the formal merit that it fits in very well with ordinary interest-discounting when, as is frequently the case, time streams of net benefits are at issue. Indeed it is a quite common presumption in economics that the rate of return payable by a risky loan or venture must be greater than the rate payable by a safe loan. For example, one would expect that normally the yields on common stocks would be greater than those on high-grade bonds. Thus in computing the present value of the benefits anticipated from a risky venture, one might apply a rate of discount of the form $r + c\sigma$, where r is the rate of interest on a very safe investment and $c\sigma$ is a risk premium which increases in proportion to σ. Of course, a more elaborate function of the standard deviation might also be used and is frequently preferred to the simple linear one. This device has the effect of sharply reducing the present value of net benefits expected with uncertainty in the remote future, while it reduces the present value of early-maturing benefits only mildly. The logic of this differential treatment (which, it should be noted, does not depend on the fact that the standard deviations of anticipated benefits may increase with distance) is open to some question.

Efforts to overcome defects of traditional approaches. The shortcoming common to all three of these traditional approaches to the problem of uncertainty is that they contain no hint of how rationally to adjust expected values for uncertainty — for example, of how to determine the parameter c of the risk-discount formula. There have been a number of attempts to fill this gap, of which we shall mention two.

In some uncertain situations there is a discontinuity in the list of outcomes such that some possible outcomes are considered catastrophically bad, while the others range from mildly bad on up. In such cases an appealing basis for choice is to select the alternative that has the smallest probability of a disastrous outcome. For example, using the data of Table 3.6, if a yield of \$130,000 or less is considered catastrophic, this principle would recommend spilling two-thirds of the contents of the reservoir; for with this decision the probability of reaching the disaster level is zero, while with either of the other alternatives it is 0.4. If the disaster level is \$140,000 or greater, however, the first two alternatives are indifferent, and the same is true if the disaster level is less than \$130,000. This example shows that this basis for decision disregards a great deal of information about each alternative; indeed, it utilizes only a single point in its probability distribution.

An approximation to the disaster-level approach can be based upon the expected value and standard deviation of the outcome distribution. Let μ_i and σ_i denote the expected value and standard deviation, respectively, of the probability distribution corresponding to the ith alternative, and let D be the disaster level. Then, generally speaking, the probability of a result as bad as the disaster level is a decreasing function of $(\mu_i - D)/\sigma_i$. Therefore one can minimize the probability of experiencing the disaster level by choosing the alternative for which this ratio is as large as possible. Under this approach the indifference curves are the loci of (μ, σ) satisfying $(\mu - D)/\sigma =$ constant. Figure 3.14 illustrates such an indifference map. Note the peculiarity that all (μ, σ) points have the same certainty equivalent, namely D. Thus the disaster-minimization approach implies and determines a gambler's indifference map on rational grounds.

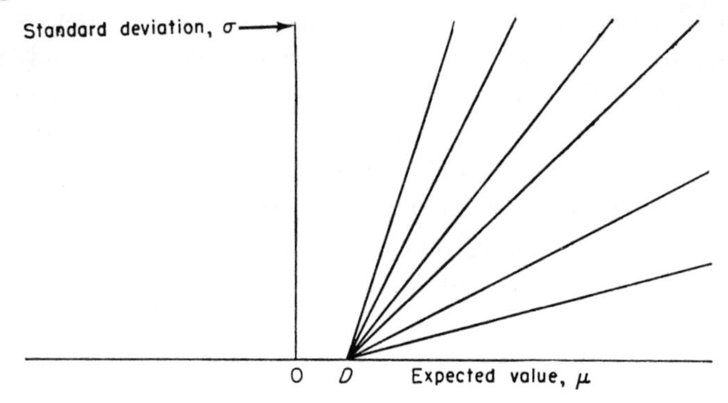

Fig. 3.14. Gambler's indifference map for minimizing the probability of disaster, where D represents the disaster level.

It can, however, lead to ridiculous results. Let us apply it, for example, to the data of Table 3.7 with a disaster level of $100,000. The critical ratio for releasing one-third of the water is 1.45, for releasing two-thirds of the water it is 1.90, but for releasing all the water it is infinite. The last alternative wins handsomely, although we know it to be inadmissible.

Another method for deducing a gambler's indifference map from rational considerations has been suggested by H. A. Thomas, Jr. specifically for application to the design of water-resource systems. This approach stems from the idea that the net benefits yielded by any installation are the present value of its gross benefits minus the present value of its costs, where the costs include the cost of any uncertainties inherent in the project. The crux of the method is the device for measuring the cost of uncertainty. Imagine that at the time the system is constructed an equalization fund is established with the understanding that in any year when actual benefits fall short of expected benefits the fund will be used to make up the difference, while in any year when actual benefits exceed expected benefits the excess will be used to replenish the fund. Thus, provided that the fund is adequate, the users of the project will receive precisely the expected benefits in every year. This device converts a risky situation into a fully insured one. The cost of the uncertainty is then just the size of the fund required to obliterate it, and this statement is true whether or not such a fund is actually established.

In estimating the size of the fund required to offset the uncertainty of any project, we cannot, of course, insist that it be large enough to sustain any imaginable string of unlucky years. All we can do is require that the probability that the fund will be exhausted within the time horizon of the project be low, say 5 percent or 1 percent. Once this probability has been established, calculation of the required size of the fund, which is the same as the cost of the uncertainty, is a purely actuarial problem, although a difficult one. When the size of the fund has been computed for each alternative design, it can be added to the construction and operating costs, and that design can be adopted for which the excess of expected net benefits over total costs, calculated in this way, is greatest.

If some plausible simplifying assumptions are made, it can be shown that the size of the fund required to hold the probability of exhaustion down to any specified level is proportional to the standard deviation of the distribution of outcomes in a single year. Specifically, the formula for the net benefits B of a project is approximately

$$B = \mu - v_\alpha \sigma / \sqrt{2r},$$

where μ is the present value of the expected benefits minus the present value of expected costs, v_α is the normal deviate with probability α of being exceeded, α is the specified probability that the fund will be exhausted, σ is the standard deviation of the single-year outcome distribution, and r is the rate of interest earned by the equalization fund. The second term in this formula is the cost of the uncertainty. Since all designs

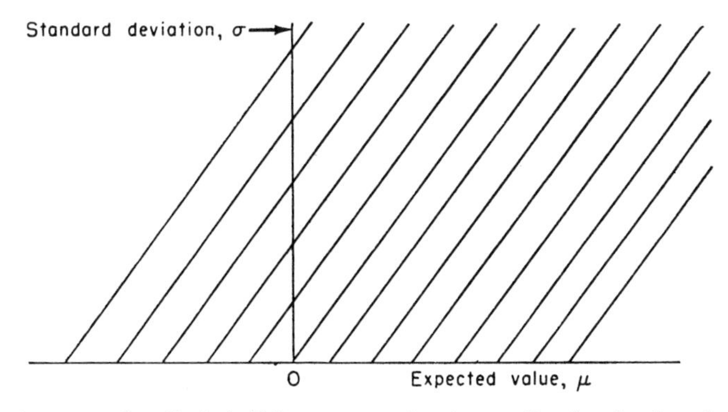

Fig. 3.15. Gambler's indifference map for the equalization-fund method.

with the same value of B are indifferent, in terms of a gambler's indifference map the indifference curves are straight lines with slope $\sqrt{2r}/v_\alpha$, as shown in Fig. 3.15.

To illustrate this approach, let us apply it to the data of Table 3.7 assuming an interest rate of 4 percent and an exhaustion probability of 0.05. Then $\sqrt{2r} = 0.283$, $v_\alpha = 1.645$, and the cost of uncertainty is 5.8σ. For the three plans considered, these costs are approximately \$765,000, \$342,000, and zero, respectively. Converting to an annual basis by taking 4 percent of each and subtracting from the annual expected values shown in the table gives approximately \$261,000, \$198,000, and \$80,000 for the annual net benefits of the three plans, allowing for the cost of uncertainty. Releasing one-third of the water in the reservoir is clearly the best policy.

MULTIPERIOD DECISION PROBLEMS

At this stage it will be useful to take account of the fact that a decision is likely to have consequences that extend over a considerable period of time. A decision regarding the capacity of a reservoir, for example, is of this nature. The consequences of such a decision are not a single yield but rather a sequence of yields, one for each year of the life of the project. These yields will not be exactly predictable, but each will be a random variable with a probability distribution that depends on the original decision (for example, as to the size of the reservoir), as well as on other factors.

Any particular stream of yields can be reduced to a present value by applying a discount factor. If r denotes the rate of interest used in discounting, y_t denotes the yield in the tth year, and T years are taken into account, the present value of the stream of yields u is

$$u = \sum_{t=1}^{T} y_t/(1+r)^t.$$

Furthermore, if $\bar{y}_t$ is the expected value of y_t, σ_t^2 is its variance, and δ_{st} is the correlation between the yields in the sth and tth years, the expected value $E(u)$ and variance σ_u^2 of u are

$$E(u) = \sum_{t=1}^{T} \bar{y}_t/(1+r)^t$$

and

$$\sigma_u^2 = \sum_{s=1}^{T} \sum_{t=1}^{T} \delta_{st}\sigma_s\sigma_t/(1+r)^{s+t}.$$

If the present value of the yields is all that matters, a plausible simplification (although somewhat dangerous, as we shall see below), then $E(u)$ and σ_u for each plan can be graphed on an indifference map and the most favorable plan chosen, just as before. If the correlations between yields in different years are small and T is sufficiently large, things are even simpler. In that case σ_u will be small in relation to $E(u)$, and $E(u)$ itself is sufficient to determine the optimal plan. This is the case of actuarial risk.

The assumption that the present value of an income stream is all that concerns the recipients, while plausible, is not innocuous. For example, even in an income stream with a high present value the variances of some of its components may be so great that a run of bad luck for a few years will endanger the solvency of the recipients. In such a case the risk of insolvency at some stage has to be taken into account, and Thomas' device for measuring it by estimating the required size of a disaster-insurance fund becomes necessary.

One characteristic of uncertain time streams that we have not yet taken into account is that they afford enough time for decisions to be revised. This simultaneously decreases their riskiness and complicates the analysis. An essential feature of uncertain time streams is that at least some of the decisions that have to be made can be postponed until relevant data have been accumulated. In the case of our irrigation district, for example, it may be necessary to decide on the capacity of the reservoir once and for all, but thereafter the amount of water to be released in each month, or in even shorter periods, need not be decided until information is available on the current contents of the reservoir, short-run predictions of inflow, and so on. Thus an element of sequential decision-making enters the problem. To be sure, these sequential decisions typically relate to operating procedure rather than to design,[22] but the criterion for appraising a design must be the yields it will afford when managed by a good operating procedure. One might even say that the best design is the one that permits use of the best possible operating procedure.

The analysis of operating procedures or sequential decision problems,

[22] Sequential decision-making often enters design problems more directly, as when the construction of a system will be spread out over a considerable period of time. Then the exact timing and design of the later units need not be decided when construction of the first units is begun, but the early units should be planned so that the later components can be added efficiently and economically when they appear desirable. For further discussion of stage construction and related possibilities, see Chapter 4.

which has a large literature under such headings as inventory theory, queuing theory, and dynamic programming, is too technical and elaborate to be discussed here at any length.[23] The central problem is whether resources available at present (water, in our context) should be used in the current period or saved for later employment. The most promising line of attack is to divide the entire future history of the undertaking into a sequence of periods, each of some convenient length like a month or a year. The number of periods may be either finite or infinite.

The analysis then proceeds by considering time horizons of different length. If the horizon is only one period long, the problem is of a familiar sort. A little notation will be helpful at this stage. Let R denote the quantity of resources available at the beginning of the period and z the quantity used. There will be some functional relationship between the benefits yielded and the quantity of resources used; denote it by $f_1(z)$. The one-period problem, then, is to choose the value of z that makes $f_1(z)$ as large as possible where the range of permissible choices of z depends on R. Now the optimal z and the corresponding net benefits are both functions of the resources available, so we may denote the maximum attainable net benefits in a single period by $f_1^*(R)$, a function of the quantity of resources available. As we shall see, this function is fundamental to the solution of the multiperiod problem.

With this background we can advance to the two-period case. Let us suppose again that R units of resources are available at the outset and z_1 units are used in the first period. Then the benefits during the first period are $f_1(z_1)$, and the resources available at the beginning of the second period will be $R - z_1$ plus the quantity of resources received during the first period. Denote the quantity of resources received by x_1, and assume that it is a random variable. Then the second period begins with resources amounting to $R - z_1 + x_1$ and, utilizing the concepts introduced in the preceding paragraph, we see that the maximum attainable benefits in the second period are $f_1^*(R - z_1 + x_1)$.

To find the present value of net benefits for both periods together, discount the second period's benefits at an appropriate rate of interest r, and add to the first period's benefits to obtain $f_1(z_1) + [f_1^*(R - z_1 + x_1)]/(1 + r)$. Unfortunately this formula cannot be evaluated at the time that the first period's utilization must be determined, since x_1 generally will not be known at that time. Its expected value can be computed,

[23] These matters, especially queuing theory, are discussed further in Chapter 14.

however, and the expected value of the benefits of the two periods together corresponding to a use of z_1 units in the first period is

$$f_2(z_1) = f_1(z_1) + [Ef_1^*(R - z_1 + x_1)]/(1 + r),$$

where E denotes expected value. If x_1 has the probability distribution $p(x_1)$, then

$$Ef_1^*(R - z_1 + x_1) = \int_{-\infty}^{\infty} f_1^*(R - z_1 + x_1)p(x_1)dx_1$$

and $$f_2(z_1) = f_1(z_1) + \int_{-\infty}^{\infty} f_1^*(R - z_1 + x_1)p(x_1)dx_1/(1 + r).$$

This formula says that the net benefits in two periods, if R units of resources are available initially and z_1 units are used in the first period, is the sum of two parts. The first part is the net benefits yielded by z_1 units used in the first period; the second is the discounted expected value of the maximum net benefits attainable in the second period. The first part is a basic datum of the problem, the one-period production function. The second part is an integral, or sum, involving a given probability distribution and a function $f_1^*(R - z_1 + x_1)$ that has been determined by analysis of the one-period case. Thus we can find the permissible value of z_1 that maximizes the sum. It will depend on R, of course, and so will the maximum value of the sum, which we can now denote by $f_2^*(R)$.

In the three-period case the only decision that has to be made at the outset is again the value of z_1, the quantity of resources to be used in the first period. As before, the expected benefits resulting from any choice of z_1 can be written:

$$f_3(z_1) = f_1(z_1) + [Ef_2^*(R - z_1 + x_1)]/(1 + r),$$

where the second term uses the function that was determined in the analysis of the two-period case. Thus we can proceed recursively and write for the t-period case with initial resources R:

$$f_t(z_1) = f_1(z_1) + [Ef_{t-1}^*(R - z_1 + x_1)]/(1 + r),$$

where $f_t(z_1)$ is the net benefit in t periods resulting from the use of z_1 units of resources in the first period and optimal use of resources in the $t - 1$ periods remaining. Since the second term was determined in the preceding step, the maximum attainable net benefits for the t periods is the maximum of this function with respect to z_1 and can be denoted by $f_t^*(R)$.

The foregoing exposition has assumed that the single-period net-benefit function is the same for all the periods considered. This assumption has served merely to simplify the notation a bit, but now we shall make a substantive use of it. We have repeatedly remarked that in each case the optimal value of z_1 is a function of the available resources R. If z_1^* is this optimal value, we can denote this relationship by $z_1^* = g_t(R)$, introducing the subscript because the function will be different, in general, for different time horizons. But if the successive periods are identical, it can be shown[24] that the functions $g_t(R)$ gradually change less as t increases, and approach some limiting function $g(R)$. This function, $g(R)$, is then the optimal operating procedure — the relationship between first-period releases and available resources — for a long or infinite time horizon. With luck, the ultimate function $g(R)$ is approached fairly quickly in the first few iterations so that it is not necessary to repeat the analysis just sketched 50 times for a 50-year horizon, or 600 times for a 600-month horizon.

In principle, this method of solution is sound and appealing; in practice, the required computations are very difficult to carry out, especially if some of the functions are stochastic. In general, therefore, this mode of analysis is more useful for formulating problems and guiding intuition than for actually deriving solutions.

Loss functions. In view of these difficulties a much simpler approach is used frequently in the following chapters to allow for some of the uncertainties that enter into designing water-resource installations for use over many periods. This approach consists in dividing the actual benefits yielded each year into two parts, a "normal benefit" and a "loss." The normal benefit is the benefit obtained when the quantity of water for which the project was designed is available and used. The loss is the discrepancy between the normal benefit and the actual benefit which results when the quantity of water available departs from the design amount. Since the consequences of a water shortage are in general more marked than those of a surplus, the emphasis in this concept is on the loss side; but it should be kept in mind that when water is more abundant than it normally is, the "loss" can be negative.

The reason for distinguishing between normal benefits and losses is that the normal amount of water and discrepancies from it affect benefits in different ways. The installations of both a water project and its cus-

[24] See R. Bellman, *Dynamic programming* (Princeton University Press, Princeton, 1957), chap. 4.

tomers are designed to make use of a certain amount of water, expected to be available in most years. Neither the project nor its customers are prepared to take full advantage of any overage that may occur, and the customers in particular stand to suffer losses when the water supply falls short of the amount required for normal operations. The relationship between the amount of the loss and the discrepancy between the actual water availability and the normal quantity will be referred to as the "loss function."

As an illustration of a loss function suppose that irrigation water is worth $6 per acre foot when supplied in the amount planned for and that, on this basis, laterals are installed, land cleared, and other measures taken. Suppose also that when installations and improvements are made and the anticipated volume of water is not forthcoming the losses amount to $12 per acre foot of deficiency. If, however, the water supply in any year is superabundant, let us say that the additional benefits are worth only $1 per acre foot. Then if Y_n denotes the normal supply of water for which the irrigation system was designed and Y_t is the actual supply in the tth year, we may summarize these data by the benefit formula:

$$\text{Benefit in year } t = 6Y_n - 12(Y_n - Y_t)^+ + (Y_t - Y_n)^+,$$

where we use the notation $(y)^+ = \max(0, y)$. In this formula $6Y_n$ is the normal benefit and $12(Y_n - Y_t)^+ - (Y_t - Y_n)^+$ is the loss function.

More generally, using the same notation, we can write

$$B(Y_n, Y_t) = N(Y_n) - L(Y_n - Y_t),$$

indicating that the normal benefits N depend on Y_n alone, the loss L depends on the difference between Y_n and Y_t, and the actual benefits B depend on both Y_n and Y_t. The expected benefits B_{exp} are then

$$B_{\text{exp}}(Y_n, Y_t) = N(Y_n) - L_{\text{exp}}(Y_n - Y_t).$$

Generally speaking, the expected loss $L_{\text{exp}}(Y_n - Y_t)$ will be an increasing function of two factors: the excess of Y_n over the expected value of Y_t, and the standard deviation of Y_t. The introduction of the loss function, therefore, has the effect of diminishing estimated benefits in response to variability and consequently makes some allowance for this aspect of uncertainty. Because of its relative simplicity, this concept will be used many times in the present volume.

One final word about uncertainty. In the latter part of this discussion we have assumed that although precise results of decisions could not be foretold, their probability distributions were known or could be estimated with tolerable accuracy. This may be the case where the uncertainty results from physical phenomena, like hydrologic conditions, concerning which adequate records and experience are available. It is not likely to be so if, for instance, the uncertainty is generated by economic or political phenomena. If the probability distributions resulting from different decisions are not known, we are thrown back on the conundrums considered earlier in this discussion. Then, as we say, it is by no means clear how to formulate an objective function or measure of merit, let alone how to find its maximum, and rational decision-making becomes very difficult indeed.

In this long discussion of decision-making under uncertainty we have been compelled to point out flaws and shortcomings in every solution that has been proposed. One would hope that this will not always be the case, and that some day satisfactory solutions will be found to this pervasive and fundamental problem. At present, however, the problem of uncertainty is clouded by uncertainty.

4 Economic Factors Affecting System Design

STEPHEN A. MARGLIN

IN this chapter some of the economic factors mentioned in the discussions of objectives and basic concepts of system design in Chapters 2 and 3 are dealt with in detail. First, the influence of the budgetary constraint, in various forms, on the optimal system design is analyzed. Next, problems raised for system design by the dynamic nature of the investment process are treated; since costs and benefits occur in time streams, the objective is to obtain the optimal investment schedule as well as design for system units. A third subject is the choice of interest rate for public investments such as water-resource development, and the method in which the opportunity costs of public investment can be calculated in a mixed private-public economy. Finally, certain problems raised by specific external alternatives to water-resource systems are examined — for example, the treatment of private projects which systems would displace.

BUDGETARY CONSTRAINTS

The problems that budgetary constraints impose on the choice of system design can be divided into two categories: those relating to the form of the budgetary constraint and those relating to the implementation of the constraint, whatever its form.

The form of the budgetary constraint depends on the social objective it is intended to serve. If, for example, the constraint is designed to limit the long-run flow of funds from the national treasury into water-resource development, the constraint properly applies to the net drain on the treasury exercised by each project, namely, the costs less the revenues and taxes stimulated by the project over its entire economic life. If, on the other hand, the purpose of the constraint is to limit participation by the national government in water-resource development, the expenditure limitation is properly applied to the portion of project costs borne by the national government. A goal of limiting the net drain on the treasury for the current year would dictate still another form of constraint: restricting the outlay of the national government for the year in question on construction of systems and operation, maintenance, and replacement of existing river-development schemes, net of revenues and taxes generated during the year. As a final example, a budgetary constraint designed to restrict the amount of foreign exchange used to import capital for the construction and replacement of project units would effect only that portion of the outlay that had foreign-currency repercussions.

From these examples it is evident that a budgetary constraint may be used reasonably to achieve any one of a wide variety of goals or any combination of them, and that there are many forms of the constraint. However, the questions to be answered in framing the constraint are basically the same in every situation: Which costs are to be included? Over how long a period of time are these costs to be constrained? How is feedback (revenues and taxes stimulated by the projects) to be incorporated into the constraint? Therefore, instead of surveying forms of the constraint applicable to a wide variety of situations, we shall examine closely only two alternative forms. For these we shall derive the appropriate marginal conditions, employ them to reveal the different effects of the two constraints on system design, and identify the conditions for which each of the constraints is appropriate. In the process we shall have enlisted the techniques by which the effects on system design of any form of the constraint can be determined. This procedure should lead to a clearer understanding of the principles by which one chooses the appropriate constraint for any given situation. We shall conclude our discussion of budgetary constraints by considering the decisions involved in allocation of a given budget among the several systems that comprise the full development program.

TWO ALTERNATIVE FORMS OF CONSTRAINT: A LIMITED INVESTMENT BUDGET VERSUS A LIMITED EXPENDITURE LEVEL

The two alternative forms that we shall consider are a constraint on construction costs alone and a constraint on construction costs plus the present value of operation, maintenance, and replacement (OMR) costs. These are chosen not so much because of a greater intrinsic interest in them than in other forms of constraint but more because of the interest and controversy they have aroused among economists.[1]

(The problem of choice by the national government between these two forms of the constraint can arise only if it undertakes both construction and OMR outlay. It does not arise if, for instance, the national government as a matter of policy undertakes only construction outlay, leaving OMR expenditure to local interests; in this case the national government's constraint applies only to construction costs, although separate constraints may be imposed on OMR outlays by local interests.)

The constraint on construction costs says to planners: "You have $\overline{K}$ dollars to spend on the construction of water-resource systems. What you spend in the future on operation, maintenance, and replacement of system units is irrelevant (provided, of course, that your expenditures pass the basic test that the incremental benefits exceed the incremental costs)." The constraint on construction-plus-OMR costs, on the other hand, says in effect: "You have $\overline{U}$ dollars to spend on the construction and operation of water-resource systems. You may spend any or all of it now for construction or at any time in the future for operation, maintenance, and replacement of system units. The portion not spent today will be made available with accrued interest in any desired time pattern."

The difference between the two forms of constraint lies in the portion of costs covered and — insofar as construction takes place in early years and OMR expenditures in later years — in the time period over which the costs are constrained. Because the two forms of the constraint cause planners to allocate expenditures between present construction costs and future OMR costs in different ways, the two forms affect system design in different ways.

In order that the two forms of the constraint may be compared at all stages, their appropriate marginal conditions will be derived simultaneously from the same ranking function. For this purpose we have selected the efficiency ranking function presented in Eq. (2.6). As described in Chapter 2, we let θ_q be the discount factor applicable to the demand period q, $E_{sq}(\mathbf{y}_{sq})$ denote the annual efficiency benefits for system s in period q as a function of the outputs in period q, $M_{sq}(\mathbf{x}_{sq}')$ denote the annual OMR costs for system s in period q as a function of the OMR inputs in period q, and $K_s(\mathbf{x}_s^0)$ denote the construction cost for system s as a function of the construction inputs. Then for each form of the con-

[1] See, for example, Otto Eckstein, *Water-resource development, the economics of project evaluation* (Harvard University Press, Cambridge, 1958), pp. 35–36.

straint we are looking for the set of designs of systems $s = 1, \ldots, S$ which maximizes

$$\sum_{s=1}^{S} \sum_{q=1}^{Q} \theta_q [E_{sq}(\mathbf{y}_{sq}) - M_{sq}(\mathbf{x}_{sq}')] - \sum_{s=1}^{S} K_s(\mathbf{x}_s^0) \tag{4.1}$$

subject to the production function for each system,

$$f_s(\mathbf{x}_s^0, \mathbf{x}_{sq}', \mathbf{y}_{sq}) = 0, \qquad \begin{array}{l} s = 1, \ldots, S \\ q = 1, \ldots, Q \end{array} \tag{4.2}$$

and to the applicable form of the budgetary constraint. In other words, the range of choice among designs to maximize Exp. (4.1) is restricted not only by the production function for each system, but also by the requirement that the sum of the inroads on the budget caused by all systems shall be less than the available budget.

Formally, for the constraint on construction costs alone we have the restriction

$$\sum_{s=1}^{S} K_s(\mathbf{x}_s^0) \leq \overline{K},$$

where $\overline{K}$ is the level of the construction budget. For the constraint that includes OMR costs as well, the restriction is

$$\sum_{s=1}^{S} K_s(\mathbf{x}_s^0) + \sum_{s=1}^{S} \sum_{q=1}^{Q} \theta_q M_{sq}(\mathbf{x}_{sq}') \leq \overline{U},$$

$\overline{U}$ denoting the construction-plus-OMR budget. The problem as it stands, therefore, is the very complicated one of allocating the budget among systems and simultaneously designing the systems. Note the way in which the budgetary constraint, whatever its form, ties together otherwise independent systems. In the absence of budgetary constraints, each system usually can be designed without regard to any other system. In the presence of a budgetary constraint, however, the planners for each system s must keep an eye on all other systems, for they must verify not only that the benefits of a proposed increment exceed its costs (as in the absence of a constraint), but also that the increments to all other systems which must be foregone if funds are allotted to system s produce less benefits per dollar than those from the proposed increment to s.

However, to facilitate exposition in this chapter and, as we shall see, to facilitate water-resource-planning by allowing designers to focus on individual systems, we separate the problem of design from the problem of allocation of the budget. This division permits us to break down the aggregate ranking function, Exp. (4.1), into S separate functions, one for each system.

The design problem for each system is to maximize the net-benefit function

$$\sum_{q=1}^{Q} \theta_q[E_{sq}(\mathbf{y}_{sq}) - M_{sq}(\mathbf{x}_{sq}')] - K_s(\mathbf{x}_s^0) \tag{4.3}$$

subject to the production function, Exp. (4.2), and the appropriate subbudget restriction, either

$$K_s(\mathbf{x}_s^0) \leq \overline{K}_s \tag{4.4a}$$

or

$$K_s(\mathbf{x}_s^0) + \sum_{q=1}^{Q} \theta_q M_{sq}(\mathbf{x}_{sq}') \leq \overline{U}_s. \tag{4.4b}$$

This restricts the choice among designs for each system to that set for which the constrained expenditures are less than or equal to $\overline{K}_s$ or $\overline{U}_s$, the subbudget allocated to the sth project.

For the time being we simply assume an arbitrary division of the over-all budget into subbudgets for individual systems. In reality, the quantities $\overline{K}_s$ and $\overline{U}_s$ for optimal allocation are related to the over-all budget in the following manner. If we express the efficiency benefits of system s, net of unconstrained costs, as functions $B_s(\overline{K}_s)$ and $B_s^*(\overline{U}_s)$ of the amount of each budget allocated to the system,[2] the optimal allocation of the over-all budget $\overline{K}$ or $\overline{U}$ is given by the values of $\overline{K}_s$ or $\overline{U}_s$ in Exp. (4.4) for which

$$\sum_{s=1}^{S} [B_s(\overline{K}_s) - \overline{K}_s]$$

and

$$\sum_{s=1}^{S} [B_s^*(\overline{U}_s) - \overline{U}_s]$$

are maximum. The values of $\overline{K}_s$ and $\overline{U}_s$ are restricted only by the constraint that the total allocation cannot exceed the budget available for the program as a whole, or

$$\sum_{s=1}^{S} \overline{K}_s \leq \overline{K}$$

and

$$\sum_{s=1}^{S} \overline{U}_s \leq \overline{U}.$$

[2] Formally, the functions $B_s(\overline{K}_s)$ and $B_s^*(\overline{U}_s)$ are defined by the maxima of the single-project efficiency net-benefit function (4.3) subject to Exp. (4.4) for successive values of $\overline{K}_s$ and $\overline{U}_s$.

At present we address ourselves to a rather detailed examination of the implications of the budgetary constraint for design. Later we shall investigate the allocation problem in corresponding detail.

DESIGN IN THE PRESENCE OF BUDGETARY CONSTRAINTS

For any given allocation of $\overline{K}_s$ or $\overline{U}_s$ for system s the design criterion is, we have observed, maximization of Exp. (4.3) subject to the constraint (4.4a) or (4.4b) and to the production function (4.2) applicable to system s. For simplicity we assume that demands for system outputs are constant over the life of the system. The Lagrangian expressions for the two design problems are therefore

$$L = \theta[E_s(\mathbf{y}_s) - M_s(\mathbf{x}_s')] - K_s(\mathbf{x}_s^0) - \mu f_s(\mathbf{y}_s, \mathbf{x}_s^0, \mathbf{x}_s') - \lambda[K_s(\mathbf{x}_s^0) - \overline{K}_s] \qquad (4.5a)$$

and

$$L^* = \theta[E_s(\mathbf{y}_s) - M_s(\mathbf{x}_s')] - K_s(\mathbf{x}_s^0) - \mu f_s(\mathbf{y}_s, \mathbf{x}_s^0, \mathbf{x}_s') - \lambda[K_s(\mathbf{x}_s^0) + M_s(\mathbf{x}_s') - \overline{U}_s]. \qquad (4.5b)$$

Equations (4.5), it will be noted, each involve two Lagrangian multipliers, the multiplier μ arising from the production function and the multiplier λ owing its existence to the budgetary constraint. Equating the partial derivatives of Eq. (4.5) to zero and denoting individual outputs and inputs by second subscripts, we obtain[3]

$$\partial L/\partial y_{sj} = \theta D_{sj}(y_{sj}) - \mu(\partial f_s/\partial y_{sj}) = 0, \qquad (4.6a)$$

$$\partial L/\partial x_{si}' = -\theta(\partial M_s/\partial x_{si}') - \mu(\partial f_s/\partial x_{si}') = 0, \qquad (4.7a)$$

$$\partial L/\partial x_{si}^0 = -\partial K_s/\partial x_{si}^0 - \mu(\partial f_s/\partial x_{si}^0) - \lambda(\partial K_s/\partial x_{si}^0) = 0, \qquad (4.8a)$$

and

$$\partial L^*/\partial y_{sj} = \theta D_{sj}(y_{sj}) - \mu(\partial f_s/\partial y_{sj}) = 0, \qquad (4.6b)$$

$$\partial L^*/\partial x_{si}' = -\theta(\partial M_s/\partial x_{si}') - \mu(\partial f_s/\partial x_{si}') - \lambda\theta(\partial M_s/\partial x_{si}') = 0, \qquad (4.7b)$$

$$\partial L^*/\partial x_{si}^0 = -\partial K_s/\partial x_{si}^0 - \mu(\partial f_s/\partial x_{si}^0) - \lambda(\partial K_s/\partial x_{si}^0) = 0. \qquad (4.8b)$$

From these we obtain the following expressions by division.

[3] As explained in Chapter 2, the demand function $D_{sj}(y_{sj})$ in the equations below is the partial derivative $\partial E_s/\partial y_{sj}$ of efficiency benefits with respect to output y_{sj}.

From Eqs. (4.6), if j and h both denote outputs:

$$\frac{\partial f_s}{\partial y_{sj}} \Big/ \frac{\partial f_s}{\partial y_{sh}} = D_{sj}(y_{sj})/D_{sh}(y_{sh}),$$

or

$$\text{MRT}_{shj} = \text{M}E_{sj}/\text{M}E_{sh}. \tag{4.9}$$

From Eqs. (4.7), if i and h both denote OMR inputs:

$$\frac{\partial f_s}{\partial x_{si}'} \Big/ \frac{\partial f_s}{\partial x_{sh}'} = \frac{\partial M_s}{\partial x_{si}'} \Big/ \frac{\partial M_s}{\partial x_{sh}'},$$

or

$$\text{MRS}_{sih} = \text{M}M_{si}/\text{M}M_{sh}. \tag{4.10}$$

From Eqs. (4.8), if i and h both denote construction inputs:

$$\frac{\partial f_s}{\partial x_{si}^0} \Big/ \frac{\partial f_s}{\partial x_{sh}^0} = \frac{\partial K_s}{\partial x_{si}^0} \Big/ \frac{\partial K_s}{\partial x_{sh}^0},$$

or

$$\text{MRS}_{shi} = \text{M}K_{si}/\text{M}K_{sh}. \tag{4.11}$$

These results are all familiar from Chapters 2 and 3. Moreover, they are the same for both forms of the constraint. For example, Eq. (4.9) asserts that if two units of output h can be obtained by sacrificing one unit of output j (that is, if $\text{MRT}_{shj} = 2$), then for the design to be optimal, the marginal benefit of output j must equal twice the marginal benefits of output h, or $\text{M}E_{sj} = 2\text{M}E_{sh}$.

A comparison of Eqs. (4.6), (4.7), and (4.8), however, reveals new conditions that show the difference between design in the absence and presence of a budgetary constraint and between the two forms of the constraint. From Eqs. (4.6) and (4.8) we obtain for both forms of the constraint:

$$\theta D_{sj}(y_{sj})/(1+\lambda)\frac{\partial K_s}{\partial x_{si}^0} = -\frac{\partial f_s}{\partial y_{sj}} \Big/ \frac{\partial f_s}{\partial x_{si}^0},$$

or

$$\text{M}E_{sj}/\text{M}K_{si} = (1+\lambda)/\theta\text{MP}_{sij}. \tag{4.12}$$

From Eqs. (4.6) and (4.7) we obtain for the constraint on construction cost only:

$$D_{sj}(y_{sj}) \Big/ \frac{\partial M_s}{\partial x_{si}'} = -\frac{\partial f_s}{\partial y_{sj}} \Big/ \frac{\partial f_s}{\partial x_{si}'},$$

or

$$\text{M}E_{sj}/\text{M}M_{si} = 1/\text{MP}_{sij}, \tag{4.13a}$$

and for the constraint on construction plus OMR costs:

$$D_{sj}(y_{sj}) \Big/ \frac{\partial M_s}{\partial x_{si}'} = -(1+\lambda)\frac{\partial f}{\partial y_{sj}} \Big/ \frac{\partial f}{\partial x_{si}'},$$

or

$$\text{M}E_{sj}/\text{M}M_{si} = (1+\lambda)/\text{MP}_{sij}. \tag{4.13b}$$

In Chapter 3, in the absence of budgetary constraints, we derived the conditions

$$\mathrm{M}E_{sj}/\mathrm{M}K_{si} = 1/\theta\mathrm{MP}_{sij} \tag{4.14}$$

relating benefits to construction outlay, and

$$\mathrm{M}E_{sj}/\mathrm{M}M_{si} = 1/\mathrm{MP}_{sij}, \tag{4.15}$$

relating benefits to OMR outlay for the special case $\mathrm{M}E_{sj} = p_j$. Thus, the impact of budgetary constraint on design is represented by the factor λ, and differences in the impact on design of the two forms of the constraint are to be found in the asymmetries with respect to λ of the corresponding equations for these forms.

To illustrate the impact on design of the two forms of the constraint, assume that an additional \$100,000 spent on construction of generators would permit an increase of 200 Mw hr annually of electric energy, worth \$20 per megawatt hour at the margin. Assume also that the value of λ, determined by the allocation made to the subbudget for system s, is 0.3 and that a $2\frac{1}{2}$-percent interest rate and a 50-yr economic life lead to a value of 28.32 for θ. Then the data for Eq. (4.12) are $\mathrm{ME}_{sj} = 20$, $\mathrm{M}K_{si} = 100{,}000$, $1 + \lambda = 1.3$, $\theta = 28.32$, and $\mathrm{MP}_{sij} = 200$. Substitution of these values gives

$$\mathrm{M}E_{sj}/\mathrm{M}K_{si} = 0.0002 < (1 + \lambda)/\theta\mathrm{MP}_{sij} = 0.00023.$$

Thus, according to our criterion for design in the presence of either form of the budgetary constraint, the extra \$100,000 spent on additional generators would not yield sufficient extra energy to justify itself. This is to be contrasted with the no-budgetary-constraint case, where substituting the same data into Eq. (4.14) yields

$$\mathrm{M}E_{sj}/\mathrm{M}K_{si} = 0.0002 > 1/\theta\mathrm{MP}_{sij} = 0.00018.$$

That is, the expenditure of the extra \$100,000 is justified by the marginal condition appropriate in the absence of the budgetary constraint. Viewed from a different angle, the budgetary constraint, through λ, tells us to forego extra benefits having a present value of \$113,280 ($\theta \times \mathrm{M}E_{sj} \times \mathrm{MP}_{sij}$) available at a cost of \$100,000 ($\mathrm{M}K_{si}$); this would, of course, be sheer folly in the absence of the budgetary constraint.

Unlike construction expenditure, the two forms of the budgetary constraint do not affect OMR outlays identically. Assume, for example, that an additional \$1000 a year spent in improving snowpack estimates (an OMR expense) would permit an annual increase of 200 acre ft in the output of irrigation water, worth \$6.00 per acre foot, and that, as before, $\lambda = 0.3$. Substituting these data into Eq. (4.15), $\mathrm{ME}_{sj} = 6$, $\mathrm{M}M_{si} = 1000$, $\mathrm{MP}_{sij} = 200$, and we have

$$\mathrm{M}E_{sj}/\mathrm{M}M_{si} = 0.006 > 1/\mathrm{MP}_{sij} = 0.005$$

for the no-budgetary-constraint case and the construction-cost-only form of the constraint. But from Eq. (4.13a) we obtain

$$\mathrm{M}E_{sj}/\mathrm{M}M_{si} = 0.006 < (1 + \lambda)/\mathrm{MP}_{sij} = 0.0065$$

for the form that includes construction-plus-OMR costs. In other words, an annual gain of \$1200 ($\mathrm{M}E_{sj} \times \mathrm{MP}_{sij}$) for an annual OMR cost of \$1000 ($\mathrm{M}M_{si}$) is justified in the absence of a budgetary constraint or when the constraint applies to construction cost only, but not when the constraint applies to OMR as well as to construction outlays.

These results are not at all unexpected. The budgetary constraint, regardless of form, reduces the levels of resource commitment to a system; the fact that the marginal benefit of a proposed addition to a system design exceeds its marginal cost is no longer sufficient to justify inclusion of the addition. Almost equally obvious is the differential impact of the two forms of the constraint. The constraint on construction costs impels planners to rely more heavily on future OMR expenditures as substitutes for present construction expenditures; for OMR outlays are required to meet only the basic test that marginal benefit exceed marginal cost. Thus an increment in construction outlay for lining irrigation canals might be foregone in favor of an increase in annual expenditure for cleaning the canals of brush growth, even though the present value of benefits per extra dollar of outlay for canal linings exceeds the benefits per extra dollar spent on cleaning. This does not happen under the form of the constraint that includes construction and OMR costs, because here the test imposed by the marginal conditions is symmetric with respect to the two types of outlay.

The difference between the tests imposed by the two forms of the constraint revolves about the Lagrangian multiplier λ, which we have yet to explore with the care that its central position commands. Suppose that $(\mathbf{y}, \mathbf{x}^0, \mathbf{x}')$ is an optimal input-output vector and that its components are subject to a set of variations that satisfy the production-function constraint. Then we have

$$\delta E = \sum_j \frac{\partial E}{\partial y_j} \delta y_j = \sum_j D(y_j) \delta y_j, \tag{4.16}$$

$$\delta M = \sum_i \frac{\partial M}{\partial x_i'} \delta x_i', \tag{4.17}$$

and
$$\delta f = \sum_i \frac{\partial f}{\partial x_i^0} \delta x_i^0 + \sum_i \frac{\partial f}{\partial x_i'} \delta x_i' + \sum_j \frac{\partial f}{\partial y_j} \delta y_j = 0. \tag{4.18}$$

For the construction-cost form of the constraint, we have, furthermore,

$$\delta \overline{K}_s = \delta K_s = \sum_i \frac{\partial K}{\partial x_i^0} \delta x_i^0. \tag{4.19}$$

By virtue of the optimality conditions (4.6a), (4.7a), and (4.8a), it follows that

$$\sum_i \frac{\partial f}{\partial x_i^0} \delta x_i^0 = -\frac{(1+\lambda)}{\mu} \sum_i \frac{\partial K}{\partial x_i^0} \delta x_i^0, \tag{4.20}$$

$$\sum_i \frac{\partial f}{\partial x_i'} \delta x_i' = -\frac{\theta}{\mu} \sum_j \frac{\partial M}{\partial x_i'} \delta x_i'. \tag{4.21}$$

and
$$\sum_j \frac{\partial f}{\partial y_j} \delta y_j = \frac{\theta}{\mu} \sum_j D_j(y_j) \delta y_j = \frac{\theta}{\mu} \sum_j \frac{\partial E}{\partial y_j} \delta y_j. \tag{4.22}$$

Substituting these expressions in Eq. (4.18), we obtain

$$\begin{aligned}\delta f &= -\frac{1}{\mu}\left[(1+\lambda) \sum_i \frac{\partial K}{\partial x_i^0} \delta x_i^0 + \theta \sum_i \frac{\partial M}{\partial x_i'} \delta x_i' - \theta \sum_j \frac{\partial E}{\partial y_j} \delta y_j\right] \\ &= -\frac{1}{\mu}[(1+\lambda)\delta \overline{K}_s + \theta \delta M - \theta \delta E] = 0. \end{aligned} \tag{4.23}$$

Therefore,
$$\theta(\delta E - \delta M)/\delta K = 1 + \lambda, \tag{4.24a}$$

that is, the ratio of the change in the present value of benefits net of unconstrained OMR costs, $\theta(\delta E - \delta M)$, to a change in the construction budget, δK, is $1 + \lambda$. Since in the absence of a constraint the value of this ratio is unity for an optimal design, the term λ measures the premium per dollar that construction outlay must earn to qualify for inclusion in the system. A value of λ equal to 0.3, for example, means that $1.00 of construction outlay must yield a stream of benefits with a present value net of operating costs of at least $1.30; in other words, it must yield a premium of $0.30. This explains why our marginal conditions led us to forego a gain of $113,280 available for a construction outlay of only $100,000; the minimum acceptable return would have been $130,000.

If we carry out the operation corresponding to Exps. (4.16) through (4.23) for the construction-plus-OMR form of the constraint, we obtain

$$\theta \delta E/(\delta K + \theta \delta M) = 1 + \lambda. \tag{4.24b}$$

In this case the ratio of the change in the present value of benefits to a change in construction *or* OMR costs is $1 + \lambda$. Not only construction outlay, as in Eq. (4.24a), but OMR outlay also, must earn $1 + \lambda$ dollars per dollar to qualify.

Equations (4.24a) and (4.24b) may not appear comparable in view of the difference in form of their left-hand sides. However, we can easily rewrite Eq. (4.24b) to fit the format of Eq. (4.24a) — that is, with the numerator containing benefits less OMR costs and the denominator containing only construction costs. Instead of Eq. (4.24b) we then have

$$\theta[\delta E - (1 + \lambda)\delta M]/\delta K = 1 + \lambda. \qquad (4.25)$$

If we compare Eq. (4.25) with Eq. (4.24a), we see that in requiring the ratio of a change in the present value of benefits net of OMR costs to a change in construction outlay to equal $1 + \lambda$, the construction-plus-OMR constraint reevaluates OMR outlay at $1 + \lambda$ dollars per dollar, whereas the construction-cost constraint values OMR outlay at its nominal cost. Thus the hypothetical snowpack expenditures pass muster under the one form of the constraint but not under the other.

The left-hand sides of Eqs. (4.24) and (4.25) both have the dimension of marginal benefit–cost ratios. However, the difference between the left-hand sides of Eq. (4.24a) and Eq. (4.25) on the one hand and Eq. (4.24b) — the traditional form of the ratio — on the other, should be borne in mind: in the traditional form, the numerator contains only benefit terms and the denominator only cost terms; in the presence of budgetary constraints, however, segregation of constrained outlay in the denominator and treatment of all other outlays as "negative benefits" in the numerator, as in Eq. (4.24a), derives naturally from the constraint.

We are now in a somewhat better position to appraise the grounds for choice between the construction-cost and the construction-plus-OMR forms of the constraint. The relative appropriateness of foregoing net benefits from construction expenditure alone, as opposed to foregoing net benefits from both construction and OMR outlays, depends on the reason underlying the government's willingness to forego benefits in imposing the constraint in the first place. If, for example, the constraint on outlay stems from a fundamental philosophical opposition to "big government," then, insofar as the government operates as well as constructs systems, the budgetary constraint properly applies to governmental economic activity over the long run, and the construction-plus-OMR form of the constraint is appropriate. If, on the other hand, the reason for the constraint is the more immediate objective of diversion of *present* resources from water-resource development to other ends, then only the immediate outlays need be limited.

This "time-horizon" principle of choice carries over to other cases of the constraint. Suppose, for example, the purpose of the constraint is to limit the cash outflow from the national treasury, rather than to limit the scale of national water-resource development. Then, as we have observed, feedback (revenue plus taxes) is properly deducted from outlay in computing the drain of each system on constrained funds. But there is still a choice to be made between a constraint on immediate outlays and a constraint on immediate plus future outlays, and it properly rests on whether governmental concern is limited to reducing pressures on the current budget or, instead, extends to the long-run outflow from the national treasury.

There are two cases, however, in which the decision between the short-horizon, construction-cost form and the long-horizon, construction-plus-OMR form of the constraint becomes unimportant. First, if the ratio of OMR expenditure to construction outlay is relatively invariable, then a constraint on construction cost alone limits OMR outlay as a by-product, and specific inclusion of OMR outlay in the constraint is superfluous. If, for example, annual OMR costs are necessarily 1 percent of construction outlay (there being, for instance, virtually no opportunity to substitute greater or smaller maintenance expenditure on generators for higher or lower quality generators), then a limitation of construction outlay to $100,000 would necessarily imply an OMR expenditure with a present value of $28,320 for an interest rate of 2½ percent and an economic life of 50 years.

Second, if the ratio of OMR costs to construction costs is very small, then regardless of the degree of variability there is little need for concern with OMR. If, for example, the annual OMR expenditure is of the order of only 0.1 percent of the construction cost, the present value of OMR outlay (for an economic life of 50 years and an interest rate of 2½ percent) is only $2.83 per $100 of construction outlay and is hardly worth our concern.

ALLOCATION OF A LIMITED BUDGET

We turn now from the design problem to the allocation problem, that is, from examination of principles for designing a single system within a given subbudget to examination of principles for dividing a budget into subbudgets for the systems that comprise the over-all development program. From now on, we no longer concern ourselves with the choice of the form of the constraint; we assume in the ensuing discussion that the effective constraint is on construction costs only.[4]

The present value of benefits net of unconstrained OMR outlays that can be obtained from each project s is dependent on the size of the subbudget $\overline{K}_s$ allotted to the project. More formally, we have represented this dependence by a function $B_s(\overline{K}_s)$. In all discussions of budget allocations in this chapter we assume this function to be strictly concave; that is, we assume that the marginal benefit-cost ratio, $dB_s/d\overline{K}_s = \theta(\delta E - \delta M)/\delta\overline{K}$, considered as a function of the subbudget, decreases with the size of the subbudget.[5] The allocation problem, as we have seen, is to divide the over-all budget $\overline{K}$ among systems in such a way that the total net benefits

$$\sum_{s=1}^{S} [B_s(\overline{K}_s) - \overline{K}_s] \tag{4.26}$$

are as great as possible subject to the budgetary constraint

$$\sum_{s=1}^{S} \overline{K}_s \leq \overline{K}. \tag{4.27}$$

The Lagrangian expression for this problem is

$$L = \sum_{s=1}^{S} [B_s(\overline{K}_s) - \overline{K}_s] - \lambda\left(\sum_{s=1}^{S} \overline{K}_s - \overline{K}\right). \tag{4.28}$$

Setting the partial derivatives of this expression equal to zero, we have

$$\partial L/\partial\overline{K}_s = dB_s/d\overline{K}_s - 1 - \lambda = 0, \qquad s = 1, \ldots, S$$

or $$dB_s/d\overline{K}_s = \theta(\delta E - \delta M)/\delta\overline{K}_s = 1 + \lambda. \qquad s = 1, \ldots, S \tag{4.29}$$

In other words, the marginal benefit–cost ratios should be the same for all systems.[6] The rationale of this condition is obvious. As long as the marginal benefit–cost ratios differ among systems, construction funds can be transferred from those systems with lower benefits per dollar of

[4] This assumption is made in order to avoid the duplication of equations required in the preceding section. The principles discussed in this and subsequent sections on budgetary constraints apply equally to any form of the constraint.

[5] Strict concavity can be identified also with a negative second derivative, that is, with the requirement $d^2B_s/d\overline{K}_s^2 < 0$. This condition appears essential to the proper functioning of the procedure for budget allocation discussed in this chapter. At least we have no guarantee that our assertions hold if the assumption of strict concavity is relaxed.

[6] More rigorously, this injunction holds only if all S systems are actually constructed. The condition $dB_s/d\overline{K}_s < 1 + \lambda$ replaces the condition (4.29) for systems not constructed.

outlay to those with higher returns, with a resulting increase in the over-all benefits of the water-resource program.

Now we can advance the final step in the explanation of our decision to forego extra energy generation that offered benefits of \$113,280 at a cost of \$100,000. Allocation of \$100,000 of the subbudget to project *s* for construction of the generators would have forced us to abandon other increments to the project with returns of at least \$130,000. Similarly, increasing the subbudget for system *s* by \$100,000 would have required curtailing outlay on some other system that could yield \$130,000 in benefits.

The design and allocation problems cannot be as neatly separated in practice as we have done here for the sake of clear exposition. The difficulty is that in our analysis of the allocation problem we use the function

$$B_s(\overline{K}_s) - {}_s\overline{K} \qquad (4.30)$$

which relates the net present value of benefits to the subbudget allocated to each system. Precise numerical definition of the present-value function would require planners to determine the net present value of benefits for each level of construction outlay, and this in turn would require them to determine the optimal design for each level of outlay. So great is the magnitude of this task that allocation of the over-all budget cannot be based on precise knowledge of the values of Exp. (4.30). Instead, allocation algorithms must be tailored to the data-gathering capacity of the planning process.

The conceptual formulation of the principles of optimal allocation of the over-all budget has not been wasted, however. The key to a workable allocation-design procedure is the Lagrangian multiplier λ. If we knew the value of λ to be, say, λ^0, we could divide the over-all budget into subbudgets for each system and design each system without further knowledge of the net present-value function (4.30). For with the value of λ known, each planner could design his system to maximize

$$\theta E_{sq}(\mathbf{y}_{sq}) - M_{sq}(\mathbf{x}_{sq}') - (1 + \lambda)K_s(\mathbf{x}_s^0) \qquad (4.31)$$

independently of all other systems; in so doing, each planner would automatically implement both the conditions of optimal design (4.6a), (4.7a), (4.8a) and the condition of optimal allocation of the over-all budget (4.29), and the budget would be exactly depleted. Design and allocation would then impose no difficulties beyond those present in the absence of a budgetary constraint. Evaluation of each dollar of construction outlay (assuming, as we are, that only these outlays are constrained) at the opportunity cost or shadow price of $1 + \lambda^0$ dollars — for example, \$1.30 for $\lambda = 0.3$ — instead of at the nominal price of \$1.00 would effectively solve the interdependence problem. Evaluation of outlay at the shadow price would substitute for a planner's "roving eye" in ensuring that no construction was undertaken on his system that would displace more worthwhile construction elsewhere; each planner would then be free to keep both eyes on the design of his own system.

Solution of the operational allocation-design problem thus requires determination of the true value of λ, that is, the value associated with a solution of Eq. (4.28). As a first step, consider the similarities and differences between the results of using the true value of λ (which we shall continue to denote by λ^0) and any other value, that is, any other nonnegative number. The marginal conditions (4.6a), (4.7a), (4.8a) and (4.29) are satisfied as well for any value of λ as for its true value λ^0. And if all planners use maximization of Exp. (4.31) as their design criterion and the same value of λ, whether or not it is in fact the true value λ^0, the requirement of Exp. (4.24) will be met since the cutoff benefit–cost ratio will be the same for each system. It will not be possible to transfer construction outlay from one system to another without reducing over-all net benefits. Indeed the only place where substitution of a wrong number for λ^0 matters is in the relationship to the available budget $\overline{K}_s$ of the funds required to construct all S projects that have been designed in accordance with criterion (4.31). If a value of λ lower than λ^0 is employed, that is, if a smaller premium is placed on construction outlay than the solution of Eq. (4.28) would dictate, the cutoff benefit–cost ratio for each system becomes less than the benefit–cost ratio associated with Eq. (4.28). As a result, systems are designed larger, and the cost of the designs exceeds the available budget for the program as a whole. Similarly, if a value of λ higher than λ^0 is used, systems are designed too small, and the projected cost of the program fails to exhaust the budget.

The term shadow price for $1 + \lambda$ (or shadow premium for λ) is not misplaced. Let us imagine allocating the over-all budget through the sale of construction funds by a central office to the field offices in charge of projects. The central office acts as a perfect competitor in "selling" any amount of construction funds to any field office at the going price. The field offices also act as competitive businesses in purchasing construction funds only as long as it is "profitable" for them to do so, that is, only as long as the net present value per dollar of construction outlay evaluated at the shadow price is positive. Indeed, the major substantive difference between our pseudo market and an actual competitive market lies in the manner in which the shadow price of construction outlay is set for the purpose of bringing "demand" (the amounts of funds required by field offices to construct the designs dictated by maximization of net present value at any shadow price) and "supply" (the amount of the over-all budget) into line. In a competitive situation, the interaction of a large number of buyers and sellers determines the price. In our case, however, whether or not the number of buyers (the field offices) is large, there is only one seller (the central office). Therefore an iterative, trial-and-error procedure like the following must replace the invisible hand of the market.

The central office places a value λ^1 (for the time being, any nonnegative number) on the shadow premium as a first approximation to λ^0. Field offices design their respective systems in order to maximize net present value with construction outlay evaluated at the shadow price of $1 + \lambda^1$. Field offices record with the central office their demands for funds for construction of designs based on the shadow price, and the central office compares the sum of these demands with the available budget. Three possibilities ensue. First, the sum of the field-office demands may just equal the available budget. The central office would then allocate to each field office the amount demanded and close up shop. The second possibility is that the sum of the demands exceeds the available budget. This indicates, as we reasoned above, that the shadow price has been set too low. Just as the price in a competitive market rises in response to a temporary excess of demand over supply, so the shadow price of construction funds must be raised by deliberate action of the central office in order to reduce the apparent benefits (that is, the benefits with outlay evaluated at shadow rather than nominal prices) of system units and thus lower the demands of field offices for funds. The third possibility is, of course, the opposite of the second, and the central office responds accordingly by lowering the shadow price in order to encourage expansion of system designs. Under either the second or third possibility, the central office estimates a new shadow price $1 + \lambda^2$, and the field offices design their systems and calculate their demands anew. This process is repeated until an estimate λ^n is sufficiently close to the true value λ^0 for the total funds demanded by field offices at the price $1 + \lambda^n$, which we denote by

$\sum_s K_s(\mathbf{x}_s{}^0; \lambda^n)$, to differ from the available budget $\overline{K}$ by less than any specified amount, for example, less than 1 percent of the budget.[7]

Thus far we have described only how the direction of change of λ^n to λ^{n+1} is determined. How can we be sure that the estimates $\lambda^1, \ldots, \lambda^n, \ldots$ will eventually get close enough to λ^0 for the total costs of designs to be essentially equal to the total budget? More formally, how do we know that the process we have described converges, λ^n approaching λ^0 and $\sum_s K_s(\mathbf{x}_s{}^0; \lambda^n)$ approaching $\overline{K}$? The answer we shall give is not wholly satisfactory, and indeed becomes even less so when we extend our discussion to the dynamic problem of multiple budgetary constraints covering future as well as present construction.

If we make our changes in λ^n according to the equation

$$\lambda^{n+1} - \lambda^n = \phi \left[\sum_s K_s(\mathbf{x}_s{}^0; \lambda^n) - \overline{K} \right]$$

where ϕ is a positive number, the λ^n's will converge to λ^0 and the budget allocations will converge to the available budget, provided only that, as assumed, the benefit function Exp. (4.30) is strictly concave and that ϕ is small enough.[8] In other words, the process converges for strictly concave benefit functions if we make each change in λ proportional to the difference between supply and demand, and if we make the constant of proportionality sufficiently small. Why need we make ϕ small? The point is that as we change λ, we may overshoot the mark, going from λ^n lower than λ^0 to λ^{n+1} higher than λ^0 or vice versa. This in itself presents no problem; indeed, it is to be expected. However, if we change λ^n too radically, not only will we overshoot the mark, but we may also end up with a worse estimate in λ^{n+1} than we had in λ^n, in the sense that the difference between the demand and supply for funds $\left| \sum_s K_s(\mathbf{x}_s{}^0; \lambda^{n+1}) - \overline{K} \right|$ may be greater than the previous difference $\left| \sum_s K_s(\mathbf{x}_s{}^0; \lambda^n) - \overline{K} \right|$. Moreover, each successive estimate may prove worse than the preceding one.

Why not make ϕ very small and be done with the problem? The

[7] Alternatively, we stop the redesign-recalculation process if λ^n falls to zero and the budget nevertheless is not depleted. This indicates that the budgetary constraint is not binding and can be ignored.

[8] See Kenneth J. Arrow et al., *Studies in linear and nonlinear programming* (Stanford University Press, Stanford, 1958), pp. 9–14 and chaps. 6–11, especially chap. 10, pp. 157–159. To take account of the possibility of a nonbinding budgetary constraint, we should actually use $\lambda^{n+1} = \max\{\lambda^n + \phi[\sum_s K_s(\mathbf{x}_s{}^0; \lambda^n) - \overline{K}], 0\}$.

reason is the desirability of speedy convergence. A large number of iterations can be expensive. By choosing a larger value of ϕ, we increase the size of the jumps between successive values of λ^n and thereby the rapidity of convergence — if we are successful in avoiding the pitfalls associated with "overshooting."

Two considerations reduce the urgency of the dilemma involved in the choice of ϕ. First, it is easy to tell whether the sequences of values of λ^n and the discrepancies between demand and supply of funds are converging or diverging for any given value of ϕ merely by inspecting successive terms $\left|\sum_s K_s(\mathbf{x}_s^0; \lambda^n) - \overline{K}\right|$ and $\left|\sum_s K_s(\mathbf{x}_s^0; \lambda^{n+1}) - \overline{K}\right|$. If the "distance" (that is, the difference) between demand and supply for λ^{n+1} is greater than the distance for λ^n, ϕ should be reduced. Second, since several iterations may be required in order to secure convergence, designs should be only gradually refined. It would be foolish for field offices to design in full detail based on the first shadow premium λ^1. Only when enough iterations have been carried out for supply and demand to come reasonably close together would field offices be justified in drawing detailed plans.

Although the iterative procedure described above will work for any choice of λ^1, the speed of convergence depends greatly on the proximity to λ^0 of the first approximation λ^1. The procedure we propose for determining the initial value λ^1 approximates the maximization of Exp. (4.26) subject to (4.27); the value of λ^1 appears as a by-product of the solution of a simple linear-programming problem, similar to the manner in which the Lagrangian multiplier is a by-product of the solution of the original allocation problem. More precisely, λ^1 appears as the direct result of the "dual problem" associated with the allocation problem by a basic theorem of linear programming.

In outline the procedure is this. Field offices make a small number of "broadbrush" estimates of the net present value $B_s(\overline{K}_s) - \overline{K}_s$ obtainable for levels of construction outlay evenly spaced over the plausible range for their systems, that is, over the range 0 to Z_s, where Z_s represents an estimate of the construction outlay dictated by optimal design in the absence of a budgetary constraint. Here estimates are even cruder than in the preliminary iterations, since they are based on rough field studies and surveys. Field offices then calculate by linear interpolation the net present value of construction outlays that lie between those selected for direct estimation. The result is a set of piecewise linear functions that can be represented mathematically to fit the linear-programming format.[9] The use of piecewise linear functions permits us to circumvent the conventional limitation of linear programming—that it is applicable only to linear functions which represent constant returns to scale—and apply it to our problem involving nonlinear functions.

These piecewise linear functions are transmitted to the central office, which then uses them to formulate the linear-programming problem. As we have observed, the solution to this problem yields the initial approximation λ^1 to the shadow premium of construction outlay. Although it may be far from the true value, the approximation λ^1 is certainly much better than pure guesswork, and the savings that result from reducing the number of iterations required for near-convergence of the sequences λ^n and $\left|\sum_s K_s(\mathbf{x}_s{}^0; \lambda^n) - \overline{K}\right|$ are undoubtedly greater than the cost of the few crude estimates and computations needed to determine λ^1 by the method proposed.

THE DYNAMICS OF WATER-RESOURCE DEVELOPMENT [10]

Water-resource-planners are accustomed to considering the relationship of the benefits of a project to its age. Most projects go through a maturation period in which economic activities expand to take advantage of the project services; during this period the project yields few benefits in comparison with its future potential. After economic activity has grown sufficiently and the project has entered its adult stage, its increasing age remains relatively unimportant until OMR costs rise to the point where the economic viability of the project is threatened.

A second determinant of the rate of benefits of water-resource developments (and, indeed, of investment projects generally) is calendar time. To illustrate the influence of calendar time, imagine three water-resource systems (for example, three TVA's), identical physically, hydrologically, and in age, but operating in 1941, 1961, and 1981. In spite of the assumed identity of the systems, their annual benefits would be different in the three years. The source of this difference is what we

[9] This is true provided we assume that the direct estimates are accurate enough for the approximating functions to reflect the assumed strict concavity, that is, the decreasing marginal benefits ($d^2B_s/d\overline{K}_s{}^2 < 0$) of the unknown "parent" $B_s(\overline{K}_s) - \overline{K}_s$ functions.

[10] The following section summarizes briefly some of the problems explored in a dynamic investment model prepared by the author of this chapter. The full study is to be published by the North Holland Publishing Co., Amsterdam, under the title *Approaches to dynamic investment planning.*

call the influence of calendar time. This is not to say that calendar time has a metaphysical influence on system benefits. Rather, it is a convenient label for describing those forces that, quite apart from the age of the project, lead to changes over time in the demands for, and hence the benefits from, identical quantities of identical outputs.

The causes of change in demand for system outputs over time fall into two broad categories, those affecting the demand for system outputs and their close substitutes symmetrically, and those stemming, by contrast, from changes in the availability and cost of substitutes and thus affecting the demand for system outputs and their substitutes asymmetrically. Examples from the first category are changes in population and in per capita income, and the emergence of new products or new processes for making old products. Examples from the second category are technologic progress, exhaustion of available sources of supply for alternatives, and discovery of new alternative sources of supply.

Consider, for example, how these factors affect the demand for a system's output of electric energy. Changes in population tend to change demand in the same direction. More people use more refrigerators, more lightbulbs, more electric ranges, and, perhaps more important, more industrial products in the production of which electricity plays an essential role. Similarly, changes in per capita income tend to produce changes in demand in the same direction. Greater incomes permit people to consume more of the products for which electric energy is essential, either as a factor of production or as a complement in consumption. The emergence of new products or new processes (or the improvement of old ones) can increase or decrease the demand for energy. The production of fuel for atomic energy, for example, has increased manyfold the demand for TVA energy. The wide use of air-conditioning and electric home-heating have had a similar impact on residential and commercial demand. On the other hand, the introduction of new products such as the fluorescent light and the transistor have tended to decrease the demand for electric energy because of their reduction in heat wastage.

The effect on demand for system energy of changes in the cost and availability of substitutes is well known. Technologic progress that lowers the cost of steam energy will clearly tend to reduce the demand for system hydroelectric energy, as would the discovery of new, relatively inexpensive, sources of the raw materials employed in the production of steam energy. By contrast, exhaustion of the available supplies of these raw materials would increase their cost of production and hence increase the demand for hydroelectric energy.

In the sections that follow we shall examine the impact on the planning of water-resource developments of changes in demand over time, or as we choose to put it, of the dependence of the benefit rate on calendar time. We shall first discuss the problem in terms of a single project, subsequently in terms of a group of projects associated with one another by virtue of a common budgetary constraint. This discussion will be by no means exhaustive. It is intended rather to point the way to dynamic planning.

DYNAMICS IN THE ABSENCE OF BUDGETARY CONSTRAINTS

To simplify the discussion, we shall begin by considering indivisible projects only, that is, projects which, if built at all, will be built to only one scale. If calendar time were irrelevant in determining benefit rates, the planning of these projects in the absence of budgetary constraints would require only that we compare the present value of benefits net of OMR costs with construction outlay. If these benefits exceeded the construction outlay, the project would be approved. The calendar-time factor, however, forces us to decide when to build as well as whether to build. Possibly the economic merit of a project can be improved by postponing its construction, and the postponement may change the net present value of a project from a negative figure for construction today to a positive value for construction at some future date.

To illustrate this numerically, assume that we wish to maximize the present value of a project with a projected construction cost of \$1000 and projected benefits of \$10 per year in the years from 1961 through

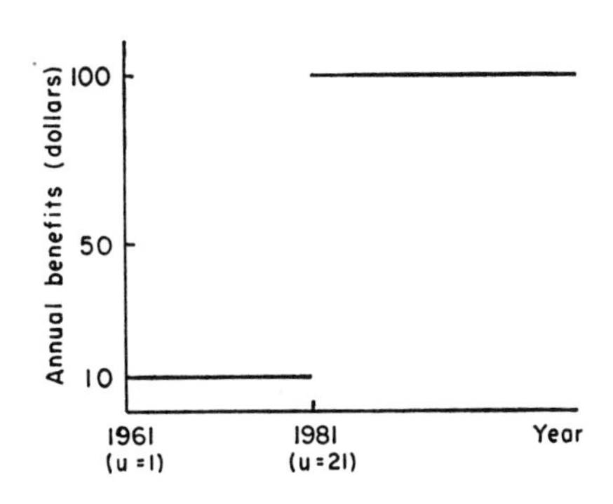

Fig. 4.1. Annual benefits of a water-resource project as a function of calendar time.

1980 and \$100 per year thereafter for the identical outputs. This benefit-rate function is presented graphically in Fig. 4.1. For greater conceptual and computational simplicity, let us assume: (1) that the construction outlay of \$1000 remains the same regardless of the year in which the project is built, and that this outlay is incurred on the first day of the year of construction (thus, the present value of outlay for construction in the tenth year, 1970, would be $\$1000 \times (1+r)^{-9}$, where r is the interest rate); (2) that project age has no influence on the benefit rate (the benefits in 1984, for example, are \$100 whether the project is one or 20 years old); (3) that the project yields benefits indefinitely;[11] (4) that all benefits accrue on the last day of the calendar year and commence in the year of construction (thus, the present value of benefits in the tenth year would be $\$10 \times (1+r)^{-10}$); (5) that the interest rate is 5 percent; and (6) that benefit rates are net of OMR costs.

If this project were constructed today, therefore, the present value of gross benefits[12] would be \$878.40; the net present value for present construction is therefore −\$121.60, and it is clear that a prudent investor would reject construction of the project for the present.

But what about construction in future years? Let us look at the cost side first. The present value of outlay for construction in year u (that is, u years from the present, counting 1961 as year one) is $\$1000 \times 1.05^{-(u-1)}$. Similarly, the present value of the cost of construction in year $u+1$ is $\$1000 \times 1.05^{-u}$. Accordingly, the difference between the present value of the cost of construction in year u and in year $u+1$ is

$$1000\,[1.05^{-(u-1)} - 1.05^{-u}]$$

$$= 1000 \times 0.05 \times 1.05^{-u} = 50 \times 1.05^{-u}. \qquad (4.32)$$

[11] If the benefit rate is independent of project age, the concept of a finite economic life of the project is without meaning. There might be some date after which no benefits would be produced, but this date would, by assumption, be independent of project age.

[12] The computations are as follows.

(*a*) The present value of benefits from 1961 through 1980 is given by the formula for the present value of a 20-yr annuity:

$$0.05^{-1} \times (1 - 1.05^{-20}) \times 10 = 124.62$$

(*b*) The present value of benefits from 1981 on is obtained by calculating the present value of a perpetuity beginning 20 years hence:

$$1.05^{-20} \times 0.05^{-1} \times 100 = 753.78$$

(*c*) The present value of total benefits is the sum of the results of (*a*) and (*b*):

$$124.62 + 753.78 = 878.40$$

This value represents the saving in the present value of cost gained by postponing construction from year u to year $u + 1$. These savings in interest cost can be viewed as the present value of the net benefits gained by placing the construction outlay, made available for one year by postponement of construction, in an alternative investment with an economic life of one year and a rate of return equal to the interest rate of 5 percent.

On the benefit side, postponement of construction from year u to year $u + 1$ results simply in the loss of the benefits from year u. The loss in the present value of gross benefits caused by postponement is thus

$$10 \times 1.05^{-u} \text{ for } u = 1, \ldots, 20 \text{ (that is, from 1961 through 1980)} \tag{4.33a}$$

and

$$100 \times 1.05^{-u} \text{ for } u = 21, \ldots \text{ (that is, from 1981 on).} \tag{4.33b}$$

It follows that the change in net present value resulting from postponement of construction from year u to year $u + 1$ is the difference between the saving in costs and the loss in benefits obtained by subtracting Exps. (4.33) from (4.32) as follows:

$$(50 - 10 = 40) \times 1.05^{-u} \text{ for } u = 1, \ldots, 20 \tag{4.34a}$$

and

$$(50 - 100 = -50) \times 1.05^{-u} \text{ for } u = 21, \ldots. \tag{4.34b}$$

From Exp. (4.34a) we see that until 1981 the marginal net present value of delaying construction is positive — the marginal saving in interest costs exceeds the marginal loss in benefits by $\$40 \times 1.05^{-u}$. Thus until 1981 the net present value of the project increases with each year that construction is postponed. In 1981, however, because of the increase in the benefit rate, the marginal net present value of delaying construction changes sign abruptly. After 1981 the annual benefits exceed the annual interest costs, and, as shown in Exp. (4.34b), net present value decreases for each year thereafter that construction is postponed. The maximum net present value is thus obtained by constructing the project in 1981, the year in which the increment in net present value changes sign from positive to negative. The project in fact shows quite a handsome payoff for 1981 construction, with a net present value of \$376.89 and a ratio of the present value of benefits to costs of 2:1.

Figure 4.2 shows the relationship of net present value in 1961 to the year of construction. We call special attention to the fact that the net present value in 1961 increases with delays in the date of construction for 17 years after the project is first "justified," that is, after it first attains a positive net present value. That postponement of construction can increase an already positive present value is as significant as that it can change a negative present value into a positive value.

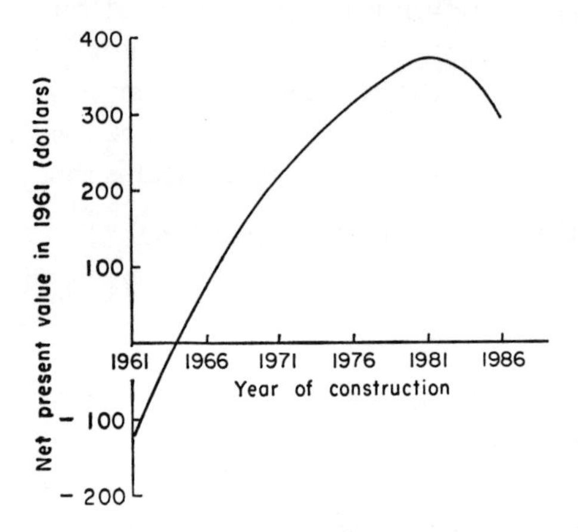

Fig. 4.2. Net present value of benefits from construction of a water-resource project as a function of year of construction.

The starkness of these results and the ease with which they were obtained are admittedly due in part to our omission of the role of project age in determining the benefit rate and to the abruptness with which we changed the calendar-time-determined benefit rate in 1981. Nevertheless, the lesson to be learned has general validity. It is possible to demonstrate analytically, although we shall not do so here, the desirability of postponing construction in a wide range of cases in which the project benefit rate depends jointly on calendar time and project age.[13] Indeed, it is only when the benefit rate depends on project age alone (or, more generally, when demand is decreasing over time so that the calendar-time component of the benefit-rate function is decreasing) that we can be sure that construction, if justified at all, is optimally undertaken in the present. In other cases, when demand is increasing over time, we cannot ignore future construction possibilities and should not, as traditional benefit–cost analysis would have us do, focus our attention on the single question of whether or not the net present value of a project is positive for immediate construction.

Use of indivisible projects in our discussion might at first appear to place a severe limitation on the applicability of our results to actual

[13] See Marglin, reference 10.

planning decisions, for an essential characteristic of water-resource developments is their divisibility. Indeed, this book owes its existence to the great variety of possible designs for potential river-development systems. However, the indivisible project of our discussion can be considered a particular increment of a project rather than the entire project. With this substitution our results are the following. First, in a wide number of cases, postponement of construction of increments, if not whole projects, will increase the present value of net benefits. Second, only in the event that the benefit rate is totally independent of calendar time or that the calendar-time component of the benefit rate is falling over time can we be sure that construction of all increments will be less desirable in the future than it is today.

Determining the optimal schedule for stage construction of a divisible project (that is, choosing construction dates for each of the increments) is, of course, far more complicated than choosing a single construction date for an indivisible project. Many of the increments of a divisible project are mutually exclusive — an increment devoted to municipal water supply, for example, cannot be employed for irrigation water since both are consumptive uses. Moreover, divisibility introduces not only the opportunity to construct the different component units of the project at different times, but also to construct individual units in stages. And the construction of individual units in stages introduces new cost considerations. Against the saving in interest cost of constructing a unit in stages must be measured not only the benefits lost by postponing construction of the increment, but the additional absolute costs of stage construction as well. It is more expensive, for example (although how much more varies greatly from case to case), to build a dam to a height of 200 ft and later raise it to 300 ft than to build it to 300 ft in one fell swoop.

The complications of determining the optimal schedule of construction of the increments of a project are rendered more imposing by the fact that scheduling cannot be separated from design. We cannot, for example, determine the optimal design for present construction and then determine the optimal sequence of construction of the component increments, for it is often the case that the optimal design for present construction does not remain optimal when we remove the restrictive assumption that the entire development must be undertaken at once.

In spite of the complexities introduced by dynamic considerations, at least one of the tools presented in Part II of this book for determining the optimal design of a system can be readily extended to determining the optimal design-plus-construction schedule. Preliminary experiments have shown that it is relatively easy to expand simulation by machine programming to allow for scheduling of separate units at different times. Inclusion of stage construction of individual units is more difficult, since it requires the introduction and manipulation of additional information relating to the costs of this type of construction; nevertheless, no insurmountable obstacles are evident.

The important point is that the dynamic aspects of design resulting from the dependence of benefits on calendar time must be introduced systematically into the planning procedure if the present value of net benefits is to be maximized. One suspects that the timing of incremental development of many large water-resource undertakings in the United States would have been modified had the role of calendar time been clearly understood rather than seen "through a glass, darkly."

DYNAMICS IN THE PRESENCE OF BUDGETARY CONSTRAINTS

Dynamic planning becomes even more complicated in the presence of budgetary constraints. In its simplest form, the problem is as follows. The investor (that is, the government agency) has a set of projects and is obliged to honor a constraint on the amount of money that can be spent today on their construction. (We assume that the budgetary constraint on immediate outlay is effective and precludes construction of all projects today, even if immediate construction is optimal for each project viewed independently.) The investor's problem, in this case, is not merely to choose which projects are to be constructed today, for it is a rare budgetary constraint that is "one shot" in nature and dooms the investor to inaction after a single burst of construction activity. In general, opportunities not selected for immediate development are not lost forever; money may be expected to become available next period and in subsequent periods to construct projects until all worthwhile development has been undertaken. Thus the investor's problem is not that of choice among projects, but rather that of determining the sequence in which projects shall be undertaken.

Traditional methods of planning in the presence of budgetary constraints fail to take systematic account of the sequential aspect of the decisions being made. For this reason they can lead to incorrect decisions, that is, sequences with net present value less than the maximum possible when the project benefit rate depends on calendar time as well as project age. As an example consider the following decision rule, adapted from the familiar "Green Book."[14] This manual tells us to rank the projects in descending order of their net present value[15] for "period-one" construction. The investor then goes down this list, constructing as many projects as the budget for the first period permits. Being a "one-shot" algorithm, the Green Book procedure gives no further explicit instructions for the remaining projects. Implicitly, however, the investor is invited to repeat the given instruction in the second period (and then in each succeeding period) for these projects.

In determining the projects to be constructed in a given period, the Green Book considers only the net present values for construction in that period. The fallacy inherent in this procedure is a familiar one to economists: the Green Book concentrates on the absolute advantage of projects in each period instead of looking at the comparative advantage of projects in different periods. A numerical example will show this.

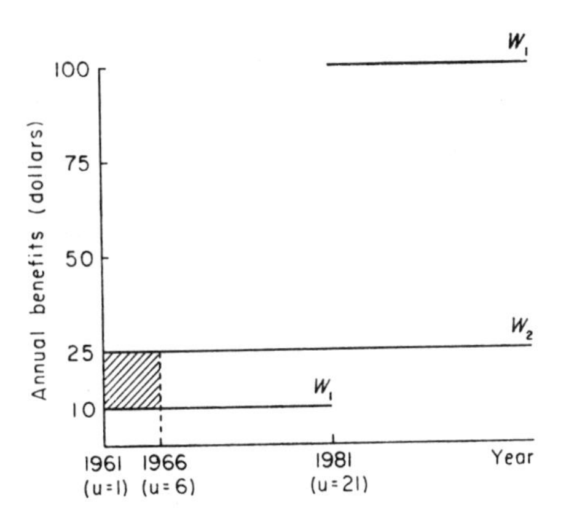

Fig. 4.3. Annual benefits of water-resource projects W_1 and W_2 as a function of calendar time. The shaded area represents the difference between the absolute benefits lost by postponing W_2 and those lost by postponing W_1.

Suppose that the group of projects consists of just two, W_1 and W_2, and that the following data and assumptions apply: (1) the projects are indivisible; (2) each costs the same amount, \$150, which remains constant

[14] *Proposed practices for economic analysis of river-basin projects*, a report to the Inter-Agency Committee on Water Resources by its Subcommittee on Evaluation Standards (May 1958).

[15] The Green Book actually suggests ranking projects in descending order of the ratio of annual benefits to annual amortized costs. Although this ratio is ambiguous when a project's benefit rate is not constant, ranking by this device is equivalent to ranking by the generally unambiguous ratio of the *present value* of benefits to the *present value* of costs in the event the projects being compared have constant benefit streams. The present-value ratio reduces to our ranking by present values alone under our assumption (see below) that the cost of all projects is the same.

regardless of the year of construction; (3) the outlay is considered to be made on the first day of the year of construction; (4) all benefits are considered to accrue on the last day of the calendar year and to commence in the year of construction; (5) the annual benefits from W_1 are the same as those from our previous example on page 179, while those from W_2 are constant at \$25; (6) the interest rate is 5 percent; and (7) the available budget is \$150 in 1961 and \$150 in 1966 (thus if one project is constructed this year, the other must be postponed for five years). The benefit-rate functions (which are net of OMR costs) are presented graphically in Fig. 4.3.

The net present value of each project, viewed in isolation, is maximized for immediate construction.[16] However, the budgetary constraints, as we observed before, permit construction of only one project now, with the other one forced to wait until 1966. The Green Book procedure tells us to construct the project now with the higher net present value for immediate construction. Straightforward computations reveal that construction of W_1 in 1961 — abbreviated (W_1, 1961) — yields a net present value of \$728.40, while (W_2, 1961) yields a net present value of only \$350. Accordingly, the Green Book procedure assigns W_1 to 1961 construction, and W_2 is residually assigned to 1966. Since the net present value of (W_2, 1966) is \$274.05, the sequence of (W_1, 1961) and (W_2, 1966) determined by the Green Book gives a total net present value of \$1002.45. On the other hand, (W_1, 1966) yields a net present value of \$718.61 which, added to the net present value of (W_2, 1961), \$350, gives a value of \$1068.61 to this alternative program, or over \$65 more than the program determined from the Green Book.

A glance at Table 4.1, which relates net present values today to

Table 4.1. Present value (1961) of net benefits of projects W_1 and W_2 construction in 1961 and 1966.

	Net present value in construction period —		
Project	1961	1966	Deferral cost
W_1	\$728.40	\$718.61	\$9.79
W_2	350.00	274.05	75.95

[16] This can be verified by direct comparison of the present value of construction in successive periods, in the manner of Fig. 4.2, or by substitution of the assumed cost and benefit data in Exps. (4.32) to (4.34). Both procedures reveal that the net present value of each project viewed in isolation decreases with postponement of construction.

construction periods in easy-to-read form, shows why. Clearly it is not the relative magnitude of the net present values of W_1 and W_2 for construction in 1961 that determines the optimal assignment of the projects to the two construction periods, but rather the relative loss in net present values caused by postponement of construction. Or, as we put it earlier, it is not the absolute advantage of one project over another at a given time, but rather the comparative advantage in different periods that determines the optimal sequence. Project W_1 loses less than \$10 in net present value if its construction is postponed from 1961 to 1966, whereas delaying construction of W_2 to 1966 causes its net present value to shrink by over \$75. Thus W_2 has a comparative advantage for 1961, W_1 for 1966.

The basis of the comparative advantage of W_2 for earlier construction is clearly expressed in Fig. 4.3. By our budgetary assumptions, both projects will be built by 1966; for this reason, and because of the additional assumptions that the benefit rate of each is determined only by calendar time and that the absolute construction cost of each remains constant over time, the benefits in the years between 1961 and 1966 alone determine the relative losses in net present values of postponing construction of the two projects. The shaded area of Fig. 4.3 represents the difference between the absolute benefits lost by postponing W_2 and the absolute benefits lost by postponing W_1. Discounted by the appropriate interest factor, this difference is the net gain (about \$65) for the program (W_2, 1961) and (W_1, 1966) over the alternative program.

Linear programming can be employed to determine the optimal assignment of projects to construction periods in problems more complicated than the two-project, two-period example just presented, that is, problems in which trial-and-error or complete-description methods for examining all feasible sequences would break down because of the number of sequences to be examined. For the present we shall not look further into the mathematics of choosing the optimal assignment of projects in the presence of budgetary constraints.[17] Once again, we shall merely state the most important results and their impact on the design of water-resource systems.

The principal lesson to be learned is the importance of a proper

[17] Our very use of the term "assignment" will suggest to readers familiar with linear programming the category of linear-programming problems into which our problem falls: the assignment of projects to construction periods is formally similar to the standard linear-programming "assignment" problem of allocating factories to locations or personnel to jobs. See Marglin, reference 10, for a more extensive discussion.

conception of the planning problem imposed by budgetary constraints. It is not, as usually formulated, one of choice among projects (or project increments). Rather it is almost invariably one of the sequence in which projects (or increments) are to be built. Thus, planning procedures that, like the Green Book procedure, are based on comparisons of the absolute advantage of projects within the same period, must give way to procedures that, like the linear-programming procedure, take comparative advantage into account in determining the sequence in which projects will be constructed.

Under certain conditions, however, the investment decision can be treated as a simple choice problem for which the Green Book procedure will lead to an optimal assignment of projects to construction periods. Stated most generally and most simply, the Green Book procedure is valid when absolute advantages among projects — the basis for the Green Book procedure — happen to coincide with comparative advantages for earlier construction — the proper basis for assignment of projects. If we arrange projects in descending order of net present value for period-one construction, it is sufficient that each project have a comparative advantage for earlier construction over projects below it on the list. If the benefit-rate functions of all projects are independent of calendar time, that is, if the demand for project outputs is constant over time and the benefit rate depends only upon project age, then this general condition is satisfied and the Green Book procedure of assigning projects to construction periods is optimal.

The parallel between this result and the corresponding result of the previous section on dynamics in the absence of budgetary constraints deserves further comment. In both cases the use of static planning procedures that ignore future construction possibilities in deciding which projects (or increments) to undertake today is justified by dependence of the benefit rate on project age only. It is likely, therefore, that the continued acceptance of traditional static planning procedures is owed to failure to appreciate the importance of calendar time in determining project benefit rates.

This discussion of dynamic planning in the presence of budgetary constraints has been limited to the planning of indivisible projects. As we have indicated parenthetically, when we wish to apply the results to divisible projects, we consider these projects as being composed of a number of indivisible project increments.

DYNAMICS AND PLANNING PROCEDURE IN IMPLEMENTING BUDGETARY CONSTRAINTS

The problem to be considered in this section is the allocation of the budgets of several time periods among systems and the design and scheduling of project development in a way that will maximize the over-all net present value of the program. The solution combines the results of our studies of the allocation of multiple-period budgets among indivisible projects and of the allocation of a single budget over divisible projects. The central office determines the appropriate shadow price for the budget of each construction period. These shadow prices are then used by field offices to design projects and schedule construction.

The determination of multiple shadow prices is, as one would expect, a more complex problem than that of determining a single shadow price. The prices are interrelated, and a change made in one in the process of convergence affects all others. Nonetheless, the procedure for implementing multiple budgetary constraints follows the procedure for a single constraint, and relies as before on the assumption of concave benefit functions. The first approximations $\lambda_1^1, \lambda_2^1, \ldots, \lambda_T^1$ to the true shadow premiums $\lambda_1^0, \lambda_2^0, \ldots, \lambda_T^0$ for T construction periods are derived by linear programming, the necessary benefit–cost data being compiled from rough field-office estimates of the net present values obtainable for selected levels of outlay in each construction period. As in the case of a single budget and a single construction period, the net present values obtainable from levels of outlay falling between those selected for direct estimate (including combinations of outlay in more than one period) are estimated by linear interpolation between the direct estimates. The form of the linear program is essentially a composite of the dynamic linear program for the assignment of indivisible projects to construction periods and the static linear program for determination of the first approximation to the shadow premium associated with a single budgetary constraint and construction period.

The procedure for determining second and successive approximations to the shadow premium is analogous to that used in the case of the single budgetary constraint. As before, designs and construction schedules are refined only as the supply and demand for funds in each period begin to converge.

The question of how much to raise or lower shadow premiums in each iteration can be answered in basically the same way as in the single-budget case. Consider the algebraic formula for deriving the $(n+1)$th approximations from the nth and the demands for funds generated by these values:[18]

$$\lambda_t^{n+1} = \lambda_t^n + \phi\left[\sum_{s=1}^{S} K_{st}(\mathbf{x}_{st}^0; \lambda_1^n, \lambda_2^n, \ldots, \lambda_T^n) - \overline{K}_t\right]. \quad t = 1, \ldots, T$$

In this equation $K_{st}(\mathbf{x}_{st}^0; \lambda_1^n, \lambda_2^n, \ldots, \lambda_T^n)$ represents the construction outlay required for project s in construction period t to maximize the shadow net present value for the approximate shadow prices $1 + \lambda_t^n$, $1 + \lambda_2^n, \ldots, 1 + \lambda_T^n$ and $\overline{K}_t$ represents the budget for period t. Limitation of ϕ to positive values ensures that the direction of change of shadow premiums properly accords with the relationship of the demand and supply of funds in the respective period, that is, that shadow premiums increase when the demand for funds exceeds the supply and decrease when the supply exceeds the demand.

The magnitude of ϕ determines the size of the change in shadow premiums in response to a given discrepancy between supply and demand. As in the single-period prototype of this planning procedure, there are two conflicting desiderata in the determination of ϕ: speedy convergence, which by itself leads to large changes in the shadow premiums and hence a high value of ϕ, and avoidance of divergence, which suggests small changes in the λ_t^n's and hence a small value of ϕ. The prospect of divergence is much more likely than in the single-constraint case. There is, as in the earlier case, the danger that use of too large a value for ϕ will overshoot the mark by so much that all values of λ_t^{n+1} become worse approximations than the previous ones in the sense that the discrepancies between demand and supply $\left|\sum_{s=1}^{S} K_{st}(\mathbf{x}_{st}^0; \lambda_1^{n+1}, \lambda_2^{n+1}, \ldots, \lambda_T^{n+1}) - \overline{K}_t\right|$ for the new approximations are each larger than the respective discrepancies $\left|\sum_{s=1}^{S} K_{st}(\mathbf{x}_{st}^0; \lambda_1^n, \lambda_2^n, \ldots, \lambda_T^n) - \overline{K}_t\right|$ for the old. But there is a more subtle danger. We might find that as we close the gap a bit between demand and supply in some periods, it becomes much larger

[18] This formula is not quite accurate. To prevent shadow premiums from becoming negative, we must employ the more precise formula $\lambda_t^{n+1} = \max\left\{\lambda_t^n + \phi\left[\sum_{s=1}^{S} K_{st}(x_{st}^0; \lambda_1^n, \lambda_2^n, \ldots, \lambda_T^n) - \overline{K}^t\right], 0\right\}$.

in others. This increases the likelihood of divergence and suggests that we should lean more to the conservative position of choosing a small value for ϕ.

As in the single-period case, the choice of ϕ is not irrevocable. If the sequences of values for λ_T^n and the discrepancies between demand and supply of funds in the various periods tend to diverge, ϕ can be reduced. And a tendency to divergence is not much more difficult to spot than in the one-dimensional case. It is indicated by an increase in the over-all "distance" between supply of and demand for funds in all periods for successive values of the λ_t^n's. This distance can be measured by the sum of the individual discrepancies

$$\sum_{t=1}^{T} \left| \sum_{s=1}^{S} K_{st}(\mathbf{x}_{st}^0; \lambda_1^n, \lambda_2^n, \ldots, \lambda_T^n) - \overline{K}_t \right|.$$

In the presence of budgetary constraints, the planning of a river development for construction today is even more closely interwoven with planning for construction in the future than if budgetary constraints were absent. The optimal design-construction schedule for each project depends not only on the benefits and costs for the project itself, but also on the time pattern of its comparative advantage with respect to other projects, and the levels of budgetary constraints in all years as well. In the planning process this greater interdependence is reflected by the dependence of the current-period shadow price on both the benefits and costs of all projects and the levels of budgetary constraints for all periods.

In a perfect certainty model this interdependence is a computational complication, but once we admit uncertainty in the projection of future benefits and costs and the level of future budgets, the interdependence between present and future planning presents a much more formidable problem. Passage of time will very likely bring different general economic conditions from those projected, affecting both benefits and costs and the level of budgets. In time a different set of attitudes toward public water-resource development may also arise, superseding those prevailing when future budgets are originally projected. And passage of time may increase the number of potential projects as new opportunities are discovered and explored; such projects will compete with those already explored.[19] Thus, shadow prices must be constantly revised with the

[19] In keeping with traditional theories of decreasing marginal productivity of capital, the average quality of these new opportunities will presumably be below that of already-explored opportunities. However, the average quality of the new opportunities may be greater than the marginal value of the already-explored opportunities, that is, greater than the shadow prices resulting from consideration of the already-explored opportunities alone.

passage of time. Of course, if the changes that time brings are not too drastic, the future shadow prices determined today can be used as the basis for new estimates appropriate to the revised budget and benefit–cost data, according to the iterative procedure described above.

It is clear, however, that while proper decisions on present development cannot be made without reference to future development, future plans, in view of the uncertainties cited above, must always be much more tentative than plans for immediate development. It may be advantageous to divide the planning process into two stages, a first stage in which project increments are divided between "present" and "future" construction, and a second stage in which the sequence of construction for increments assigned to the "present" is determined.

In the first stage, the plans formulated for development of water resources are broad in scope. For this purpose one might conceive of a "present" of as much as five years' duration, an "immediate future" of the next five years, and a "distant future" of the next 10 or 15 years. At this stage in the planning process budgets for all the years within a period would be lumped together, so that only three budgetary constraints and hence three shadow prices would need to be considered. The broad development plans formulated by means of the iterative shadow-price procedure described above would thus differentiate particular segments of river-development schemes according to "present," "immediate-future," or "distant-future" development.

In the second stage, detailed project designs and construction schedules would be formulated, once again by means of shadow prices, but only for the portions of projects assigned to the "present." Thus in the second stage the number of budgets and shadow prices to be considered would be small — only five in our tentative outline, one for each year — and these budget levels, occurring in the near future, can be known with relative certainty.

GENERAL AND SPECIFIC EXTERNAL ALTERNATIVES [20]

In Chapter 2 we first discussed efficient design in a simplified context, that of the competitive model. We then relaxed several of the assumptions of this model to account for certain situations that planners are

[20] The model presented in this section draws heavily on Peter O. Steiner's article, "Choosing among alternative public investments in the water resource field," *Am. Econ. Rev.* *49*, 893 (1959). As noted in Chapter 1, Steiner was associated with the Harvard Water Program in 1956–57.

likely to encounter. But we did not discuss modifications made necessary by impediments in the free flow of resources between sectors of the economy. Where such impediments exist, public investment of any kind may displace private investment or public investment of some other kind. The measure of gain from the public investment is not then its total net benefits, but rather the incremental gain over the displaced investment. Furthermore, we did not concern ourselves in Chapter 2 with complications stemming from the government's pre-emptive rights in water-resource development. Public water-resource development opportunities may represent attractive private investment opportunities as well, opportunities the realization of which would be denied by public development. Once again, the real gains of the public undertaking are limited by the gains possible in its absence, namely, the potential gains of the private alternative. Finally, we have not given any attention to the possible creation of private opportunities.

To analyze these factors, we shall develop a multiple-sector model of the economy and trace through all of its sectors the consequences of decisions in the sector over which water-resource-planners have direct control.

Formally, we define the government water-resource-development sector as the set of *internal alternatives*, that is, the set of alternative system designs for all projects from which planners choose the designs to be constructed. We label this sector, to which all of our discussion in Chapters 2 and 3 referred, $S1$. A second sector contains the run-of-the-mill, marginal investments throughout the economy that would be displaced in part by assignment of resources to government river-basin development. We call this sector *general external alternatives* and give it the label $S2$. The third sector, $S3$, consists of the *specific external alternatives* to government water-resource development — the substitute plans of private industry, municipalities, special districts, and the like for fulfilling the wants that development by the national government would meet (economic alternatives) or for utilizing the same sites that the national government's projects would occupy (technologic alternatives). The set of *private opportunities created* by government river developments is called $S3'$. The close relationship between $S3$ and $S3'$ will be seen when these sectors are discussed.

RELATIONSHIP BETWEEN GOVERNMENT WATER-RESOURCE DEVELOPMENT (S1) AND RUN-OF-THE MILL PRIVATE INVESTMENT (S2)

Why be concerned, one might ask, if government water-resource development displaces run-of-the-mill investment alternatives? The very fact that the displaced investments are run-of-the-mill indicates that their loss is of little moment; their marginal net return must be zero or they would not be marginal. The answer is that investments in *S*2 may be marginal from the standpoint of the *private* entrepreneur and at the same time have high return from the public point of view (and exploitation of worthwhile investment in the *public* sector may be prevented by budgetary constraints). As we pointed out in Chapter 2, the rate of interest applicable to the evaluation of public investment may well be different from the market rate. Symmetrically, this "social" rate of interest is the proper rate for evaluating private investment from the public point of view. In the next section we explore one important reason why this social rate of interest may be different from the market rate of interest which guides investment decisions in the private sphere of the economy.

The choice of interest rate.[21] In Chapter 2 we suggested that planners of public investment should not accept the "solution" offered by the market to the problem of choice of interest rate even if the investment market were perfectly competitive. This may seem a strong assertion, for if the investment market is competitive, the interest rate represents a "price" ratio of present consumption to future consumption to which every individual (through borrowing and lending) adjusts his consumption policies over time so as to make the market interest rate his own marginal rate of substitution between present and future consumption (that is, his marginal time preference in consumption) as well as the marginal rate of return on all investments (that is, on investments in *S*2).

The objection to the market solution is that individuals may have preferences that, although an integral part of their attitudes toward consumption now versus consumption later, are inexpressible in the market place. In particular, none of us is able to put into effect in the market his preferences with regard to other people's consumption. I may

[21] The argument set forth in this section builds on the work of William J. Baumol, *Welfare economics and the theory of the state* (Harvard University Press, Cambridge, 1952), p. 92, and A. K. Sen, "On optimizing the rate of saving," *Econ. J.*, *71*, 479 (1961).

well place less of a premium on my own consumption now as opposed to the consumption of an unknown member of a future generation at some specified date in the cooperative context of public investment, in which I know a sacrifice on my part will be matched by sacrifices by all other members of the community, than in an individualistic market arrangement in which I have no such assurance.

For example, if provision of $1.00 of consumption to a future individual gives me satisfaction equivalent to $0.10 of my own consumption today, in a unilateral market context my time preference for my own present consumption over a future individual's consumption is 1/0.1 or 10. If investment of $1.00 in the present will at the margin provide $2.00 for a future member of the community at a specified date, I would not undertake investment for future generations on my own, for each dollar of such investment would "cost" me $0.80 more than the psychic gain I derive from it (2 × $0.10). This does not mean that I would undertake no investment at all; merely that I would undertake none specifically on behalf of future generations. And if for simplicity we assume that all have identical tastes, none would undertake investment unilaterally to provide for future generations.

In the context of public investment, however, where each individual knows that his sacrifice will be accompanied by sacrifices by all other members of the community, each properly compares the psychic value of the future individual's consumption not with his own current consumption alone, but with his own portion of the sacrifice entailed in public investment plus any psychic pleasure he would forego because of the sacrifices of his contemporaries.

If, for example, each values $1.00 of a contemporary's consumption as equivalent to $0.15 of his own consumption, then as long as all are required to undertake an equal amount of extra investment to provide for future generations, this marginal time preference is the ratio of the sum of $1.00 of one's own consumption and the psychic value of $1.00 of consumption by each of the other $n - 1$ members of the community to the psychic value of future consumption; that is to say, $[1 + (n - 1)(0.15)]/n(0.1)$. In our example if n is equal to or greater than 17, this ratio is less than 2. Since a present investment of $1.00 yields $2.00 for a future individual, the investment accordingly is held to be desirable by each individual if undertaken on a cooperative basis, although it is far from being so as a private, unilateral venture. (In fact, in our example, no one finds a private investment yielding less than $10.00 for $1.00

attractive.) As n becomes very large, one's own sacrifice becomes an insignificant consideration, and in our example one's public-investment time-preference ratio, $[1 + (n - 1)(0.15)]/n(0.1)$ reduces approximately to 0.15/0.1, the ratio of the psychic value one receives from $1.00 of consumption by a contemporary to the psychic value of consumption by a future individual.

It is neither inconceivable nor inconsistent therefore that, while our time preference for present consumption over the consumption of a future individual in the individualistic context of the market may be too high to justify investment solely for future generations, our premium on present consumption in the context of public investment, in which each individual knows that others share the investment burden, may be sufficiently small to justify public investment for future generations above and beyond that intended to provide for the future needs and wants of the present generation.

Of course, the world is not quite so simple as assumed in this discussion. Individuals value consumption on the part of others, present and future, differently, and these value parameters (assumed above to be 0.15 and 0.1) change with the level and distribution of the costs and proceeds of investment. Also, the marginal return in the future from present investment (assumed above to be $2.00 per dollar) depends on the level of investment. Nevertheless, the impact of the argument remains even after relaxation of the simplifying assumptions. The degree of concern for future generations in the abstract (as distinct from specific concern for one's children and near relations) is generally too low to be operative in the market; hence the market interest rate is determined only by the interaction of hedonistic and familial utility and the productive possibilities of investment. However, one's preferences with respect to consumption by his contemporaries as well as by future generations can be expressed in his attitude toward public investment. Therefore, marginal time preferences with regard to public investment may well differ from the market rate of interest.[22]

Two important characteristics of this argument should be noted. First, it strikes at the heart of the competitive model and thus is more telling than criticisms of market interest rates that are directed at imperfections

[22] This is not to say that agreement among the members of the community upon a rate of discount for public investment different from the market rate of interest will be unanimous. For example, were a few individuals to derive no pleasure from consumption by others, present and future, unanimity would be impossible, for these individuals would resent any personal sacrifice whatsoever.

in markets for investment funds. The prescription called for by imperfections is a distillation by public planners of the spread of market interest rates brought about by these imperfections into an average rate appropriate to public investment.[23] Thus, if the spread of interest rates at which individuals borrow and lend funds in an imperfect market were between 4 and 12 percent, the public planners could be sure that, insofar as imperfections are the issue, the appropriate rate of interest lies somewhere between these extremes. The argument presented above, however, does not justify limiting the public planners' attention to the range of market rates of interest assumed to lie between 4 and 12 percent; it is quite conceivable that time preferences of individuals for public investment will be, say, 2 percent. The range of private interest rates has no normative significance for public investment.

Our argument differs in a second fundamental way from previous criticisms of market interest rates. The English economists A. C. Pigou and M. N. Dobb[24] have rejected market interest rates on grounds both that future generations are unrepresented in their determination and that individuals simply do not know what is good for themselves when it comes to planning for the future. Whatever else their merits, these arguments imply an authoritarian rejection of individual preferences with which we are unwilling to associate ourselves. Thinking somewhat differently, Gerhard Colm[25] has suggested that underlying preferences of individuals are not uniquely defined, differing in the political arena from what they are in the market place simply because the considerations that enter one's mind in the two contexts differ. This argument is sufficiently similar to ours that it is only by bearing in mind the following crucial difference that they can be distinguished: whereas the argument we advance stresses the difference between the market and the political arena with respect to the preferences an individual can express effectively, Colm's argument stresses the differences in an individual's preferences themselves that a change in context occasions.

[23] See John Krutilla and Otto Eckstein, *Multiple purpose river basin development: studies in applied economic analysis* (Johns Hopkins University Press, Baltimore, 1958), chap. 4, for an exposition of this approach.

[24] A. C. Pigou, *The economics of welfare* (Macmillan and Co., London, ed. 4, 1932), pp. 24–30; M. N. Dobb, *Political economy and capitalism* (International Publishers, New York, 1945), pp. 295–298, and *Economic theory and socialism* (International Publishers, New York, 1955), pp. 70–74.

[25] Colm's argument is summarized, on the basis of cited references and personal conversation with Colm, by R. A. Musgrave, *The theory of public finance* (McGraw-Hill, New York, 1959), pp. 87–88.

The opportunity costs of public investment. Rejection of the market rate of interest in favor of a social rate of interest means that the optimal level of investment is that for which the marginal productivity of investment equals the social rate of interest, rather than the level at which the marginal productivity of investment equals the market rate of interest. If the social rate of discount is lower than the market rate, the impact of the result in a frictionless competitive model is simply that the community in its collective political capacity undertakes investments with a future return, evaluated at the market rate of interest, too low to justify their exploitation privately. That is, the community invests in *S*2 until further investment becomes marginal from the public as well as the private point of view. This action is not, however, directly applicable to actual investment decisions, for it takes no heed of the different kinds of investment opportunities likely to be open to public and to private development. In a mixed-enterprise economy like that of the United States, for example, the government cannot itself exploit all opportunities in *S*2 left untouched by the market for which the future gain, when viewed collectively, justifies the present sacrifice; for some of these opportunities will lie in the sector that in the present institutional structure has been marked off as "private."

Of course, if the government exercised absolute control over private investment, its inability itself to undertake all the socially desirable investment opportunities in *S*2 would be of little moment. It could through fiscal and monetary policy induce private enterprise to exploit all such opportunities, that is, all opportunities for which the present value of net benefits is positive at the social rate of interest. Appropriate use of monetary operations to ensure plentiful and cheap credit, coupled with subsidies and differentiated tax rates, would make socially desirable opportunities privately desirable as well.[26]

In actuality the U. S. economy, and especially the capital market, is not sufficiently competitive to permit application of this counsel of perfection; moreover, the government does not, by virtue of the institutional structure, exercise the degree of control over private investment through fiscal and monetary policy that would be necessary in this

[26] This is in fact the alternative to public investment advanced by Jack Hirschleifer, James C. De Haven, and Jerome W. Milliman in their recent study, *Water supply: economics, technology, and policy* (University of Chicago Press, Chicago, 1960), chap. 6. By means of monetary policy they would seek to drive the market rate down to the social rate of interest. Thus, while the social rate would not be determined by the market rate, the two would be equal, for the market rate would be determined by the social rate. Fiscal policy would be necessary only to avoid the inflation that increased private investment would otherwise set off.

country's economy to ensure private development of all socially desirable opportunities in the private sector.

On the other hand, the boundary between the public and private investment sectors, *S*1 and *S*2, is not absolute in modern capitalistic economies. Government fiscal and monetary policies can and do affect private investment. And public investment, as a result of deviations from competitive assumptions in the actual determination of private investment, can in certain circumstances prevent private development of socially desirable opportunities in *S*2. The extent to which the rate of private investment declines in response to increases in the rate of public investment will depend on the level of employment in the economy and the method of financing the public investment.

Thus, in situations of less than full employment of resources — especially long-run, structural underemployment typical of underdeveloped countries — increased public investment, insofar as it can be arranged to take up the slack in the economy, need disturb both private investment and consumption somewhat less than in situations of full employment. Also, financing public investment by taxes on business profits would reduce private investment more than an equal amount of taxation on luxury goods, because of the importance of the availability of funds supplied internally by firms in determining their investment programs.

The potential displacement of private by public investment poses the problem of this section: How should the planning of those investments that the government does undertake be affected by the existence of private investment opportunities in *S*2 that are (1) socially desirable by virtue of the discrepancy between the social interest rate and higher rate(s) of interest governing private investment, and (2) displaced by public investment by virtue of the institutional structure?

This is a problem of "second best." The optimal solution, undertaking all public and private investment that has a positive present value at the social rate of interest, is precluded by the institutional structure, and we are consequently faced with the necessity of choosing the best of inferior combinations of public and private investment in *S*1 and *S*2.[27] Clearly the goal in planning public investment should be avoidance of displacement of "better" opportunities in the private sector, *S*2.

This answer, however, simply shifts the ground of the question. The

[27] For a general discussion of "second best," see R. G. Lipsey and R. K. Lancaster, "The general theory of second best," *Rev. Econ. Stud. 24* (*1*), 11 (1956–57). The argument is further developed in M. McManus, "Comments on the general theory of second best" and Lancaster and Lipsey "McManus on second best," *Rev. Econ. Stud. 26* (*3*), 209 and 225 (1959).

problem becomes how to decide if an alternative private investment represents a better opportunity than public investment. "Better" is easy to define if, of two alternatives entailing the same initial investment, the benefits less operating costs for one are greater in every year than for the other. For example, if projects *A* and *B* both have the same capital cost and their respective time stream of annual benefits less OMR costs are as depicted in Fig. 4.4, we have no trouble identifying project *A* as the superior investment. The question of superiority becomes difficult to

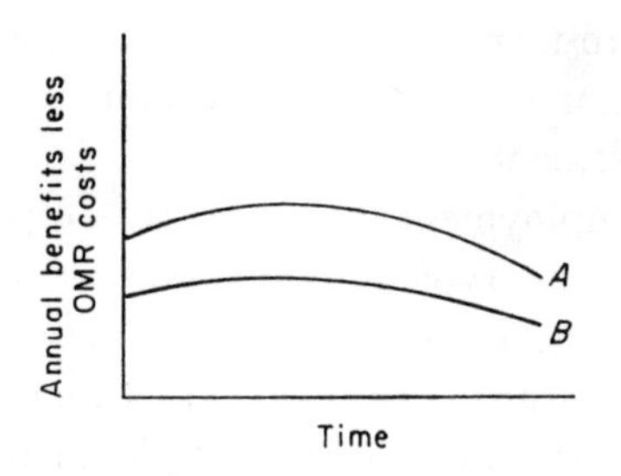

Fig. 4.4. Time streams of benefits of two investment projects: nonintersecting streams.

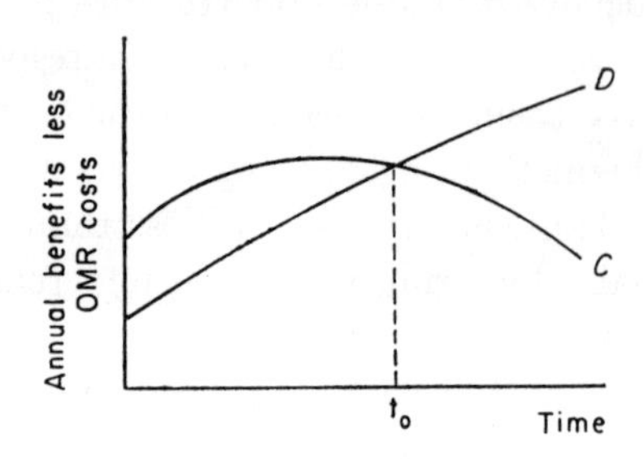

Fig. 4.5. Time streams of benefits of two investment projects: intersecting streams.

resolve only if the time streams intersect, as in the example of projects *C* and *D* in Fig. 4.5. Here *C* is better only until time t_0, after which *D* becomes superior. As a special case of the intersection of benefit streams, neither of the two alternative investments with different economic lives will be unambiguously superior if one of them produces more benefits during the period in which their life spans overlap, and the other continues to provide benefits after the first has long ceased to exist economically.

The solution to the problem of comparing alternative public and private investment in a mixed-enterprise economy, originated in embryonic form by Otto Eckstein[28] and developed in a masterful paper by Peter O. Steiner,[29] is straightforward enough. Since the social rate of interest reflects the community's weight on consumption at different times, the appropriate basis for comparison of alternative public and private investment is the present value of net benefits evaluated at the social rate of interest.[30] The general guide for the planning of water-resource de-

[28] Eckstein, reference 1, pp. 101–103.

[29] Steiner, reference 20.

[30] An alternative basis for comparing alternative investments is by their rates of return, that is, the rates of interest at which the present value of each becomes zero. The "rate-of-

velopments — adding increments to systems to the point where the present value of benefits (net of OMR costs) equals the construction cost — remains formally the same; however, the construction "cost" now includes the present value of private investment in $S2$ that construction of the system increment prevents. In other words, cost becomes once again, as in the presence of other deviations from the assumptions of the competitive model, an opportunity rather than a money concept.

Much of interest about the determinants of opportunity cost can be learned by examining a particularly simple class of alternative private investments. Suppose that marginal private investment in $S2$, that is, the private investment displaced by public investment, yields annual benefits equal to ρ_1 percent of the investment outlay in perpetuity, and that ρ_1 is constant over the relevant range, namely, the range of investment alternative to water-resource development. Then ρ_1 has a simple interpretation: it is the marginal productivity of investment in the private sector of economy.[31] We assume that the benefits of private investment are consumed as they become available; otherwise the real alternative is not the perpetuity of ρ_1 percent annually, but rather the benefit stream of the nominal alternative plus reinvestment. Finally, we denote the amount of private investment in $S2$ displaced by each dollar of outlay on water-resource development by α_1, this quantity being restricted by the inequality $0 \leq \alpha_1 \leq 1$.

With these assumptions, the opportunity cost of water-resource development in view of private investment foregone for each dollar of construction outlay on water-resource development is

$$[\alpha_1\rho_1 + (1 - \alpha_1)r]/r, \tag{4.35}$$

where r represents the social rate of discount. The terms in the brackets may be viewed as representing the annual value of an "investment" of one dollar divided (in the proportions α_1 and $1 - \alpha_1$) between the private alternative with a value of ρ_1 dollars per year and consumption with

return criterion" was in vogue for many years in the literature of business-investment-planning. See, for example, Joel Dean, *Capital budgeting* (Columbia University Press, New York, 1951). Within recent years it has been demonstrated conclusively that this criterion is not a generally valid basis of comparison, because the alternative with the higher rate of return will not necessarily have a higher present value at the relevant rate of interest, in our case the social rate of interest. See J. Hirschleifer, "On the theory of optimal investment decision," *J. Pol. Econ. 66*, 329 (1958); J. H. Lorie and L. J. Savage, "Three problems in rationing capital," *J. Business 28*, 229 (1955); E. Solomon, "The arithmetic of capital budgeting decisions," *J. Business 29*, 124 (1956); A. Alchian, "The rate of interest, Fisher's rate of return over cost, and Keynes' internal rate of return," *Am. Econ. Rev. 45*, 938 (1955).

[31] It is also the internal rate of return on marginal private investment. See footnote 30.

a value of r dollars per year. The annual return divided by r is of course the present value of the perpetuity.[32]

Expression (4.35) defines the minimum present value (net of OMR costs) that system increments must provide in order to qualify for construction; in the language peculiar to the planning of water-resource development, this expression represents the cutoff (marginal) benefit–cost ratio which replaces the ratio 1/1 in the presence of deviations of opportunity from money costs. The true cost of system construction is thus

$$[\alpha_1\rho_1 + (1 - \alpha_1)r][K(\mathbf{x})]/r, \tag{4.36}$$

where $K(\mathbf{x})$ represents the money cost of system inputs, denoted by the vector $\mathbf{x}$. If, for example, the social rate of interest r is 2 percent; the perpetual annual return of displaced private investment ρ_1, 5 percent; and the amount of private investment displaced by each dollar of public investment α_1, \$0.75, then Exp. (4.35), the opportunity cost per dollar of public investment or the cutoff benefit–cost ratio, is \$2.125. The cost of system construction, Exp. (4.36), becomes $2.125K(\mathbf{x})$, and no system increments are properly undertaken unless they provide at least \$2.125 of present value (net of OMR costs) per dollar of construction outlay.[33]

We noted briefly in defining the alternative private investment that it is not the nominal benefit stream of the private alternative that is important, but the actual stream of benefits that alternatives provide, taking reinvestment into account. The same holds true of the water-resource system itself. Reinvestment of system benefits tends to offset any loss of benefits from alternative investment in $S2$ which resulted from directing the original outlay to water-resource development. If, as system benefits become available, a constant proportion α_2, $0 \leq \alpha_2 \leq 1$, of benefits (net of OMR costs) is reinvested in the exploitation of op-

[32] Dr. A. K. Sen has pointed out that this formula is valid only in the event that resources are fully employed and public investment consequently displaces private investment and consumption on a dollar-for-dollar basis. In the event of unemployment, the private consumption displaced per dollar of public investment need not be a dollar minus the private investment displaced, $1 - \alpha_1$, but something less. To take account of this possibility, we replace $1 - \alpha_1$ by the coefficient β_1 and replace the full-employment restriction $\alpha_1 + \beta_1 = 1$ by the more general restriction $\alpha_1 + \beta_1 \leq 1$.

[33] Note that if $\alpha_1 = 1$, that is, if displacement of private investment is "one for one," the second term of Exp. (4.35) drops out and the opportunity cost per dollar of system construction outlay becomes simply ρ_1/r, the present value of the displaced benefits of private investment. On the other hand, if $\alpha_1 = 0$, that is, if public investment entails no displacement of private investment, Exp. (4.35) reduces to 1, the value of present consumption displaced by each dollar of investment in water-resource development. Expression (4.36) then reduces to $K(\mathbf{x})$, the money cost of system construction.

portunities with a yield of ρ_2 percent of the investment outlay annually in perpetuity, then the opportunity cost per dollar of system construction outlay becomes

$$\frac{1}{[\alpha_2\rho_2 + (1 - \alpha_2)r]/r} \times [\alpha_1\rho_1 + (1 - \alpha_1)r]/r$$

where the first quotient represents the influence of the reinvestment of system benefits. This expression can be written more simply as

$$\frac{\alpha_1\rho_1 + (1 - \alpha_1)r}{\alpha_2\rho_2 + (1 - \alpha_2)r}.$$

Thus the opportunity cost of system construction outlay becomes

$$\frac{\alpha_1\rho_1 + (1 - \alpha_1)r}{\alpha_2\rho_2 + (1 - \alpha_2)r} K(\mathbf{x}). \tag{4.37}$$

It is clear that if ρ_2 equals ρ_1, and if α_2 equals α_1, the opportunity cost reduces simply to the nominal cost of $1 per dollar. That is, if the productivity of reinvestment equals the productivity of alternative investment in $S2$, and the proportion of benefits reinvested equals the ratio of displaced alternative investment to the original outlay on system construction, then the opportunity cost of system construction becomes the money cost. Displacement of private investment and reinvestment in this case cancel each other out, and both can be ignored. By the same token, the opportunity cost of system construction can be less than the nominal cost of $1 if water-resource development leads to proportionally more reinvestment of benefits than to original displacement of investment ($\alpha_2 > \alpha_1$), or if the reinvestment opportunities are superior to the displaced investment opportunities ($\rho_2 > \rho_1$).

Determination of the values of the ρ's and the α's is of course an empirical matter. Once we drop the simplifying assumptions that the marginal productivity of investment and the marginal propensity to invest are constant throughout the private sector of the economy, it is clear that the values of ρ_1 and α_1 depend on the distribution of the costs, and ρ_2 and α_2, on the distribution of the benefits of water-resource development. The values of ρ_1 and ρ_2 will be determined respectively by the investment opportunities open to those who bear the costs and those who enjoy the benefits of system construction. Similarly, the values of α_1 and α_2 will be positively correlated with the disposition to investment relative to consumption and negatively correlated with the ability and willingness to borrow funds to finance exploitation of the investment opportunities open to those who bear the costs and enjoy the benefits of water-resource development.

What is the importance of the willingness and ability to borrow of cost-bearers and beneficiaries? Simply that to the extent that this willingness and ability are present, cost-bearers and beneficiaries are not dependent on decreases in the level of public investment or increases in benefits to supply funds needed for the exploitation of investment opportunities.[34]

In focusing our attention upon general private alternatives ($S2$) to water-resource development and private reinvestment of water-resource-development benefits, we have been assuming implicitly that all socially desirable investment opportunities in the public area of the economy are exploited, that is, that $S2$ consists solely of private investment opportunities. In actual fact, institutional constraints may prevent the government from carrying out the dictates of efficiency in the public area as well as in the private sector.

Insofar as investment in water-resource development displaces alternative public investment (in highways, for instance), the cost of system construction must be divided into three rather than two segments; instead of a fraction displacing private investment, α_1, and a fraction displacing private consumption, $1 - \alpha_1$, we now have a fraction displacing private investment, α_1^{pri}, a fraction displacing consumption, α_1^{con}, and a fraction displacing alternative public investment, α_1^{pub}. The opportunity cost of water-resource development, taking into account only displacement of alternatives, then becomes

$$[\alpha_1^{\text{pri}}\rho_1^{\text{pri}} + \alpha_1^{\text{pub}}\rho_1^{\text{pub}} + \alpha_1^{\text{con}}r]K(\mathbf{x})/r \tag{4.38}$$

and the relationship between the α's is[35]

$$\alpha_1^{\text{pri}} + \alpha_1^{\text{pub}} + \alpha_1^{\text{con}} = 1. \tag{4.39}$$

[34] It should be observed that the existence of the opportunity cost discussed in this section may affect not only the scale of water-resource systems but the combination of outputs as well. Insofar as different outputs go to different groups among which ρ_2 and α_2 differ, system design under the efficiency objective will tend to be biased in favor of those outputs for which reinvestment of benefits is highest. For example, if the benefits of irrigation water tend to go to low-income farmers with low marginal propensities to invest and meager investment opportunities, whereas the benefits of electric energy tend to go to industrial entrepreneurs with high marginal propensities to invest and excellent investment opportunities, electric energy will be relatively more attractive than irrigation.

[35] This assumes that the economy is operating at full employment. For the general case, the following relationship (discussed in footnote 32) among the weights replaces that in Eq. (4.39): $\alpha_1^{\text{pri}} + \alpha_1^{\text{pub}} + \alpha_1^{\text{con}} \leq 1$.

The marginal productivity of alternative public investment, ρ_1^{pub} (again we assume that the alternative is a perpetuity), need not equal the marginal productivity of investment in the private sector in the imperfect world in which we live and for many types of public investment, such as education, will be much more difficult to measure.

Finally, insofar as benefits from water-resource development recaptured by the government are reinvested, that is, insofar as new public investment is actually increased as a result of recapture of system benefits by the government,[36] we must similarly expand the reinvestment factor of the opportunity-cost coefficient. Instead of Exp. (4.38) we have, after simplification,

$$\frac{\alpha_1^{\text{pri}}\rho_1^{\text{pri}} + \alpha_1^{\text{pub}}\rho_1^{\text{pub}} + \alpha_1^{\text{con}}r}{\alpha_2^{\text{pri}}\rho_2^{\text{pri}} + \alpha_2^{\text{pub}}\rho_2^{\text{pub}} + \alpha_2^{\text{con}}r} K(\mathbf{x}) \tag{4.40}$$

as the true cost of water-resource development. The α_2's are related in the same manner as the α_1's in Exp. (4.39).

Since we have introduced the concept of opportunity cost (shadow price) earlier with regard to budgetary constraints, we should not conclude this section without a word on the relationship of that opportunity cost to opportunity cost in the present section. The opportunity cost $1 + \lambda$ in the context of a budgetary constraint is an internal concept; viewed from the standpoint of any individual system, this opportunity cost measures the value of benefits foregone from all other systems for each dollar spent on the system in question. The opportunity cost in the present context is an external concept; it represents the net value of benefits foregone from investments alternative to water-resource development. There is no problem in deciding which opportunity cost is valid in the presence of both a budgetary constraint and alternative investments: it is simply the higher of the two. That is, if the external opportunity cost exceeds the internal, the budget should not be fully utilized, whereas if the internal opportunity cost exceeds the external, alternative investments are irrelevant.

[36] It is probably true that in the United States the level of new federal investment is little affected by the extent to which the benefits of earlier investment are recaptured. On the other hand, such recapturing well may be a crucial factor in determining the level of state and local public investment in the U. S., and of all public investment in underdeveloped countries, since government investment in those countries depends much more on the extent to which the government is able to pre-empt resources from the consumption sector of the economy.

RELATIONSHIP BETWEEN GOVERNMENT WATER-RESOURCE DEVELOPMENT (S1) AND SPECIFIC EXTERNAL ALTERNATIVES (S3)

We have noted that the specific external alternatives which comprise the third sector, *S*3, can be either economic or technologic in nature; they can be alternative means for meeting demands to be fulfilled by a project or alternative uses for the sites to be occupied by the project. For example, a new or enlarged road or railroad that would provide transport facilities in lieu of a navigation system proposed by the federal government would be an economic alternative. By contrast, the construction of a hydroelectric power plant at a particular reservoir site bv a private power company constitutes a technologic alternative to development by the government of this site for irrigation storage.[37] If the government also were to develop this site for power, thus pre-empting the market for electric energy and precluding private power development on these grounds, the private alternative would be both economic and technologic.

For both of these classes of alternatives it is essential that the water-resource-planner be able to select from among the many potential alternatives those which are relevant to system design. How is relevancy to be determined in any given instance? Generally stated, the relevant alternative is the one that would actually be constructed in the event of inaction on the part of the national government. This does not mean that only a single alternative need be considered for each economic demand and for each site included in the system under study; the relevant alternatives are not limited to those that would replace the *full* system being planned. It is as important for the planner to know the external alternative to increasing the size of the system, say, a hydroelectric power system, from one yielding 3×10^6 Mw hr per year to one producing 4×10^6 Mw hr per year, as to know the relevant external alternative if no system were constructed at all. Just as the consideration of internal alternatives in system design is based on incremental analysis, so must the consideration of external alternatives be similarly based. Specific external alternatives of *S*3 may be redefined, therefore, as the set of actual responses, considered incrementally, to decisions on the part of

[37] Technologic alternatives will often be alternatives only on an "all-or-nothing" basis; that is, any development whatsoever by the planning unit will often preclude all construction of technologic alternatives at the site.

the planning unit to forego corresponding increments of a proposed system.

How can the national-government planning unit determine the response of government units or private firms to a decision by the planning unit to forego a particular economic or technologic alternative? As for the economic alternative, the only general basis on which this question can be answered is by reference to the benefits and costs of the alternative from the point of view of the government unit or private firm that would build the alternative, given the objective of this unit or firm. We can reasonably assume that, like the planning unit, the other unit or firm will make its decisions so as to maximize the present value of net benefits. The alternative that maximizes the present value of net benefits based on the latter's objectives (rather than on those of the planning unit) is thus the relevant alternative to any proposed system increment.

For example, the relevant economic alternative to the system increment that increases annual energy output from 3 to 4×10^6 Mw hr is, if constructed by a private firm, the steam plant that would maximize the present value of net profits to the private firm, given the existence of a public hydroelectric system generating 3×10^6 Mw hr per year.[38] The relevant economic alternative, if constructed by a state or local government with the objective of, say, state or local income redistribution, would be the steam plant that would maximize the present value of net benefits with respect to the given redistribution objective, again taking into account the existence of the nationally-constructed system generating 3×10^6 Mw hr per year.

For technologic alternatives, the basis for determining relevancy is somewhat different. The national government not only may have "first refusal" of an opportunity for water-resource development; in addition, it may be able to control the choice of technologic alternative to a certain extent through its licensing power and through voluntary agreements with the government unit or private firm constructing the alternative. If it once turns down an opportunity, the relevant technologic alternative is the one that would be licensed or negotiated. If no license would

[38] Determination of relevancy by means of profit maximization must be qualified in the presence of governmental regulation of private economic activity (such as public utilities) that might compel deviations from profit maximization. In the event of such control, the relevancy of private economic alternatives is determined by the alternative that would be imposed by the regulating authorities. The alternative need not supply the same levels of output as the proposed system increment. In the present example the alternative steam plant may, but need not, supply 1×10^6 Mw hr of electricity per year.

be applied for, and no alternative would be constructed by voluntary agreement, the alternative to construction of the system increment is no (incremental) development whatsoever of the site.

The criterion for determining the relevant alternative from *S*3 has an important result: hypothetical, as distinct from real, alternatives are excluded from consideration. If, for example, the costs of the benefit-maximizing alternative steam plant exceed its benefits,[39] it would not be constructed, and the alternative to the planning unit's 1×10^6 Mw hr increment would be no (incremental) action at all. At the same time, existing alternative facilities, although certainly real, are not relevant alternatives in our sense; if an existing railroad provides an alternative means of transport to that of a proposed navigation system, the effect of the existing facilities on system benefits should be reflected fully in the demand for the navigation system. Only to the extent that enlargement of the railroad, requiring new expenditures, is alternative to a navigation system (to meet a growing demand) is the railroad an alternative in the sense in which we use the term.

The relevancy of economic alternatives depends also on the objective sought by the planning unit. For example, privately produced steam energy to meet industrial demands might be an alternative to system energy if the objective for system energy is economic efficiency. If, on the other hand, an objective of income redistribution is sought by providing system energy to farmers whose demands for electricity are such that private utilities do not consider them profitable, there would be no economic alternative to the system.

The objective also determines the relevancy of technologic alternatives to the extent that the government will, through its licensing powers or through negotiation, tailor the alternative to fit the objective — as may be done, for example, in requiring private builders to design their power facilities so that they can be connected with regional transmission networks.

In separating relevant from irrelevant alternatives, we emphasize that the primary criterion of externality is framed in terms of jurisdiction: alternatives belong to *S*3 rather than *S*1 only if they fall outside the purview of the planning unit. There is, for instance, no good reason for treating a steam plant as an external alternative to hydroelectric capacity if both can be constructed by the planning unit; steam energy should be

[39] Benefits and costs as used here mean those evaluated in accordance with the objective of the alternative investor.

considered as simply another internal alternative, along with hydroelectric energy at sites x, y, and z, and the optimal mix chosen from the set of combinations of energy from all four sources. In other words, our definition of $S1$ is flexible, expanding in any given case to include all variables subject to direct decision by the planning unit. These variables need not necessarily themselves represent measures of water-resource development, but may be complementary to such measures in carrying out the objectives for which the water-resource development is undertaken.

Do we mean by "planning unit" the national government as a whole or the agency actually preparing water-resource and related development plans? Since the undertakings of all national agencies are financed ultimately from the same budget, we believe it preferable to view determination of the best among competing proposals submitted by the agencies as a problem of internal maximization by the national government as a whole. We are interested here essentially in evaluating national development in the presence of alternatives to such development, and we consider coordination among the subordinate agencies of the national government as "given," that is, achieved separately.[40]

The effect of external alternatives on system benefits. The planning goal does not change when we take into account the specific-external-alternative sector; it remains the design of water-resource systems to make the maximum contribution to a predetermined objective. But if maximum system benefits are to represent this maximum contribution in the presence of external alternatives, the concept of benefits must be amended to reflect the potential contribution of the alternatives to the objective. We therefore redefine system net benefits as the sum of the net benefits of the system proper and those of external alternatives not displaced.

"Net benefits of alternatives" then has two meanings. In determining relevancy of alternatives in the first place, it refers to the benefits and costs from the standpoint of the alternative producer. In evaluating the contribution of an external alternative to the government's objective, however, it is not the net benefits from the point of view of the unit constructing the alternatives that count, but rather the net benefits

[40] How is the set of external alternatives affected if the planning unit is not the national government but a local government or private firm instead? With technologic alternatives, the structure of government may determine that a planning unit will have the chance for development only if higher units all pass their turn. Thus from the point of view of lower units, "alternatives" in the hands of higher units are not alternatives at all. Economic alternatives in the hands of higher units are also likely to be irrelevant, since decisions on their construction often are made independently of the decisions of lower units.

with respect to the government's objective. Thus if the objective is efficiency, the maximum willingness to pay for outputs of alternatives (rather than, for example, actual revenues) constitutes the benefits. If the objective is income redistribution, the actual amount paid by the intended beneficiaries must be subtracted from the maximum amount they would pay to arrive at the redistribution benefits of the alternative.

Costs, too, must be assessed from the social point of view. Taxes, although they represent a cost to a private concern, are not a cost from the social point of view. Consequently they are properly left out of the benefit–cost analysis of alternatives. And, more generally, as in evaluation of the costs of a system itself, money costs are meaningful only as long as factor prices reflect opportunity costs. Otherwise (as, for example, in evaluating labor costs under conditions of less than full employment), costs must be revalued on the basis of shadow prices reflecting true social valuations of the factors employed in the construction of alternatives.

Furthermore, interest payments as such are not costs from the social point of view. Rather the benefits and costs of alternatives should be discounted at the rate of social time preference with respect to the given objective. As for the system itself, insofar as this discount rate is below the marginal productivity of capital in *S*2, and insofar as construction of a specific external alternative in *S*3 displaces a run-of-the-mill investment in *S*2, the opportunity cost of this displacement must be included in the computation of costs of the alternative from *S*3. In other words, if marginal investments throughout the economy have a positive net present value when discounted at the rate of social time preference, the net present value that would have been earned by the marginal investments must be included as an (opportunity) cost of the specific alternative to the extent that construction of a specific alternative actually diverts funds from these marginal investments. We shall assume that the run-of-the-mill alternatives to specific alternatives come from the same source as the general alternatives to government water-resource development, that is, that *S*2 is the pool of alternatives to both *S*1 and *S*3.

In showing how this revised definition of system net benefits can be applied in maximizing the value of an objective function, we first examine a relatively simple case with but one external alternative and no budgetary constraint. Assume that the objective is economic efficiency.[41]

[41] Once again, this assumption is not really restrictive; for any other objective we simply replace benefits and costs, as defined with respect to the efficiency objective, with their appropriate counterparts.

If external alternatives are disregarded, this objective is to choose the levels of electric-energy production, flood protection, irrigation water, or other system outputs, and the levels of reservoir capacities, power plants, irrigation canals, and other inputs so as to maximize

$$\sum_{q=1}^{Q} \theta_q[E_q(\mathbf{y}_q) - M_q(\mathbf{x})] - aK(\mathbf{x}) \tag{4.41}$$

subject to the production-function restriction $f(\mathbf{x}, \mathbf{y}_q) = 0$.[42] The coefficient a represents the opportunity-cost coefficient of Exp. (4.40); that is, the term $aK(\mathbf{x})$ represents Exp. (4.40).

Let us assume that the external alternative is steam energy, which is an economic alternative to system energy since it would meet the same demand (although not necessarily to the same extent). However, because steam plants would not occupy system hydroelectric sites, steam energy is not a technologic alternative. First, we must identify the relevant alternatives, as defined by the benefits and costs to the alternative producer (modified by the licensing and regulatory powers of government units). This set of relevant alternatives can be summarized in a functional relationship in which the quantity of external steam energy actually produced is taken as dependent upon the quantity of system energy actually produced. If we denote steam energy by the variable A^1 and system energy by the variable y_1, this function $A^1(y_1)$ can be depicted graphically as in Fig. 4.6.

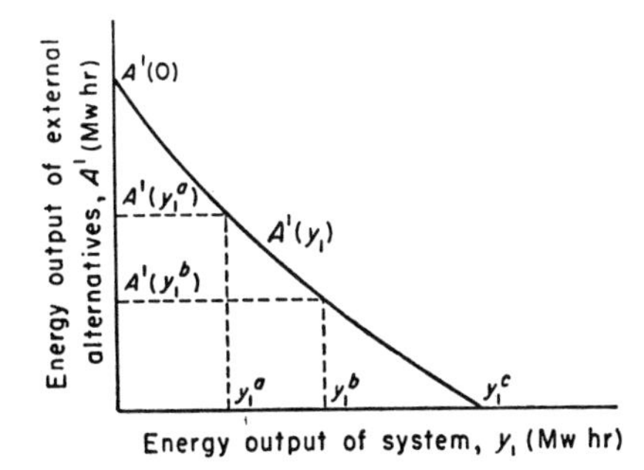

Fig. 4.6. Relationship between energy output of system and that of external alternatives (steam plants).

If no system energy is provided, $A^1(0)$ Mw hr of external energy will, by assumption, be produced. As system-energy output is expanded, the demand for energy is gradually pre-empted and the production of external energy decreases. When system-energy output reaches y_1^c units, external-

[42] We need not for present purposes distinguish between construction inputs ($\mathbf{x}^0$) and OMR inputs ($\mathbf{x}'$).

energy production is no longer profitable at all (in that the cost to the external producer presumably exceeds the benefits to him), and for increments of system-energy output beyond y_1^c there is consequently no external alternative. The external alternative to any increment of system energy between zero and y_1^c Mw hr may be found directly from Fig. 4.6. The alternative to the increment from y_1^a to y_1^b Mw hr, for example, is the increment from $A^1(y_1^b)$ to $A^1(y_1^a)$ Mw hr of external energy; that is, the increase in system-energy output from y_1^a to y_1^b Mw hr displaces $A^1(y_1^a) - A^1(y_1^b)$ Mw hr of external energy.

The next step is the computation of the additional net benefits provided by each level of external energy produced. Emphasis is properly placed on the word "additional"; for each quantity of external energy, we wish to determine the net benefits over and above those of the accompanying quantity of system energy. We write these net benefits as follows:

$$\sum_{q=1}^{Q} \theta_q\{E_q^1[A^1(y_1)] - M_q^1[A^1(y_1)]\} - a'K^1[A^1(y_1)]. \tag{4.42}$$

Here the term $E_q^1[A^1(y_1)]$ represents the additional benefit of external energy during demand period q; $M_q^1[A^1(y_1)]$ the OMR cost during demand period q; and $K^1[A^1(y_1)]$ the construction cost, all expressed as functions of the level of external energy. The parameter a' is given by the opportunity-cost coefficient of Exp. (4.40) but with the values of the α's appropriate to displacement of investment in $S2$ by specific alternatives substituted.

Total net benefits are the sum of expressions (4.41) and (4.42):

$$\sum_{q=1}^{Q} \theta_q[E_q(\mathbf{y}_q) - M_q(\mathbf{x})] - aK(\mathbf{x}) + \sum_{q=1}^{Q} \theta_q\{E_q^1[A^1(y_1)] - M_q^1[A^1(y_1)]\} - a'K^1[A^1(y_1)]. \tag{4.43}$$

Expression (4.43) expresses the efficiency objective as maximization of the net efficiency benefits of the system proper plus the additional net efficiency benefits of the quantity of external energy that remains economically feasible for each given level of system energy. This is equivalent to designing the system to maximize the difference between the net benefits of the system proper and the net benefits of the displaced alternative source of energy, or designing the system so as to maximize the incremental efficiency of the system over and above the efficiency of alternative production of energy.[43] Whereas formulation of the objective as maximization of the superiority of system outputs over external sources of outputs points up the alternative relationship of external to system energy, it disguises the true objective expressed in Exp. (4.43), namely, maximization of the net benefits from both sectors $S1$ and $S3$. Nonetheless, the second formulation is intuitively more useful in certain circumstances, as we shall see.

If the external alternative is technologic rather than economic, there is little change in the manner of determining the effect on system benefits. We begin by expressing the set of relevant alternatives as a function of the level of the corresponding system element, but in this case it is an *input* instead of an *output* that displaces the alternative. Suppose that a private hydroelectric development is an alternative to a government flood-control reservoir at site m with a capacity denoted by the variable x_m. Assume, further, that the government is not interested in direct development of the hydroelectric potential of the site itself. As we have pointed out before, the relevancy of technologic alternatives may be determined more by the licensing and other powers of the government and less by dollar-benefit considerations than is the case for economic alternatives. Nonetheless, we have a function, $A^m(x_m)$, relating the level of external hydroelectric-energy production at site m, A^m, to the capacity of the flood-control reservoir, x_m. This function may be discontinuous, since any flood-control capacity, no matter how small, displaces the private hydroelectric storage reservoir altogether. This possibility is illustrated by the function represented in Fig. 4.7(a), in which A^m is positive for $x_m = 0$ and zero for all positive levels of x_m. But joint private and government development of the site is also a possibility; in such a case successive increments of flood-control protection simply make private hydroelectric-energy production more expensive, as marginal units of energy more quickly cease to cover their costs. The result is that the function $A^m(x_m)$ decreases gradually, as shown in Fig. 4.7(b).

Figure 4.7 shows two possibilities, but others, such as multiple discontinuities and sharp corners, are equally likely to occur. Whatever the shape of the function $A^m(x_m)$, the next step is to relate the net benefits of the alternative to the level of external energy produced. This is done by the function

[43] This can be shown by subtracting from Exp. (4.43) the net benefits from external energy accompanying the zero output level of system energy. Subtraction of this constant term does not affect the choice of benefit-maximizing system design.

$$\sum_{q=1}^{Q} \theta_q\{E_q^{m}[A^m(x_m)] - M_q^{m}[A^m(x_m)]\} - a'K^m[A^m(x_m)]. \tag{4.44}$$

Just as for economic alternatives, these benefits are added to the net benefits of the system proper, Exp. (4.41). We point out once again that the form of maximand (the net benefits from system and external outputs, or the net benefits of the system less the benefits of the alternative displaced) is optional.

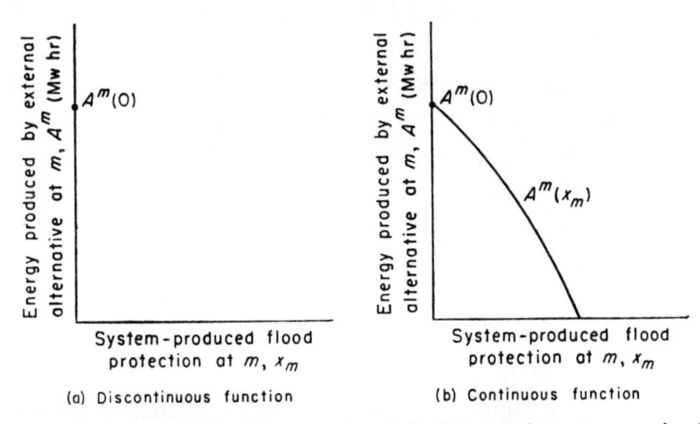

Fig. 4.7. Relationship between system-produced flood protection at reservoir site m and an external alternative consisting of hydroelectric-energy production at m.

If an alternative is simultaneously economic and technologic, its relationship to the system is expressed by writing the A-function as dependent on both the output and the input that displace the alternative. For example, if the system in which the flood-control reservoir x_m figures can also produce electric energy (at site m or elsewhere) for which the private hydroelectric development at m is alternative, we write the level of external-energy production as a function of the capacity of the system reservoir at m, x_m, and the level of system-energy output, y_1, thus: $A^{1,m}(y_1, x_m)$. Apart from this modification, benefit-counting proceeds as above, with

$$\sum_{q=1}^{Q} \theta_q\{E_q[A^{1,m}(y_1, x_m)] - M_q[A^{1,m}(y_1, x_m)]\} \\ - a'K^{1,m}[A^{1,m}(y_1, x_m)] \tag{4.45}$$

replacing Exp. (4.42) or (4.44) as the term to be added to Exp. (4.41).

When, as will often be the case, there is an alternative for more than one output or input, we must go down the list of system outputs and inputs, determining first the alternatives that are both economic and technologic, and then the alternatives that are either economic or technologic, but not both simultaneously. With the A-functions for all alternatives thus defined, benefits and costs are computed as functions of the levels of the external outputs (the A's) provided. Net system benefits are obtained by adding the resulting benefit functions to the net-benefit function for the system proper, Exp. (4.41).

The literature of water-resource-planning places considerable emphasis on a somewhat different use of data relating to the specific-external-alternatives sector: the use of alternative costs as a measure of system benefits.[44] We should state explicitly what the relationship of this is to our exposition of the influence of alternatives on system net benefits. The concept of alternative costs of outputs relates of course to economic alternatives. It has been argued that an upper limit on the benefits that can be claimed for any output of a water-resource system is given by the cost of providing the output by alternative means. For this to be true, the following condition must hold: the incremental net *benefits* of the alternative must equal the *benefits* of the quantity of system output that would displace the alternative. It is not sufficient (and theoretically not even necessary) for the alternative to provide the identical *quantity* of output that the system would provide.[45] Thus, to return to an earlier example, whereas it would not be necessary that an external alternative source of steam energy provide exactly the same quantities of energy as the system sources it rivals, the benefits of the quantities it provides must be identical with the benefits of the corresponding amounts of system energy if alternative costs are to be used as a measure of system benefits.

The use of alternative costs as a measure of benefits is justified by some as a matter of convenience. If the planner wishes to employ the cost of an (external) alternative as the measure of the benefits of a system output, he must be sure that the gross benefits of the alternative are equal to the gross benefits from system outputs, for alternative costs in themselves have no ultimate significance.

The significance of S3 alternatives for system design. The effect of $S3$

[44] See, for example, Eckstein, reference 1, pp. 52–53 and 69–70; and Inter-Agency Committee on Water Resources, reference 14, pp. 40–42.

[45] Of course, it is difficult to see how, in practice, the benefits from the alternative could equal those of the system output unless the alternative quantity of energy were at least as great as the corresponding system quantity. Indeed, even if it were as great, inefficient (from the social point of view) distribution practices might drive the benefits of alternative production below those of the same amount of system energy.

alternatives on system design depends on the relationship of the benefits of system outputs and inputs, net of all opportunity costs, to the social benefits of displaced alternatives, net of their opportunity costs. The greater opportunity costs of government expenditure in the presence of budgetary constraints that effectively limit the size of the government water-resource development program mean that external alternatives will, by and large, be more attractive in the presence of budgetary constraints than in their absence. This accords with one's common-sense expectation. When funds are limited, aspects of federal water-resource systems for which no external alternatives exist are relatively more desirable than otherwise, since the exploitation of the other opportunities by external alternatives will in general produce substantial social net benefits, although perhaps not as much as would government development. On the other hand, those opportunities for which no *S*3 alternatives exist will, by definition, not be developed at all in the absence of government action. Since the two cases — budgetary constraint and no budgetary constraint — differ, they must be treated separately. Accordingly, we turn first to the effects on system design in the absence of budgetary constraints of the change in the objective function outlined above.

External alternatives are designed to maximize the net benefits to the external entrepreneur, whom we assume to be represented by a profit-seeking organization in the discussion that follows. To this entrepreneur social benefits from income redistribution, employment, regional growth, and the like are in general of no interest, and design, operation, and distribution of outputs of alternatives will not be influenced by such considerations except to the extent imposed by the government through licensing and other means. Hence, for such nonefficiency objectives external alternatives will generally be less effective than a government system in achieving the desired goal.

With respect to the efficiency objective, the case is somewhat more complicated. In some ranges of output, economic external alternatives may be more efficient than their rival system outputs, even though the goal of the external entrepreneur is maximization of private rather than social net benefits. The social benefits of an alternative may be almost as large as those from an equivalent output produced and distributed by the government; but the costs of the external alternative may be considerably less than the costs of providing the output within the system

For example, the marginal costs of external production of steam energy may be much less than internal production of hydroelectric energy after a certain level of this energy output is reached, and the benefits of externally produced energy may be almost as great as those of internally produced energy. If the production of steam energy internally (that is, in $S1$) is not possible, the external alternative may be superior to the production of more internal hydroelectric energy. And even if the marginal net benefits of system hydroelectric energy are greater than those of external steam energy, the net benefits of a competitive system output such as flood control might, when added to the social net benefits of external energy, exceed the net benefits foregone on system energy. In other words, deleting some hydroelectric energy from a proposed system, leaving it to be produced by an external entrepreneur, might increase over-all social net benefits by making possible the use of scarce capacity for flood protection. Although an increment in system energy might yield greater net benefits than either alternative energy or system flood control, the combination of external energy with system flood control might be the best solution of all in terms of the efficiency objective function.

In the presence of budgetary constraints technologic, economic-technologic, and economic alternatives are likely to affect optimal system design more significantly — and this, to some extent, at least, for all objectives. In the face of an internal opportunity cost $1 + \lambda$ (imposed by a budgetary constraint) higher than the external opportunity cost a, the extra social benefits that government development brings in comparison with an external alternative may not justify the marginal cost of government development. This is shown graphically in Fig. 4.8, which shows the marginal benefit–cost ratios for two economically independent systems A and B as functions of construction outlay, K_A and K_B.

The curves labeled A_1 and B_1 represent the respective functions when no external alternatives are considered — that is, the curves are drawn up in terms of the objective function, Exp. (4.41). The horizontal line at the value $1 + \lambda^1$ represents a hypothetical cutoff benefit–cost ratio (an internal opportunity cost) determined by the level of the budget and the attractiveness of all potential national water-resource developments. If we ignore external alternatives, the optimal levels of expenditure on the two systems are $K_A{}^1$ and $K_B{}^1$, at which points the marginal benefit–cost ratios fall to $1 + \lambda^1$. If we consider external alternatives, however, their availability reduces somewhat the pressure on the government budget, so that a lower cutoff benefit–cost ratio, $1 + \lambda^2$ (also represented in Fig. 4.8), will replace the value $1 + \lambda^1$.

But the system marginal benefit–cost-ratio functions will not be affected identically. If system A has many alternatives and system B has none, the marginal benefit–cost-ratio function of system A will be reduced, say, to the broken curve labeled A_2, whereas the curve of system B will remain unchanged. The optimal outlay for system A is reduced from $K_A{}^1$ to $K_A{}^2$, whereas the optimal outlay for system B is increased

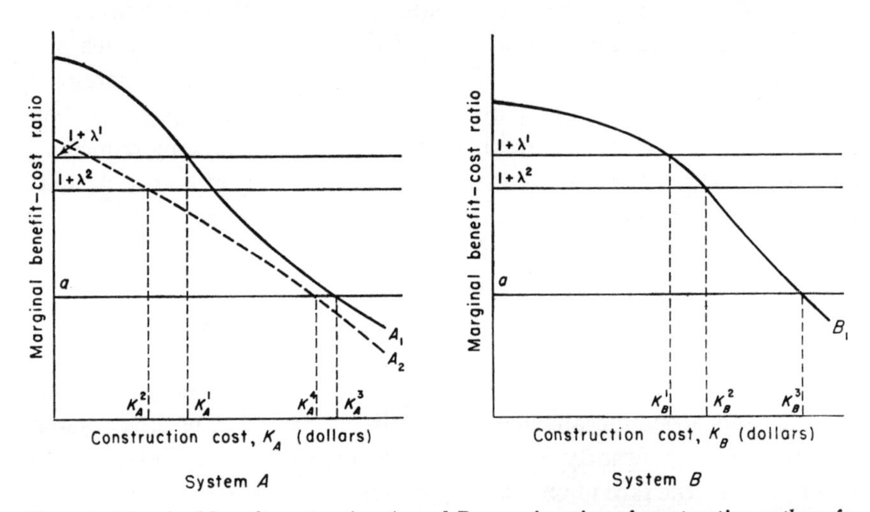

Fig. 4.8. Marginal benefit–cost ratios A_1 and B_1 as a function of construction outlays for two economically independent water-resource systems A and B. Curves A_1 and B_1 represent these functions when no external alternatives are considered. Curve A_2 represents the marginal benefit–cost ratio for system A in the presence of external alternatives; the ratio for system B remains the same as when there are no external alternatives. Lines $1 + \lambda^1$ and $1 + \lambda^2$ represent hypothetical cutoff benefit–cost ratios (internal opportunity costs) in the presence of a budgetary constraint without external alternatives and with them, respectively, and line a represents the external opportunity cost when there are no budgetary constraints.

to $K_B{}^2$ because funds become less scarce. In the presence of alternatives, the expenditure on system A will be limited to those increments for which there are no alternatives and those aspects for which the alternatives perform most poorly in relation to the planning objective, so poorly indeed that the marginal benefit–cost ratio of system increments, even in the presence of these alternatives, is greater than $1 + \lambda^2$. Because of the fall in the cutoff benefit–cost ratio, these latter inputs and outputs (like those of system B) will be developed more intensively than would have been the case if alternatives had been ignored.

This considerable effect of alternatives in the presence of budgetary constraints is to be compared with their almost negligible effect (for the efficiency objective at least) in the absence of constraints. This is illustrated in Fig. 4.8 by the reduction of optimal outlay for system A from $K_A{}^3$ to $K_A{}^4$. Of course, there is no effect on system B in the absence of budgetary constraints; the optimal outlay remains $K_B{}^3$.

It is well to remember what the drop in the marginal benefit–cost-ratio function of system A represents: the increments of system A do not become absolutely less valuable, but only relatively so. In terms of the objective viewed as maximization of net benefits of all system increments and all undisplaced alternatives, the difference between the marginal benefit–cost-ratio function A_1 without alternatives and the function A_2 adjusted for alternatives occurs because of the opportunities for external alternatives created by the absence of system development.

RELATIONSHIP BETWEEN GOVERNMENT WATER-RESOURCE DEVELOPMENT (S1) AND THE CREATION OF EXTERNAL OPPORTUNITIES (S3)′

In the preceding section dealing with the specific-external-alternatives sector, we were concerned with the effects on system design of alternatives that would be displaced, and therefore with benefits that would be lost, by the construction of system increments. The obverse of such displacement is the creation of external opportunities by water-resource development. The social benefits from such external opportunities are often referred to as secondary benefits.[46] A government hydroelectric system suitable for production of peaking power may, for example, create an opportunity for an externally constructed, base-load steam plant that would not be economically feasible without the peaking power of the government system. And a large-scale government irrigation development, in creating entire new communities or making old ones more prosperous, usually generates a wide range of external opportunities to supply the goods and services that the beneficiary farmers require. How do the benefits of external opportunities created by a government river development, that is, investments in $S3'$, affect the objective function, and in turn system design? Do the benefits of all such opportunities "count," that is, do they influence system design?

[46] For a fuller discussion of this subject, see Krutilla and Eckstein, reference 23, chaps. 2 and 3; and Julius Margolis, "Secondary benefits, external economies, and the justification of public investment," *Rev. Econ. and Stat. 39*, 284 (1957).

The answers to these questions depend principally on the objective. We shall therefore examine this problem of secondary benefits in relation to each objective individually. In doing so, we shall find the concepts developed in the preceding section to be of considerable help because of the obverse relationship between the creation of external opportunities in $S3'$ and the displacement of external opportunities in $S3$. In fact, we have previously viewed alternatives as external opportunities created by government inaction, so that the only formal difference between the two types of external opportunities is that we are now dealing with opportunities created by *positive* government action.

The efficiency objective. Let us take as an example a government hydroelectric system well suited for peaking purposes. Suppose that the level of base-load private steam energy that could be profitably developed, $\hat{A}(y_i)$, where $\hat{A}$ represents energy output of the external alternative created and y_i represents system-energy output, increases with the level of system energy as shown in Fig. 4.9. A comparison of Figs. 4.6 and 4.9

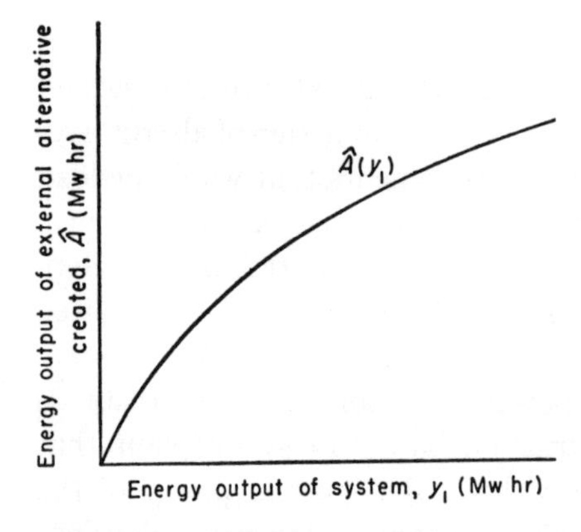

Fig. 4.9. Relationship between energy output of system and that of external alternative created.

reveals quite clearly the fundamental relationship between external alternatives and positive external opportunities created: the slope of the A-function in Fig. 4.6 (giving the amount of external alternative output not displaced for each level of system output or input) is negative, whereas the slope of the $\hat{A}$-function in Fig. 4.9 is positive. In Fig. 4.6, therefore, the amount of external alternative output not displaced decreases with increases in system outputs or inputs, whereas in Fig. 4.9 the amount of outputs of positive external opportunities created increases with increases in system outputs.

For any level of system-energy output, over-all social net benefits are the sum of system benefits and the additional social net benefits contributed by the private steam-energy development. The system benefits are given by the sum of Exps. (4.41) and (4.45), and the additional social benefits by

$$\sum_{q=1}^{Q} \theta_q[\hat{E}_q\hat{A}(y_i) - \hat{M}_q\hat{A}(y_i)] - a\hat{K}A(y_i). \tag{4.46}$$

Inclusion of Exp. (4.46) in the objective function means that system-energy output is optimally expanded until the marginal social benefit of system energy together with the additional marginal (net) social benefit of the external steam energy made possible by the government system falls to the marginal cost of system-energy output (provided that there is no effective budgetary constraint and that second-order conditions of maximization are satisfied at this point). If the additional marginal social benefit of external energy is positive, this net-benefit-maximizing level of system-energy output will be greater than the net-benefit-maximizing level without the external opportunity for steam-energy production created by the generation of system energy. There is thus both a scale and a substitution effect: secondary benefits tend to increase the over-all scale of systems and to increase the development of outputs (such as system energy in the present example) that create secondary opportunities at the expense of outputs that do not.

The method for reflecting the benefits from external opportunities in the objective function then is to determine $\hat{A}$-functions for every such opportunity,[47] and to add their net-benefit functions, each in the form of Exp. (4.46), to the objective function (4.41) plus (4.45).

However, our work is complete in a formal sense only. Are all external opportunities created (including, for example, opportunities for services such as barbershops and moving-picture houses in newly irrigated areas) of equal significance in determining optimal design? The answer is implied in Exp. (4.46): we need to evaluate explicitly only those external opportunities created by water-resource development for which the social benefits, net of all opportunity costs, are nonzero.[48] In other words, only the nonmarginal opportunities are of importance for system design. The

[47] The $\hat{A}$-functions may be more complicated than the simple one represented in Fig. 4.9. Creation of external opportunities may depend, for instance, on the level of many system elements rather than only one as in Fig. 4.9.

[48] In the paragraph that follows we shall be concerned with potentially supramarginal external opportunities. External opportunities created for which the social net benefits are negative must of course also be analyzed explicitly. Their effects on system design are the opposite of the effects of opportunities with positive net benefits.

nonmarginality of a broad class of external opportunities depends in turn on whether or not the economy is operating at full employment.

At full employment in a competitive economy, opportunities of the barbershop or moving-picture house variety are simply marginal with respect to the efficiency objective, for such opportunities created by the demands of the system beneficiaries are offset by the displacement of similar opportunities throughout the rest of the economy. The resources required to construct and operate such consumer services for the system beneficiaries presumably would be used in the absence of the system to construct and operate similar services elsewhere in the economy, and there is no reason to suppose that the system-created opportunity is superior to the displaced opportunity. Thus, on balance, the system-created opportunity adds no social net benefits. In other words, social benefits and costs need be included in the objective function for only those opportunities that represent external economies of the government system.

Redistribution objectives. In this section we cover three aspects of the relationship of secondary opportunities to redistribution objectives. First, the effect of secondary opportunities on redistribution benefits depends on whether the objective is a group or regional one, and we shall briefly explore this asymmetry. Second, the impact on design of these changes in redistribution benefits depends on which of the methods outlined in Chapter 2 for implementation of a redistribution objective is chosen. (The question of choice of methods arises, it will be recalled, because implementation of redistribution objectives involves both efficiency and redistribution considerations; the composite nature of redistribution objectives creates a degree of freedom as to whether the efficiency and redistribution components both appear in the function to be maximized by system design or one of them appears as a constraint on system design.) The third aspect of the relationship between secondary opportunities and redistribution objectives can be disposed of quickly: in addition to scale effects, secondary opportunities lead to substitution effects, just as in the case of a pure efficiency objective. But these are independent of the method chosen to implement the redistribution objective. The existence of disproportionate secondary opportunities among outputs always tends to encourage substitution of outputs for which the secondary opportunities created for redistribution of income are large for outputs for which these opportunities are small, just as on the efficiency side of the coin.

We turn now to the first point, the asymmetry between group and regional redistribution objectives in the impact of secondary opportunities on redistribution benefits.

(1) With respect to the benefits of income redistribution to a selected *group* of beneficiaries, secondary opportunities require no special attention; the net benefits of such opportunities to the intended beneficiaries, insofar as they are conscious of these benefits, are fully reflected in their willingness to pay for system outputs. For example, the supramarginal advantages to the intended beneficiaries of a secondary opportunity for steam energy will be reflected, insofar as the beneficiaries are aware of them, in the extra willingness of the beneficiaries to pay for the system outputs created by the supramarginal opportunity.

(2) The objective of redistribution to a particular *region*, unlike redistribution to a particular group, imposes no restrictions on the individuals whose gains are to be counted as benefits. Thus we rightly look beyond the direct beneficiaries of the system. For this objective the contribution of secondary opportunities is measured by the amount of income they generate within the region. Even in conditions of full employment, therefore, we count not only those opportunities representing external economies of the system, but all secondary opportunities created in the region — whether or not they are marginal from the point of view of the economy as a whole, that is, whether or not they are marginal with respect to the efficiency objective. And not only the incomes generated within the region from the further processing of goods produced by direct beneficiaries, but also the incomes generated in the region by the provision of consumer goods and services to the direct beneficiaries, are properly counted under a redistribution objective. The fact that ten opportunities for consumer services created by water-resource development in Arizona are offset by the displacement of ten similar opportunities in New England or California is of no consequence if the objective is redistribution of income to Arizona.

It would appear then that, besides the income within the region from processing and similar activities, the regional redistribution benefits of all consumer services — every barbershop, moving-picture house, bank, and fruitstand — that owe their existence to a water-resource system must be computed and that, in turn, opportunities created in the region by these services must also be counted, and so on. Fortunately, we can cut through this chain by means of the familiar multiplier analysis of economic theory: neglecting the particular kinds of economic activity created by these secondary, tertiary, and later-round opportunities, we seek instead to measure the aggregate income generated within the region by the expenditure of the direct beneficiaries of the system.

The redistribution benefits to the direct beneficiaries within the region in year t are given by

$$E_t(\mathbf{y}_t) - \sum_j p_{y_{tj}} y_{tj}, \tag{4.47}$$

where $E_t(\mathbf{y}_t)$ represents the maximum willingness of beneficiaries to pay and $\sum_j p_{y_{tj}} y_{tj}$ their actual payment in year t, both expressed as functions of the output $(\mathbf{y}_t)$. Expression (4.47), as we have seen in Chapter 2, represents the income of direct beneficiaries in year t creditable to the water-resource system. If b is the marginal propensity to spend income within the region and $1 - b$ is the marginal "leakage" (including savings and expenditures outside the region) per dollar of income, and if the multiplier chain is assumed to work itself out completely within the year, then the total secondary income generated within the region by the expenditure of direct beneficiaries in year t is

$$[b/(1 - b)][E_t(\mathbf{y}_t) - \sum_j p_{y_{tj}} y_{tj}].$$

The sum of direct benefits and secondary benefits from consumer services[49] is

$$[1/(1 - b)][E_t(\mathbf{y}_t) - \sum_j p_{y_{tj}} y_{tj}].$$

Whether implicitly, as in the case of a group-redistribution objective, or explicitly, as in the case of a regional objective, the creation of secondary opportunities adds to the redistribution benefits of any design. But, as we indicated, the impact on design of this addition to benefits depends on which of the methods of Chapter 2 is employed to implement the objective. More specifically, it depends on whether redistribution enters the picture as a constraint on system design (method I, pp. 70–78); as part of a weighted sum with efficiency in the function to be maximized by system design (method II, pp. 78–81); or as the sole maximand subject to a constraint on performance of the system in terms of efficiency (method III, pp. 81–82).

If redistribution is a constraint on system design, that is, a minimum

[49] A similar multiplier analysis should be made for the further generation of income by expenditures out of the incomes from processing and other immediate secondary opportunities within the region, including supramarginal opportunities (like the steam plant discussed in relation to the efficiency objective) which represent external economies of the system.

level of annual income is specified for the intended beneficiaries, then the existence of secondary opportunities means that a smaller system will permit the intended beneficiaries to attain the specified level of income. If, on the other hand, income to the intended redistribution beneficiaries is given an explicit weight relative to efficiency, and the design goal is formulated as maximization of the weighted sum of redistribution and efficiency benefits, then for any given value of the weight the extra redistribution benefits arising from secondary opportunities will make larger developments more attractive, and the result will be larger systems. This result is reinforced insofar as the secondary opportunities are external economies and thus also add to efficiency benefits. Finally, if the composite objective is cast as maximization of the redistribution benefits subject to a specified minimum level of performance in terms of efficiency, the additional redistribution benefits from secondary opportunities produce no scale effects. The component of the composite objective that limits the over-all scale of development in this case is specification of a minimum level of efficiency net benefits; consequently, secondary opportunities relax this limitation only insofar as they represent external economies and thus improve system performance in terms of efficiency.

CONCLUSION

An essential element of the process of system design is to determine those economic factors which have major relevance. Each planner must ask the following questions: Are budgetary constraints, in fact, applicable to the design problem? In view of the dynamic nature of the design problem, is a static analysis adequate? What is the appropriate rate of discount for use in design of the system? Are there general and specific external alternatives and external opportunities created that must be considered? To the extent that budgetary constraints, dynamics, and external alternatives must be taken into account explicitly, the design problem becomes increasingly complex and difficult. In this chapter we have presented certain techniques for reducing the adverse effects of complexity — by use of shadow prices when there are budgetary constraints, for example. But these and other problems will require much further exploration as this general framework of analysis is applied to actual planning of water resources.

5 Application of Basic Concepts: Graphic Techniques

MAYNARD M. HUFSCHMIDT [1]

WE have suggested in Chapter 3 that it is possible to develop optimal designs for very simple water-resource systems by graphic techniques that establish the relationships between certain basic economic and technologic concepts. The nature of these techniques is discussed in this chapter, and their use is illustrated for the system shown in Fig. 5.1, which contains three potential structures (two reservoirs and

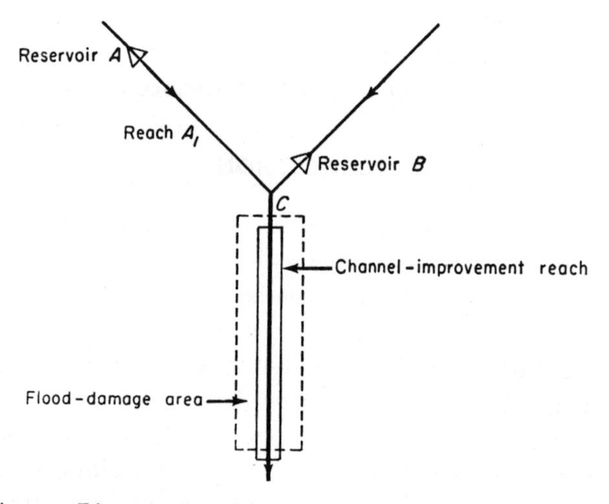

Fig. 5.1. Diagrammatic sketch of a simple water-resource system.

a channel improvement) that, in combination, serve the single purpose of flood control. Application of graphic techniques to this system will be deferred, however, until we have discussed in general terms how certain basic concepts of Chapter 3 — the production function and the net-benefit function — will be used.

[1] The model presented in this chapter was developed by Joseph G. Polifka while on leave from the U. S. Soil Conservation Service to Harvard in 1957–58.

GRAPHIC TECHNIQUES

As stated in Chapter 3, the production function can be split into two parts, one concerned with the relationship between the resources consumed (more specifically, their costs) and the capacity of the resultant structures, and the other with the relationship between the structures and the useful outputs. Cost–capacity curves are simple and straightforward, capacity–output curves more complex. Whereas the previous discussion focused on storing water for withdrawal uses, the present chapter is concerned with reservoir storage for reducing flood peaks and with channel capacity for safely handling residual peak flows. Because these capacity–output relationships are stochastic, they require probabilistic analysis, which in this chapter takes the form of flood-frequency analysis.

The concept of the net-benefit function discussed in Chapter 3 finds application in the present chapter in the following ways.

(1) The conditions that must be met if a design is to maximize net benefits — conditions that relate marginal rates of substitution and transformation — are applied to the three structures and single output of our simple example. Accordingly, Eq. (3.2) is solved for least-cost combinations of the two reservoirs and the channel improvement that will provide protection against floods of several degrees of magnitude and frequency. The condition that marginal system benefits must equal marginal system costs is then used to find the optimal combination of structures and optimal level of flood protection.

(2) Because the production function is stochastic, the net benefit to be maximized is itself a random variable; hence we can use the expected value of net benefits. This we derive by constructing a function showing the expected value of gross flood-control benefits for protection from floods of various degrees of magnitude and frequency and deducting the pertinent minimum costs.

(3) In solving for the least-cost combinations of reservoirs, we use curves of equal cost (iso-cost) and of equal flood-peak reduction that are analogous to the curves of equal output of Fig. 3.6.

The graphic techniques that illustrate the basic concepts in this chapter are familiar to many engineers and have been employed in the design of a number of developments. In planning the flood-retarding reservoirs for the Miami River Basin in Ohio before 1920, for instance, the engineering staff of the Miami Conservancy District developed a technique for simultaneous adjustment of the heights of several proposed dams and the scales of channel improvements that would achieve, at minimum cost, a desired degree of protection at flood-damage centers. To apply this technique, called "balancing" the flood-control system, the staff devised a graphic method.[2] However, the technique did not deal specifically with the problem of finding the optimal scale of development.

Both least cost and optimal scale were dealt with by J. S. Ransmeier, who demonstrated how the balancing procedure could be applied to a simple reservoir and levee system for flood control.[3] To do this, he first constructed a function showing the least-cost means of combining reservoirs and levees for preventing flood damage over a range of stream-channel capacities; then he showed how to select the scale of development of reservoirs and levees at which annual net benefits are maximized.

THE MODEL

In our model of two flood-retarding reservoirs on separate tributaries and one channel improvement in a flood-damage area below the confluence of the tributaries, the site of reservoir B (as shown in Fig. 5.1) is immediately above the confluence at C; the site of reservoir A is farther above the confluence. Possible measures for stabilizing and increasing the capacity of the channel include realigning and revetting its banks and enlarging its cross section by dredging. The objective is to find the combination of the two reservoirs and one channel improvement that will maximize the expected value of net benefits.

FLOOD FLOWS: MAGNITUDE AND CHARACTERISTICS

Because flood control is the sole issue, data on flood runoff are our main interest. Table 5.1 shows for each of four selected probability levels (1, 5, 10, and 20 percent) a set of assumed peak flows at the sites of reservoirs A and B and the point of confluence C. For example, there is a specific probability of one in a hundred that the peak flood flow of 31.0×10^3 cu ft/sec at C will be equaled or exceeded in any one year. We

[2] Sherman M. Woodward, *Hydraulics of the Miami flood control project*, Miami Conservancy District Technical Reports, pt. 7 (Dayton, Ohio, 1920), pp. 286–308.

[3] Joseph Sivera Ransmeier, *The Tennessee Valley Authority: a case study in the economics of multiple purpose stream planning*, (Vanderbilt University Press, Nashville, 1942), pp. 399–403. More complex systems are discussed on pp. 404–421. Ransmeier indicates as the source for his presentation a paper by Calvin Davis on "Reservoir economics," given before a seminar on the economics of multiple-purpose projects, conducted in 1940–41 by the Training Division of the TVA Personnel Department. The substance of this paper is contained in Calvin Davis, *Handbook of Applied Hydraulics* (McGraw-Hill, New York, ed. 1, 1942).

Table 5.1. Assumed peak flood flows for a simple water-resource system at selected frequencies of occurrence or exceedance.

Frequency of occurrence or exceedance (percent)	Peak flood flows (10^3 cu ft/sec)			
	At site of reservoir *A*	At site of reservoir *B*	Contribution of reach A_1 at *C*	At confluence point *C*[a]
1	20.0	15.0	4.0	31.0
5	16.0	12.0	3.2	24.8
10	12.0	9.0	2.4	18.6
20	10.0	7.5	2.0	15.5

[a] Peak flow at $C = 0.6 \times$ peak flow at $A +$ peak flow at $B +$ contribution of reach A_1.

assume that peak flows at *B* are translated to *C* with no diminution, that those at *A* are reduced by 40 percent, that the contributions of reach A_1

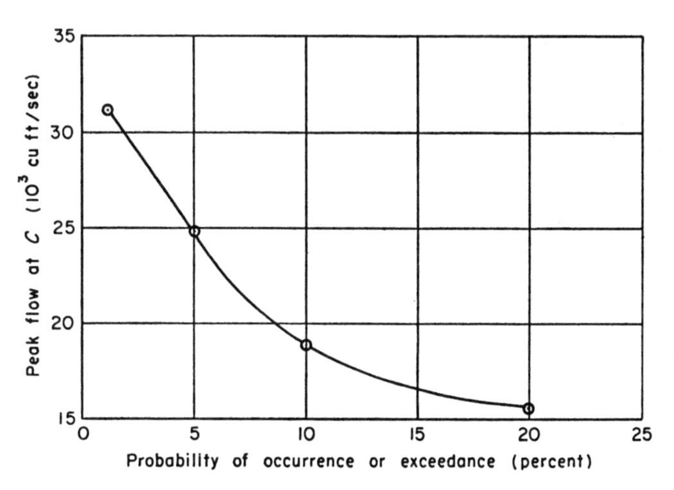

Fig. 5.2. Peak flood flow at confluence point C for stated probability of occurrence or exceedance.

are known and directly additive in forming peak flows at *C*, and that peak flows at given probability levels occur together at all three points of the system.[4] Based on the information presented in Table 5.1, we

[4] These relationships have been simplified and may not be consistently applicable over a range of actual cases. For example, a peak flow of given probability may occur at *A* along with a peak flow of higher probability at *B*. For small watersheds, however, the assumption of consistent correlation of flows is not unreasonable. The interrelationships of reservoirs and intermediate reaches in affecting flood peaks downstream also are usually more complex than we have assumed. However, flood-routing techniques for determining these effects are highly

can construct a continuous function of peak flood flows at C measured against probabilities of their occurring or being exceeded. This is done in Fig. 5.2.

ECONOMIC DATA

Assumed capital costs of reservoirs A and B and of the channel improvement over the relevant ranges of reservoir capacity and safe channel capacity, respectively, are shown in Figs. 5.3 and 5.4. Annual operation,

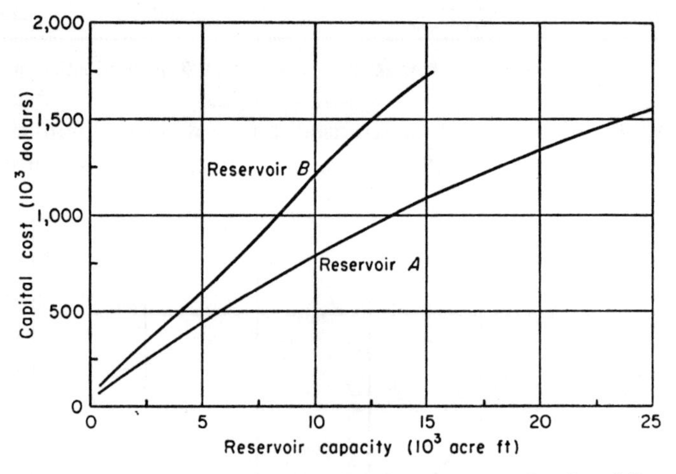

Fig. 5.3. Capital-cost–capacity functions for reservoirs A and B.

maintenance, and replacement (OMR) costs are assumed to be a fixed proportion of capital costs (0.25 percent for reservoirs A and B and 2.0 percent for the channel improvement).[5]

The benefit data for the system are based on a more complex type of analysis. For each peak flood flow of given probability shown in Fig. 5.2, there is an associated flood-damage value. These can be calculated over

developed and well-known to river-system-planners. We have assumed that by use of such techniques, and because of the simplifying assumptions previously stated, the effects of any combination of sizes of reservoirs A and B on reducing flood peaks at C can be determined readily.

[5] The assumption of fixed proportions between capital costs and average annual OMR costs covering all scales of development is less tenable for channel improvements than for reservoirs. Large-scale channel improvements by definition pass high flood flows, and the channel works are subject to major damage only by the relatively few floods with still higher peaks; smaller-scale channel improvements, designed to pass lower flows, are subject to damage from the more numerous floods that exceed the lower flows. Generally speaking, therefore, the smaller the scale of the channel improvement, the greater the ratio of OMR costs to capital costs.

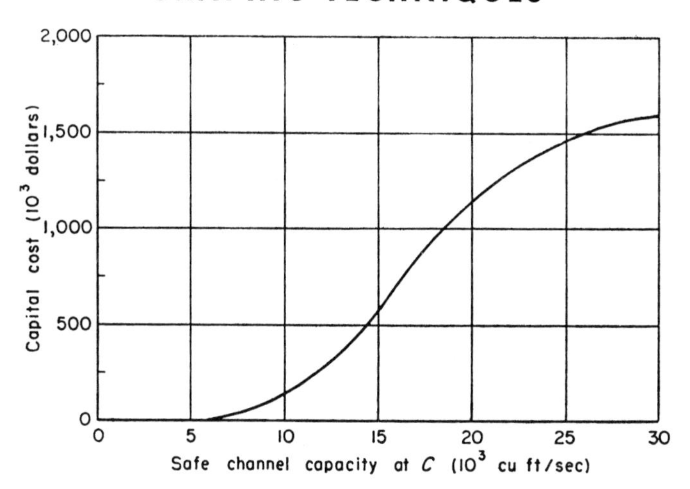

Fig. 5.4. Capital cost of channel-improvement works that will assure a safe channel capacity of given magnitude.

the entire range of flood flows by the conventional techniques of determining stage-discharge and stage-damage relationships. By conventional techniques, too, a flood-damage–frequency function can be constructed, as in Fig. 5.5, to show the chance (with uncontrolled conditions) of any

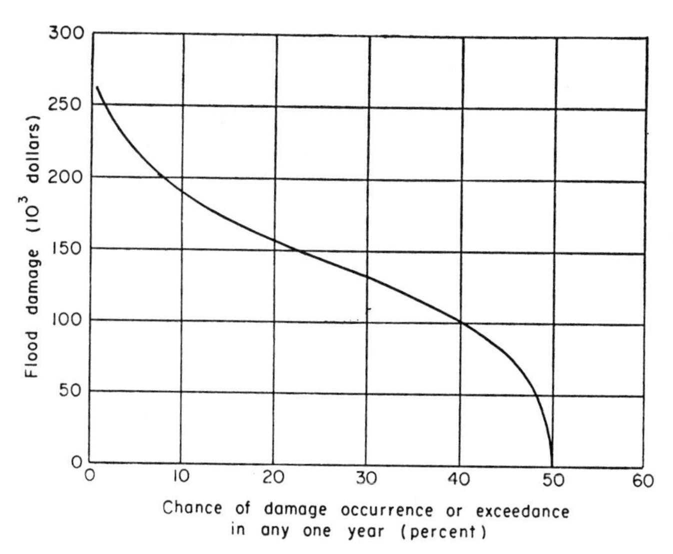

Fig. 5.5. Flood-damage–frequency function with uncontrolled conditions. Average annual damage, or expected value of annual flood damage, $= \Sigma$ *(damage* $\times$ *frequency expressed as a decimal)* $=$ *area under function/100* $= 72 \times 10^3$ *dollars.*

specific level of flood damage being equaled or exceeded in any one year.[6] From Fig. 5.5, for instance, there is a 40 percent chance that flood damages will be at least as high as $100,000 in any one year, and a 10 percent chance that they will be at least $190,000.

From this function the expected value of annual flood damage can be derived by weighting the damage values according to their probabilities of occurring or being exceeded. This is equivalent to integrating the curve shown in Fig. 5.5 or measuring the area under it in units of 10^3 dollars. The expected value of annual flood damage determined in this way is $72,000.

THE ANALYSIS

The primary goals in analyzing the information so far obtained are: (1) determination of least-cost combinations of reservoirs to give complete protection against floods of given magnitudes and frequencies; (2) determination of least-cost combinations of reservoirs and channel improvements to protect against such floods; and (3) determination of the optimal scale of development. In the terminology of Chapter 3, this amounts to locating certain efficient values on a stochastic production function, bringing the production and net-benefit functions together, and seeking the member of the production-function locus that maximizes the expected value of net benefits.

DETERMINING LEAST-COST RESERVOIR COMBINATIONS

Starting with reservoirs A and B, the isoquants of the two-unit, single-purpose example in Chapter 3 give direction to our analysis. However, we shall express the size of the reservoirs at A and B in terms of capacity rather than cost, as was done in Fig. 3.6. Curves of equal capital cost are then drawn analogous to the linear curves of equal output of Fig. 3.6 for the two reservoirs, as in Fig. 5.6. Each curve is constructed from the individual cost–capacity curves of Fig. 5.3 by selecting combinations of capacities with *specified* combined capital costs. The curve in Fig. 5.6 labeled $2{,}500 \times 10^3$ dollars, for instance, records all combinations of capacities of reservoirs A and B that can be provided for that capital

[6] The method used by the Corps of Engineers to compute the damage–frequency function is described by Otto Eckstein, *Water-resource development, the economics of project evaluation* (Harvard University Press, Cambridge, 1958), pp. 121–124. See especially Figs. 12 and 13.

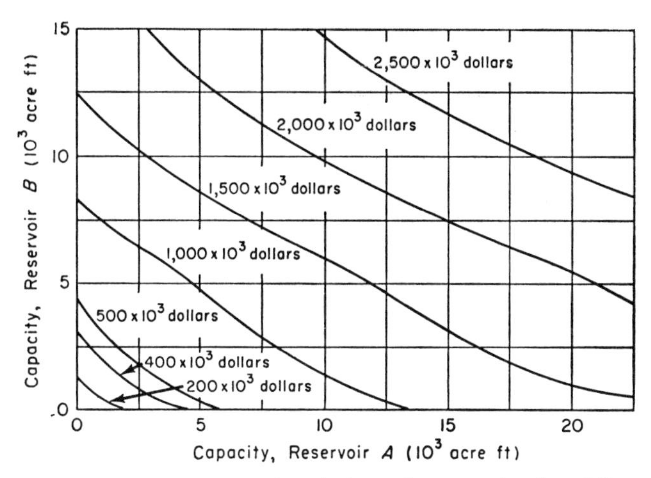

Fig. 5.6. Curves of equal capital cost for reservoirs A and B.

expenditure. Since OMR costs are assumed to be single-valued functions of capital costs, they need not be considered specifically at this point in the analysis.

Following the example in Chapter 3 further, we construct curves of equal flood-peak reduction for floods with various frequencies of occurring or being exceeded. These are analogous to the convex equal-output curves in Fig. 3.6 and are shown in Figs. 5.7 to 5.10 for flood-peak frequencies of 1, 5, 10, and 20 percent. Each curve is a trace of all combina-

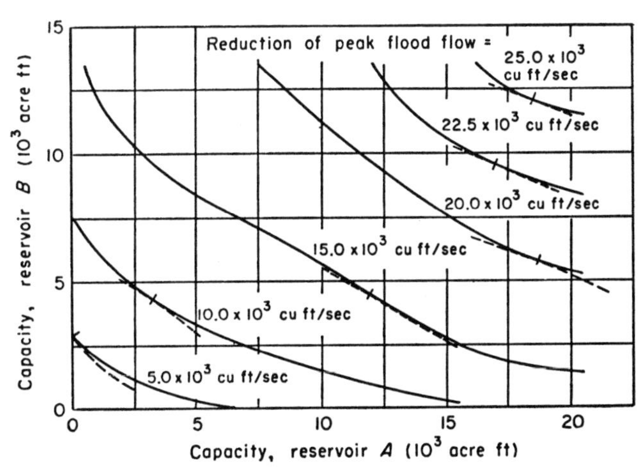

Fig. 5.7. Curves of equal flood-peak reduction for floods with 1 percent probability of occurrence or exceedance. Dotted lines indicate points of tangency of equal-cost curves.

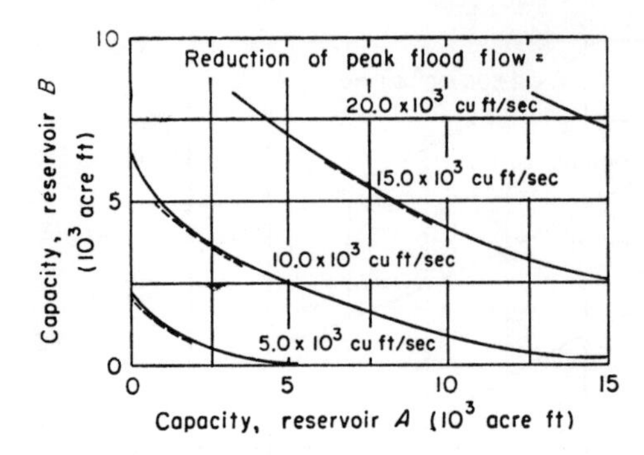

Fig. 5.8. Curves of equal flood-peak reduction for floods with 5 percent probability of occurrence or exceedance. Dotted lines indicate points of tangency of equal cost curves.

tions of capacities of the two reservoirs that will reduce a designated flood peak by the same amount. The uppermost curve in Fig. 5.7, for example, records all combinations of capacities of reservoirs A and B that will reduce by 25.0×10^3 cu ft/sec the flood peak with a 1 percent frequency of occurring or being exceeded.

To construct these curves, we route the flood flows corresponding to the four frequencies through the reservoirs to C, using the simplifying flood-routing assumptions discussed earlier. The flood hydrographs themselves will not be presented here; but a flood hydrograph typically possesses a single sharp peak, as shown in Fig. 12.3. The total volume of water that runs off during a specific flood can be estimated from the hydrograph. If the reservoir storage space is known, a revised hydrograph reflecting the withholding capabilities of the reservoir can then be constructed. In Fig. 5.7, for instance, storage of about 2.6×10^3 acre ft in reservoir B (and none in reservoir A) reduces by 5.0×10^3 cu ft/sec the

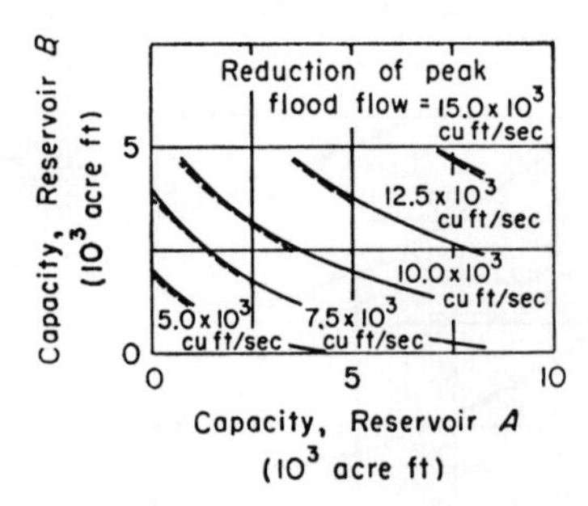

Fig. 5.9. Curves of equal flood-peak reduction for floods with 10 percent probability of occurrence or exceedance. Dotted lines indicate points of tangency of equal-cost curves.

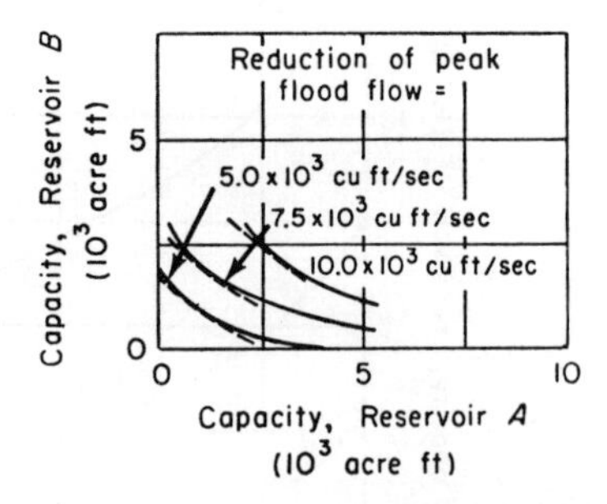

Fig. 5.10. Curves of equal flood-peak reduction for floods with 20 percent probability of occurrence or exceedance. Dotted lines indicate points of tangency of equal-cost curves.

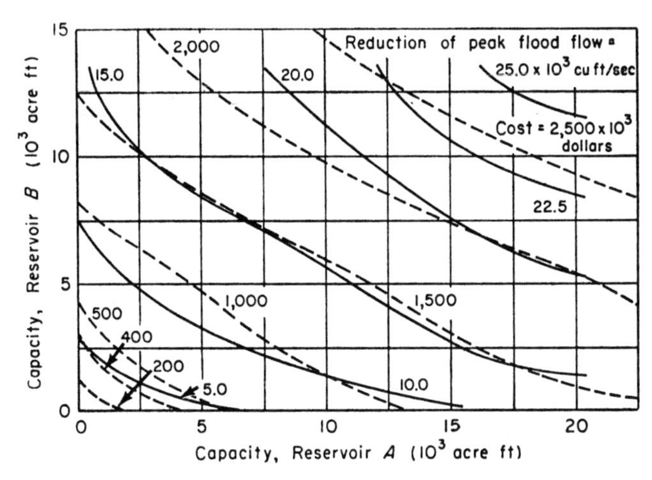

Fig. 5.11. Equal-cost curves (dotted lines) superimposed on curves of equal flood-peak reduction for floods with 1 percent probability of occurrence or exceedance. Values for equal-cost curves are in units of 10^3 dollars; for curves of equal flood-peak reduction, in units of 10^3 cu ft/sec.

flood peak with 1-percent probability of occurring or being exceeded. Table 5.1 puts the natural peak flow at point *C* for the 1 percent probability flood at 31.0×10^3 cu ft/sec. If we assume simple triangular shapes for both the natural and controlled flood hydrographs and a 12-hr duration for the 1 percent probability flood, it follows that its peak flow of 31.0×10^3 cu ft/sec and total water volume of $31.0 \times 10^3 \times 12/24 = 15.5 \times 10^3$ acre ft are translated by effective use of 2.6×10^3 acre ft of reservoir storage into a controlled flow, also of 12-hr duration, with a peak of 26.0×10^3 cu ft/sec and a total water volume of about 13.0×10^3 acre ft.[7]

Again, as in the example in Fig. 3.6, the curves of equal cost of Fig. 5.6 are superimposed on the curves of equal flood-peak reduction of Figs. 5.7 to 5.10 for each of the four selected flood-peak frequencies. An example for the flood peak with 1 percent probability of occurring or being exceeded is presented in Fig. 5.11. The superimposed curves identify the points on the six curves of equal flood-peak reduction that

[7] If the hydrographs of all floods, whatever their probabilities of occurrence or exceedance, had the same shape (not necessarily the simple triangular shape assumed for descriptive purposes in this chapter), only one set of curves of equal flood-peak reduction would be needed, because the task of reducing the peak flood flow by a designated amount would be the same in all instances. However, the shapes of actual flood hydrographs do differ and the curves of equal flood-peak reduction in Figs. 5.7 to 5.10 are unique, although they have the same general configuration. See Chapter 12 for further discussion of flood hydrographs.

Table 5.2. Least-cost combinations of reservoirs *A* and *B* required to provide certain reductions in flood peak at point C for floods with stated probabilities of occurrence or exceedance.

Natural flood flow				Reservoir *A*		Reservoir *B*			
Probability of occurrence or exceedance (percent)	Peak at *C* (10^3 cu ft/sec)	Reduction in flood peak at *C* by reservoir regulation (10^3 cu ft/sec)	Reduced peak at *C*, after regulation (10^3 cu ft/sec)	Capacity (10^3 acre ft)	Capital cost (10^3 dollars)	Capacity (10^3 acre ft)	Capital cost (10^3 dollars)	Total capital cost (10^3 dollars)	Average annual cost (10^3 dollars)[a]
1	31.0	5.0	26.0	0.00	0	2.60	360	360	13.6
1	31.0	10.0	21.0	3.25	313	4.25	525	838	31.5
1	31.0	15.0	16.0	11.80	910	4.50	550	1460	55.2
1	31.0	20.0	11.0	18.70	1275	5.75	680	1955	73.9
1	31.0	22.5	8.5	17.00	1190	9.50	1160	2350	88.9
1	31.0	25.0	6.0	18.50	1270	11.90	1450	2720	101.6
5	24.8	5.0	19.8	0.00	0	2.25	325	325	12.3
5	24.8	10.0	14.8	2.00	200	4.00	500	700	26.4
5	24.8	15.0	9.8	8.30	675	5.00	600	1275	48.2
5	24.8	20.0	4.8	13.50	1010	8.00	970	1980	74.9
10	18.6	5.0	13.6	0.00	0	2.10	315	315	11.9
10	18.6	7.5	11.1	0.50	65	3.00	400	465	17.6
10	18.6	10.0	8.6	3.00	280	3.00	400	680	25.7
10	18.6	12.5	6.1	3.50	325	4.80	590	915	34.5
10	18.6	15.0	3.6	7.40	610	4.80	590	1200	45.3
20	15.5	5.0	10.5	0.70	75	1.00	175	250	9.4
20	15.5	7.5	8.0	1.15	125	2.00	300	425	15.7
20	15.5	10.0	5.5	2.40	240	2.80	380	620	23.0

[a] Capital cost amortized over a 50-yr period at $2\frac{1}{2}$-percent interest rate plus OMR costs taken at 0.25 percent of unamortized capital cost.

are likely to be tangent to the appropriate equal-cost curves. The points of tangency, indicated in Figs. 5.7 to 5.10, are found by trial, equal-cost curves being drawn in the regions of apparent tangency. With the general shape of the surfaces of equal cost and equal flood-peak reduction determined, these trial solutions need not take long.

At points of tangency, the ratio of the marginal cost of the two reservoirs is equal to the ratio of their marginal productivities; or, expressed differently, the ratio of the marginal costs of the two reservoirs is equal to their marginal rates of substitution in accordance with the least-cost condition defined by Eq. (3.2). Therefore, the points of tangency in Figs. 5.7 to 5.10 identify the least-cost combination of reservoirs required to reduce the flood peak of the specified frequency.[8]

Least-cost combinations of reservoirs and their performance are summarized in Table 5.2. In the last column, costs of the various reservoir combinations are given on an average annual basis. These were computed under the assumption of a 50-year amortization period with $2\frac{1}{2}$ percent interest and, as stated previously, with annual OMR costs assumed to be 0.25 percent of total capital costs.[9]

DETERMINING LEAST-COST COMBINATIONS OF RESERVOIRS AND CHANNEL IMPROVEMENT

Having found the reductions in flood peaks and the annual costs of a number of reservoir combinations, we can now construct functions that will determine the minimum annual cost of controlling floods to specific peak levels, for floods of various frequencies of occurring or being exceeded. These are the curves labeled *a* in Figs. 5.12 to 5.15. Next, we plot on these figures the curves labeled *b*, which show the annual cost of channel-capacity improvement as a function of safe channel capacity. The annual costs are computed from the capital costs shown in Fig. 5.4, again using an amortization period of 50 years with $2\frac{1}{2}$ percent interest, and assuming that annual OMR costs are 2 percent of the total capital

[8] In several cases (the 5.0×10^3 cu ft/sec curve in Figs. 5.7, 5.8, and 5.9) the least-cost combination is a "corner solution," rather than a tangency solution. The marginal conditions, then, do not apply. Neither would they be applicable if sharp breaks or discontinuities were to occur in the functions.

[9] The alternative formulation, used in Chapter 3 and in other chapters of this book, is to express all costs and benefits in terms of present values. Average annual costs and benefits are used in this chapter because uniform time streams of OMR costs and of benefits are assumed. The basic concepts apply in exactly the same way, however, and the same optimal design is obtained, no matter which formulation is used.

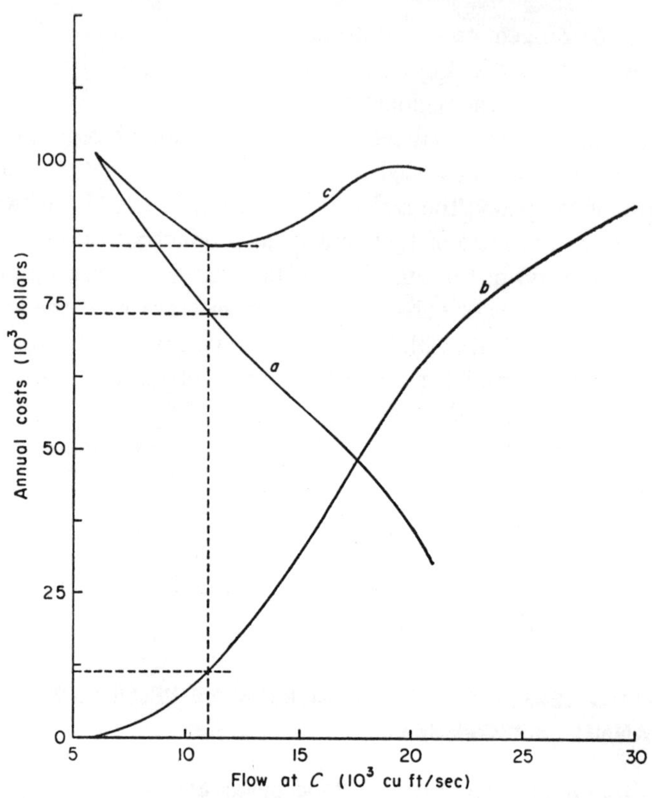

Fig. 5.12. Least-cost combination of reservoirs and channel improvement for control of floods with 1 percent probability of occurrence or exceedance. Curve a shows the annual cost of providing floodwater-retarding reservoirs to reduce the peak flow to the indicated level, b the annual cost of improving the channel to carry indicated flows safely, and c the combined annual cost of eliminating losses from floods of 1 percent probability.

costs. The functions are then combined as *c*, the cost of both reservoirs and channel improvement, in order to identify the least cost of controlling floods of the four designated probabilities. In each case the slopes of curves *a* and *b* are equal but of opposite sign at points of minimum cost. In Fig. 5.12, for example, a small shift in flow from 11.0×10^3 cu ft/sec, at which the cost of combined flood control is least, makes the incremental costs for retarding reservoirs greater than the incremental savings for channel improvement, or the converse, depending upon the direction of the shift. Stated in another way, the ratio of the marginal productivity to the marginal cost of channel improvement just equals the ratio of marginal productivity to the marginal cost of floodwater-retarding reservoirs at this point.

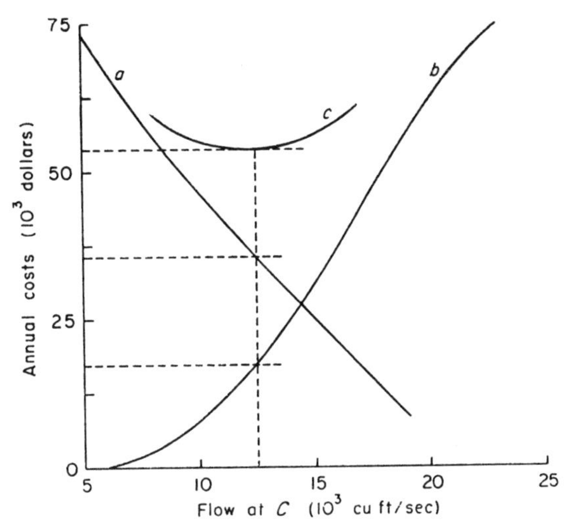

Fig. 5.13. Least-cost combination of reservoirs and channel improvement for control of floods with 5 percent probability of occurrence or exceedance. Curve a shows the annual cost of providing floodwater-retarding reservoirs to reduce the peak flow to the indicated level, b the annual cost of improving the channel to carry indicated flows safely, and c the combined annual cost of eliminating losses from floods of 5 percent probability.

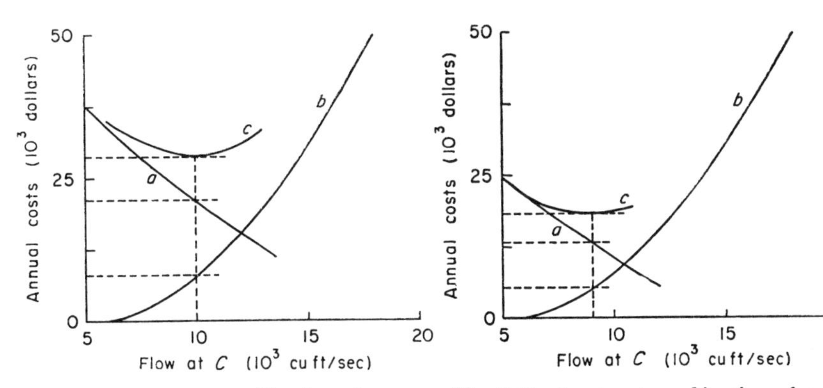

Fig. 5.14. Least-cost combination of reservoirs and channel improvement for control of floods with 10 percent probability of occurrence or exceedance. Curve a shows the annual cost of providing floodwater-retarding reservoirs to reduce the peak flow to the indicated level, b the annual cost of improving the channel to carry indicated flows safely, and c the combined annual cost of eliminating losses from floods of 10 percent probability.

Fig. 5.15. Least-cost combination of reservoirs and channel improvement for control of floods with 20 percent probability of occurrence or exceedance. Curve a shows the annual cost of providing floodwater-retarding reservoirs to reduce the peak flow to the indicated level, b the annual cost of improving the channel to carry indicated flows safely, and c the combined annual cost of eliminating losses from floods of 20 percent probability.

DETERMINING THE OPTIMAL SCALE OF DEVELOPMENT

To determine the scale of development at which net benefits are maximized, the flood-damage–frequency function of Fig. 5.5 can be combined with functions of the residual damages associated with fully controlled floods of given frequencies. Figure 5.16 shows such a combination for the

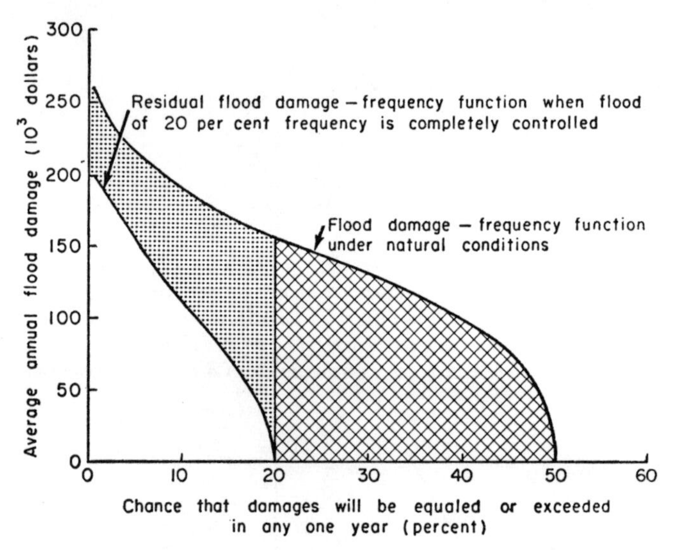

Fig. 5.16. Flood-damage–frequency functions under natural conditions and with complete control of floods of 20 percent frequency. Crosshatched area measures damages prevented by complete control of floods of 20 percent frequency or greater. Dotted area measures damages prevented by less than complete control of floods of less than 20 percent frequency. Remaining area under curve measures residual flood damages.

20 percent frequency flood. The physical data that identify the damages prevented come from flood-routing studies, in which representative floods for the full frequency range are routed through the selected system to establish the reduced river stages at the point of confluence *C*. The controlled and residual damages for a specific least-cost combination of structures then are: (1) damages from floods with specified or greater frequencies completely *prevented* by the specific least-cost combination of reservoirs and channel improvement; (2) damages from floods with less than the specified frequency *reduced* by the specific combination of structures; and (3) residual damages from floods with less than the specified frequency in the presence of the specific structures. For the

20 percent frequency flood, for example, (1) above is measured in Fig. 5.16 by the crosshatched area; (2) by the dotted area; and (3) by the remaining area under the curve.

In accordance with Fig. 5.16, therefore, complete control of the flood with 20 percent chance of occurring or being exceeded will reduce average annual damages from $72,000 under uncontrolled conditions (measured by the sum of the three areas under the uncontrolled flood-damage–frequency curve) to $23,000 (measured by the unmarked area under the residual-damage–frequency curve). Hence, average annual flood-protection benefits for this degree of flood protection are the difference, or $49,000. Corresponding totals of flood-protection benefits for full control of floods with frequencies of 1, 5, and 10 percent are $72,000, $67,000, and $60,000 respectively.

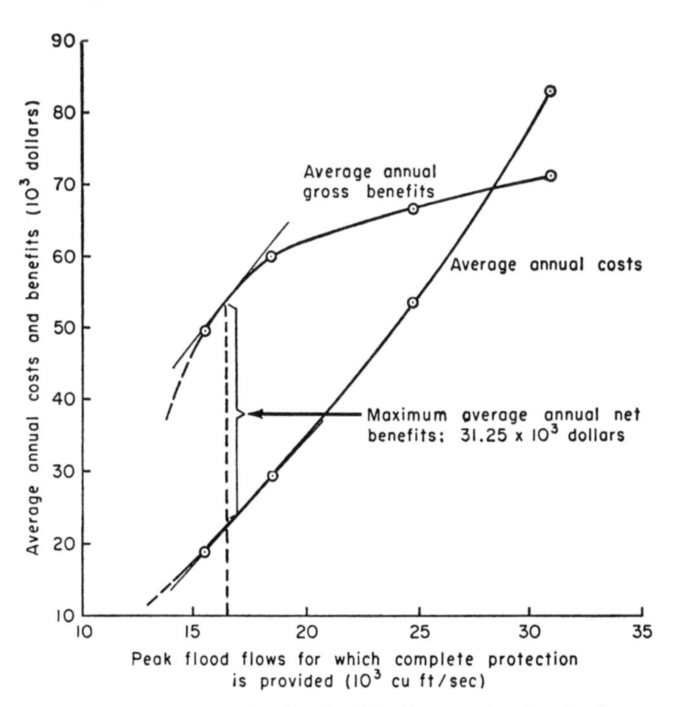

Fig. 5.17. Determination of the optimal scale of development for the simple water-resource system.

The two curves in Fig. 5.17 give average annual costs and gross benefits of providing complete protection against floods of different magnitudes and frequencies, and are derived from the benefit figures given above and the costs indicated in Figs. 5.12 to 5.15. The curves are analogous to the functions in Fig. 3.7, output being expressed in terms of peak flood flows against which complete protection is provided.

As in Fig. 3.7, we can now find the points of parallel tangents for the two curves in Fig. 5.17. These points of equality of marginal costs and marginal gross benefits identify the scale of development for which net benefits are at a maximum; namely, full control of a peak flood flow of 16.5×10^3 cu ft/sec and associated annual gross benefits of $53,500, annual costs of $22,250 and annual net benefits of $31,250.

Turning back to Fig. 5.2, we find the flood-probability level that corresponds to a peak flood flow of 16.5×10^3 cu ft/sec is 15 percent. The corresponding curves of equal flood-peak reduction for reservoirs *A* and *B* are drawn in Fig. 5.18 in accordance with the development of Figs.

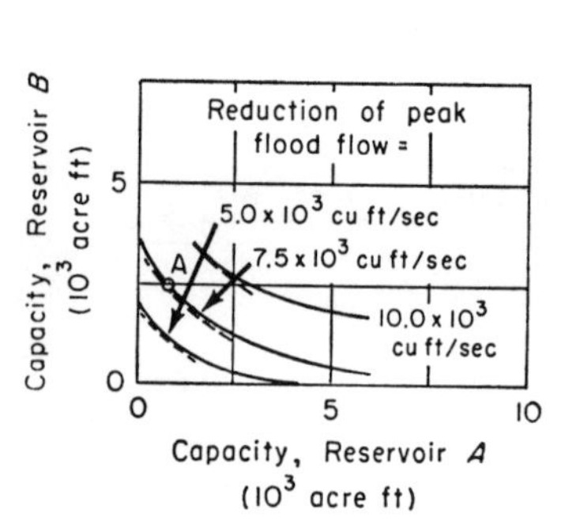

Fig. 5.18. Curves of equal flood-peak reduction for floods with 15 percent probability of occurrence or exceedance. Dotted lines indicate points of tangency of equal-cost curves. Point A defines the least-cost combination of the two reservoirs that will reduce the peak flow from 16.5×10^3 cu ft/sec to 9.0×10^3 cu ft/sec.

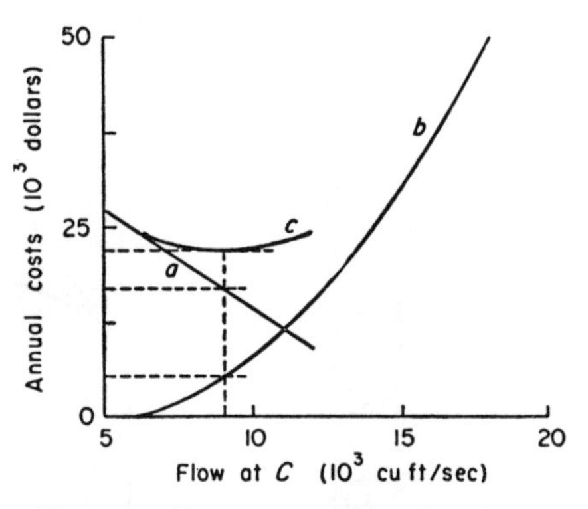

Fig. 5.19. Least-cost combination of reservoirs and channel improvement for control of floods with 15 percent probability of occurrence or exceedance. Curve a shows the annual cost of providing floodwater-retarding reservoirs to reduce the peak flow to the indicated level, b the annual cost of improving the channel to carry indicated flows safely, and c the combined annual cost of eliminating losses from floods of 15 percent probability.

5.7 to 5.11. The points of tangency with the appropriate curves of equal cost are included. Figure 5.19 finally shows the pertinent least-cost combination of reservoirs and channel improvement in these circumstances and permits determination of the least-cost means of providing the optimal level of flood protection.

The completed analysis has led to the following results.

(1) At the optimal scale of development of our simple example of a river-basin system, complete protection is provided against a flood with a peak flow of 16.5×10^3 cu ft/sec and a frequency of occurring or being exceeded of 15 percent.

(2) As shown in Fig. 5.19, optimal design calls for channel improvements that will safely accommodate 9.0×10^3 cu ft/sec of flood flow, and floodwater-retarding reservoirs that will reduce to 9.0×10^3 cu ft/sec the peak flow of the flood with a frequency of occurring or being exceeded of 15 percent. As indicated in Fig. 5.18, the least-cost combination of reservoirs to achieve this reduction is a capacity of 0.7×10^3 acre ft at reservoir *A* and 2.5×10^3 acre ft at reservoir *B*.

(3) Resulting system costs and benefits as collected or calculated from Figs. 5.17, 5.18, and 5.19 are listed in Table 5.3.

Table 5.3. Optimal design of the simple water-resource system.

		Cost (10^3 dollars)		Average annual benefits (10^3 dollars)	
Unit	Capacity (10^3 acre ft)	Capital	Average annual[a]	Gross	Net
Reservoir *A*	0.7	90.0	3.35	—	—
Reservoir *B*	2.5	360.0	13.40	—	—
Channel improvement	—	100.0	5.50	—	—
Total		550.0	22.25	53.50	31.25

[a] Annual OMR costs plus capital costs amortized over a 50-yr period at an interest rate of $2\frac{1}{2}$ percent.

CONCLUSION

The basic concepts of Chapter 3 can be used directly in a graphic solution of a simple river-basin design problem. As we move to more complex problems, however, graphic analysis becomes too difficult. Use of the simulation techniques and mathematical models described in Chapters 6 to 14 would then seem to be in order.

A large amount of basic information is needed to solve even the relatively simple problem considered in this chapter. Although the information is of the kind with which water-resource-planners are familiar, it is larger in amount and requires more extensive analysis than that with which they are accustomed to working. For the reservoir cost–capacity curves to be representative, for example, the coordinates of a fairly large number of points must be found, and a sizable number of simplified flood-routing analyses may be needed to construct an adequate number of curves of equal flood-peak reduction that are of good quality. In some instances relatively simple and quick methods of assembling the required data in sufficient quantity are available; in other instances new methods must be developed.

It follows that the work of defining the cost–input, input–output, and output–benefit functions in adequate detail for optimal design of simple systems is manageable, and that the procedure described in this chapter can be applied in a straightforward way to many different combinations of system units and purposes. Nevertheless, the relationships among the variables can be identified properly only with skill and imagination. A major advantage of graphic over nongraphic techniques is that the trail to the optimal design is not difficult to follow. The effects of applying marginal analysis, for instance, are readily recognized in the identification of least-cost designs and the selection of the optimal scale of development. The flexibility of the procedure commends it: it can be applied for the definitive solution of simple cases involving at least five variables,[10] and it can be used, as shown in Chapter 10, in the final determination of the optimal combination for a complex system that has been subjected to preliminary analysis by other means.

[10] A procedure applicable to four reservoirs and a channel improvement for the single purpose of flood control was devised and tested successfully in January 1959 by Sheldon G. Boone and Alan S. Payne, then on leave from the U. S. Soil Conservation Service as Water-resource Fellows at Harvard University.

Part II METHODS AND TECHNIQUES

6 Methods and Techniques of Analysis of the Multiunit, Multipurpose Water-Resource System: A General Statement

ARTHUR MAASS AND MAYNARD M. HUFSCHMIDT

WATER-RESOURCE systems, even relatively simple ones, may be created in almost infinite variety through different combinations of system units, levels of output, and allocations of reservoir capacity to various uses. The aim of system design is to select the combination of variables that maximizes net benefits in accordance with the requirements of the design criterion. This criterion itself is a function of the objectives which, as we have said before, are chosen by the community from among a number of alternatives.

This seemingly unlimited freedom of selection among system components can be circumscribed by using techniques that enable us to identify readily those combinations of variables that lie on the production function and are, therefore, technologically neither infeasible nor inefficient. These combinations must include the one that has the highest possible value of net benefits.

It is the purpose of the present chapter to describe briefly and in general terms the methods and techniques that we have employed for analysis of a multiunit, multipurpose river-basin system, using a model system as the test vehicle. Broadly speaking, we have employed two types of techniques. The first one simulates the behavior of a water-resource system on a high-speed digital computer, observes its response over extended periods of time to alternative combinations of structures and levels and purposes of development, and selects the best combination. The second technique adapts the design problem so that it can be solved by certain mathematical methods which proceed automatically to the optimal solution for relatively simple problems. Both techniques are designed to allow simultaneous consideration of a large number of alternatives, conjoining of engineering and economics, and adaptation to any reasonable objective and institutional constraint. The extent to which they succeed and their collateral advantages and limitations will be discussed in this and subsequent chapters.

THE SIMPLIFIED RIVER-BASIN SYSTEM

Because we discovered no river-basin system in the United States that would adequately serve our purpose of testing ways and means of system analysis, we chose to create a simplified river-basin model of our own. To be of broad use, the model system had to fulfill the following requirements: (1) it had to be simple enough to allow for easy experimentation, yet sufficiently complex to be representative of actual multiunit, multipurpose river developments; (2) we had to know enough about its unit costs and benefits to permit a full examination of relevant combinations of facilities and outputs; (3) it had to include the most important kinds of water-resource outputs or purposes, namely, withdrawal-consumptive uses, nonwithdrawal-nonconsumptive uses, and retardation or withholding uses; (4) its hydrology had to be typical of a major region of the United States; and (5) it had to lend itself to detailed and thorough analysis by high-speed digital computers, at relatively small cost in money and manpower.

The simplified river-basin system we developed contains four reservoirs that serve three purposes: irrigation, electric power, and flood protection. Power is generated in two hydropower plants, and a large area of land is irrigated through a diversion dam, irrigation canals, and distributing works. All reservoirs can contribute in some degree to flood control. In all, there are 12 design variables. Streamflow data for the system were selected from the U. S. Geological Survey records for the Clearwater River and its tributaries, in Idaho. Although all other basic data (including costs of reservoirs, power plants, and irrigation works, and benefits for the three system purposes) are hypothetical, the values and relationships used were based on actual experience. A further simplification is that specified monthly distributions of annual target outputs for irrigation water and firm energy were assumed to be optimal. Whereas we were able to analyze the full system by simulation on a high-speed digital computer, we could use only parts of it in our experiments with mathematical models. The full system is shown in Fig. 7.1.

Although water-resource developments may include more than a single river basin, ground- as well as surface-water flows, and nonstructural as well as structural measures, we chose to analyze a single river basin with surface flows (except for irrigation return flows) and structural measures only, because these comprise the heart of the multiunit, multipurpose problem. If our techniques can deal with this problem, it should be possible to develop them further for ultimate application to situations involving many factors omitted from our analysis.

CONVENTIONAL METHODS OF ANALYSIS

We have applied conventional design techniques to the model system to solve for its optimal combination of design variables, and have used the results to compare with those obtained by what we believe to be more useful and informative techniques. By conventional techniques we mean those currently in use by U. S. water-resource planning agencies. We realize that no single technique of river-basin analysis is common to all agencies, although they generally agree on certain principles of plan formulation. Disparities in practice from agency to agency and between some field offices within a single agency can be ascribed in part to differences in the problems with which they are concerned, in part to differences in agency traditions, assigned responsibilities, availability of funds, and background and experience of personnel.

The task of analyzing the simplified river-basin system by conventional techniques was assigned to three groups of water-resource engineers, drawn from the major federal water-resource agencies and associated with the Harvard Water Program at different times. They used the same economic assumptions that were used to test the simulation technique, namely, an interest rate of $2\frac{1}{2}$ percent, an economic life of 50 years, no budgetary constraints, and static rather than dynamic conditions.

The groups examined the system separately for each of the three purposes (irrigation, power, and flood control) and developed roughly optimal single-purpose designs for them. With these as a nucleus, they formulated two optimal two-purpose designs, and subsequently a third two-purpose design from the information provided by the first two. From the optimal two-purpose plans they developed a tentative three-purpose plan, which was subsequently optimized by applying techniques of marginal analysis. Desk computers were used and short-cut methods employed to cut down the work of examining the many possible designs. Storage–yield relationships, for instance, were based on operation studies of short, critical periods rather than on the entire 32 years of hydrologic record that were available.

Out of this work emerged a best design for the simplified river-basin system. This included all three purposes, but retained only two of the four possible reservoirs, which supplied water to two power plants and to irrigation diversion and distribution works. The resulting design served as a basis for comparison with the optimal design obtained by the simulation technique.

SIMULATION

The availability today of high-speed digital computers of large capacity makes it possible to simulate the performance of relatively complex river-basin systems for periods of any desired length, whereas conventional methods using only desk computers permit operating only selected parts of the system during a limited period of hydrologic record. In effect, simulation on high-speed digital computers has become a new technique of river-basin analysis; accordingly, it was made an important part of our studies.

The essence of simulation is to reproduce the behavior of a system in every important respect. Simulating our simplified river-basin system on the computer, therefore, involved representing all of the inherent characteristics and probable responses of the system to control by a model that was largely arithmetic and algebraic in nature but included also some nonmathematical logical processes. Control was exercised through an operating policy or procedure that took the form of instructions to the computer and had as its objective the conversion of natural runoff into flow complexes that would generate useful outputs. Our 12 design variables consisted of reservoirs, power plants, irrigation works, target outputs for irrigation and energy, and specified allocations of reservoir capacity for active, dead, and flood storage. Our hydrology was composed of monthly data on normal inflow and 6-hr data for months of flood. These flows were routed through the reservoirs, power plants, and irrigation system for a 50-yr period according to a specified operating procedure, and the magnitudes of the physical outputs and the resulting gross and net benefits were determined.

Preparing the information on the simplified river-basin system for simulation on the computer involved a number of operations. First, the basic data for the system (including the hydrologic data) had to be assembled and arranged in a form easily handled by the computer. Next, an operating procedure had to be formulated to serve as the fundamental control for the simulation. Finally, the basic data for the system and the operating procedure had to be coded for use on the computer.

Our simulation reproduced the behavior of the simplified river-basin system for flood control as well as for irrigation and power generation. Conventional operating studies, in contrast, normally perform flood-control operations separately. The simulation program started with a monthly operating procedure having monthly inflows, releases, and target outputs; when a flood month was encountered, the program shifted to a flood-routing procedure based on a 6-hr interval. At the end of the flood month, the program returned to the monthly operating procedure. Although more complicated, the simulation was more realistic, as flood flows and their consequences became an integral part of the sequential computations.

The results of the analysis, furthermore, were stated in economic terms by including cost and benefit functions in the simulation program, and evaluating deficits and surpluses in outputs as well as target outputs themselves. These results were converted into present values of net benefits at a number of alternative interest rates. Economics and engineering were thereby conjoined in the design process.

One simulation program was coded for an IBM 704 computer by the Fortran II automatic coding technique. This allowed the use of symbols, not unlike those of elementary algebra, to describe the quantities represented. The resulting simplified form of the program was then converted by the computer into a more complex form for use in actual simulation runs.

Testing a system design by simulation proceeded as follows. After selecting a trial value for each of our 12 design variables, the computer, following its Fortran II instructions, produced the desired physical and economic results as output. Included in the information for the simulation period were the total firm and dump energy generated; the number of monthly energy deficits and the cumulative deficit over the entire simulation period; the number of annual irrigation deficits and their accumulated value; the minimum and maximum contents of the reservoirs; the number of damaging floods and the largest flood flow experienced; the gross irrigation, energy, and flood-control benefits; the sums of capital costs and of operation, maintenance, and replacement (OMR) costs; and, finally, the net benefits.

LIMITATIONS OF SIMULATION

Simulation as just described has important limitations. For example, it does not yield an immediate optimal answer. Each answer pertains merely to the operation study of the selected combination of variables. However, the computer's capacity to perform repetitive calculations with speed is so great that the chance of including in its answers a design close to the theoretical optimum is much greater than when conventional techniques are employed. Nevertheless, there remains the problem of using the computer most effectively. Assuming that significant increments are chosen for each variable in a given river-basin system, the computer need not be asked to examine all possible combinations of variables. The combinations to be simulated can be selected, using a method that justifies confidence that the best result obtained approximates the theoretically best combination. How this can be done is discussed later in the chapter.

A second limitation of simulation results from the lack of flexibility in the operating procedure of the system. In a sense, each design is a combination of structures (or nonstructural measures like flood-plain zoning), levels of development for the several target outputs, and the operating procedure. In simulation there is great freedom to test different combinations of structures and target outputs, but not of operating procedures. In the simulation of the simplified river-basin system, for example, the computer is instructed first to release water for irrigation from reservoir *A* and then from reservoir *B*, and to release water from reservoirs *C* and *D* for energy production at power plant *G* according to a certain order of priorities. These instructions cannot be varied readily. Indeed, every major change in operating procedure requires writing and testing a new computer program. Although operating procedure should be treated as a variable, therefore, the preparation of needed instructions and the additional computer operations required are so time-consuming that we are compelled to consider operating procedure as substantially constant.

A third limitation of simulation derives from the use of historical streamflow records. These are run through the computer in the order of their occurrence — for the simplified river-basin system, over a period of 32 years. Yet this record may be a very limited representation of the possible configurations of flows. There is no reason, for example, to expect that in the Clearwater River, on which the hydrology of the simplified river-basin system is based, the next 32 years of streamflow will be the same as the last 32.

PARTIAL ANSWERS TO THESE LIMITATIONS

To overcome these limitations of simulation, we have introduced sampling methods for the efficient selection of combinations of variables, flexible operating procedures for testing them, and synthetic streamflows as adjuncts to the historical records.

Sampling. The net-benefit response to simulating combinations of the 12 design variables in the simplified river-basin system creates, in concept, a 12-dimensional surface that includes at least one point with the highest net benefits. The only practical means of finding the optimal combination or combinations of the 12 design variables responsible for this point or series of points is to employ a suitable method of sampling the variables and thereby to explore the net-benefit response surface.

Sampling, as we shall see, can be systematic or random. In systematic sampling, the magnitudes of the variables are selected in accordance with some ordering principle; in random sampling, purely by chance from an appropriate population. Among systematic sampling methods, we have tested the uniform-grid or factorial method, the single-factor method, the method of incremental or marginal analysis, and the method of steepest ascent. These we have used singly or in combination, together with random sampling, to determine their suitability for analysis of the simplified river-basin system, and by inference for analysis of other river developments as well.

Our studies have shown that a reasonable approximation of the optimal design can be obtained for our system in less than 200 trial simulations. This number stands in contrast to a total of over 10^{23} possible combinations when individual increments of reservoir capacity and irrigation output are set at 10^5 acre ft and the magnitudes of the other system components are varied proportionately. Sampling, therefore, has proved to be a powerful means for reducing to manageable proportions the task of examining the wider range of possible designs.

More specifically, we learned that an effective sampling strategy could be devised for river-basin systems such as ours by selecting, to begin with, a random sample of moderate size from the full range of all system variables — following it, if need be, by a second random sample covering a more restricted range of these variables. Regions of high net benefits could then be examined systematically by the methods of steepest ascent and uniform grid. Finally, we could proceed to marginal analysis of the high-value combination, making successive adjustments until there was no further gain in net benefits.

Proceeding in this way, we were able to develop a design for the simplified river-basin system that yielded net benefits of $811,000,000 against $724,000,000 for the best design by conventional methods. The two designs are compared in Table 6.1. Since exploration of the response surface by sampling showed but a single major net-benefit hill, we are reasonably sure that the best combination obtained by our sampling methods is approximately optimal for the operating procedure and hydrology used in the simulation.

Operating procedures. In developing operating procedures for simulation, we aimed, on the one hand, to simplify their formulation and programming so that a designer would be free to choose among alternative procedures and thereby treat the operating procedure more nearly as a design variable; on the other hand, we wanted to establish principles of optimal operating procedure.

To facilitate the programming of operating procedures, we formulated several, programming one for the Univac I and two for the IBM 704 computer. Although it is possible to develop and program many operating procedures in this way, it would be tedious to do so, and we should still not be certain that the optimal procedure for a particular combination of design variables had been established. It would be more systematic to develop a master operating procedure, the different components of which could be varied within wide limits. In effect, we should then be able to test a number of discrete operating procedures by combining the components of the master program in different ways. While no reprogramming would be required, the initial writing and programming of such a master procedure would be a formidable task. Each of the discrete procedures, furthermore, would remain relatively inflexible, because the component rules would stay the same throughout the simulation period. A procedure that does not take account of the state of the system periodically during the 50 years of simulation is likely not to be optimal.

To formulate principles and programming methods for optimal procedures, we can view the procedure as a means for apportioning releases among reservoirs, purposes, and time intervals so as to optimize the value of the objective function (in our case, the expected value of net benefits). While it is difficult to formulate and program a sufficiently flexible pro-

Table 6.1. Comparison of best system designs obtained from conventional methods and from simulation techniques.

	Conventional methods	Simulation techniques
Variable	*Value*	
Reservoir capacities and allocations of storage (10^6 acre ft)		
A, total storage	0.0	0.0
B, active storage for irrigation and power generation	2.8	10.0
B, dead storage	5.7	3.0
B, flood storage	0.6[a]	0.4[b]
B, total capacity	9.1	13.0
C, flood storage	0.0	0.0
C, total storage	0.0	0.0
D, active storage for power generation	3.6	2.2
D, flood storage	0.2[a]	0.0
D, total storage	3.8	2.2
Power-plant capacities (Mw)		
B	257	325
G	298	300
Target outputs		
Irrigation (10^6 acre ft)	4.2	4.2
Energy (10^9 kw hr)	1.77	2.4
Costs (10^6 *dollars*)		
Capital	467.7[c]	493.6
OMR	90.9[c]	93.7
Total	558.6[c]	587.3
Benefits (10^6 *dollars*)[d]		
Gross, irrigation	555.5[e]	558.4
energy	375.4[f]	488.0
flood control	352.0[g]	352.0
total	1282.9	1398.4
Net	724.3	811.1

[a] Allocated exclusively for flood control.
[b] Included in active storage for irrigation and power generation.
[c] Slightly higher than comparable values in Table 8.6. All costs (and benefits) shown here were obtained from simulation of the best conventional-methods design on the computer; values in Table 8.6 were estimated from hand calculations, as described in Chapter 8.
[d] Present value, based on 50-yr economic life with interest rate of $2\frac{1}{2}$ percent.
[e] Slightly less than comparable value in Table 8.6, for the reason given in note *c* above.
[f] Higher than comparable value in Table 8.6 by $\$24 \times 10^6$, because dump-energy benefits are counted here but not in Table 8.6.
[g] Higher than comparable value in Table 8.6 by $\$33.8 \times 10^6$, because a greater number of floods were assumed in simulation analysis of the 50-yr period than in the analysis described in Chapter 8.

cedure, we can record some success. The initial steps of necessity were focused on physical, rather than economic, objectives. Examples are our flexible rules for allocating necessary releases to reservoirs and to successive time intervals, to either avoid or minimize spills. The rules change each month in accordance with the state of the system, decisions being aided by forecasts of probable future flows. It should also be possible ultimately to develop rules for allocating releases among purposes in conjunction with releases among structures and time intervals. Then substitution of economic objectives for physical ones would bring us close to a fully flexible operating procedure governing releases among structures, purposes, and time intervals and thereby maximizing the expected value of net benefits. As yet, however, the problems connected with operating procedures are largely unsolved.

Synthetic streamflow sequences. As stated before, the 32-yr record of streamflows upon which simulation of the simplified river-basin system was based is in some ways an inadequate expression of the possible flows and, as such, an unreliable base for precise system design. Critical sequences of years of low and high runoff inherent in the statistical population of river flows may be missing. But no matter how poorly a brief record may identify the time frequency of years or seasons of unusually low and high flows, unless it is very short indeed, it will permit fairly reliable estimates of mean annual and seasonal flows and their variances. As we shall see, these statistical parameters, together with a few assumptions about the population of flows (which may be based on experience with near-by rivers), can make it possible to construct statistical models that will generate synthetic flows of any desired length.

We designed a model to synthesize, with the aid of a high-speed computer, a record of 510 years. Since the economic design period for our river-basin system was 50 years, we divided the 510-yr record into ten replicate 50-yr periods or hydrographs (plus an initial 10-yr transient or "warm-up" period). This technique enabled us to test our system designs more exhaustively than by utilizing the available historical flows alone.

Largely because of limitations on computer time, we did not use the synthetic streamflow sequences in sampling the net-benefit response surface for the simplified river-basin system. However, we did make use of the synthesized record of streamflows to examine three system designs, including one that was near-optimal for the historical hydrograph itself. According to these three studies, the gross benefits for irrigation and energy were quite stable and close to the results obtained with actual inflow data. Not unexpectedly, flood-control benefits proved to be more variable. Indeed, they accounted for most of the variance in the net benefits measured by the replicate 50-yr sets.

Obviously these results for a single stream and its tributaries do not justify drawing far-reaching conclusions. For the simplified river-basin system, however, the average net benefits derived from the ten sets of synthetic hydrologic records must offer a considerably more reliable estimate of the expected net benefits than can the net benefits based on the actual hydrology. The variation in the ten sets of results, furthermore, should be a valuable guide to a system-planner in selecting the best design in the light of hydrologic uncertainty.

MATHEMATICAL MODELS

The methods we have just discussed for selection of the optimal combination of system units and outputs, treatment of operating procedure as a variable, and stochastic manipulation of streamflows are reasonably satisfactory, but there may be even better ways of going to the heart of the multipurpose, multiunit problem and isolating the optimal combination or combinations of units, outputs, and operating procedures. The existence of mathematical techniques that can make optimal decisions for relatively complex objective functions and sets of constraints, in combination with high-speed digital computers that can perform the required repetitive computations, has encouraged us to adapt our design problem to solution by these techniques. To this end further simplification of our river-basin system has been necessary. Whereas our simulation operations involved the equivalent of many thousands of equations, the number had to be reduced drastically to use mathematical models that optimize.

We have experimented with several kinds of mathematical models but present only two of them in this book. Both employ data derived from the simplified river-basin system and use techniques of linear programming. The basic difference lies in their method of handling the stochastic component of streamflow. We have chosen to call one the multistructure model and the other the stochastic sequential model. The limits of application of the models were not tested, for lack of time and because it seemed desirable for expository purposes to keep the examples relatively uncomplicated. However, we did go far enough to exemplify the essence of the techniques; and we shall be quite specific in pointing out the limitations of their application to complex multiunit, multipurpose river systems, and in suggesting necessary simplifications that will make river-system problems amenable to solution by the mathematical optimizing techniques in question.

THE MULTISTRUCTURE MODEL

Three versions of the multistructure model were of special interest. In the first and simplest one, it was assumed that the life of the project is composed of a succession of identical and typical years and that a first approximation to design can be reached by considering a single year in isolation. The model deduced simultaneously the optimal set of structures for putting the flows to use, the optimal set of outputs that can be obtained with the flows and structures, and the optimal operating procedure for achieving these outputs. To do this, we identified certain decision variables, such as active- and dead-storage capacities of reservoirs, capacities of fixed- and variable-head power plants, and target outputs for irrigation and energy, and sought magnitudes for all of the variables that confer the maximum value on an objective function while satisfying certain technologic constraints.

There is considerable freedom in defining the objective function. In our example it included many terms or coefficients for the benefits and costs of the chosen variables. The terms representing marginal benefits of irrigation and capital costs of reservoirs and power plants were nonlinear, but since they could be approximated by linear segments, they could be accommodated in the solution.

Examples of technologic constraints in our system were that flows in all reaches of the system must be nonnegative, that power plants and the flows reaching them in each season must be large enough to generate energy outputs, and that none of the decision variables can be negative. Since these constraints are inequalities rather than equalities, they require the use of programming, rather than Lagrangian analysis, for maximization of the value of an objective function which they restrain. As in the case of the objective function, it is not necessary that all of the terms or coefficients of these constraints be linear so long as they can be approximated by piecewise linear functions.

The most severe simplifications or limitations of the first version of the multistructure model were that each year is treated like all other years, no overyear storage therefore being provided, and that inflows are treated as determined or certain. A second and third version of the multistructure model were developed to deal with these simplifications.

In general, overyear storage can be accounted for in one of three ways: (1) by using the historical or a synthetic streamflow record, which requires a full simulation operation and takes us outside the realm of simplifying mathematical models; (2) by analyzing the "critical period" (or periods) of an historical or synthetic record; and (3) by introducing the probability distribution of the inflow record to determine probabilities of outputs for reservoirs of different sizes. Although defining the critical period of a record is always difficult and inexact, hydrologists have had considerable experience in dealing with this matter, and it is a customary part of conventional techniques. The multistructure model employs the second and third approaches; the stochastic sequential model, the third one only.

The second version of the multistructure model illustrates the use of critical-period analysis to estimate the need for overyear storage. As a matter of abstract principle, the methods used in the single-year case can be applied to a critical period covering any number of years, but as a matter of practice, this extension cannot be carried very far because the number of equations that have to be solved simultaneously increases about in proportion to the number of years, and the difficulty of solution increases as the cube of the number of equations. The second version of the multistructure model showed how this rapid growth of a problem toward unmanageable size could be retarded by first dealing with the equations for each year or other period separately and later revising the results of the preliminary analysis in the light of the performance of the system in other years. These revisions determined, for example, how much water should be retained in storage at the end of each year for use in later years, and therefore the reservoir capacity that should be provided for overyear storage. This so-called "decomposition" version of the multistructure model is illustrated in Chapter 13 by application to a four-period problem; with little additional difficulty it could have been applied to a much longer sequence of elementary periods.

In the third version of the multistructure model, inflows were treated stochastically, probability distributions of flow being substituted for mean seasonal flows. This is one way to handle the problem of overyear storage; but even when overyear storage is not involved, it is important to treat the inflows stochastically, because the probability of attaining a given target output depends on the inflow probabilities. When water availability is random rather than fully predictable, it is unrealistic to insist that output commitments be met in every period of a system's life. Instead we must be prepared to accept shortages in adverse periods in order to reap the benefits of the water supply in normal and bountiful years. Therefore, when random variability is taken into account, it is advisable to omit commitments for firm energy, irrigation supply, and the like from the list of constraints imposed on the system design, and to replace those constraints by loss functions that measure the cost of shortages.

Even with this simplification, the multistructure model was too difficult to solve in the random-variability case. Still another simplification had to be invoked: that in each period of operation the system would start with a normal quantity of water in its reservoirs. When this simplification, or rather falsification, was made, fairly complicated multistructure-multipurpose-multiperiod systems could be analyzed without excessive difficulty. A few pilot studies, summarized in Chapter 13, have indicated that such simplified analyses provide very instructive first approximations to optimal designs.

THE STOCHASTIC SEQUENTIAL MODEL

The stochastic sequential model was also developed for the purpose of finding that combination of reservoir capacity, level of development or target output, and operating procedure which will maximize the value of an objective function subject to technologic constraints. Unlike the multistructure model, its central focus is on the stochastic nature of streamflows. Using random variations in flows, the model derives a stochastic storage-yield relationship for a single reservoir capacity in the form of probability distributions of the draft that maximizes the expected value of the objective function. The result obtained can be interpreted in terms of the optimal combination of target output and operating procedure for that output, including overyear storage, assuming a given reservoir capacity. Finally, by methods of sampling and iteration, a reservoir capacity and its associated output and operating procedure can be selected that confer a high value on the objective function.

Because of the complex calculations necessary to develop its elaborate stochastic storage–yield relationship, this model requires considerably more simplifications in other respects than does the multistructure model. The most important can be summarized as follows.

(1) At the present stage of development the stochastic sequential

model cannot handle serial correlation of seasonal inflows nor, for the same reasons, more than one reservoir when they are in series.

(2) Because it permits only one decision on draft in any one interval, the benefit function in the model is a function of a single variable, namely the draft, although this variable can represent a combination of purposes. Thus the allocation of a draft among purposes so as to maximize the value of an objective function is a suboptimization problem external to the model, whereas it is integral to the performance of the multistructure model.

(3) The stochastic sequential model optimizes for target output and operating procedure; the choice of reservoir size, however, must be by sampling. The multistructure model optimizes for all three in combination.

The greater limitations of the stochastic sequential model, it will be realized, result from its greater advantages in handling overyear storage and the probabilities of attaining given target outputs.

RECAPITULATION

We have simulated the operation of a moderately complex river-basin development on a large digital computer. Apart from the early conjoining of economics and technology, we have done little more than follow existing planning procedures. Instead of examining a few alternative schemes by hand, however, with the same expenditure of time and effort we have examined hundreds of alternatives on a computer. Even with powerful techniques of sampling to select the combinations of variables to be examined, we were not satisfied that we were reaching our objective efficiently. So we simplified our requirements further and developed mathematical models in the hope of finding in a single operation that combination of size of structure, operating procedure, and output of a river-basin system that would maximize the value of an objective function. We have concluded that when mathematical models of this kind are perfected, we shall be in possession of an adequate strategy for the design of simple river-basin systems. For more complex developments, however, perhaps even for the simplified river-basin system, the mathematical models will probably continue to give only an approximate answer or a good first fit. With this we can turn to simulation. The operating procedure that we program will then be optimal or near-optimal, because of knowledge obtained from the mathematical model; similarly, the combinations of variables that we select for computer studies will be in the region of the optimum, because that region will have been identified by the model. Accordingly we shall have improved upon the partial answers to the limitations of simulation given earlier in this chapter.

7 A Simplified River-Basin System for Testing Methods and Techniques of Analysis

BLAIR T. BOWER

In searching for a river basin that could serve as a model for our studies, we examined a number of U. S. drainage basins in which the development of water resources, existing and proposed, had been carried forward to varying degrees of completeness. Since no system was uncovered that met our objectives sufficiently well to be used directly, our alternative was to devise an artificial river-basin system satisfying these objectives. The present chapter describes the resulting system and explains our reasons for adopting its component parts and attributes.

CONSIDERATIONS IN DEVISING THE SYSTEM

In devising an artificial system that would fit our purposes, five factors appeared to be of particular importance.

(1) The system had to be sufficiently simple to keep the necessary calculations from becoming cumbersome, yet complex enough to be reasonably representative of the "real world." Specifically, the system had to include physical facilities and purposes in sufficient number to create competitive as well as complementary interrelationships.

(2) The system had to lend itself to the study of all relevant combinations of facilities and outputs. This required that cost and benefit data cover (a) costs at all possible scales of development of the facilities, and (b) benefits at all levels of output for all pertinent purposes. It required further that no possible combination be excluded prematurely or arbitrarily because of institutional constraints. These qualities, a search showed, were not possessed by any existing or proposed water-resource development in the United States.

(3) It had to be possible to develop at least one of each major kind of output of a water-resource system: a withdrawal-consumptive use; a nonwithdrawal, essentially nonconsumptive use; and a retardation or

withholding use. For these purposes we selected respectively the irrigation of crops, the development of water power, and the reduction of flood damage. Irrigation was specified because it is more consumptive than most other withdrawal uses; water power, because it is a function of power head as well as of water volume; and flood-damage reduction, because it is concerned with the withholding rather than the releasing of water. These three uses, it should be noted, comprise not only the major types of output, but probably also the most complicated.

(4) The hydrology of the proposed system had to be typical for some major region of the United States. Although artificial runoff series could have been developed, it proved simpler to make use of recorded streamflows that possessed the following characteristics: they were free from the influence of major storage and diversion works already within the basin; they were large enough to justify power development; they rose to peak flows sufficiently high and frequent to cause flood damages; and they fell to low values during periods long enough to require annual holdover, or more than seasonal, storage.

(5) The system had to lend itself to exploration by modern high-speed digital computers in order to permit broad as well as deep analysis at relatively low cost in money and manpower, and in a short period of time.

The simplified river-basin system developed for study is sketched in Fig. 7.1. The assumed combination of four reservoirs and three purposes as shown appeared to create an adequate number of interrelationships to typify the complexities of an actual system of substantial proportions. Adding further reservoirs or purposes, on the one hand, probably would not have altered the basic system interrelationships significantly, but it would have increased very markedly the labor of data preparation, computer programming, and detailed analysis. Reduction of the number of reservoirs and purposes, on the other hand, would have stripped the system of complexities essential to our aims.

The hydrologic information available for a dozen U. S. river basins was examined before the U. S. Geological Survey record of the runoff of the Clearwater River Basin in Idaho was selected as most nearly meeting the fourth consideration described above. This record was unhampered by river regulation or by significant depletions from water use. Further, the Clearwater stream system had, as an important feature, two upstream branches uniting in a main stem that received a major tributary downstream. This made possible the study of reservoirs placed both in parallel and in series.

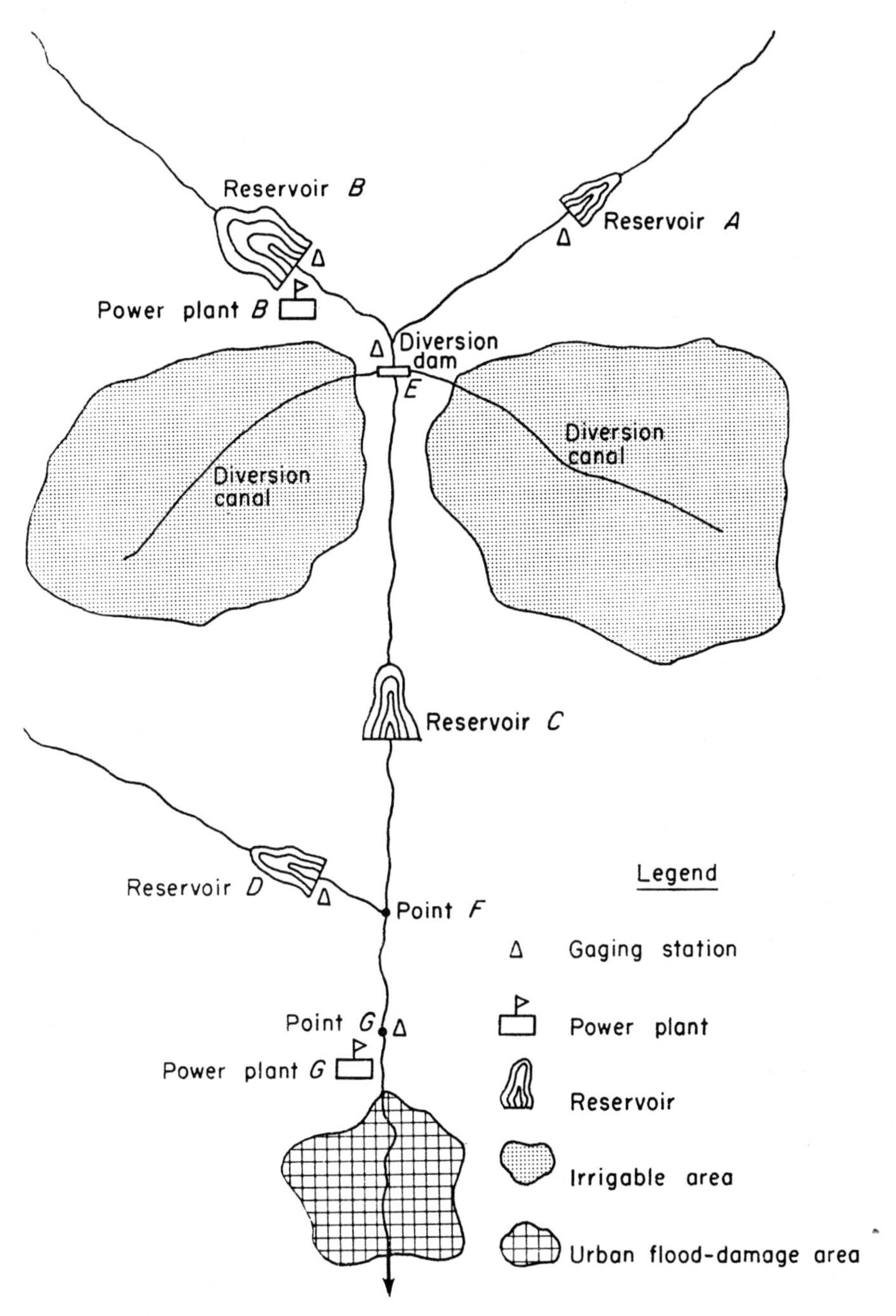

Fig. 7.1. Sketch of the simplified river-basin system (not to scale).

Some features of water-resource development, although important in themselves, were excluded from the simplified system. The most significant of these are touched upon briefly in the following paragraphs.

(1) Water quality was neglected, although it plays a part in almost all uses of water. Explicitly, this attribute of water could have been considered in a number of ways. Positing a certain pattern of flows for dilution of waste waters entering the streams and adding to the diversion requirements for irrigation to ensure adequate leaching of salt residues from the soil are examples. Implicitly, water quality could have been considered as affecting the sale price of water drawn from the river-basin system for municipal or industrial purposes. Nevertheless, it was felt that the problems of water quality were not sufficiently unique in their implications to warrant consideration in our studies.

(2) Water supply for municipal and industrial needs was not made one of our purposes because it is a withdrawal use quite similar to irrigation. Moreover, as long as the supply is not diverted from the drainage basin, this use of water is much less consumptive and less variable in monthly requirements throughout the year than is irrigation. As a system output, therefore, irrigation poses the same basic problems as water supply but with desirable complicating factors.

(3) The recreational use of water was excluded from consideration because it is essentially a nonwithdrawal-nonconsumptive use.[1] Recreational use of reservoirs for example, demands that suitable water surfaces be maintained during the outdoor seasons of the year by keeping the reservoirs relatively full. Operating reservoirs for recreation is thus similar to operating them for flood control by means of rule curves. Recreational use of streams themselves necessitates the maintenance of water quality and volume by controlled release of water from storage.

(4) Also excluded from the system to be studied was the use of alternative sources of water and of alternative ways of satisfying water needs. Water drawn from the ground for irrigation of crops, by itself or in conjunction with surface supplies, offers an example of an alternative source.[2] Groundwater recharge, recycling of industrial water, reclamation of waste

[1] As is true for substantially all water uses or purposes, recreation is not completely nonconsumptive. Water is lost by evaporation from recreational bodies of water, and there are consumptive personal needs for water in recreation areas. Moreover, water may be degraded in quality by recreational activities and this may have adverse effects on other uses.

[2] See, for example, R. L. Nace, S. W. West, and R. W. Mower, "Feasibility of ground-water features of the alternate plan for the Mountain Home Project, Idaho," *U. S. Geol. Survey Water Supply Paper No. 1376* (1957).

water, and flood-damage reduction by flood-plain zoning, flood-warning systems, and the like exemplify alternative ways of meeting water requirements and conjunctive features of actual river-basin systems.

Substitution of any one of these four aspects of water-resource planning and development in the artificial system, it was reasoned, would not add as much to its complexity as would the three purposes selected. If successful methods could be devised for analysis of a river-basin system that included those three purposes, therefore, it seemed safe to assume that such methods would be applicable also to systems both larger and smaller than the one selected and including other combinations of purposes and facilities, both in greater and in lesser number.

PHYSICAL LAYOUT OF THE SYSTEM

As shown in Fig. 7.1, the chosen development comprises four impounding reservoirs *A*, *B*, *C*, and *D*; two power plants, one at reservoir *B*, the other at point *G* where pondage is provided by a low weir; and one irrigation-diversion dam at point *E*. The site for each of these structures is assumed to be the only one feasible for the desired system purposes. The four impounding reservoirs offer both flood-control storage and power releases. However, only reservoirs *A* and *B* can provide releases for irrigation. Reservoir *D* lies on a tributary stream which joins the main stem at point *F*.

Power plant *B* is typical of a variable-head station operating in conjunction with a major storage reservoir, whereas pondage at power plant *G* reconciles only daily and weekly load variations. No other power developments are assumed to be possible, a situation not likely to occur in reality but which simplifies the operation and analysis of the system. Irrigable areas lie on both banks of the main river downstream from reservoirs *A* and *B*. Return flow reenters the river channel below *E* but upstream from reservoir *C*; none of it can be reused within the irrigated areas. A flood-damage zone is situated just below point *G*.

An actual river basin would usually contain more than one or two irrigable areas and flood-damage zones. But the addition of more areas for these purposes would have complicated unduly the operation studies, the computer programming, and the general analysis, without significantly clarifying the involved interrelationships existing in a four-reservoir, three-purpose system.

HYDROLOGY OF THE SYSTEM

STREAMFLOW DATA

As the streamflow data for our system were taken from the U. S. Geological Survey record of the Clearwater River, a tributary of the Snake River in the Columbia River Basin, the gaging stations and records at sites *A*, *B*, *D*, *E*, and *G* each have their counterparts in the Clearwater River Basin. These are given below with the dates of available streamflow data.[3]

Point *A*: The south fork of the Clearwater River near Grangeville, Idaho; June 1923 through October 1955.

Point *B*: The sum of runoffs recorded for the Selway River near Lowell, Idaho, 1929–55, and the Lochsa River near Lowell, Idaho, 1929–55.

Point *D*: The north fork of the Clearwater River near Ahsahka, Idaho; August 1929 through October 1955.

Point *E*: The Clearwater River at Kamiah, Idaho; June 1923 through October 1955.

Point *G*: The Clearwater River at Spalding, Idaho; June 1925 through October 1955.

The natural flows within the river reaches between reservoirs *A* and *B* and point *E* are the differences between those at *E* and the sum of those at *A* and *B*. For convenience it is assumed that the incremental runoff between point *E* and reservoir *C* is small enough to be neglected. This makes the flows at reservoir *C* and point *E* identical. The inflow to reservoir *C* consists of the return flow from irrigation diversions and the surplus flow past *E*. The combined natural flows in the stretches of river between reservoirs *C* and *D* and point *G* are the differences between the measurements at *G* and the sum of those at *E* and *D*.

Table 7.1 identifies the mean monthly distribution of runoff as a percentage of the mean annual runoff for the Clearwater River at Kamiah (point *E*) for the period of record. The observed pattern is typical of the hydrology of a catchment area in which the melting of winter snows

[3] Available runoff data through September 1950 are contained in "Compilation of records of surface waters of the United States through September 1950, Part 13, Snake River Basin," *U. S. Geol. Survey Water Supply Paper No. 1317*; data for the water years 1951 to 1955 are contained in Papers No. 1217, 1247, 1287, 1347, and 1397.

Table 7.1. Recorded mean monthly runoff, Clearwater River at Kamiah, Idaho (point *E*), 1923–55.

Month	Percentage of mean annual runoff
November	3
December	4
January	3
February	3
March	6
April	16
May	32
June	22
July	6
August	2
September	1
October	2
	100

produces large spring runoffs, followed by low summer and fall discharges. River basins of this kind are found in wide regions of the western United States. Their behavior is paralleled in other parts of the world in quite different circumstances. In India, for example, the monsoons also concentrate much of the annual runoff into a few months.

All streamflow data were tabulated to the closest 10^2 acre ft, although the recorded measurements were not that precise. This was done for consistency and to ensure a sufficient number of figures, or "guard digits" in computer terminology, for rounding off the machine computations without producing inconsistent results.

Where simulation techniques were employed, the traditional monthly periods for runoff and water requirements were used, except for flood-routing where the time intervals had to be reduced to 6 hours. In the analysis of some mathematical models quarterly, semiannual, and annual periods were chosen. Examples of calculations based on these longer time intervals are found in Chapters 13 and 14.

ADJUSTMENT OF STREAMFLOW DATA

Since the streamflow records at all stations were not as long as those at Kamiah (*E*), missing monthly flows were interpolated by simple correlation of the runoff observed at a given station with that at Kamiah (see Chapter 12 for techniques and results). Two further adjustments were made in the basic streamflow data, primarily for convenience in computer programming and system analysis. Since two sets of flow data, those between reservoirs *A* and *B* and point *E* and those between reservoirs *C* and *D* and point *G*, were obtained by subtraction, some negative values inevitably resulted. The first adjustment in basic data, therefore, was to set these small negative values equal to zero and to modify the positive differences accordingly. The second adjustment was to convert all monthly flows to equivalent runoffs for standard months of 30.44 days. This standardization simplified computer programming appreciably without introducing significant errors, because the proportions of the monthly flows involved were less in volume than were known errors in the streamflow measurements themselves.

In the ordinary course of simulation studies of the simplified system, a 50-yr period of analysis was preceded by a preliminary 10-yr run to compensate for any possible bias introduced by assuming that the reservoirs were full at the start of each study. The 60-yr runoff record needed for this purpose was obtained by appending to the observed 32-yr sequence of monthly flows the first 28 years of the same series.

The 6-hr flood flows required for simulation studies were read from hydrographs plotted from recorded mean daily flood flows and, in some instances, from instantaneous peak flows.

For more extensive simulation studies, 510 years of monthly runoff and flood-flow data were synthesized from the observed streamflow record by statistical methods (see Chapter 12).

ASSUMPTIONS GOVERNING SYSTEM OUTPUTS

The assumptions made in developing the simplified river-basin system are indicated below in reference to the three selected purposes of irrigation, water power, and flood-damage reduction.

IRRIGATION

Consumptive use and diversion requirement. The consumptive use and diversion requirement for irrigation were based on climatological data for and irrigation practices in the Lewiston region of Idaho. The resulting values are typical of irrigation in semiarid regions with a moderately long growing season of about 205 days. Table 7.2 shows the crop distribution and associated farm-irrigation efficiencies assumed in the simplified sys-

Table 7.2. Assumed crop distribution and farm-irrigation efficiencies.

Crop	Percentage of irrigated area	Farm-irrigation efficiency (percent)
Alfalfa	35	60
Pasture grass	15	50
Sugar beets	20	55
Potatoes	10	50
Small grains	10	55
Deciduous orchard	5	60
Small truck[a]	5	50
	100	

[a] Assumed to produce two crops during the irrigation season.

tem.[4] For the sake of simplicity, the crop pattern was held constant regardless of the level of irrigation output, but moderate changes in the distribution of land classes were made. In a real situation both crop pattern and land-class distribution change as the total area under irrigation increases. However, regardless of crop pattern and land-class distribution, the end results in the computation of an irrigation output are a net consumptive use and a diversion requirement for the crops to be irrigated.

The net consumptive use during the irrigation season was calculated by the Blaney-Criddle method,[5] and the effective precipitation during the irrigation season was assumed to be constant. For the crop pattern presented in Table 7.2 the unit net consumptive use is 2.0 ft (or 2.0 acre ft per acre) in each irrigation season.

In order to simplify computation of the unit irrigation-diversion requirement, we assumed that the consumptive use of the cropped area during the nonirrigation season was equaled by the average precipitation during that part of the year; accordingly, no allowance had to be made for soil moisture. Conveyance losses were assumed to average 30 percent of the annual diversion requirement irrespective of its level. This is essentially true when main conveyance canals are lined.

With these assumptions the unit irrigation-diversion requirement was computed to be 5.0 ft yearly, distributed by months as shown in Table 7.3.

[4] Farm-irrigation efficiency is defined as the percentage of water delivered at the farm headgate that is consumptively used by crops.

[5] H. F. Blaney and W. D. Criddle, "Determining water requirements in irrigated areas from climatological and irrigation data," *U. S. Soil Conservation Service Technical Publication 96* (1952).

Table 7.3. Assumed monthly distribution of annual irrigation-diversion requirement.

Month	Percentage of total annual diversion requirement
April	12.4
May	14.6
June	16.6
July	19.0
August	18.0
September	12.4
October	7.0
November–March	0.0
	100 0

The resulting monthly diversions were assumed to represent the most efficient time-spacing of the irrigation water for the postulated climate and crop pattern.[6]

Irrigation return flow. In estimating the return flow from irrigation diversions, we assumed for simplicity that no water would be lost by evaporation from drainage and wasteway channels, that no return flow

Table 7.4. Assumed monthly distribution of annual irrigation return flow with full irrigation supply.

Month	Percentage of total annual return flow
November	8
December	7
January	5
February	4
March	4
April	6
May	8
June	10
July	12
August	14
September	12
October	10
	100

[6] In economic terms, this assumes that the marginal value of the irrigation diversion has been equated for each month of the irrigation season and that the farm-irrigation system has been designed at the optimal economic point, or the point where the cost of facilities, labor, and other resources needed to increase the irrigation efficiency just equals the saving effected by reducing the gross diversion requirement at the farm headgate.

would be consumed on nonirrigated land, and that no water would escape into neighboring basins. Return flow would therefore equal the difference between the irrigation-diversion requirement and the net consumptive use of irrigation water, or 3.0 ft annually. The monthly distribution of this return flow in years of full irrigation supply, shown in Table 7.4, was selected to accord with observations in several Bureau of Reclamation projects or study areas.

For years in which the full irrigation-diversion requirement could not be satisfied, the amount of the return flow was reduced by six-tenths of the amount of the diversion shortage. In the absence of quantitative information about the effects of shortages on return flows, monthly return flows were modified in the following simple way. Beginning with the month succeeding the recorded shortage, the normal return flows from irrigation were reduced by the proportions of the shortage shown in Table 7.5. The sum of the reduction factors for the seven months is

Table 7.5. Assumed reduction factors for monthly irrigation return flows in the presence of irrigation shortages.

Month succeeding shortage month	Reduction factor
1st	0.24
2nd	0.12
3rd	0.09
4th	0.06
5th	0.03
6th	0.03
7th	0.03
	0.60

seen to be 0.6. The time distribution of the drop in irrigation return flows is similar to an infiltration curve for irrigation water, although this curve can vary greatly from area to area. The time distribution is similar in shape also to the normal return-flow pattern, in that the largest return flows occur at times closest to the largest diversions and, in the case of shortages, the largest reductions in return flow take place closest to the time of shortage.

Irrigation benefits. To be of significant use in the study of methods for planning water-resource systems, the benefits derived should involve problems of economic choice among designs in terms of costs, purposes served, and amounts of water available for utilization. Accordingly the maximum annual irrigation output of the system was set at 6×10^6 acre ft, for 1.2×10^6 acres. This is sufficiently high to have the available water rather than the available acreage limit the effective physical development. Unit annual gross irrigation benefits were taken as decreasing from about $6.50 per acre foot at low levels of development to about $5.00 per acre foot at maximum development. This drop in unit benefits results from the falling off of over-all land quality commonly associated with an expansion of the area to be placed under irrigation. Figure 7.2 shows the unit gross irrigation benefit function adopted.

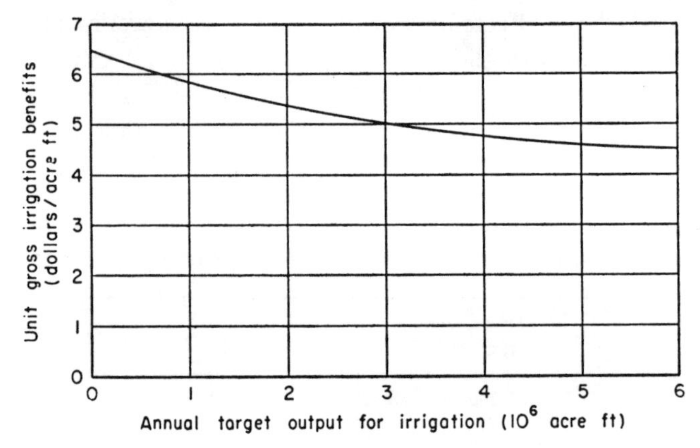

Fig. 7.2. Assumed unit gross irrigation benefit function.

When irrigation shortages occur, it stands to reason that gross irrigation benefits must decrease. However, the detailed economic effects of irrigation shortages are a function of their time, frequency, and intensity of occurrence, serial correlation among the occurrences, and flexibility of the irrigation operation. If, for instance, water is in short supply at a time when moisture is needed to germinate small grains, the entire crop may fail; whereas, for other crops such as alfalfa, shortages occurring during the last two months of the irrigation season may merely lessen the last cutting of the crop. To give another example, frequent shortages of water for irrigated pasture may so reduce yields and benefits that operation of the irrigation system is no longer economical, particularly if the deficits are severe and long drawn out. However, if the irrigator conducts a flexible operation — if, for instance, he can draw on labor and machinery to apply the available water more effectively to his crops — he may run up his costs and still cut his losses, because the gross benefits otherwise not received are likely to be greater than the increased costs, at least for small shortages. With farm irrigation efficiencies of 50 to 60 percent, as in the system under study, some margin does indeed exist for irrigators to cut economic losses connected with water shortages.

In the absence of quantitative information on the response of irrigators to water shortages and the associated net economic losses,[7] the irrigation loss function sketched in Fig. 7.3 was based on judgment and practical

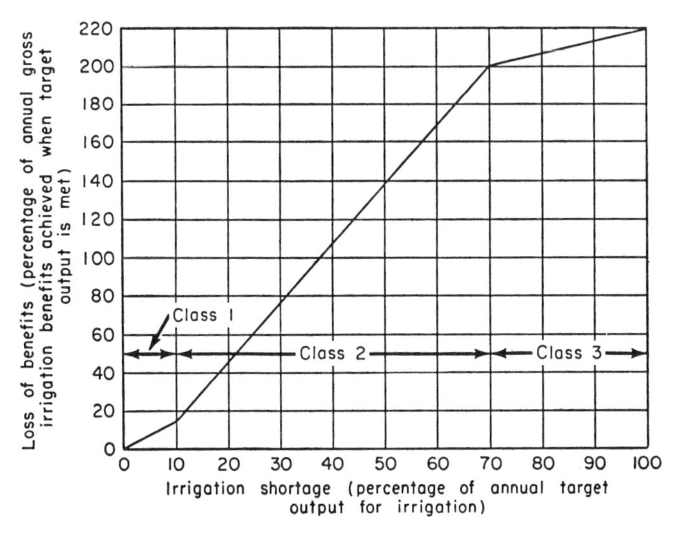

Fig. 7.3. Assumed irrigation loss function showing loss of benefits when target output is not met.

experience. The relation between losses and water shortages was developed only for annual, rather than also for monthly, deficiencies.[8]

The loss function comprises three straight lines that identify class 1 shortages as lying between 0 and 10 percent, class 2 shortages between 10 and 70 percent, and class 3 shortages between 70 and 100 percent of the annual diversion requirement. The reasons underlying this classification are stated briefly below.

Within the range of class 1 shortages, the rise in benefit losses is judged

[7] However, see the discussion of the irrigation loss function in U. S. Bureau of Reclamation, *Seedskadee Project, Wyoming,* Preliminary Plan Formulation Appendix (Washington, May 1957), p. 61.

[8] Developing the relationship between losses and monthly deficiencies was not considered justified for system-design purposes in view of inadequate data, the wide variation in monthly responses to water shortages, and the presence of serial correlation among monthly losses (the fact that the loss from a given shortage in a given month depends on the magnitude of shortages in preceding months).

to be slow because it is possible to improve distribution and farm irrigation at little increased cost, and crop yields may fall but slightly. Gross benefits may decline little or not at all, but production costs are likely to swing upward in the process of achieving increased efficiencies of water utilization. The loss function shown in Fig. 7.3 incorporates the impact of water shortages on both benefits and costs.

Within the range of class 3 shortages, losses may be relatively complete. At some level of shortage, indeed, not enough water is available to permit a crop to mature. The supply of water may even be insufficient to ensure the survival of fruit trees and perennial crops such as alfalfa. More than one year's harvest may then be lost, since it may take five or more years for a new orchard to come into commercial production and two to three years for a good stand of alfalfa to be reestablished. Associated with this level of shortages, therefore, are not only the losses for the shortage year itself, but also losses extending beyond it over additional years.

Just where the physical level for virtually complete loss should be set is an open question.[9] For some crops and irrigation operations, complete loss might follow a 50 percent shortage in the annual diversion requirement; for others, an 85 percent shortage. It should be explained in this connection that a given percentage reduction in the diversion requirement does not necessarily imply the same percentage reduction in consumptive use or water made available to and used by the crop. For example, a unit diversion requirement of 5.0 ft to which is attached a 30 percent conveyance loss, as has been assumed for our system, results in a unit farm input of about 3.5 ft, whereas the unit consumptive use is but 2.0 ft in an irrigation season. Even with a constant percentage conveyance loss, therefore, the irrigator has some leeway, and the percentage reduction in consumptive use may be kept below the percentage reduction in the diversion requirement. To accomplish this, the irrigator can check ditches, gates, and pipes for leaks, obstructions, and weeds; time water applications so as to prevent waste; change his method of cultivation; and, if sufficiently forewarned, alter the crop pattern.

Between the boundaries of class 1 and class 3 shortages, the reduction in benefits is assumed to vary linearly with the degree of shortage, in the absence of information to the contrary. Within this range losses increase with water shortage in spite of the best efforts of the irrigator.

[9] U. S. Bureau of Reclamation, reference 7.

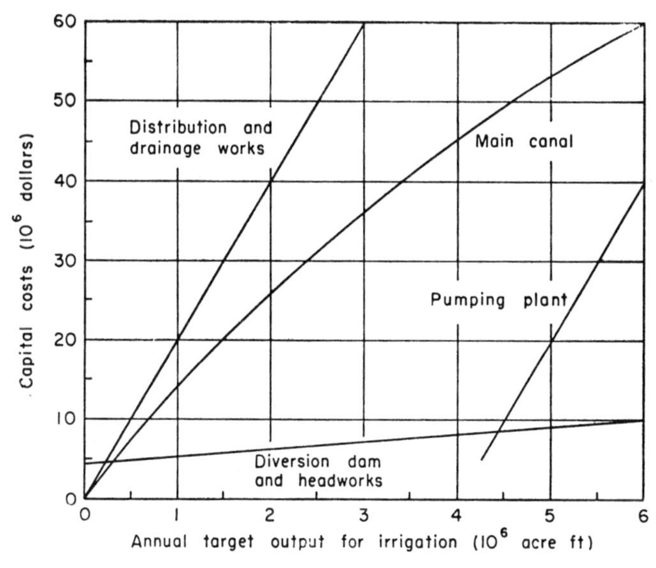

Fig. 7.4. Assumed capital costs of irrigation-diversion, distribution, and pumping works.

This lengthy discussion of the water-shortage–irrigation loss function is believed to be justified by its importance in actual practice, as a conceptual problem, and in relation to our objectives. If economic choice is not to be restricted, some of the alternatives that must be studied are sure to include combinations of scales of development and levels of target output for irrigation that will result in shortages. The loss function was

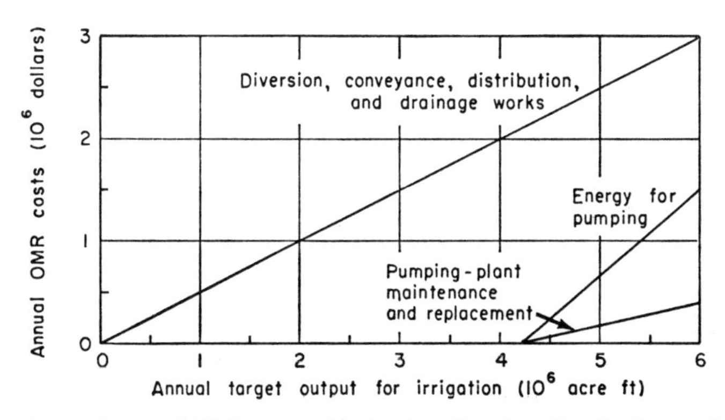

Fig. 7.5. Assumed annual OMR *costs of irrigation-diversion, distribution, and pumping works.*

introduced to make possible an economic evaluation of such combinations,[10] which otherwise would be ruled out arbitrarily.

Irrigation costs. Estimates of capital costs and costs of operation, maintenance, and replacement (OMR) for the irrigation-diversion works and distribution and pumping facilities were based on data from Bureau of Reclamation projects. They are shown in Figs. 7.4 and 7.5, respectively. Costs are given for the most efficient design and operation at a given level of development and reflect economies of scale. As indicated previously, the main conveyance canal was assumed to be lined. It was also assumed that the first 840,000 acres in the irrigable area can be served by gravity, whereas any further acreage requires pumping of diversions.

WATER POWER

Energy requirement. The electric-power requirement of our system was selected to be representative of a diversified agricultural-industrial, rural-urban economy. No extremely high demands were included for any single purpose (such as heavy, continuous, in-season irrigation pumping or large electrochemical industries). Energy requirements were assumed to be for general industrial and commercial demands, some summer air-conditioning needs, typical domestic requirements, and varied agricultural uses, including some pumping for irrigation diversions, for drainage, and for sprinkler irrigation. The resulting composition of the energy requirement, or target output for energy, is shown in Table 7.6. The assumed

Table 7.6. Assumed composition of annual energy requirement.

Load class	Percentage of total load
Domestic and rural	30
Industrial	35
Irrigation pumping and air conditioning	15
Commercial	10
Losses, internal utility use, and miscellaneous	10
	100

[10] Combinations with many large shortages extending beyond some "tolerable" limit would be questioned in any event, as is done now in agency practice. A steeply sloping loss function penalizes such combinations so greatly that they are, in effect, rejected on economic grounds. The loss function can thus serve as an alternative to an arbitrary cutoff level. For a specific example of a tolerable shortage-cutoff level, see Chapter 8.

monthly distribution of the load and the associated load factors are listed in Table 7.7.

Table 7.7. Assumed monthly distribution of annual energy requirement.

Month	Percentage of total annual energy requirement	Monthly load factor
November	8.1	0.63
December	8.3	0.61
January	8.2	0.62
February	7.5	0.60
March	7.3	0.58
April	7.7	0.60
May	8.3	0.62
June	8.9	0.64
July	9.1	0.66
August	9.3	0.68
September	9.1	0.68
October	8.2	0.64
	100.0	
Annual load factor		0.60

It was assumed that the values shown in Table 7.7 do not respond to changes in the magnitude of the total annual target output for energy. In reality, changes in the total load are associated with shifts in the composition of energy requirements and consequent changes in the monthly energy distribution and load factors. In practice, a power-market survey would determine the specific magnitudes and times of peak loads for significant load classes during daily, weekly, and monthly periods. Whether the target output would be realized by power plant B or G was assumed to be immaterial.[11] This simplifying assumption implies either that energy losses between the two plants are negligible or that the two plants are equidistant from the centers they serve.

Power-plant capacity. With the monthly target output for energy and load factors established, the installed capacity that is required could be readily computed. In order to eliminate much detailed data preparation and to decrease the difficulty of computer programming, we made the following simplifying assumptions.

(1) No thermal power plants were to be built.

(2) The power facilities were not to become parts of a larger integrated

[11] The target output is the energy needed to meet the given load. It is measured as the energy output of the generators at the low-voltage side of the switchyard. Additional energy, produced with surplus water and capacity, is considered to be nonfirm or dump energy.

power network. The complexities of analyzing and designing hydrothermal power systems were thereby avoided; they have been studied by others.[12] Some interconnection with other systems was posited, and some energy was therefore assumed to be available to compensate for system shortages. The matter is discussed further in a later section of this chapter.

(3) Continuous overloading of units was not permitted; only normal overloading was allowed in response to short-term peak demands.

(4) No specific allowance was made for reserve capacity. Rounding the computed required capacity to the nearest higher 5,000 kw resulted in a significant increment in the power reserve only for low installed capacities. In practice, provision is generally made for a reserve capacity equaling 10 to 15 percent of the total peak demand.

The requisite installed capacity was found to reach a maximum in the month of June, when the assumed energy requirement is 8.9 percent of the total annual requirement and the load factor is 0.64. This follows from the relationship

$$c = pE/100lh.$$

where c is the requisite installed capacity, p is the percent of total annual energy requirement, E is the total annual energy requirement, l is the monthly load factor, and h is the number of hours in a month.

In our studies of different levels of target output for energy and different scales of development, the associated minimum required capacities of power plants B and G were ascertained by considering power-plant capacities as variables along with other input variables.

The maximum yearly target output for energy for our simplified system was set at 4×10^9 kw hr. As is the selected maximum target output for irrigation, this level is high enough to make the amount of available water the effective physical limit of development. Between 0 and 1.6×10^9 kw hr the desired energy output can be supplied by either power plant B or G. Above that figure both installations are needed.

Energy benefits. Unit energy benefits were assumed to be constant for all levels of output of the simplified system, namely, 7 mills per kilowatt hour of firm energy and 1.5 mills per kilowatt hour of nonfirm energy, at both power plants B and G. The firm-energy benefit is an average

[12] See, for example, L. S. Wing and R. H. Griffin, "Selection of installed capacity at hydroelectric power plants," *Proc. Am. Soc. Civil Engrs. 81*, Paper 697 (1955); A. P. Fugill, "Principles and practices of modern system planning," *Trans. Am. Inst. Elec. Engrs. 74*, 1323 (1956); and Philip Sporn, *The integrated power system* (McGraw-Hill, New York, 1950).

value resulting from a combination of dependable installed capacity and firm energy produced, and no attempt was made to compute its component parts. Average rates of this kind are often employed by water-resource agencies during the early stages of planning.

It was assumed that occasional energy shortages could be tolerated. This is realistic, since many loads can actually be interrupted or reduced for short periods of time. However, the economic responses to energy shortages — for example, in "brown-out" periods in the Pacific Northwest — are not quite so evident as are the responses to irrigation shortages. Although the same factors are involved as in irrigation shortages (time, frequency, intensity, serial correlation, and flexibility of use), a different loss function was needed for the economic evaluation of energy shortages. In the absence of specific information, the same type of simplified three-class loss function was assumed as for irrigation shortages (Fig. 7.3). However, the losses were related to monthly rather than annual shortages, and class 1 shortages were defined as those between 0 and 10 percent of the monthly requirement; class 2, between 10 and 25 percent; and class 3, over 25 percent.

For class 1 or 2 energy shortages, supplementary energy was assumed to be available through an intertie at a constant price of 9 mills per kilowatt hour. At this price the loss to the system would be 2 mills per kilowatt hour. Because interconnections among power systems are becoming increasingly common, such supplementary energy may be expected to be available. For our system it was limited to 25 percent of the monthly target output for energy. Above this level the amount to be supplied from the "outside" was considered so large as to affect the basic design of the outside power system. The shortage within our system would then logically become part of the "normal" load of the outside system. In these circumstances the two systems would have to be designed for integrated operation, and this would drastically alter the scope of our study. In order to avoid this, designs with energy shortages greater than 25 percent were not considered feasible.[13]

Power-plant costs and characteristics. Assumed capital costs and OMR costs for the power plants and their appurtenant facilities were based on experience recorded by Creager and Justin,[14] Doland,[15] and the Federal

[13] An arbitrary cutoff was applied as before. As discussed in Chapter 10, however, the task of finding an optimal combination by systematic sampling would be greatly eased if a steeply sloping energy loss function were used to weed out combinations with high energy shortages.

[14] W. P. Creager and J. D. Justin, *Hydroelectric handbook* (John Wiley, New York, ed. 2, 1950), pp. 238–241.

[15] J. J. Doland, *Hydro power engineering* (Ronald Press, New York, 1954), chap. 7.

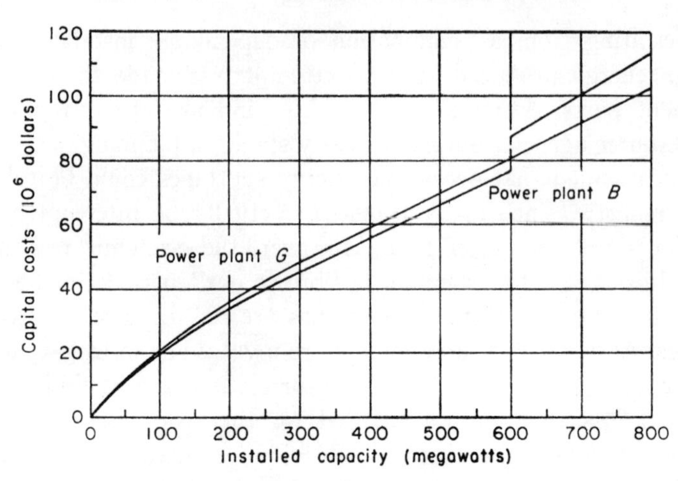

Fig. 7.6. Assumed capital costs of power plants B and G. Capacity of power plant B above 600 Mw requires more complicated, and thus more expensive, structure.

Power Commission.[16] The two categories of costs are shown in Figs. 7.6 and 7.7. The best design was assumed to have been developed at each capacity in terms of penstocks, turbines, inlet structures, surge tanks,

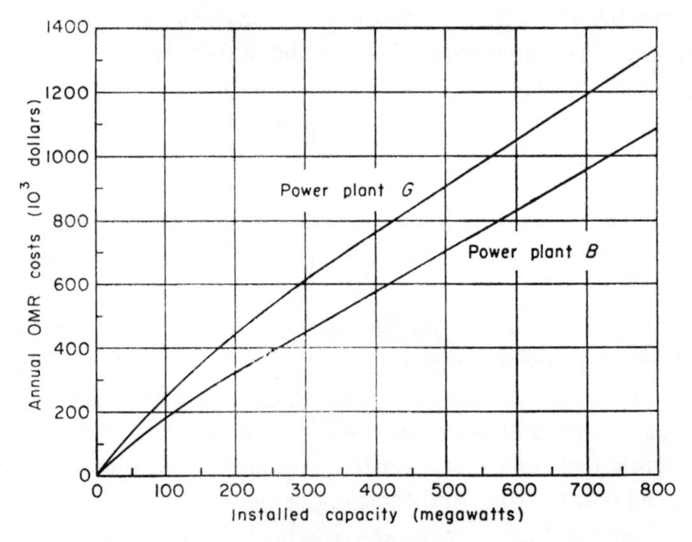

Fig. 7.7. Assumed annual OMR *costs of power plants B and G.*

[16] U. S. Federal Power Commission, "Information for staff use in estimating power costs and values," *Tech. Mem. 1, Bur. of Power* (1955), Table 6; and U. S. Federal Power Commission, *Hydroelectric plant construction cost and annual production expenses, 1953–1956,* F.P.C. S-130, with first and second annual supplements, 1957 and 1958, F.P.C. S-135 and 139.

and other appurtenances. A further assumption was that economies of scale were realized as plant size increased. Cost estimates were carried no farther than the low-voltage side of switchyards.

The geology and topography of the site for power plant *B* were assumed to be such that plants with capacities above 600 Mw require a more complicated structure than plants with lower capacities. This causes a discontinuity in the cost curve at 600-Mw capacity. Penstock costs and costs of dam modification identified with changes in power-plant capacity were allocated to the power plant rather than to the reservoir. At *G* the costs of pondage were included in full in the power-plant costs.

Power plant *B* was assumed to be a variable-head plant in which the net effective head varied in accordance with the curve drawn in Fig. 7.8;

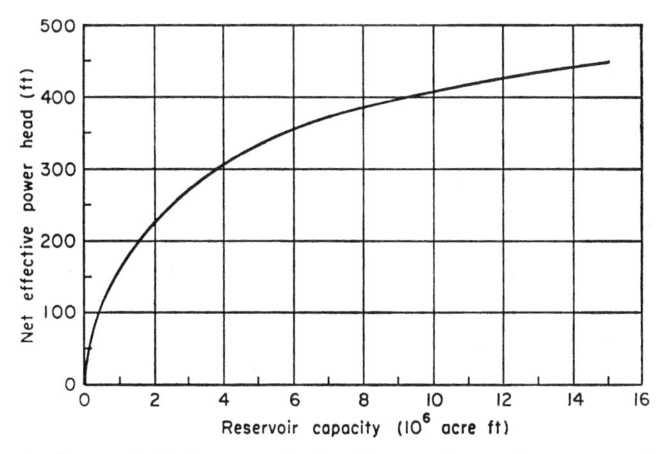

Fig. 7.8. Assumed effective power head and reservoir capacity at power plant B.

the tailwater effect on the net head was negligible; and the over-all plant efficiency from forebay to tailwater remained constant at 85 percent, regardless of head or installed capacity. The variation in head was limited, however, by the requirement that the maximum power head be no more than twice the minimum. This is a practical assumption dictated by turbine performance. The others are simplifying assumptions that greatly ease the preparation of data, machine programming, and system analysis.

Power plant *G* was assumed to be a constant-head plant in which the fixed net head was 160 ft; the tailwater effect on the net head was negligible; and the over-all plant efficiency from forebay to tailwater remained constant at 88 percent, regardless of installed capacity. In reality, a power plant in which a separate power canal or tunnel forms a relatively constant headwater elevation and which possesses a relatively constant tailwater elevation,[17] or a run-of-the-river plant with relatively small variation in head, would be the closest approximation to this kind of plant.

FLOOD-DAMAGE REDUCTION

Flood characteristics. Our system was assumed to contain but one flood-damage center, located just below the gaging station at *G*. Examination of the historical streamflow record at *G* suggested 105×10^3 cu ft/sec as a zero flood-damage flow useful for our purposes.[18] Accordingly, nine damaging floods would have been experienced during the 32-yr period of record. Eight of these were spring snowmelt floods; one was a winter flood. Such a distribution is typical of the Columbia River Basin as a whole. The nine flood hydrographs, together with information on peak flows between 75×10^3 and 105×10^3 cu ft/sec, were later used in identifying the number, magnitude, and time of occurrence of floods to be expected in a synthesized runoff record of 510 years (see Chapter 12).

Flood-control capacity. For the simulation studies described in Chapter 10 flood-control space was specifically set aside in reservoirs only for the months of April, May, and June in conformance with the operation of multiple-purpose reservoirs in the Columbia River Basin system for flood control.[19] In this region multiple-purpose reservoirs for irrigation, power, and flood control are generally at low levels in the fall and winter, because of summer and early fall irrigation releases and fall and winter power drafts. Usually this leaves room to store fall and winter floods without allocating specific space for flood control. Because the April floods in the chosen hydrology are usually smaller and less frequent than the May and June floods, the flood-control capacity of the reser-

[17] These conditions are approximated in the Swift No. 2 plant of the Cowlitz County Public Utility District on the Lewis River, Washington, and in several of the Southern California Edison plants on the upper San Joaquin Basin.

[18] In the Clearwater River Basin flood damages actually occur at much lower flows. But we wanted to limit the number of floods in the simplified system because of the large amount of information that had to be developed for each flood.

[19] See *Report on Columbia River and tributaries, northwestern United States,* House Doc. 531, 81 Cong., 2 Sess., *VII,* 2747.

voirs was made only half as large for April as for May and June, when full capacity was allocated.[20]

Flood-control capacity can be set aside in any one or more of the reservoirs. How such capacity should be distributed between them is partly a function of the relative contribution that each subbasin makes to the flood volumes at the damage center,[21] and partly a function of the multipurpose character of the storage reservoirs in the system.

Over the historical period of record the relative contributions of the two major tributaries, recognized as *B* and *D* in the simplified system, were reasonably constant. However, reservoir *B* can provide not only flood-control storage, but also storage for power production and irrigation, and so should logically retain as much water as possible. This is the direct opposite of releasing water to make room for the control of floods. Hence the amount of flood-control capacity allocated to *B* had to be tempered to its storage of water for power and irrigation.

Reservoir *A* serves an area that contributes only small amounts of water to flood flows. Because of drawdowns for irrigation and power, this reservoir is generally low at the start of the spring flood season. Moreover, its drainage area is amenable to some control by reservoir *C*. In our simulation studies, therefore, we assigned no flood-control capacity to reservoir *A*.

Reservoir *C* lies downstream from tributary *B* and because of this can exert a measure of control over watershed *B* as well as over watershed *A* and some additional area. It follows that regulation of flood flows issuing above *C* can be provided by reservoir *B* or *C*, or by some combination of the two. Since *C* lies below the irrigation diversion and serves the main purpose of supplying water to power plant *G*, multiple-purpose considerations are somewhat less cogent at *C* than at *B*. The same is more or less true for reservoir *D*, which supplies water to power plant *G*. Because of its location on a major tributary, however, reservoir *D* can regulate some contributions to major flood flows.

It was assumed that the reservoirs could be drawn down in one month

[20] Under certain conditions this type of flood-control rule curve may be unsuited to snowmelt basins and to other climatic areas. Whenever energy and irrigation outputs are zero or relatively small, insufficient drawdown may not leave enough room for retention of fall and winter floods. For river systems in humid areas such as the northeastern United States, where floods may occur in every month of the year, specific flood-control capacity would have to be available each month.

[21] See, for example, U. S. Army Engineer Division, North Pacific, *Water resource development of the Columbia River Basin*, Report Brochure (1958).

to desired flood-control levels. In order to be ready for April floods, therefore, releases had to be started on March 1. The assumed channel capacities were large enough so that released flows could be carried without causing flood damage of their own. Because snowmelt floods can be forecast reasonably well for basins such as ours, choice of a one-month "evacuation time" was realistic and in reasonably good accord with existing practice.[22]

Flood-routing. The routing of flood waters through multiple reservoirs and stream channels, an intricate problem in itself, was complicated further by our objective of integrating flood-routing into our over-all simulation studies. Specifically this meant that whenever a flood occurred in a given month, the computer had to "shift gears," so to speak, in order to route the flood through the system on the basis of 6-hr flows before shifting back to monthly information.

Although it complicates computer programming, flood-routing was made part of each study for three important reasons.

(1) Flood control is a fundamental factor in our multipurpose, economic-choice approach to river-basin analysis, and we wished to analyze the flood-control purpose along with other purposes in the same simulation study. The common practice of making flood-routing a separate procedure is avoided.

(2) The hydrologic conditions prevailing in the system at the start of a flood are determined intrinsically by the monthly study. In separate flood-routing studies, by contrast, they may be arbitrarily chosen.

(3) Correlatively, the hydrologic situations established in the system at the end of the flood are determined, as they should be, by the flood-routing results. These terminal conditions then become the initial conditions for resumption of the monthly information.

Let us summarize the advantages of this method. Each flood and the response of the system to it are made part of the normal sequence of hydrologic events. A better estimate is obtained of the effects of all flood flows, not merely of some of them, nor of the so-called "standard project flood." Flood-routing in itself was not of concern to us;[23] either a simpler or a far more sophisticated method of flood-routing would have been conceptually acceptable.[24]

[22] See *Report on Columbia River and tributaries*, reference 19.

[23] Numerous refined methods of flood-routing have been developed elsewhere. For examples, see E. Isaacson, J. J. Stoker, and A. Troesch, "Numerical solution of flow problems in rivers," *Proc. Am. Soc. Civil Engrs.* Paper 1810 (1958), and D. M. Rockwood, "Columbia Basin streamflow routing by computer," *Proc. Am. Soc. Civil Engrs.* Paper 1874 (1958).

[24] It is recognized that flood-routing would be difficult to combine with a monthly procedure

An examination of the flow records for floods at the various gaging stations in the Clearwater Basin showed that valley storage was small.[25] Accordingly, for ease of simulation, a simple time-lag method of flood-routing was adopted rather than some variant of the more usual general-storage equation.[26] The travel time from either A or B to G was estimated to be roughly 12 hours; from C to G, about 6 hours; from D to G, virtually negligible. The 6-hr flows at A and B were therefore advanced by 12 hours and those at C by 6 hours. For any particular interval of time, the natural flow at G then fell into line with the natural or regulated flows at A, B, C, and D.

Flood flows were routed through the system for an entire month even when the flood was of shorter duration. In order to simplify computations, we made no allowance for reservoir surcharge, although storage above spillway level would have reduced outflows and improved flood regulation. The timing of flood flows of the Clearwater River that occurred toward the end of a given month was shifted sufficiently to contain the flood hydrograph within the month.

Flood-control benefits. Damage caused by flood waters within the simplified system are delineated in Fig. 7.9. Dollar costs were selected to bring flood control within the range of monetary benefits derived by developing the simplified system for irrigation and water power. In this way a problem of economic choice among the three purposes was created.

Flood damages were assumed to be a function of the peak flow only. In real situations velocity of flow, time of year, quality of flood waters, and duration and depth of inundation may also be important.[27] The flood-damage curve shown in Fig. 7.9 was derived as follows. Up to a flow of 105×10^3 cu ft/sec, there is no damage. Once the associated

---

for very large basins, such as the Missouri or the Ohio, where the travel time of the flood wave from the upstream to the downstream end of the basin may be a matter of weeks rather than days. Incorporating the proper travel times into the monthly procedures might then become impossible. Instead, some form of segmental analysis of the basin would be required.

[25] See *Report on Columbia River and tributaries*, reference 19, *IV*, 2806: "Flood flows of the Columbia River and its major tributaries are largely confined to regular channels throughout their entire reaches east of the Cascade Range. Since there is a relatively small amount of overbank flow, valley storage is small in comparison with the volume of runoff." Also, p. 2807: "In view of the fact that the effect of valley storage in reducing the peak of floods on the Columbia River is small, the routing procedure developed for reproducing natural flood hydrographs and the effect of controlled flows from upstream reservoirs was based on transposing flows from upstream to downstream stations by the lag time derived from an analysis of past floods, and adding the local inflow from tributary streams."

[26] See R. K. Linsley, M. A. Kohler, and J. L. H. Paulhus, *Applied hydrology* (McGraw-Hill, New York, 1949), chap. 19.

[27] See S. W. Wiitala, K. R. Jetter, and A. J. Somerville, *Hydraulic and hydrologic aspects of flood-plain zoning*, U. S. Geol. Survey Open File Report (1958), 85 and 89.

river stage is reached, the banks are overtopped; water sweeps over a portion of the flood plain, and a block of damages results. With rising stage the area of inundation is widened, and damages mount until the interior flood plain bounded by a terrace formation is under water. Thereafter, there is no additional damage until the rising flood waters are able to wash into the higher-lying outer flood plain. Then, once again, damages increase rapidly. Eventually all the valley lands have been inundated and damages reach a second plateau.

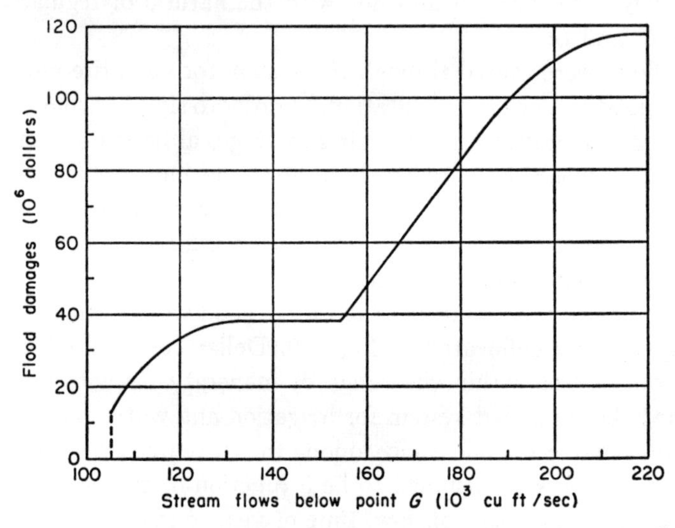

Fig. 7.9. Assumed relationship of flood damages to streamflows below point G. Up to a flow of 105 × 10^3 cu ft/sec no damage occurs.

In our system the flood-control benefits achieved by suitable storage were calculated as the difference in damages associated with the uncontrolled and the regulated flows at *G*. If these benefits were to be received, some explicit flood-storage capacity had to be specified. In practice this may be as great as the volume of water discharged during the maximum flood beyond the zero damage level. For our system this volume was found to fall well below the sum of the capacities of the four reservoirs. Storage of 2 × 10^6 acre ft in combinations of reservoirs *B*, *C*, and *D* appeared to control the largest flood likely to be encountered — even the largest flood developed by our 510-yr trace of synthetic streamflow.

Flood-control costs. In readying their specified flood-control capacity, we assumed that our reservoirs would normally be drawn down through turbines and irrigation outlets and that separate flood-release outlets would not be needed. No specific costs for flood-control facilities were therefore assumed in our studies.

RESERVOIR COSTS AND CHARACTERISTICS

The curves shown in Figs. 7.10 and 7.11 trace the assumed capital costs and the assumed annual OMR costs for increasing the capacities

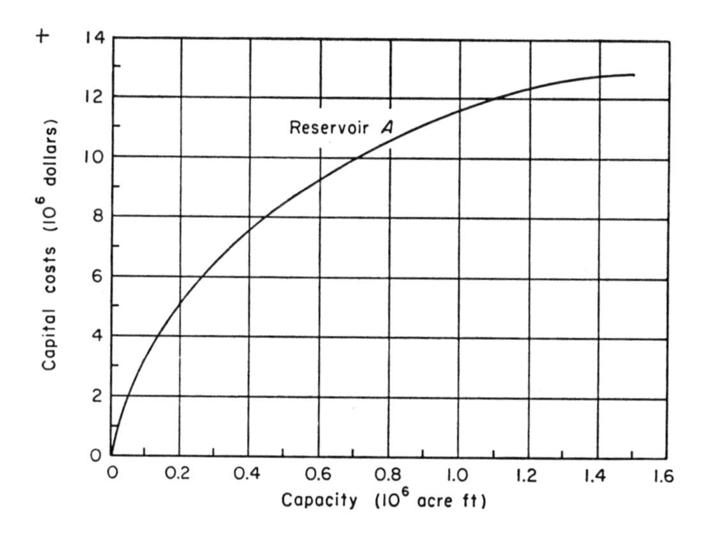

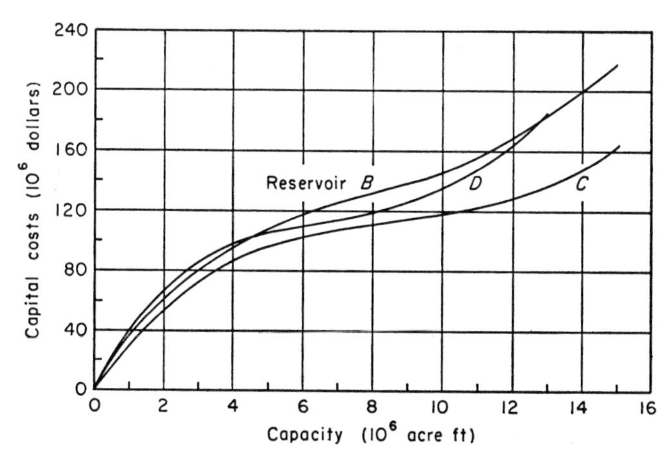

Fig. 7.10. Assumed capital costs of reservoirs A, B, C, and D.

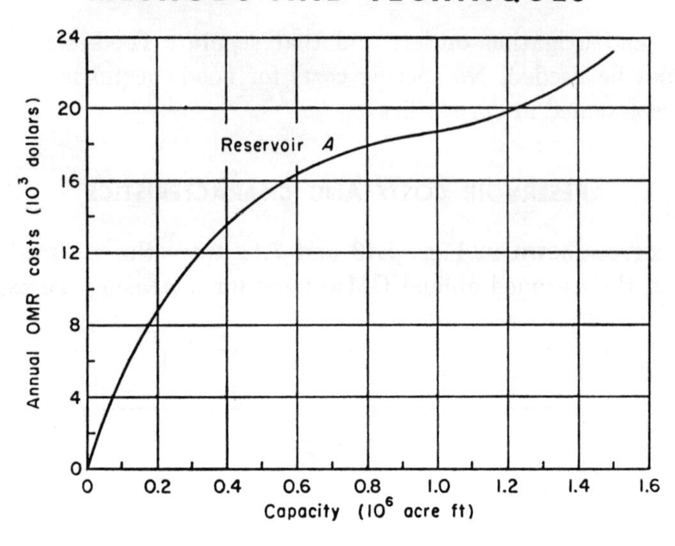

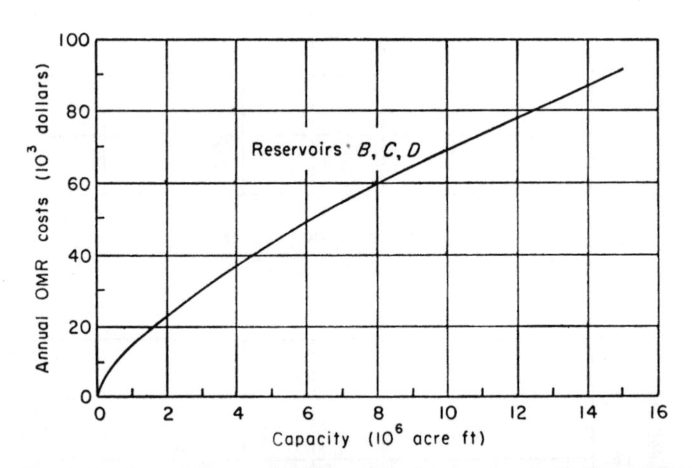

Fig. 7.11. Assumed annual OMR *costs of reservoirs A, B, C, and D.*

of the four reservoirs of our system. The assumed relationships are believed to be realistic for such structures, but they are more comprehensive than the information ordinarily collected by water-resource planning agencies,[28] which is restricted to one or, at most, a few scales of reservoir development. In order to provide a real problem in economic design,

[28] For the Clearwater River Basin, only limited cost data were available on dams actually proposed for construction, such as the Bruces Eddy and Kooskia dams. See *Report on Columbia River and tributaries*, reference 19, *IV*, 1429.

furthermore, we had to balance the costs of the structures in the system with the benefits to be obtained. The following assumptions were made in setting the capital costs of the reservoirs.

(1) Recorded reservoir costs are the sum of the costs of constructing the dam, its spillway, and outlet works; necessary relocation of highways and other works; clearing the reservoir area; and similar related undertakings. Costs are assumed to include interest during construction, but at a constant level throughout all studies.

(2) The reservoir cost curves are the same for all purposes of use, although in reality this is not strictly true. If power is to be produced, for example, penstocks must be added. To exclude differences of this kind, we assigned costs of structures other than those common to all reservoir functions to the development of the specialized function. The cost of penstocks, for instance, was charged to power-plant construction.

(3) The upper limit of reservoir storage was set at approximately three times the mean annual flow at the given sites. Ordinarily, this is high enough to make water resources and economics the determining factors rather than an arbitrarily imposed physical limit of reservoir size.

The following general assumptions relating to reservoirs were made in analyzing the system.

(1) The hydrologic factors of evaporation and seepage from the reservoirs were neglected, because the work involved in searching out reasonable information and incorporating it in the simulation seemed disproportionately greater than the benefits to be derived.

(2) Loss of storage by sedimentation was not taken into account, although this is a major problem in the arid and semiarid portions of the United States. In a full study of water-resource development, both the quantity of sediment and its pattern of deposition in reservoirs would be considered explicitly.

(3) No allowance was made for maintaining permanent pools for recreation, fish culture, and the like, because they were not among the purposes selected for our system. At reservoir *B*, to be sure, allocations for dead storage were permitted, but this was done for power production.

OPERATION OF THE SYSTEM

As described in detail in Chapters 9 and 10, the performance of the simplified river-basin system was studied for many different combinations of system variables. Performance, in terms of physical outputs and economic benefits achieved, was measured by simulating the behavior of the system on a digital computer. We thus were required to construct a set of rules for storing and releasing water in reservoirs — a specific operating procedure — in a form suitable for the computer. After several trials an operating procedure was developed. The reasons for adopting this particular procedure derive from (1) the hydrology of the basin, (2) the physical configuration of the system, including location of the potential structures, and (3) the nature of the system outputs. These salient characteristics are summarized briefly in the following section. A general discussion of the problems of devising operating procedures for simulation studies is contained in Chapter 11.

UNIQUE CHARACTERISTICS OF THE SIMPLIFIED SYSTEM

Hydrologic properties of the basin. Hydrologically, the system possesses certain general characteristics which are indicated below.

The adopted hydrology is typical for a western snowmelt area, with most of the annual runoff and flooding occurring in April, May, and June. Winter floods are rare, and low summer flows follow the high spring runoffs.

The mean annual runoff at reservoir site *B* is about eight times that at site *A* and about the same as that at site *D*.

Natural inflow between sites *A* and *B* and diversion point *E* as well as between sites *C* and *D* and point *G* are relatively small in comparison with the runoff at *B* and *D*.

The basin tributaries on which are located dam sites *B* and *D* are equally responsible for most of the flood runoff.

Physical configuration of the system. The following physical characteristics of the basin, as illustrated in Fig. 7.1, are of significance in building an operating procedure:

Sites for but four reservoirs, at *A*, *B*, *C*, and *D* (however, a smaller number may be considered in any one simulation study).

Sites for but two water-power plants, a variable-head plant at *B* and a fixed-head plant at *G*.

A single irrigation area (lying on both banks of the stream) that can be served by only two reservoirs, *A* and *B*.

A single downstream flood-damage center at point *G*, where the flood damages can be reduced by storage of flood waters in any or all of the four upstream reservoirs.

Generation of energy at power plant *B* by water supplied from reservoir *B* only, but at power plant *G* by water from all upstream reservoirs.

Regulation in reservoir *C* of spills and releases from reservoirs *A* and *B*, unregulated inflows between reservoirs *A* and *B* and diversion point *E*, and return flows from the irrigated area.

System outputs. The characteristics of the system outputs for irrigation, water power, and flood-damage reduction relevant to the operating procedure are given below.

Water is supplied to the single complex of irrigable lands by diversion works at point *E*. Irrigation diversions begin in April and end in October, with the maximum diversion during August. No return flows are used on the irrigated area.

Electric power is produced in plants *B* and *G*, energy output being measured at the bus bars. The load can be met from either plant. Energy in excess of a specified amount is marketed as dump energy; internal energy deficits up to 25 percent of monthly target outputs are made up by purchase from connected outside systems. Monthly target outputs show a small but significant variation throughout the year.

Reservoir space for the storage of flood flows need be set aside only in April, May, and June.

RATIONALE FOR THE ADOPTED OPERATING PROCEDURE

With these characteristics of the simplified system in mind, the operating procedure is based on the following reasoning.

Insofar as possible, target outputs are met by unregulated flows in the system and water that would otherwise spill or be released from reservoirs in order to draw them down to flood-storage levels.

Because the only variable-head power plant at a storage reservoir is at *B*, the water level in reservoir *B* is kept as high as possible consistent with meeting target outputs. Releases from storage for irrigation are therefore made first from reservoir *A*. Similarly, withdrawals from storage for water power are made from other reservoirs before reservoir *B* is lowered.

The flood flows at reservoir site *A* are relatively small. Since flood storage at *A* will not reduce damages at *G* significantly, none need be allocated to reservoir *A*.

Reservoirs C and D are downstream from the irrigation area, so that specific withdrawals for energy are made first from C and D, next from reservoir A, and last, if needed, from reservoir B.

Because energy deficits can be made up by purchase from outside the system, whereas irrigation deficits cannot, and because water power is generated by releases for irrigation from B, water is drawn from A and B to meet the target output for irrigation before any system releases are made to meet the target output for energy.

Energy generated at B by drafts for other purposes is at a maximum in the spring and summer because of high flood-control and irrigation releases. Hence, withdrawals of reservoir waters specifically for energy production are greatest in the fall and winter. Because inflow into reservoir D is highest in April, May, and June, reservoir D is drawn down in advance of that period. Accordingly, heavy fall and winter drafts for energy are made preferentially from reservoir D, so that spills from it may be minimized.

Because reservoir C operates largely as a reregulating reservoir by storing spills from A and B and the return irrigation flow, it is not easy to formulate a rule for minimizing spill from C. However, it seems sensible to release water for energy in late winter, spring, and summer from reservoir C in order to free some of its space for catching spills from reservoirs A and B during the months of high flow. By February or late winter, it should be noted, reservoir D will normally have been drawn down through heavy fall and early-winter releases.

Since major spring floods in the adopted system are caused by snowmelt, and techniques for forecasting such floods are relatively accurate, space for flood storage need be allocated only until the major flood hazard has passed.

Because of the assumed safe channel capacity at G, which is high in relation to flood-storage allowances, reservoirs can be safely lowered to flood-storage level in a single month.

Some system designs incorporate large reservoirs; others, much smaller ones. This gives significance to the question of initial reservoir contents. If the reservoirs are assumed to be full at the start of a simulation run, large reservoirs are given an advantage over smaller ones. For removal or reduction of any possible bias a 10-yr transient or "warm-up" period of operation is made to precede actual simulation. The reservoir contents at the end of this period then become the starting volumes.

In view of the many possible combinations of system units, operating parameters, and outputs, the procedure indicated below does not necessarily take full account of all contingencies. Discussion of the difficulties of formulating "optimal" operating procedures is deferred until Chapter 11.

OPERATING INSTRUCTIONS

Based upon the above reasoning, operating instructions were scheduled as follows.

(1) Use spills[29] from reservoirs *A* and *B*, if any, together with inflow between *A*, *B*, and *E* to provide the target output for irrigation.

(2) If the target output for irrigation is not realized, draw water from storage in reservoir *A* until the target output is met or *A* is emptied.

(3) If *A* is emptied and the target output for irrigation still remains unsatisfied, draw water from reservoir *B* until the target output is reached or *B* is drawn down to dead-storage level.

(4) Route both spills and releases from reservoir *B* through power plant *B* up to its maximum water capacity. Compute the energy generated and note its contribution toward meeting the target output for energy.

(5) Use spills from reservoirs *C* and *D* and the inflow between reservoirs *C* and *D* and point *G* for additional energy production at power plant *G*.

(6) If the total energy yield of power plants *B* and *G* by these means does not meet the target output, release water in accordance with the following schedule as far as need be.

(*a*) September through January:

(i) Release water from reservoir *D* until the target output is met, the water capacity of power plant *G* is reached, or reservoir *D* is emptied.

(ii) Draw water from reservoir *C* until the target output is attained, the water capacity of power plant *G* is reached, or reservoir *C* is emptied.

(iii) Make additional releases from reservoir *A*, if this is still possible, until the target output is realized, the water capacity of power plant *G* is reached, or reservoir *A* is emptied.

(iv) Draw additional water from reservoir *B*, if this is still possible, until the target output is met, the water capacities of power plants *B*

[29] The term "spill" in this context means water which would flow over the spillway if no releases were actually made.

and *G* are both reached, or reservoir *B* is drawn down to dead-storage level. The additional releases from reservoir *B* may generate energy at both power plants *B* and *G*. Because power plant *B* is a variable-head plant, calculation of the required draft from reservoir *B* involves a trial-and-error procedure.

(*b*) February through August:

(i) Release water from reservoir *C* until the target output is met, the water capacity of power plant *G* is reached, or reservoir *C* is emptied.

(ii) Draw water from reservoir *D* until the target output is attained, the water capacity of power plant *G* is reached, or reservoir *D* is emptied.

(iii) Make additional releases from reservoir *A*, if this is still possible, until the target output is realized, the water capacity of power plant *G* is reached, or reservoir *A* is emptied.

(iv) Draw additional water from reservoir *B*, if this is still possible, until the target output is met, the water capacities of power plants *B* and *G* are both reached, or reservoir *B* is drawn down to dead-storage level. As explained before, a trial-and-error procedure must be followed in computing the required draft from reservoir *B*.

(7) Make specific provision for flood-storage capacity during the months of April, May, and June in any combination of reservoirs *B*, *C*, and *D* as follows.

(*a*) During March, draw down each reservoir until one-half the total specified flood-storage capacity has been provided by April 1.

(*b*) If the snowmelt flood occurs in April, make no releases during the month. If it does not, lower each reservoir during April until the total specified flood-storage capacity has been provided by May 1.

(*c*) If the snowmelt flood occurs in May, make no releases during the month. If it does not, empty each reservoir until the required flood-storage-capacity has been provided by June 1.

(8) During March, April, and May route flood-storage releases through the turbines at power plants *B* and *G* up to their turbine capacities, and use the waters issuing from reservoir *B* to satisfy the target output for irrigation. (In effect, therefore, instructions for flood-storage releases are the same as instructions (1) and (5) for spills.) Fulfill any remaining requirements for irrigation and energy output in accordance with instructions (2) and (3) for irrigation and (4) and (6) for energy.

(9) If the input hydrology indicates that the month is one in which a flood occurs, follow the flood-routing instructions given below. The procedure is essentially the same as the monthly one described in instructions

(1) to (6), but it differs in several details. First, a 6-hr interval is specified. Second, if the required energy cannot be generated by initial releases from reservoir *B* and by drawing down reservoirs *A*, *C*, and *D*, no additional water is taken from reservoir *B*; instead, the energy deficit is computed and the procedure moves to the next interval. Third, 12 hours elapse while flows from *A* and *B* reach point *G*, and 6 hours while flows from *C* reach *G*; no time lag is applied to flows from *D*.

Instructions for routing flood flows during each 6-hr period are as follows.

(*a*) Release to keep contents of reservoirs *B*, *C*, and *D* down to flood-storage levels, subject to the condition that the total flow at *G* must not exceed the safe channel capacity of 52×10^3 acre ft in a 6-hr interval.

(*b*) If there is danger of exceeding the safe channel capacity at *G*, some of the flood flow must be stored. Therefore, draw water first from reservoirs *C* and *D* in proportion to the contents above their flood-storage levels.

(*c*) Once the contents of reservoirs *C* and *D* have been safely lowered to flood-storage levels, release water from reservoir *B* at a rate up to the remaining safe channel capacity at *G* and up to a volume equal to the contents of the reservoir above flood-storage level.

When the flood occurs, flows in excess of the safe channel capacity at *G* are held in the reservoir space provided for this purpose. However, water is purposely discharged from the reservoirs throughout the flood period up to the safe channel capacity at *G*.

SUMMARY OF SYSTEM PROPERTIES

The essential elements and assumptions of our simplified river-basin system are categorized in Table 7.8. The component units, allocations of reservoir capacity, and outputs constitute the 12 variables that, along with the assumed operating procedures, were considered in our studies.

It should be repeated that the specific characteristics and cost and benefit values assumed for the system were not of concern in themselves so long as they were consistent with reality. Our main concern has been with methods of analysis, and applicability of the results that can be obtained by these methods.

It is well to remember also that our system comprises purposes and conditions of water-resource development at least as complex as those encountered in other regions of the United States as well as in other

Table 7.8. Essential elements and assumptions of the simplified river-basin system.

Element	Assumed range of capacity or output
Components of system	
Reservoir A[a]	0–1.5 $\times$ 10^6 acre ft
Reservoir B	0–15.0 $\times$ 10^6 acre ft
Reservoir C[a]	0–15.0 $\times$ 10^6 acre ft
Reservoir D[a]	0–13.0 $\times$ 10^6 acre ft
Power plant B[a]	0–0.8 $\times$ 10^6 kw installed capacity[b]
Power plant G[a]	0–0.8 $\times$ 10^6 kw installed capacity
Allocations of reservoir capacity	
Active storage, reservoir B[a]	0–15.0 $\times$ 10^6 acre ft[c]
Dead storage, reservoir B[a]	0–15.0 $\times$ 10^6 acre ft[c]
Flood storage[d]	
Reservoir B[a]	0–2 0 $\times$ 10^6 acre ft
Reservoir C[a]	0–2 0 $\times$ 10^6 acre ft
Reservoir D[a]	0–2.0 $\times$ 10^6 acre ft
Purposes with annual target outputs for —	
Irrigation[a]	0–6 0 $\times$ 10^6 acre ft
Energy[a]	0–4 0 $\times$ 10^9 kw hr

[a] One of the 12 system variables.

[b] Subject to the condition that when reservoir $B = 0$, power plant $B = 0$.

[c] Subject to the conditions that maximum power head $\leq$ twice minimum power head, and that when power plant $B = 0$, dead storage $= 0$.

[d] The objective of flood storage is to reduce the flood flow at point G to no more than 105×10^3 cu ft/sec, the maximum safe channel capacity.

countries of the world. Examples of such purposes and conditions are the long holdover-storage period, the semiarid climate, the wide monthly variation in the target output for irrigation, and the intricate nature of the water-power and flood-control developments. Other areas, to be sure, may possess appreciably greater ratios of maximum-to-minimum annual flow, and peak flood to mean annual discharge. In other geographic regions groundwater, recreation, and water quality may be of primary concern. Conceivably the methods of analysis developed for our system would be less effective in such localities. How broadly applicable such methods of analysis are to other systems is a question that should be kept clearly in mind in evaluating the usefulness of our procedures.

In succeeding chapters the simplified river-basin system is made the test vehicle for the methods and techniques of analysis that are central to our purpose.

8 Conventional Methods of Analysis

WILLIAM W. REEDY

In the design of large multiunit, multipurpose river developments by U. S. Government agencies,[1] certain methods and techniques of analysis are widely used. These are presented in general terms below, along with a description of how they were adapted to analysis of our simplified river-basin system.

System design, known as "project formulation" or "plan formulation" in agency terminology, is essentially a procedure for evaluating all pertinent evidence and preparing a plan of river-basin development. Such a plan includes the purposes the development is to serve, the physical means for meeting these purposes, the sizes of needed facilities and service areas, and the levels of output. System plans are formulated to achieve the national objectives of water-resource development set forth in federal legislation and in statements of departmental policy and to do so within the framework of state water laws, interstate compacts, and established water rights. The planning process in turn may suggest modifications of state or federal laws and policies, or of court decrees, in order to resolve unusual situations disclosed in the course of system design.

Within the framework of these objectives agencies formulate their plans to promote economic efficiency[2] in water-resource development; their goal is to create the maximum excess of total system benefits (efficiency and otherwise) over total system costs.

[1] These federal agencies comprise the Bureau of Reclamation, Department of the Interior; Corps of Engineers, Department of the Army; and Soil Conservation Service, Department of Agriculture. The Tennessee Valley Authority, which planned and executed the first large-scale multiunit, multipurpose river development in the world, has substantially completed its major water-development tasks.

[2] Federal agencies differ on how objectives other than economic efficiency are to be reflected in the analysis of benefits and costs. The Bureau of Reclamation prefers to include regional redistribution benefits along with efficiency benefits in its economic analysis; other federal agencies generally prefer to separate these classes of benefits to preserve a pure efficiency benefit–cost comparison. For a recent statement of the differing views, see the so-called Green Book, *Proposed practices for economic analysis of river basin projects*, a report to the Inter-Agency Committee on Water Resources by its Subcommittee on Evaluation Standards (May 1958).

METHODS AND TECHNIQUES

THE CONVENTIONAL PLANNING PROCESS

Stated briefly, the conventional planning process of federal agencies comprises the collection of needed information, the preparation of a tentative plan based upon analysis of this information, and the search for an optimal plan by modification of the tentative plan through incremental analysis.[3] In the sections that follow, each of these steps is explained in general terms.

COLLECTION OF INFORMATION

The river basin or study area is delineated, and a comprehensive inventory of its needs and resources is compiled. Efforts are made to identify all significant problems of river behavior (such as the recurrence of damaging floods, soil erosion, and poor water quality) and potential uses of water (such as full irrigation for potentially irrigable lands and supplementary irrigation for inadequately irrigated areas, municipal and industrial water consumption, generation of hydroelectric energy, maintenance of minimum river stages for navigation, and preservation and enhancement of recreational opportunities and useful aquatic life). In-

[3] More complete descriptions of the process can be found in agency and planning manuals and in reports of independent observers and special water-resource study groups. Specific sources are as follows:

(*a*) U. S. Bureau of Reclamation, *Reclamation instructions*, pt. 115, Power investigations (1959) and pt. 116, Economic investigations (1959); also *Reclamation manual*, vol. 3, Basin and project development (1949), vol. 4, Water studies (1948), vol. 5, Irrigated land use (1947), vol. 7, Associated multiple uses (1947), and vol. 8, Surveying and mapping (1949).

(*b*) U. S. Army Corps of Engineers, *Engineer manuals*, Engineering and design, ser. 1110 (in part); Survey investigations and reports, ser. 1120; and Cost allocations for multiple-purpose projects, ser. 1160.

(*c*) U. S. Soil Conservation Service, *Watershed protection handbook* (1957); *Economics guide of watershed protection and flood prevention* (1958).

(*d*) United Nations Economic Commission for Asia and the Far East, *Multiple-purpose river basin development*, pt. 1, Manual of river basin planning (1955).

(*e*) United Nations Department of Economic and Social Affairs, *Integrated river basin development*, Report by a panel of experts (1958).

(*f*) Report of the U. S. President's Water Resources Policy Commission, vol. 1, *A Water policy for the American people* (U. S. Govt. Printing Office, Washington, 1950), chaps. 3, 4, and 7.

(*g*) Alfred R. Golze, *Reclamation in the United States* (McGraw-Hill, New York, 1952), chaps. 5–8.

(*h*) Luna B. Leopold and Thomas Maddock, Jr., *The flood control controversy* (Ronald Press, New York, 1954), chap. 7-11.

(*i*) Otto Eckstein, *Water-resource development, the economics of project evaluation* (Harvard University Press, Cambridge, 1958), chaps. 5–8.

formation is gathered on all regional water sources, both surface and ground, on existing water uses, and on water quality. Sites for projected structures are located on maps and by field reconnaissance. Where necessary, new basic data are collected and gaps in information are filled by installing stream-gaging stations, by tracing and measuring groundwater flows and resources, by topographic mapping and geologic exploration of sites for structures such as dams, water conduits, and power plants, and by locating and assessing suitable construction materials.

PREPARATION OF A TENTATIVE PLAN

As suggested before, a preliminary analysis of all available data is made and a tentative or pilot plan for effective development of the available water resources is prepared with due regard for regional or basin-wide needs. The objectives of the plan are often stated in physical terms. Attaining a particular degree of flood protection, irrigating a given acreage of land, or supplying a specified amount of water to municipalities and industries are common examples. Engineering analyses and designs then look to providing the target outputs at least cost.

Hydrologic studies identify available yields or outputs for each chosen purpose, the operation of reservoirs and other contemplated hydraulic structures being simulated by using the historical record of streamflows and without attempting to estimate future streamflow. Necessary calculations were formerly made almost entirely on hand-operated calculating machines; today an increasing proportion of the work is being programmed for electronic digital computers.

Estimates of capital costs are based on the engineering works incorporated in the plan, as are estimates of the cost of operating and maintaining the selected works and replacing structures and equipment expected to wear out during the period of analysis.

For calculation of the benefits created by the plan, each purpose is studied by itself. The principle underlying this evaluation is that system benefits can be measured in terms of the difference between forecasts of future conditions in the presence and absence of a given project. There are significant disparities in the policies and techniques followed by the various agencies in estimating system benefits. Discussion of these differences, however, is not considered to lie within the province of this chapter.[4]

[4] For a discussion of these differences see Eckstein, reference 3(*i*).

When the tentative plan has been prepared, justification for the required expenditure of money is tested. Development of the system is said to be justifiable when its estimated costs are lower than its estimated benefits, and lower too than the costs of the most likely alternative means for attaining the same level of benefits.

MODIFICATION OF THE TENTATIVE PLAN BY INCREMENTAL ANALYSIS [5]

Improvement of the tentative plan by subjecting it to incremental analysis is next sought, with the aim of increasing net benefits. Through this incremental or marginal analysis the planner seeks to isolate the maximum excess of benefits over costs that can be achieved. Incremental analysis thereby becomes the principal technique in the final stages of plan formulation or system design.

The incremental-analysis procedure consists of adding to or subtracting from the tentative plan substantial segments of purpose, levels of output, or sizes of structure, and determining through operation studies the resulting physical effects on the system as a whole and on the consequent benefits and costs. The differences between the benefits and costs of the system before and after modification constitute respectively the incremental gross benefits and the incremental costs attributable to each increment. When the incremental gross benefits run higher than the incremental costs, the increment is normally incorporated in the plan of development.

An increment may be a separable unit of a water-resource plan, such as a reservoir or a water-power plant, or a separable purpose, such as irrigation or flood control. It may also be a part of a unit or purpose; for example, an increment of reservoir capacity or power-plant capacity, or an increment in level of output, such as irrigation or power.

Increments are so dimensioned that their inclusion in or omission from the plan exerts a measurable effect or involves a practical choice. Thus an increment that changes a total active reservoir capacity of 100,000 acre ft by as little as 1,000 acre ft is often too small to be significant.

[5] This subject is discussed in greater detail and more theoretically in Chapters 2 and 3. The rationale for incremental analysis as seen by the federal agencies is discussed in the Green Book of reference 2. A recent description of the application of incremental analysis by the Bureau of Reclamation is contained in Floyd E. Dominy, "Methods of investigating hydroelectric energy potential on multiple-purpose reclamation projects," paper presented at World Power Conference Sectional Meeting, Madrid, Spain, June 1960; U. S. Dept. of the Interior, Information Service (mimeographed).

Resulting benefits and costs cannot be estimated with the needed accuracy. Similarly, a parcel of 5,000 acres of uniformly productive irrigable land that lies at some distance from a main block of 75,000 acres would be added to an irrigation scheme in its entirety or not at all. In these circumstances a few preliminary computations might well prove that the cost per acre of lengthening the irrigation canal to serve only a portion of the distant area is so great in comparison with the per-acre cost of canal lengthening to serve the entire area that there would be little point in breaking the 5,000-acre parcel into smaller increments.

Incremental analysis of the plan progresses until all reasonable increments have been tried and there is no further growth of net benefits. The procedure therefore develops a series of alternative plans. Economically optimal is the plan that includes all increments to which higher benefits than costs are attached and that excludes all increments for which the converse is true. In the course of the analysis the tentative plan may be modified substantially; in some instances, indeed, earlier plans may be reconsidered in new combinations.

DIFFERENCES IN AGENCY PRACTICES

Although federal agencies agree in general on the principles of plan formulation, practice varies significantly from agency to agency, and from office to office within some of the agencies. Because differences in degree of precision and uncertainty surround the evaluation of benefits and, to a lesser extent, that of costs, some offices place less weight than others on precise and detailed calculations or other quantitative considerations. Instead, they stress the inclusion in the final plan of elements that derive from general objectives and qualitative considerations.

There are other reasons for disparities in practice. Among the most important are the size and type of the projects studied. The problems involved in planning for the development of power, irrigation, flood control, and navigation in the Columbia River Basin, for instance, are substantially different in kind and order of magnitude from those connected with the reduction of flash floods on a small watershed in Texas or a single-purpose flood-control development in New England.

Differences in agency tradition and assigned responsibilities, in availability of funds, and in background and experience of personnel also make for disparities in practice. Thus, in formulating system designs, the Bureau of Reclamation uses certain criteria in addition to maximizing net benefits. Reclamation law requires that costs allocated to irrigation be repaid to the federal government. It would be inadmissible, therefore, to formulate a plan under which reimbursible costs could not possibly be repaid, even though the plan showed the greatest net benefits.

LIMITATIONS IN AGENCY APPROACHES TO SYSTEM DESIGN

There are two fundamental types of limitation to the approaches of federal agencies to system design. These are indicated below with examples of each type.

(1) The objectives for development, such as economic efficiency and income redistribution, are often not specified properly and translated into adequate design criteria for river-basin planners.

(*a*) Existing institutional constraints may be accepted as binding without adequate consideration of their effect on design criteria. When constraints imposed by existing law, administrative policy, or interstate agreement are accepted uncritically, a plan may be overlooked that will provide greater benefits in terms of the design criteria than will any plan honoring these constraints.

(*b*) At the outset of the planning process certain physical objectives may be specified arbitrarily rather than being developed from design criteria that are in consonance with the basic objectives of the plan. An *a priori* decision may be reached that a certain amount of flood protection must be provided, or that water must be supplied in given volumes for municipal and industrial purposes. In effect, this practice sets up physical objectives in lieu of economic ones, although (as discussed in Chapter 2) physical objectives alone cannot form an adequate frame of reference for acceptable design criteria.

(*c*) Tolerable shortage levels for irrigation, municipal and industrial water supply, and hydroelectric energy may be assumed without a study of their implications in terms of the design criteria such as could be obtained through use of an economic loss function. This is another form of arbitrary physical constraint. For instance, the tolerable limits of irrigation deficits may be set in accordance with considered judgment and experience. But while a limit of this type may be reasonable, in an economic sense, for one design, it may be uneconomical for another in which the savings obtained by building smaller reservoirs would exceed the losses from the larger irrigation deficits generated.

(*d*) Insufficient attention may be paid to the dynamic aspects of system design. Too often, systems are designed rigidly to fit a single pattern of demands for a combination of uses over the entire economic life of the system. Yet, as discussed in Chapter 3, estimates of future demands are subject to a high degree of uncertainty. It is often possible, at reasonable cost, to incorporate flexibility into the design, so that operation of the constructed system thereby can be adjusted readily to conform to changes in the demand patterns of the several uses. Furthermore, the incorrect assumption is often made that maximum physical development of a system, a site, or the available water must be attained at once or in a single stage of construction. To the contrary, it is frequently physically possible and economically advantageous to construct projects in several stages that accord with increases in demands as time goes on. Within the adopted design criteria the economic desirability of development by stages should be established by comparing the savings in capital and operating expense through deferring construction with the sum of the increased cost of building the project in stages and the benefits withheld because of the deferral.

(*e*) Budgetary constraints are almost never assumed. Federal planning agencies are not asked to set limitations on their budgets, and higher governmental levels do not set guiding constraints on plan formulation. Nevertheless, such constraints are applied during the construction stage. If a limit were imposed on the allowable total cost of building a system and this were reflected properly in the design criteria, the optimal system design might be significantly different from the optimum arrived at without budgetary constraints.

(2) The techniques for project formulation may be ineffective in finding the best design, even though the objectives and design criteria are stated properly.

(*a*) Too few of the many physically feasible designs that have economic merit may be examined. If this is the case, the probability is high that the plan selected will create significantly lower net benefits than are actually attainable. In the past the major restraint on operation studies has been the time required to perform the numerous and repetitive computations needed for proper assessment of each design. This restraint can be greatly lessened today by using electronic digital computers and applying sampling methods to choose the designs to be tested.

(*b*) System units may be eliminated without fully considering their costs and benefits. A reservoir site may be excluded from further study for the sole reason that a railroad will have to be relocated when the reservoir is built, although this site may, in fact, be competitively economical both because of unusually favorable topographic and foundation conditions and because gross benefits may be high.

(*c*) The technique of selecting a tentative plan and applying incremental analysis to it may be inadequate. If several widely different designs yield high net benefits on test, and the pilot plan itself, as modified by incremental analysis, does not produce close to the highest net benefits, incremental analysis might fail to isolate the optimal design. In these circumstances exercise of even the best technical judgment may be of little avail.

(*d*) Hydrologic information may be inadequate. If the normal procedure of running operation studies for the period of recorded or correlated runoff — no matter how short — is followed, or if only years of critically low flow are examined, the dimensions of physical structures and their purposes both will be based, more often than not, on a relatively short trace of a hydrologic past that can but inadequately represent the vagaries of the future. Because of this basic handicap, hydrologists are giving much thought to the synthesis of hydrographs that will include a larger sample of patterns of streamflow likely to be encountered during the life of the system.

(*e*) Nonstructural measures (flood-plain zoning, flood-warning systems, and water-quality controls, for example) may not be incorporated in the project formulation. Such measures, either of themselves or when combined with suitable engineering structures, may provide equivalent benefits at less cost than would structural measures alone. Substitution or addition of such alternatives in the system design may change the "mix" of the inputs substantially.

APPLICATION OF CONVENTIONAL TECHNIQUES TO THE SIMPLIFIED RIVER-BASIN SYSTEM

In order to relate conventional techniques of river-basin planning to our own research, we asked three groups of graduate students in our water-resource seminar[6] to develop a plan for the simplified system de-

[6] Membership of the three groups was as follows:

Group 1: Dante A. Caponera, United Nations Food and Agriculture Organization; Clyde W. Graham, U. S. Soil Conservation Service; Kenneth E. Ristau, U. S. Army Corps of Engineers; and Robert E. Whiting, California Department of Water Resources.

Group 2: William H. Davis, U. S. Public Health Service; James S. King, U. S. Army Corps of Engineers; Joseph G. Polifka, U. S. Soil Conservation Service; and John M. Wilkinson, U. S. Bureau of Reclamation.

scribed in Chapter 7. No restrictions were imposed on the methods or techniques to be used. Most of the men were engineers who could count years of experience with federal water-resource agencies. The first two task groups conducted their studies during the academic year 1957–58, the third a year later. The third group made use of some of the results obtained the year before, but in modified form because certain basic data had been revised for a clearer definition of the problem. The work discussed here is essentially that of the third group.

DESIGN OBJECTIVE AND CONDITIONS

The assignment of the three groups was to develop the optimal plan for our simplified system with economic efficiency as the design criterion. The objective, therefore, was to select the combination of purposes, levels of output, and sizes of structures that would maximize net benefits under the following assumptions: a specific interest rate, no budgetary constraints, and static rather than dynamic conditions. Conformance with current agency practice called for an interest rate of 2.5 percent and a 50-yr period of analysis.

DESIGN APPROACH

The study groups approached their assignment in the following general manner and sequence.

(1) The three purposes of the accepted system (irrigation, power, and flood control) were studied separately. The smallest practicable level of development at which individual benefits exceeded costs was used as a point of departure, and increments were added until there was no further rise in net benefits. The addition of grossly uneconomical increments was avoided by careful inspection of possible schemes, exercise of engineering judgment, and shortcut methods of investigation. The procedure, designated the "edges" approach, is illustrated graphically in Fig. 8.1. Each separate purpose is represented as the edge of a block. Points *A*, *B*, and *C* on the edges identify respectively the specific levels of irrigation, power generation, and flood control resulting in maximum net benefits for single-purpose developments.

(2) The optimal single-purpose system that produced the greatest

Group 3: Howard R. Bare, U. S. Army Corps of Engineers; Howard J. Mullaney, U. S. Army Corps of Engineers; Robert F. Wilson, U. S. Bureau of Reclamation.

surplus of benefits over costs was selected as the base plan for two-purpose designs. The other two purposes were added to this system individually, and were also combined with each other. Three sets of two-purpose designs were investigated in this way. Optimal combinations were arrived at by adding output and structural increments to each two-purpose system until no further growth of net benefits occurred.

(3) The optimal two-purpose plans, finally, were made the frame of reference for deciding on the nature of a tentative plan in which all three chosen purposes of water-resource development were combined. This three-purpose design was optimized, like its forerunners, by adding and subtracting significant increments. From Fig. 8.1 the optimal level of

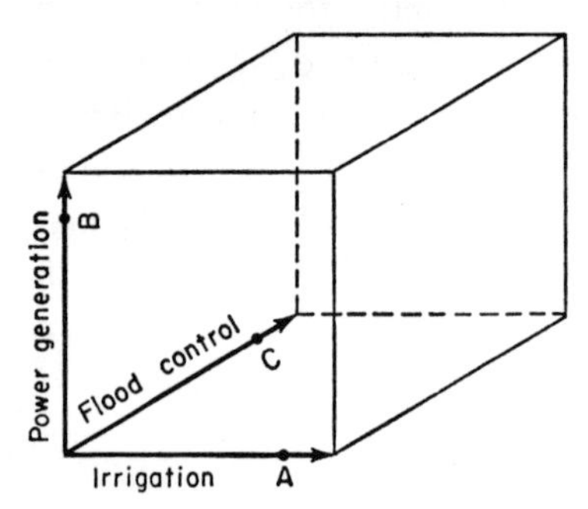

Fig. 8.1. The "edges" approach to system design. Points A, B, and C on the edges identify respectively the specific levels of irrigation, power generation, and flood control resulting in maximum net benefits for single-purpose developments.

development of a two-purpose system is seen to lie in one of the three boundary planes; that of a three-purpose system, somewhere within the block itself. Increments were made large enough to create significant changes and avoid unnecessary computations. Although small increments in the region of the optimal design might have given it more precise definition, the additional work required did not seem to be justified. In practice, suitable dimensions would be suggested by physical and economic factors. A physical factor related to irrigation diversion, for instance, is the configuration of the irrigable lands. Examples of economic factors associated with output level and size of structure are the changes in slope of the respective benefit and cost curves.

The analyses of the task groups were strictly sequential only within the compass of the three individual sets of single-purpose developments and two-purpose developments. Studies of these sets were pursued in parallel, and the results were pooled for evolution of the tentative three-purpose plan.

Desk computers were employed in all operation studies. Because this was tedious, only critical periods of not more than a few years were simulated for irrigation and power.

Operating rules for choosing the reservoir from which water was withdrawn for system use were based on judgment alone. No alternative procedures were tested. The operating rules adopted were simple and did not hedge against future deficits: (1) all reservoir inflow was stored unless it was needed to meet a target output; and (2) whenever there was usable water, it was released from storage to fulfill immediate target outputs without conserving storage for future use. All damaging floods of record were studied in the flood-control analysis.

PROCEDURES IN DETAIL

SINGLE-PURPOSE IRRIGATION

Irrigation shortage was kept below the following limits: (1) in any one year, the total irrigation shortage was not to exceed 50 percent of the total annual target output for irrigation; and (2) in a 50-yr period, the sum of all shortages was not to exceed 150 percent of this output. The shortages allowed by federal agencies are of similar dimension.

Since operation studies were confined to short periods of time, the second criterion was complied with by inspection and judgment. Restrictions such as these are not fixed by experimental information; they are based on experience acquired over many years in planning and operating irrigation projects.

The irrigation loss function shown in Fig. 7.3 was not used, and water shortages did not reduce irrigation benefits in any study. Average deficits were small and of about the same magnitude in all plans; hence their effect was negligible.

Diversion without storage. Our first study looked for the maximum amount of irrigation that could be accomplished by diversion in the absence of storage. This was accepted as a reasonable lower limit of irrigation. Below it benefits would diminish in proportion to output, whereas costs would shrink but slightly because of the relatively high constant cost of the needed diversion works; net benefits accordingly would be less.

It was found that 0.5×10^6 acre ft of water could be delivered annually without exceeding the allowable shortages. Analysis of benefits and costs gave a value of net benefits of $56,500,000.[7] This value is listed in Table 8.1 as one of the elements of study 1. Suitable increments were

Table 8.1. Summary of operation studies for single-purpose irrigation.

Study	Target output for irrigation (10^6 acre ft)	Reservoir capacity[a] (10^6 acre ft) for —		Costs (10^6 dollars)		Benefits (10^6 dollars)	
		A	B	Capital	Total[b]	Gross[c]	Net[d]
1	0.5	0	0	23.5	30.6	87.1	56.5
2	1.0	0.2	0	45.2	59.6	166.2	106.6
3	1.5	0.5	0	64.8	86.5	238.5	152.0
4	3.0	0	1.0	136.3	179.3	425.4	246.1
5	4.2	0	1.8	194.2	254.4	558.4	304.0
6	4.7	0	3.0	248.1	329.8	613.5	283.7
7	3.0	0.4	0.8	138.0	181.3	425.4	244.1
8	5.0	0.4	3.5	278.8	373.7	647.2	273.5
9	4.2	0.4	1.4	191.6	252.0	558.4	306.4

[a] Active storage.
[b] Capital costs plus annual OMR costs capitalized at 2.5 percent over a 50-yr period.
[c] Present value of annual gross benefits on a 50-yr, 2.5 percent basis, assuming that full benefits begin with first year of operation.
[d] Gross benefits minus total costs.

added to this design to maximize net benefits for the single purpose of irrigation.

Storage in reservoir A. As shown in Table 8.1, the second and third studies tied target output for irrigation to reservoir capacity at site *A*. The target was specified first, then trial operations were run to find the minimum reservoir capacity needed to supply the output without going beyond the allowable shortages. An annual target output of 1.5×10^6 acre ft was found to be close to the maximum figure attainable with the aid of storage at site *A* alone. Raising the capacity of this reservoir above 0.5×10^6 acre ft had no advantage. As shown in Table 8.1, net benefits of $152,000,000 were attained in study 3.

Storage in reservoir B. Three studies (4, 5, and 6) were aimed at determining the effects of storage in reservoir *B*. The discontinuity in the irrigation capital-cost function, Fig. 7.4, at an annual output of 4.2×10^6 acre ft suggested this as a logical scale of development. An annual target output of 4.7×10^6 acre ft (listed in Table 8.1) was the maximum for storage at site *B* alone. Reservoir capacities in excess of 3.0×10^6 acre

[7] All benefits and costs are expressed in terms of present value, obtained by discounting to the present, at an annual interest rate of $2\frac{1}{2}$ percent, the relevant streams of benefits and costs during the 50-yr period of analysis.

ft could afford little additional regulation. Study 5, with net benefits totaling $304,000,000, was optimal largely because of the discontinuity in the cost function for irrigation works.

Storage in reservoirs A and B. With both individual studies of reservoirs *A* and *B* exhibiting positive net benefits, it became possible to investigate the effect on net benefits of operating the two reservoirs as a system. This was done in studies 7, 8, and 9. A target output of 4.2×10^6 acre ft was selected once again (for study 9) because of the discontinuity in the irrigation capital-cost function.

It is evident that the two reservoirs could be operated in many different ways to supply a given target output. Some of them would make efficient use of available storage capacity; others would not. One reservoir could be forced to spill while space in the other was left unfilled. Simple operating rules were adopted, therefore, with the objective of reducing spill to a minimum. Since the capacity of reservoir *A* was larger in proportion to its storable or mean inflow than was the capacity of reservoir *B*, *A* was less likely to fill each year than was *B*. Thus the operating rule asked that no release of water be made from reservoir *A* until it was full or until reservoir *B* was empty. Although a better rule might have been found for operating over a long period of study, the adopted rule resulted in little or no waste during the critical periods examined.

The system operation studies showed that a capacity above 0.4×10^6 acre ft for reservoir *A* was of little assistance in reducing shortages, particularly when such capacity was combined with large capacities at reservoir *B*. In years of no shortage, additional capacity in reservoir *A* was not needed; in years of low flow with accompanying shortages, a reservoir at *A* greater in size than 0.4×10^6 acre ft would not have filled. An annual target output of about 5.0×10^6 acre ft was about the maximum that could be expected without exceeding allowable shortages, no matter how much storage capacity there was in reservoirs *A* and *B*. In study 9, the best of the three studies with reservoirs *A* and *B*, net benefits rose to a high figure of $306,400,000.

As recorded in Table 8.1, the best target output for irrigation was 4.2×10^6 acre ft. Above this the excess of costs over benefits for increments of output is triggered primarily by the discontinuity of the capital-cost and the operation, maintenance, and replacement (OMR) cost functions at this point. A pumping plant is needed when target outputs rise above 4.2×10^6 acre ft.

The net benefits realized in studies 5 and 9 are so close that the studies would have to be analyzed carefully to decide which was really the better. Because the system was to be developed for purposes other than irrigation alone, it appeared preferable to assign all the irrigation storage to reservoir *B*, to take advantage of the low marginal costs attached to such use considering the large sizes of reservoir *B* likely to be optimal for multipurpose development. Hence study 5 was considered to be the most desirable single-purpose design for irrigation and to mark a useful starting point for combining irrigation with other purposes.

SINGLE-PURPOSE FLOOD CONTROL

The objectives of the operation studies for flood control were (1) to determine the amounts by which reservoir storage would lower the flood peaks at point *G*, and (2) to estimate the amount of storage needed to keep flood flows at *G* either at or below 105×10^3 cu ft/sec, the assumed channel capacity of the river as it passes through the downstream urban area sketched in Fig. 7.1. Once again, costs and benefits were estimated in searching out the plan that would generate the largest net benefits.

Nine hydrographs for each of the gaging stations at reservoirs *A*, *B*, and *D* and at point *E* (the inflow to reservoir *C*) and point *G* were plotted for floods of record in excess of 105×10^3 cu ft/sec at point *G*. Eight of these floods came from spring snowmelt; the ninth was apparently caused by a December rain that deluged the area above reservoir *D*.

Each reservoir was assumed to be empty at the start of the flood hydrograph. Floods did not have to be routed through the reservoirs, because reservoir outlets were assumed to be large enough to pass all inflows that did not have to be held back. No forecast or foreknowledge of the shape and peak of the hydrographs was presumed. During the early flood stages free reservoir capacity was filled as necessary to keep the peak at *G* down to 105×10^3 cu ft/sec. As a result, small reservoir capacities were pre-empted before they could lower the peaks of large floods. Forecasts of the shape and height of the peaks would have permitted more effective use of reservoir capacity. Only waters within the flood peaks themselves would have been retained. All peaks would have been lowered, whereas with our procedure there was no reduction in some instances. Nevertheless, application of the more refined and difficult type of operation did not seem warranted. Furthermore, no allowance was made for potential surcharge in the reservoirs when passing floods.

Examination of the recorded hydrographs indicated that, because of size limitations, a reservoir at site *A* would reduce flood peaks at point *G* but slightly. Since studies showed that maximum net benefits would be obtained with almost complete control of all floods of record, no detailed studies were made for reservoir *A* operating alone or in combination with other reservoirs.

Initially, single-purpose flood-control calculations were made separately for reservoirs *B*, *C*, and *D* to determine the capacity required at each to keep the peak flow at point *G* down to 105×10^3 cu ft/sec in the nine floods of record. The results are summarized in Table 8.2, the last

Table 8.2. Summary of operation studies for single-purpose flood control. I.

	Unregulated peak flow at point *G*		Capacity required at each reservoir operating alone to control flows to 105×10^3 cu ft/sec at point *G* (10^3 acre ft)			
Flood	Date	Amount (10^3 cu ft/sec)	Reservoir *B*	Reservoir *C*	Reservoir *D*	Damage with no regulation (10^6 dollars)
1	June 1927	109	4	4	4	20.3
2	May 1932	121	21	21	21	34.5
3	June 1933	136	58	58	58	38.3
4	Dec 1933	172	83[a]	91[b]	124	69.3
5	May 1936	107	1	1	1	17.0
6	Apr 1938	134	49	49	49	38.3
7	May 1947	114	13	13	13	27.5
8	May 1948	177	786	786	764[c]	78.0
9	May 1949	123	73	73	73	35.8
Average annual damage in 32 years						11.2

[a] Capacity required to achieve maximum possible flood reduction to 138×10^3 cu ft/sec.
[b] Capacity required to achieve maximum possible flood reduction to 131×10^3 cu ft/sec.
[c] Capacity required to achieve maximum possible flood reduction to 123×10^3 cu ft/sec.

column of which shows the estimated damages that would be caused by each flood of record in the absence of flow regulation by storage. It is seen that neither reservoirs *B* nor *C* alone could completely control the December 1933 flood, nor could reservoir *D* completely control the flood of May 1948. In each of these instances, marked by superscripts in Table 8.2, the chosen reservoir capacity was sufficient to store all flows passing the particular reservoir site.

The amounts by which the nine floods could be reduced in single reservoirs of different size at each site were also determined, and the damages

Table 8.3. Summary of operation studies for single-purpose flood control. II.

Study	Reservoir	Reservoir capacity (10^3 acre ft)[a]	Costs (10^6 dollars)		Benefits (10^6 dollars)	
			Capital	Total[b]	Gross[c]	Net[d]
10	*B*	50	1.9	2.1	136.1	134.0
11	*B*	100	3.6	3.9	215.0	211.1
12	*B*	500	17.2	17.6	215.0	197.4
13	*B*	790	26.5	26.9	284.2	257.3
14	*C*	50	1.7	1.9	136.1	134.2
15	*C*	100	3.2	3.5	215.0	211.5
16	*C*	500	17.0	17.3	215.0	197.7
17	*C*	790	26.2	26.6	284.2	257.6
18	*D*	50	2.2	2.4	125.4	123.0
19	*D*	100	4.2	4.5	215.0	210.5
20	*D*	500	20.8	21.1	249.0	227.9
21	*D*	770	31.0	31.5	286.5	255.0
22	*D*[e]	200	8.3	29.2	318.2	289.0
	C[f]	600	20.2			
	Both	800	28.5			

[a] Active storage.
[b] Capital costs plus annual OMR costs capitalized at 2.5 percent over a 50-yr period.
[c] Present value of annual gross benefits on a 50-yr, 2.5-percent basis.
[d] Gross benefits minus total costs.
[e] Capacity of 124×10^3 acre ft in Table 8.2 raised to 200×10^3 acre ft to allow reservoir *D* to control the December 1933 flood by itself.
[f] Capacity of $(764 - 124) \times 10^3$ acre ft in Table 8.2 rounded to 600×10^3 acre ft to allow reservoir *C* along with reservoir *D* to control all floods.

caused by channel overflows were calculated by relating flood damage to streamflows below point *G* in accordance with Fig. 7.9. Table 8.3 brings together the pertinent physical information and the allied costs and benefits. When each reservoir was operated alone, maximum net benefits were correlated with reservoir capacities providing the maximum possible reduction in flood damages, as demonstrated by studies 13, 17, and 21. In study 22 reservoir *D* was made sufficiently large to control the December 1933 flood by itself. This capacity combined with capacity at reservoir *C* was adequate to control all floods. In terms of net benefits this combination of flood-storage capacities proved to be optimal for single-purpose flood control.

SINGLE-PURPOSE POWER GENERATION

In the power studies the assumptions outlined in Chapter 7 for the simplified river-basin system were modified as follows.

(1) The over-all efficiency of power plant *G* was reduced from 88 to 85 percent.

(2) The allowable energy shortage in any one month was dropped from 25 percent to zero.

(3) No estimates were made of nonfirm energy generation or benefits. Although the benefits in question are small in comparison with those derived from the sale of firm energy, they become more important when target outputs are low, because more water then becomes available for nonfirm energy. Nevertheless, in this case plan formulation is not affected significantly when nonfirm energy benefits are neglected.

In the first study power plant *G* was operated as a run-of-the-river unit with no upsteam regulation by storage. As shown in Table 8.4, with an

Table 8.4. Summary of operation studies for single-purpose power generation.

Study	Capacity of reservoir *B* (10^6 acre ft) Active storage	Dead storage	Total storage	Target output for energy (10^9 kw hr)	Power-plant capacity (Mw)	Costs (10^6 dollars) Capital	Total[a]	Benefits (10^6 dollars) Gross[b]	Net[c]
23	0	0	0	0.165	32[d]	8.6	11.1	32.8	21.7
24	2.7	0.4	3.1	0.666	129	106.6	113.9	132.2	18.3
25	2.7	0.7	3.4	0.710	137	112.9	120.6	141.0	20.4
26	2.7	1.7	4.4	0.858	166	131.2	140.3	170.3	30.0
27	2.7	2.7	5.4	0.955	185	144.5	154.5	189.6	35.1
28	2.7	3.7	6.4	1.056	204	155.2	166.1	209.7	43.6
29	2.7	4.7	7.4	1.128	218	163.9	175.4	223.9	48.5
30	2.7	5.7	8.4	1.185	230	172.0	184.1	235.3	51.2
31	2.7	6.7	9.4	1.238	240	180.3	193.0	245.8	52.8
32	2.7	7.7	10.4	1.280	248	189.8	202.8	254.1	51.3
33	2.3	6.7	9.0	1.165	226	175.7	187.7	231.3	43.6

[a] Capital costs plus annual OMR costs capitalized at 2.5 percent over a 50-yr period.
[b] Present value of annual gross benefits on a 50-yr, 2.5-percent basis, assuming that the entire target output for energy is sold each year.
[c] Gross benefits minus total costs.
[d] Power plant *G*; other indicated capacities pertain to power plant *B*.

installed capacity of 32 Mw it was found possible to generate 165×10^6 kw hr of electricity annually, the resulting net benefits being $21,700,000. Run-of-the-river operation, therefore, proved to be economical and an acceptable starting point for incremental analysis.

It should be noted that there is little complementarity of use between irrigation and the generation of firm energy at power plant *G*. Attention thus was turned to single-purpose generation of electricity at power plant *B*, possibly to be combined later with irrigation storage in the hope of effecting significant economies.

Historically, the most critical time for power drawdown during a fairly short holdover period was observed to be from July 1936 through March 1938. It was found that an active reservoir capacity of about 2.7×10^6 acre ft would completely regulate the stream during this period. The reservoir would fall to its lowest level at the end of March 1938 and would refill by the end of June 1939. Nine single-purpose operation studies were carried out with this reservoir capacity and with varying amounts of dead storage. The results are summarized in Table 8.4. Maximum net benefits were obtained in study 31, for a dead storage of 6.7×10^6 acre ft.

In study 33 active storage was reduced to 2.3×10^6 acre ft and combined with 6.7×10^6 acre ft of dead storage, to determine whether or not net benefits would be improved if the stream were not fully regulated during the critical period. As shown in Table 8.4, net benefits were found to be substantially less than for the comparable study, no. 31.

Single-purpose power studies were not pushed to the optimal scale of development. The net benefits noted in studies 24 through 33 could have been improved, for example, by including the contributions of power plant *G*, operating on regulated flows from reservoir *B* and unregulated runoff below *B*. Furthermore, energy generation — and probably net benefits, too — could have been increased by building a reservoir at site *D* for further regulation of flows at power plant *G*. Consideration of increments such as these was left for later multiple-purpose study.

TWO-PURPOSE STUDIES COMBINING IRRIGATION AND FLOOD CONTROL

Single-purpose development of the chosen system for irrigation provided greater net benefits than did either flood control or power production by themselves. The conditions noted for study 5 in Table 8.1, namely, an annual target output for irrigation of 4.2×10^6 acre ft and a storage capacity of 1.8×10^6 acre ft in reservoir *B*, were therefore selected as a frame of reference into which other purposes could be fitted. Since the next highest net benefits accrued with flood control, it was selected as the first purpose to be added to irrigation in the creation of a multipurpose development.

Table 8.5. Summary of multipurpose studies.

Study	Purpose[a]	Capacity allocated at reservoir B (10^6 acre ft): Active storage for — Irrigation and power	B: Active storage for — Flood control	B: Dead storage, power head	B: Total storage	Capacity allocated at reservoir C (10^6 acre ft): Active storage for — Power	C: Active storage for — Flood control	C: Total storage	Capacity allocated at reservoir D (10^6 acre ft): Active storage for — Power	D: Active storage for — Flood control	D: Total storage
34	I, F	1.8	0.8	0	2.6	0	0	0	0	0	0
35	I, F	1.8	0.6	0	2.4	0	0	0	0	0.2	0.2
36	I, P	2.8	0	6.7	9.5	0	0	0	0	0	0
37	I, P	2.8	0	0	2.8	0	0	0	0	0	0
38	I, P	2.8	0	5.7	8.5	0	0	0	3.6	0	3.6
39	I, P	2.8	0	5.7	8.5	1.3	0	1.3	3.6	0	3.6
40	F, P	2.2	0.6	7.5	10.3	0	0	0	0	0.2	0.2
41	I, P, F	2.8	0	5.7	8.5	1.3	0.8	2.1	3.6	0	3.6
42	I, P, F	2.8	0	5.7	8.5	1.3	0.6	1.9	3.6	0.2	3.8
43	I. P. F	2.8	0.6	5.7	9.1	0	0	0	3.6	0.2	3.8

Study	Purpose[a]	Power: Annual target output for energy (10^9 kw hr)	Power: Installed power-plant capacity (Mw), Power plant B	Power: Installed power-plant capacity (Mw), Power plant G	Irrigation: Annual target output (10^6 acre ft)	Costs (10^6 dollars): Capital	Costs: Total[b]	Benefits (10^6 dollars): Gross[c] Irrigation	Gross[c] Flood control	Gross[c] Power	Gross[c] Total, gross	Net[d]
34	I, F	0	0	0	4.2	212.2	272.6	558.7	284.2	0	842.9	570.3
35	I, F	0	0	0	4.2	216.3	276.9	558.7	318.2	0	876.9	600.0
36	I, P	0.657	55	88	4.2	306.2	376.9	558.7	0	130.5	689.2	312.3
37	I, P	0.355	0	69	4.2	232.0	297.4	558.7	0	70.3	629.0	331.6
38	I, P	1.770	257	298	4.2	456.8	547.6	558.7	0	351.4	910.1	362.5
39	I, P	2.030	300	345	4.2	507.0	602.0	558.7	0	403.0	961.7	359.7
40	F, P	1.238	240	0	0	196.3	209.3	0	318.2	245.8	564.0	354.7
41	I, P, F	2.030	300	345	4.2	525.2	620.3	558.7	284.2	403.0	1245.9	625.6
42	I, P, F	2.030	300	345	4.2	523.2	618.4	558.7	318.2	403.0	1279.9	661.5
43	I, P, F	1.770	257	298	4.2	463.3	554.2	558.7	318.2	351.4	1228.3	674.1

[a] I = irrigation; F = flood control; and P = power generation.
[b] Capital costs plus annual OMR costs capitalized at 2.5 percent over a 50-yr period.
[c] Present value of annual gross benefits on a 50-yr, 2.5 percent basis, assuming that full benefits begin with first year of operation.
[d] Gross benefits minus total costs.

In all studies in which flood control was important, reservoir space earmarked for flood-control storage was left empty throughout the year. To save time and effort, the effects of joint use of reservoir space for flood control and irrigation, or for flood control and power generation, were not examined. To ensure optimal use of the available storage, some system of flood-forecasting would have had to be formulated, a time-consuming undertaking. A simpler operation would have been to reserve the necessary space for flood control until the critical season had passed, and afterward to fill as much of the space as possible for conservation use. Undoubtedly, flood peaks were lowered incidentally by operating the reservoirs for single-purpose irrigation or power generation, or for the two purposes together. None of these possibilities was actively studied, however. Since rigid preservation of the flood-control space calls for more space than is actually needed to regulate the floods of record, the calculated net benefits of multiple-purpose plans that include flood control are less than would otherwise be true.

In order to take advantage of the decrease in marginal costs associated with increases in the size of reservoir *B*, the 0.8×10^6 acre ft of required flood-control space shown in Table 8.3 (study 22) were added to the 1.8×10^6 acre ft of required irrigation storage listed in Table 8.1 (study 5). The elements of this study, no. 34, are summarized in Table 8.5 (with the results of other multipurpose studies). Net benefits totaled $570,300,000. Because flood regulation could be improved by adding flood-control space in reservoir *D*, study 35 was designed to transfer 0.2×10^6 acre ft of such space from reservoir *B* to reservoir *D*. This resulted in raising net benefits to $600,000,000.

TWO-PURPOSE STUDIES COMBINING IRRIGATION AND POWER GENERATION

Power generation was added as an increment to irrigation in order to take advantage of the complementarity of use of the storage capacity of reservoir *B*. As shown in Table 8.5, study 36 was based upon 2.8×10^6 acre ft of active and 6.7×10^6 acre ft of dead storage in this reservoir, or on about the same conditions found to be optimal for the single-purpose power development in study 31 of Table 8.4. Energy was generated at power plants *B* and *G* both. Most of the active storage at reservoir *B* was drawn upon for irrigation, resulting in a large potential energy output during the irrigation season as irrigation releases were run through the turbines, and a much smaller one during the winter. Since there was no way of raising the winter output without dropping irrigation output below its optimal value, only a small portion of the energy developed during the irrigation season could be marketed as firm energy. It was doubtful, therefore, whether the cost of dead storage for boosting power head could be justified. For this reason dead storage and power generation at reservoir *B* were not included in study 37. Instead, power was developed at site *G* only, and the expected slight increase in net benefits did result.

By drawing water from reservoirs *C* or *D* to raise the winter output of power plant *G*, it stood to reason that a larger portion of the energy that could be generated during the irrigation season under the conditions of study 36 would become firm power. Preliminary operation studies suggested, however, that the amount of dead storage in reservoir *B* should be dropped below the original value of 6.7×10^6 acre ft, because energy generated by heavy summer drafts for irrigation would remain higher than firm-energy requirements. Lowering the power head by affording less dead storage could not affect the generation of firm energy during the summer; it could only reduce nonfirm energy. On the other hand, the amount of power that could be generated at *B* during the winter was relatively so small that it could not be altered significantly by any change in dead storage.

In accordance with this reasoning, dead storage in reservoir *B* was reduced to 5.7×10^6 acre ft in study 38. At the same time, 3.6×10^6 acre ft of capacity were allotted to reservoir *D* for the purpose of supplying water to power plant *G* and thereby firming up the winter output of power plant *B*. To allow for a longer holdover period, this and subsequent studies were performed for a period of operation extending from July 1936 through June 1943, rather than through March 1938 as previously.

Reservoir *C* with a capacity of 1.3×10^6 acre ft was next added to the system, further to firm up the output of power plant *B*, but, as the results of study 39 show, this was done to no advantage. The conditions of study 38 gave higher returns with net benefits of $362,500,000.

The two-purpose studies of irrigation and power proved that their complementary use of water from reservoir *B* increased net benefits only slightly over the combined value of $356,800,000 recorded for study 5 in Table 8.1 and study 31 in Table 8.4, which were optimal for single-purpose irrigation and power generation, respectively. This is explained by the fact that, without reservoir *D*, there was little complementary use of water from reservoir *B* in the two-purpose studies. The cost of reservoir *D*, on the other hand, would nearly equal the gross benefits derived from its addition to the system.

TWO-PURPOSE STUDIES COMBINING FLOOD CONTROL AND POWER GENERATION

In study 40 flood control was combined with power generation. Power-plant and associated reservoir capacities were taken, with only slight modification, from study 31 in Table 8.4, because they were optimal for single-purpose power generation. However, advantage was taken of the decreasing marginal costs of reservoir *B* by including in it the 0.6×10^6 acre ft of flood-control space allocated to reservoir *C* in study 22 of Table 8.3, which was optimal for flood control. The remaining 0.2×10^6 acre ft of needed flood-control space were left in reservoir *D*. Reservation of flood-control space throughout each year for reduction of flood peaks precluded complementary use of this space for other purposes. Power plant *G* was taken into account in this study, because it had not been in the single-purpose power studies. As shown in Table 8.5, net benefits in study 40 amounted to $354,700,000.

THREE-PURPOSE STUDIES COMBINING IRRIGATION, POWER GENERATION, AND FLOOD CONTROL

In study 41 flood control was added to irrigation and power as a final increment. Although the construction of reservoir *C*, as discussed in the preceding section, was not justified as an increment for power generation by itself, it was thought that economies of scale might justify its development for a combination of power generation and flood control. Therefore, 0.8×10^6 acre ft of flood-control space was added in reservoir *C*. There followed a large rise in net benefits over study 38, the best plan for irrigation and power, and a slight increase over study 35, the best plan for irrigation and flood control.

Study 42 showed that net benefits could be further improved by transferring 0.2×10^6 acre ft of flood-control space from reservoir *C* to reservoir *D* in order to regulate the damaging flood of December 1933.

Finally, in study 43 reservoir *C* was left out and its flood-control space of 0.6×10^6 acre ft was added to reservoir *B*. As shown in Table 8.5, a slight increase in net benefits over those of study 42 was obtained.

Net benefits of $674,100,000 are shown in Table 8.5 for study 43. These are higher than for any other design analyzed. It was considered that the design parameters used for this study probably were close to the optimum possible for our simplified system; no further studies therefore were made.

A summary of reservoir and power-plant capacities, target outputs for irrigation and energy, costs, and benefits for the assumed best system design is presented in Table 8.6.

Table 8.6. Best combination of system variables developed by conventional methods, with associated costs and benefits.

Variable	*Value*
Reservoir capacities (10^6 acre ft)	
B, active storage for irrigation and power generation	2.8
B, active storage exclusively for flood control	0.6
B, total active storage	3.4
B, dead storage	5.7
B, total capacity	9.1
D, active storage for power generation	3.6
D, active storage exclusively for flood control	0.2
D, total capacity	3.8
Target outputs	
Irrigation (10^6 acre ft)	4.2
Energy (10^9 kw hr)	1.77
Power-plant capacities (Mw)	
B	257
G	298
Costs and benefits (10^6 *dollars*)	
Costs	
Capital costs	463.3
OMR costs[a]	90.9
Total costs	554.2
Gross benefits[a]	
Irrigation	558.7
Power generation	351.4
Flood control	318.2
Total gross benefits	1228.3
Net benefits[b]	674.1

[a] Present value of costs or benefits capitalized at 2.5 percent over a 50-yr period.
[b] Gross benefits minus total costs (present values).

The approach to optimal system design described in this chapter was chosen specifically with our simplified system in mind, and proceeds from a study of purposes rather than of projects. The purpose that yields the greatest net benefits becomes the starting point for multipurpose plan formulation. A different configuration of runoff, dam sites, or purposes might have suggested an entirely different procedure. It might have been better to select as the starting point a project, such as a reservoir and power plant, that creates more multipurpose net benefits than any other project. Increments could then have been added to determine the optimal system design. Alternative approaches could also be considered. The important point to remember is that no single approach can be universally effective.

In our plan-formulation procedures, furthermore, the studies were not carried as far as federal agencies normally would do in an actual situation. Only three of the studies, indeed, included all three purposes. Further incremental analysis of the best plan might have proved the possibility of developing still greater net benefits.

The most promising change appears to be a reduction in dead storage at reservoir *B* in order to bring power generation at power plant *B* during the irrigation season more into line with power generation during the winter at this plant, and at power plant *G* through water released from reservoir *D*. Raising the active capacity of reservoir *B* should be studied also, although this incremental procedure does not appear overly promising. Examples of additional increments are changes in the levels of target output for irrigation and flood control, and further adjustments of target output for energy. Because of the discontinuities in the assumed irrigation cost functions, increases in the target output for irrigation probably would not improve net irrigation benefits; however, the possibility should be checked. Storage space for flood control probably could be decreased without losing flood-control benefits, because flood peaks are lowered concomitant with operating reservoirs for other purposes.

The precision of incremental analysis is governed, in part, by increment size. Comparatively large increments were purposely chosen. This is an appropriate decision for the early steps of incremental analysis; however, the additional studies that have been suggested for the three-purpose systems call for smaller increments if more refined results are to be obtained.

Standard procedures for plan formulation require that the completed plan be tested by assuming that each purpose in its turn is the last increment. The benefits created should naturally be greater than the costs of adding the purpose. Although this was not done for study 43, the best combination found, inspection does suggest that it would pass this test.

In the studies of our simplified system minimum costs of attaining equivalent benefits for each purpose by alternative means, as discussed in connection with the principles and procedures of plan formulation, were not calculated. Instead, it was assumed that the cost of the alternative would have been greater than the cost of including the purpose in the system plan.

As a final note, it is interesting that, even with a severe constraint on capital expenditures, almost all of the maximum net benefits were obtainable. According to Table 8.5 the net benefits of the best two-purpose plan for irrigation and flood control are almost as large as those of the optimal three-purpose plan. But in further comparing studies 35 and 43, we may well ponder the fact that a 114 percent increase in capital expenditures yields no more than a 12 percent gain in net benefits.

9 Analysis by Simulation: Programming Techniques for a High-Speed Digital Computer

DEWARD F. MANZER AND MICHAEL P. BARNETT

ANALYSIS of a river-basin system is but one example of the diverse and complex problems that can be solved efficiently through simulation by modern computers. Other examples are simulations of vehicular traffic networks, combinations of warehousing and marketing, and constructs of the U. S. domestic economy.[1] Because it is the essence of simulation to reproduce, in some cognate way, the behavior of a system in every important detail, simulation problems typically are large-scale and complex. Masses of input data, as well as the results of intermediate calculations, must be kept at hand. In digital-computer practice they are preserved in the internal storage of the computer or, if necessary, in backing storage — for example, on magnetic tapes.

Until digital computers of the magnitude of the Model No. 650 of the International Business Machines Corporation or the Univac I of the Remington Rand Corporation were constructed, the solution of large-scale simulation problems was not feasible. Even the storage capacity of these computers was not commodious enough to handle efficiently an analysis of the size and complexity of our simplified river-basin system. That had to await elaboration of the very large computers of the IBM 700 class, with an internal magnetic-core storage capacity of up to 32,768 words.

Prior to the introduction of the IBM 650 computer in 1954, no major river development had been simulated on a digital computer. Perhaps the first enterprise of this kind was the simulation of the Nile Valley plan in 1955 by Morrice and Allan.[2] This involved the analysis of as

[1] For examples of the use of simulation in the solution of industrial, military, and management problems, see D. G. Malcolm, "Bibliography on the use of simulation in management analysis," *Operations Research 8*, 169 (1960).

[2] For a detailed account of the Nile Valley simulation, see H. A. W. Morrice and W. N. Allan, "Planning for the ultimate hydraulic development of the Nile Valley," *Proc. Instn. Civil Engrs. 14*, 101 (1959).

many as 17 reservoirs or power sites, and the simulation of their behavior over a 48-yr period. The objective of the study was to find, with the help of an IBM 650 computer, the particular combination of reservoirs, control works, and operating procedure that would maximize the volume of useful irrigation water. No economic optimization was involved.

In the United States, large-scale simulation experiments were begun by the Corps of Engineers in 1953 for reservoir management on the main stem of the Missouri River.[3] The operation of six great reservoirs was simulated in such a way as to maximize power generation, subject to constraints imposed by specified requirements for navigation, flood control, and irrigation. Again, no economic analysis entered directly into the study. Absent, too, was the element of design. The structures were assumed to be fixed in size and place. Procedures were formulated and programs were written for the Univac I, using the mathematical techniques of the calculus of variations along with linear programming or Lagrangian multipliers.

A trial translation of this approach to the more complex Columbia River Basin proved to be beyond the capacity of computers then available to the Corps of Engineers. A simpler program was written for an IBM 650 computer which merely calculated the physical consequences of a single operating procedure for a single month of time. The power outputs of a large number of alternative system designs on the Columbia River analyzed in this way lie at the base of the Corps of Engineers' revised plan of development for this basin.[4]

Water-resource agencies have been slow to take full advantage of the significantly larger capacities of computers in the IBM 700 class, but a start has been made. The Tennessee Valley Authority, for example, installed an IBM 704 computer in 1958, and the Bonneville Power Administration experimented that same year with an IBM 704 simulation program for power operations on the Columbia River. No one, however, has yet completed simulation of a large and complex river-basin system

[3] This was followed in 1954 by simulation of a two-reservoir system on the Rio Grande by the International Boundary and Water Commission, U. S. and Mexico, using an IBM 701 computer. A detailed description of the Missouri River simulation is contained in DATAmatic Corporation, *Report on use of electronic computers for integrating reservoir operations*, vol. 1, prepared in cooperation with Raytheon Manufacturing Company for the Missouri River Division, Corps of Engineers, U. S. Army (January 1957).

[4] F. S. Brown, "Water resource development — Columbia River basin," in *Report of meeting of Columbia Basin Inter-Agency Committee*, Portland, Oregon (December 1958). See also D. J. Lewis and L. A. Shoemaker, "Columbia River system power analyses by use of digital computers," a paper presented at the Hydraulics Division Conference, Am. Soc. Civil Engrs., at Fort Collins, Colorado, July 1–3, 1959.

on such computers for purposes of system design. Provocative, too, is the pronouncement by Swain and Reisbol that the most important advantage of computers to planners is the facility with which these machines can be used to compare the economic as well as the physical effects of alternative schemes of development.[5]

STATEMENT OF THE PROBLEM OF RIVER-BASIN SIMULATION

The core of the simulation procedure employed by the Harvard Water Program is the representation of the simplified river-basin system, in all its inherent characteristics and probable responses to water management, by a largely arithmetic model. The information contained in this model is marshaled in a program of instructions for the computer. Physical displacements, changes in condition, and other time-dependent variations in the status of water within the system become components of the information as responses to the control introduced by the dams, reservoirs, power plants, and irrigation canals proposed for construction.

The purposeful conversion of the natural runoff into flow complexes that generate useful outputs establishes an operating policy or procedure. The streamflow is predictable only in a stochastic sense, as will be discussed at length in Chapter 12. Conceptually, nevertheless, for each particular combination of measures investigated, it is possible to decide upon an operating procedure that is optimal in terms of the maximum present value of net benefits that can issue from the developed system.

We can state our immediate simulation problem briefly as follows. Given: (1) a certain combination of reservoirs, power plants, and irrigation-diversion and distribution facilities; (2) target levels of irrigation and energy outputs; (3) specified allocations of reservoir capacity for active, dead, and flood storage; (4) a representative series of monthly runoff values, and 6-hr flows for flood months; and (5) a specified operating procedure — then by routing the available flows through the reservoirs, power plants, and irrigation system for an extended period of years, such as 50, determine the physical outputs and the magnitude of the net benefits created.

The following successive steps must be taken to solve this problem.

(1) The parts of the system must be identified schematically, as illustrated for our river basin in Fig. 9.1, and their internal functional

[5] F. E. Swain and H. S. Reisbol, "Electronic computers used for hydrologic problems," *J. Hydraulics Div., Am. Soc. Civil Engrs.*, 85 (November 1959).

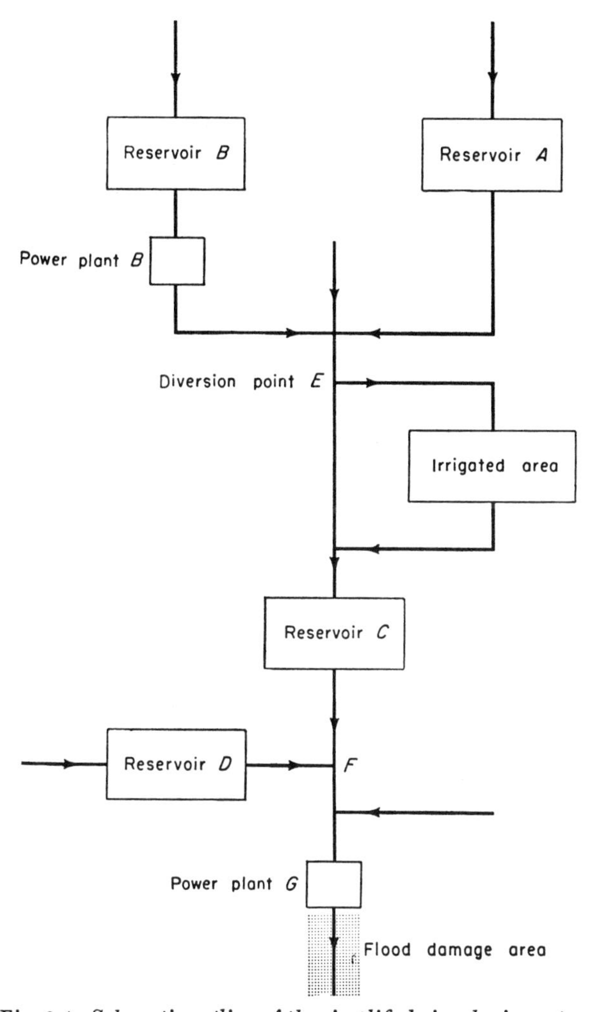

Fig. 9.1. Schematic outline of the simplified river-basin system.

relationships, such as the rise and fall of power head with reservoir capacity, the variation of costs with available inputs, and the changes in benefits resulting from different outputs, must be defined (see Chapter 7, particularly Figs. 7.2 to 7.11).

(2) The runoff information must be prepared in suitable statistical form. As stated in Chapter 7, the availability of records of monthly flow and 6-hr flood-flow data was predicated in our example for five key points in the system, and 60 years of observed or synthetic data were incorporated in each simulation study.

(3) An operating procedure for system management must be formulated. A specific operating procedure for simulation analysis is discussed in Chapter 7, with a discussion of the more general selection of appropriate operating procedures in Chapter 11.

(4) The formulated operating procedure must be coded for the computer.

(5) The rules for converting into monetary worth the computed system outputs obtained by implementing the operating procedure must be coded.

(6) The physical and economic responses of simulation must be arranged in a form suitable for printing as computer output.

The present chapter is concerned with steps (4), (5), and (6) — namely, the coding of operating procedure, the conversion of realized system outputs into dollars, and the presentation of final results.

CONSTITUENT SIMULATION PROCEDURES

The coding and recording operations required for our system necessitate the following simulation procedures: (1) a monthly operating pro-

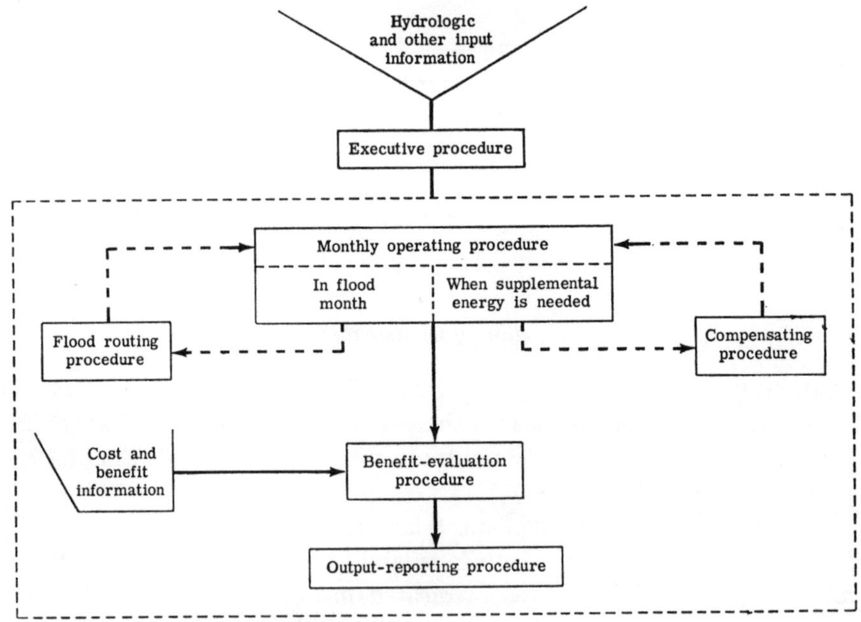

Fig. 9.2. Schematic representation of the major simulation procedures applied to the simplified river-basin system.

cedure, (2) a flood-routing procedure, (3) a compensating procedure, (4) a benefit-evaluation procedure, and (5) an output-reporting procedure. As illustrated schematically in Fig. 9.2, the first three procedures use hydrologic information. They are tied together integrally and are followed sequentially by the remaining procedures. The benefit-evaluation procedure is supported by economic information, and an executive or overall management procedure guides the entire simulation.

At the start of the simulation the volume of water stored in the reservoirs of the system is specified. Reservoirs are assumed to be either full or at levels recorded at the end of a preparatory "warm-up" run, extending over a first 10-yr period in our case. The monthly operating procedure then routes, or continues to route, the water through the system month by month with the objective of meeting target outputs. When predesignated flood months are encountered, the 6-hr flood-routing procedure displaces the monthly procedure. The compensating procedure is called when needed to make supplementary releases, at reservoir *B* in our case, for the generation of energy.

After simulation has progressed through the preparatory period (which, as suggested, may be zero) and through the subsequent active period, the benefit-evaluation procedure is asked to determine the dollar value of outputs achieved for irrigation, energy, and flood control, including such deficits as may be recorded. Finally, the output-reporting procedure selects and arranges for printing on tape the physical and economic replies requested.

Each of these procedures is discussed in this chapter. A general, guiding statement of objectives or policy is followed in each case by a considerably more detailed specification of the adopted strategy.

THE FORTRAN II METHOD

In programming our simplified system for simulation on the IBM 704 computer,[6] we were able to take advantage of the Fortran II automatic coding technique. This permits writing the simulation program as a series of statements that individually summarize algebraic, logical, or organizational processes in a symbolism akin to that of elementary algebra. These statements are represented simply and clearly in a type of shorthand, by mnemonic letter symbols that give the engineer or planner

[6] This work was done in large part at the Computation Center, Massachusetts Institute of Technology.

meaningful descriptions of the quantities they denote. The programmer then presents to the computer this package of statements — known as the source program — in Fortran language. The computer, by following an elaborate Fortran compiler program, translates the language of the source program into a different language that the computer subsequently uses in actual operations. The programmer is thus freed from the detailed and tedious task of himself writing the program in the complex computer language.

Two important advantages accrue from this and other features of the Fortran II method.

(1) Programming is greatly accelerated because of the ease of coding and revising a program in Fortran language. Moreover, the program is readily broken into separate subroutines that can be checked and modified independently. Several programmers can be put to work simultaneously, each on a separate part of the task. While they are so engaged, they can communicate effectively with one another and with outsiders in the simple Fortran language. Finally, certain diagnostic facilities built into the compiler program help to correct errors in the source program.

(2) Programs coded in Fortran II language are not restricted to any one type of computer. Many modern, large IBM units are equipped with compilers that can handle Fortran source programs effectively. However, the programming techniques we developed for simulating the behavior of our river-basin system were developed specifically for the IBM Series 700 computers.

APPLICATION TO THE SIMPLIFIED RIVER-BASIN SYSTEM

In discussing the procedures applied to our simplified river-basin system, we present the individual steps of each procedure first in descriptive terms, then in more formal mathematical terms that are set in smaller type. All symbols used are listed alphabetically, with their definitions, in the Appendix, located at the end of this chapter. As indicated previously, the notation commonly employed in the equations that define the successive operating steps differs from that of conventional algebra. In each equation the single symbol to the left of the equality sign is balanced by a single number or by a collection of symbols and numbers to the right. The equation is activated by (1) introducing the values previously associated with each of the symbols on the right-hand side of the equation, (2) computing the resultant, and (3) assigning its value to the symbol on the left-hand side of the equation.

Much of the basic numerical information referred to in the various operating procedures is presented in the tables and figures of Chapter 7. To aid the reader, some specific cross references are included in the steps that comprise the procedures. The power and energy formulations used for power plants *B* and *G* are presented in Table 9.1.

MONTHLY OPERATING PROCEDURE

Basic monthly policy. The basic policy of the monthly operating procedure for computer simulation of our simplified river-basin system is essentially that outlined in Chapter 7, and can be stated as the following series of instructions or commands.

If possible, meet the target outputs for irrigation and energy (1) with water that would otherwise spill from the reservoirs, and (2) if need be, with supplementary releases of stored waters. Do not draw upon storage specifically for the generation of dump energy.

Create the specified flood readiness in reservoirs *B*, *C*, and *D* at the start of April, May, and June by necessary releases during the preceding months.

To reach the target output for irrigation, make releases (1) from reservoir *A* until it is emptied, and (2) if required, from reservoir *B*.

Use all drafts from reservoir *B*, including releases for irrigation and drawdowns for flood readiness, to generate energy at power plant *B* up to the water capacity of its turbines. Generate such additional energy as may be required to achieve the target output by withdrawing water from reservoirs *C* and *D*, insofar as this is possible. From September through January, draw water (1) from reservoir *D*, and (2) from reservoir *C* should reservoir *D* be emptied before the target output has been reached. If the target output for energy remains unsatisfied, make supplementary releases (1) from reservoir *A*, and (2) if necessary, from reservoir *B* by introducing the compensating procedure for supplementary releases from reservoir *B*. From February through August, use water (1) from reservoir *C*, and (2) from reservoir *D* if forced to do so because reservoir *C* is emptied before the target output has been met.

If the input data portend a potentially dangerous flood in the month under simulation, regardless of what month it is, by-pass the normal

Table 9.1. Power and energy relationships used in the simulation program.

Element of system	Dependent variable	Independent variables and constants	Relationship assumed for simplified river-basin system
Power plants B and G	Energy output, E (Mw hr)	Water flow Q (acre ft) Net head, h (ft) Plant efficiency, e (percent/100) Conversion factor, 0.001024, converts from ft $\times$ acre ft to Mw hr	$E = 0.001024\ Qhe$
Power plant B	Energy output per month, E_m (Mw hr/month)	Average head, AVHEAD (ft) Effective power flow through turbines, EFPOFL (10^2 acre ft/month) Conversion factor, PREFFB = 0.0871, involves constants 0.001024 and $e = 0.85$, and converts from ft $\times$ 10^2 acre ft to Mw hr	$E_m = 0.0871$ AVHEAD $\times$ EFPOFL
Power plant B	Maximum energy output per month, E_{max} (Mw hr/month)	Maximum head, HDMX (ft) Maximum water capacity of turbines, $FLCPB_{max}$ (10^2 acre ft/month) Conversion factor, PREFFB = 0.0871	$E_{max} = 0.0871$ HDMX $\times$ $FLCPB_{max}$
Power plant B	Maximum energy output per month, E_{max} (Mw hr/month)	Rated power capacity of plant, PPB (Mw) Conversion factor, 730.56 (number of hours in standard month of 30.44 days)	$E_{max} = 730.56$ PPB
Power plant B	Water capacity of turbines, FLCPB (10^2 acre ft/month)	Rated power capacity of plant, PPB (Mw) Average head, AVHEAD (ft) Maximum head, HDMX (ft) Conversion factor, 8387.6, converts from Mw/ft to 10^2 acre ft/month	FLCPB = 8387.6 PPB $\times$ $AVHEAD^{1/2}/HDMX^{3/2}$
Power plant B	Maximum water capacity of turbines, $FLCPB_{max}$ (10^2 acre ft/month)	Rated power capacity of plant, PPB (Mw) Maximum head, HDMX (ft) Conversion factor, 8387.6	$FLCPB_{max} = 8387.6$ PPB/HDMX

Reservoir B	Head, y (ft)	Total reservoir capacity, x (10^6 acre ft)	$y = 43.1366 + 153.369x - 42.6982x^2 + 5.39929x^3 + 0.552778x^4 - 0.240631x^5 + 0.0247446x^6 - 0.000851325x^7$
Power plant G	Energy output per month, E_m (Mw hr/month)	Water flow at G, FLOWG (10^2 acre ft) Water capacity of turbines, WCAPG (10^2 acre ft/month) Conversion factor, PRFCTG = 14.4, converts from 10^2 acre ft to Mw hr	If FLOWG < WCAPG, E_m = PRFCTG × FLOWG; if FLOWG ≥ WCAPG, E_m = PRFCTG × WCAPG
Power plant G	Water capacity of turbines, WCAPG (10^2 acre ft/month)	Rated power capacity of plant, PPG (Mw) Conversion factor, 50.73, converts from Mw to 10^2 acre ft/month	WCAPG = 50.73 PPG

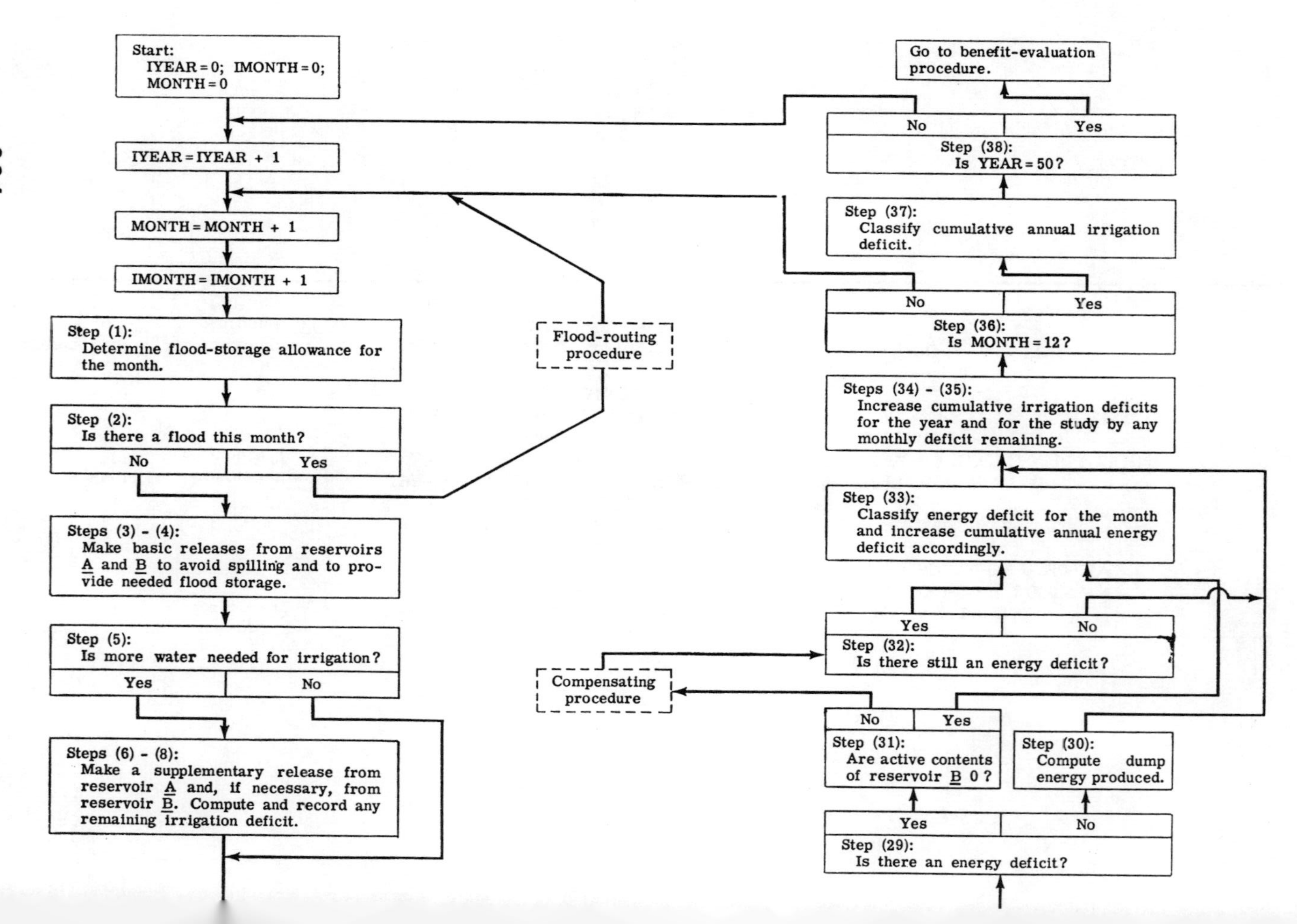
Start:
IYEAR = 0; IMONTH = 0;
MONTH = 0
IYEAR = IYEAR + 1
MONTH = MONTH + 1
IMONTH = IMONTH + 1
Step (1):
Determine flood-storage allowance for the month.
Step (2):
Is there a flood this month?
No
Yes
Flood-routing procedure
Steps (3) - (4):
Make basic releases from reservoirs A and B to avoid spilling and to provide needed flood storage.
Step (5):
Is more water needed for irrigation?
Yes
No
Steps (6) - (8):
Make a supplementary release from reservoir A and, if necessary, from reservoir B. Compute and record any remaining irrigation deficit.
Go to benefit-evaluation procedure.
No
Yes
Step (38):
Is YEAR = 50?
Step (37):
Classify cumulative annual irrigation deficit.
No
Yes
Step (36):
Is MONTH = 12?
Steps (34) - (35):
Increase cumulative irrigation deficits for the year and for the study by any monthly deficit remaining.
Step (33):
Classify energy deficit for the month and increase cumulative annual energy deficit accordingly.
Yes
No
Step (32):
Is there still an energy deficit?
Compensating procedure
No
Yes
Step (31):
Are active contents of reservoir B 0?
Step (30):
Compute dump energy produced.
Yes
No
Step (29):
Is there an energy deficit?

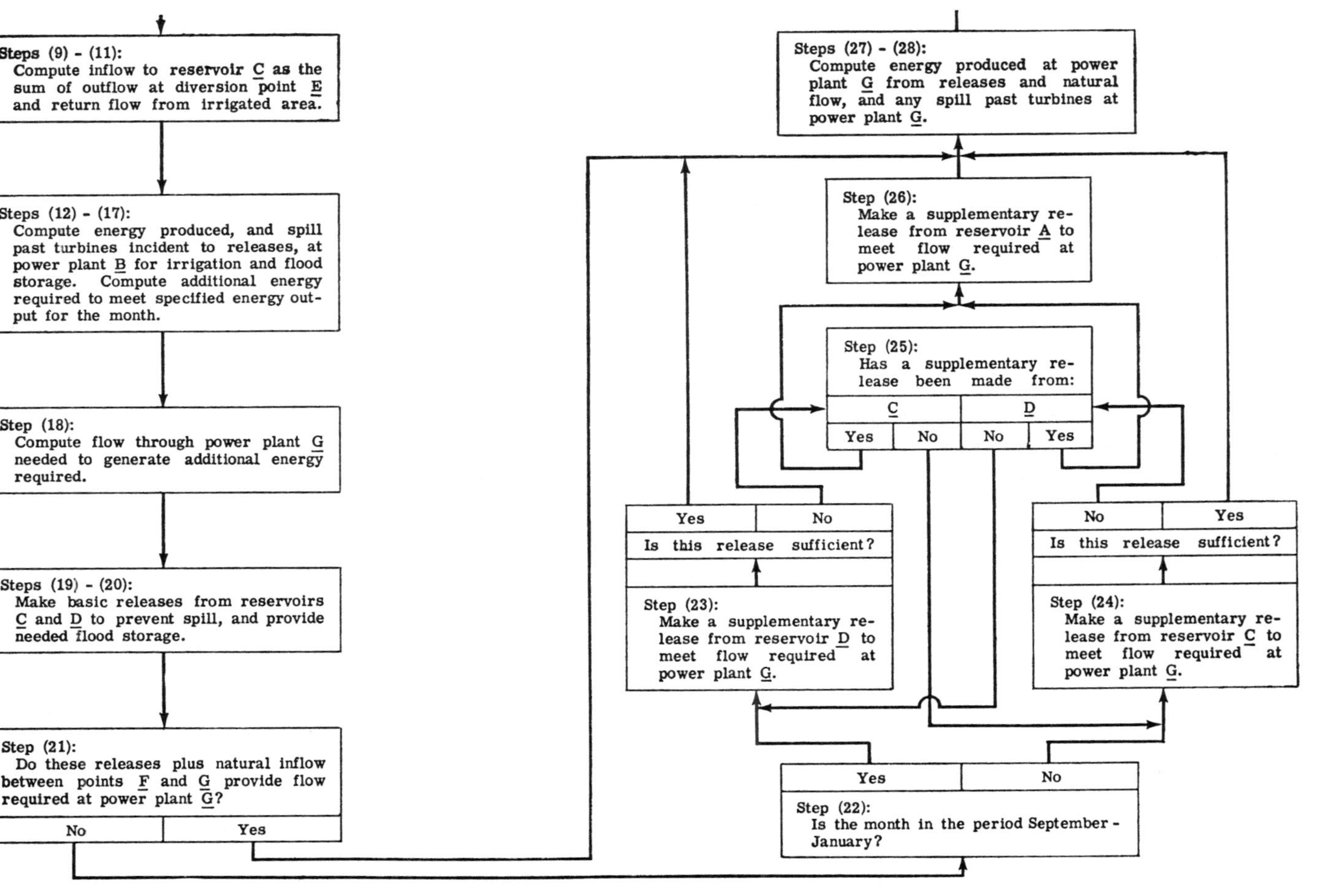

Fig. 9.3. Flow chart for the monthly operating procedure.

monthly routine and call upon the flood-routing procedure to prevent as much flood damage as possible.

Elaboration and formal statement of the monthly procedure. The following sequence of instructions establishes the monthly procedure, which is illustrated in the flow chart of Fig. 9.3.

(1) Determine from the input data the allocation for flood storage and its distribution among reservoirs B, C, and D. The allocation is zero for all months except April, May, or June; for April, it is one-half that specified for May or June. Make allocations for each April, but do so in May only if the sum of the April runoffs at reservoirs A, B, and D is less than that for May or June; make the allocation in June only if the sum of the April or May flows is less than that for June. Draw down reservoirs in the month preceding that of the expected flood in order to reach the allocation for flood storage at the beginning of the critical month. Then go to step (2).

To help in handling the arrays of quantities used in the simulation, enumerate successive months of the hydrologic year by the subscript MONTH which assumes the values 1 to 12 for July through June.[7] Enumerate successive months of the entire simulation by the subscript IMONTH, which takes the values 1 to 600 for a 50-yr simulation, and successive years of the simulation by the subscript IYEAR. Denote the maximum monthly allocation for flood storage given as input data by FDSTAL and the allocation for flood storage for the next consecutive month by FLDALW.

Let FLOWA(IMONTH), FLOWB(IMONTH), and FLOWD(IMONTH) be the monthly flows at points A, B, and D, respectively, as obtained from the hydrologic input data, and let FLOAPR, FLOMAY, and FLOJUN denote the sum of the flows into reservoirs A, B, and D for the months of April, May, and June, respectively.

Initially the subscripts MONTH, IMONTH, and IYEAR are equal to zero. To begin the procedure for the first month, set MONTH = MONTH + 1, IMONTH = IMONTH + 1, and IYEAR = IYEAR + 1, where MONTH + 1 denotes the month of the year following MONTH, IMONTH + 1 denotes the month of the study following IMONTH, and IYEAR + 1 denotes the year of the study following IYEAR.

(a) If MONTH is March, set FLDALW = $\frac{1}{2}$ FDSTAL and go to step (2).

(b) If MONTH is April, test for flood conditions by summing the flows into reservoirs A, B, and D during April, May, and June:

[7] In order to show the reader the appearance of the finished Fortran program, the following Fortran convention has been adopted and used in the equations throughout this chapter. To represent a one-dimensional array or a subscript variable A_I, Fortran uses the symbol or variable followed by the subscript within parentheses, $A(I)$. The subscript may consist of up to six letters but must begin with one of the letters I, J, K, L, M, or N. If a particular element of the array is wanted in computation, the subscript I is set equal to the number of the element in the array; thus, to get the tenth element of the array $A(I)$, I is set equal to 10 and the element is called for as $A(I)$, which the computer interprets as $A(10)$.

FLOAPR = FLOWA(IMONTH) + FLOWB(IMONTH) + FLOWD(IMONTH).
FLOMAY = FLOWA(IMONTH + 1) + FLOWB(IMONTH + 1)
+ FLOWD(IMONTH + 1),
FLOJUN = FLOWA(IMONTH + 2) + FLOWB(IMONTH + 2)
+ FLOWD(IMONTH + 2).

(i) If FLOAPR $\leq$ FLOMAY, set FLDALW = FDSTAL so as to create the necessary flood readiness for May and go to step (2).

(ii) If FLOAPR > FLOMAY, compare FLOAPR with FLOJUN.

(*a*) If FLOAPR > FLOJUN, set FLDALW = 0, because the flows point to April as the critical month rather than May or June. Then go to step (2).

(*b*) If FLOAPR $\leq$ FLOJUN, set FLDALW = FDSTAL to furnish the necessary flood-storage space for May and go to step (2).

(c) If MONTH is May, compare FLOMAY with FLOJUN.

(i) If FLOMAY > FLOJUN, set FLDALW = 0 and go to step (2).

(ii) If FLOMAY $\leq$ FLOJUN, set FLDALW = FDSTAL to provide the necessary flood-storage space for June and go to step (2).

(d) If MONTH is neither March, April, nor May, set FLDALW = 0 and go to step (2).

(2) Examine the runoff data to determine whether or not a flood is recorded for IMONTH. If it is, shift from the monthly to the flood-routing procedure; if not, continue with the monthly procedure.

(3) Compute the basic release from reservoir *A* that will keep it from spilling, and set the two supplementary releases (for irrigation and for firm energy) to zero.

Denote the starting contents, the basic and supplementary releases for the current month, and the capacity of reservoir *A* by CONTSA(IMONTH), RLS1A, RLS2A, RLS3A, and CAPA, respectively. Denote any spill that will occur at reservoir *A* by EXCESA.

Let EXCESA = CONTSA(IMONTH) + FLOWA(IMONTH) − CAPA. If EXCESA > 0, then

RLS1A = EXCESA;

if not, RLS1A = 0,

RLS2A = 0,

and RLS3A = 0.

(4) Compute the basic release from reservoir *B* that will prevent spill and create the flood readiness specified in step (1), and set the supplementary release (for irrigation) to zero.

Denote the starting contents, the basic and supplementary releases for the current month, and the active capacity of reservoir *B* by CONTSB(IMONTH), RLS1B, RLS2B, and CAPB, respectively, and let FRACB represent the fraction of the total flood storage allocated to reservoir *B*. Denote any spill that will occur at reservoir *B* by EXCESB.

Let EXCESB = CONTSB(IMONTH) + FLOWB(IMONTH) + FRACB × FLDALW − CAPB. If EXCESB ≥ 0, then

$$\text{RLS1B} = \text{EXCESB};$$

if not,

$$\text{RLS1B} = 0$$

and

$$\text{RLS2B} = 0.$$

(5) Compute the irrigation deficit incurred by keeping releases from reservoirs *A* and *B* down to their basic values as computed in steps (3) and (4), and go to step (10), (9), or (6) if the deficit is negative, zero, or positive, respectively.

The monthly target output for irrigation and any associated deficiency are denoted by REQIR(MONTH) and DEFIR(IMONTH), respectively. Then

$$\text{DEFIR(IMONTH)} = \text{REQIR(MONTH)} - \text{FLOWE(IMONTH)} - \text{RLS1A} - \text{RLS1B},$$

where FLOWE(IMONTH) is the monthly flow entering between reservoirs *A* and *B* and diversion point *E*.

$$\text{If DEFIR(IMONTH) is } \left.\begin{matrix}\text{negative}\\ \text{zero}\\ \text{positive}\end{matrix}\right\} \text{ go to step } \left\{\begin{matrix}(10).\\ (9).\\ (6).\end{matrix}\right.$$

(6) Decrease the irrigation deficit as much as is possible and necessary by drawing supplementary water from reservoir *A*, and go to step (9) or (7) if the remaining deficit is zero or positive, respectively.

Set the supplementary release RLS2A = DEFIR(IMONTH), if the contents of reservoir *A* permit; that is, if

$$\text{DEFIR(IMONTH)} \leq \text{CONTSA(IMONTH)} + \text{FLOWA(IMONTH)} - \text{RLS1A}.$$

If not,

$$\text{RLS2A} = \text{CONTSA(IMONTH)} + \text{FLOWA(IMONTH)} - \text{RLS1A}.$$

Then reduce the irrigation deficit by the supplementary release, or

$$\text{DEFIR(IMONTH)} = \text{DEFIR(IMONTH)} - \text{RLS2A}.$$

$$\text{If DEFIR(IMONTH) is } \left.\begin{matrix}\text{zero}\\ \text{positive}\end{matrix}\right\} \text{ go to step } \left\{\begin{matrix}(9).\\ (7).\end{matrix}\right.$$

(7) Decrease the irrigation deficit as much as is possible and necessary by drawing supplementary water from reservoir *B*, and go to step (9) or (8) if the remaining deficit is zero or positive, respectively.

Set the supplementary release RLS2B = DEFIR(IMONTH), if the contents of reservoir *B* permit; that is, if

$$\text{DEFIR(IMONTH)} \leq \text{CONTSB(IMONTH)} + \text{FLOWB(IMONTH)} - \text{RLS1B}.$$

If not,

$$\text{RLS2B} = \text{CONTSB(IMONTH)} + \text{FLOWB(IMONTH)} - \text{RLS1B}.$$

Then reduce the irrigation deficit by the supplementary release, or

$$\text{DEFIR(IMONTH)} = \text{DEFIR(IMONTH)} - \text{RLS2B}.$$

$$\text{If DEFIR(IMONTH) is} \left.\begin{matrix}\text{zero}\\ \text{positive}\end{matrix}\right\} \text{go to step} \left\{\begin{matrix}(9).\\ (8).\end{matrix}\right.$$

(8) Increase the cumulative irrigation deficit for the current study by the monthly irrigation deficit remaining after supplementary releases have been made from reservoirs A and B.

Denote the total irrigation deficit for the study by TODFIR. Then

$$\text{TODFIR} = \text{TODFIR} + \text{DEFIR(IMONTH)}.$$

(9) Set the flow past diversion point E to reservoir C to zero, because all water passing E has been diverted for irrigation. Then go to step (11).

Denote the monthly flow past diversion point E to reservoir C by OUTFLE(IMONTH). Then

$$\text{OUTFLE(IMONTH)} = 0.$$

Go to step (11).

(10) Set the flow past diversion point E to reservoir C equal to the difference between (a) the releases from reservoirs A and B plus flows entering between reservoirs A and B and diversion point E, and (b) the target output for irrigation. Set the irrigation deficit for the month to zero.

$$\text{OUTFLE(IMONTH)} = -\text{DEFIR(IMONTH)}$$

and

$$\text{DEFIR(IMONTH)} = 0.$$

(11) Record the total releases from reservoirs A and B and the reservoir volumes left intact. Set the flow into reservoir C equal to the flow past diversion point E plus the normal return flow from irrigation minus any decrease in return flow because of irrigation deficits in previous months. The normal irrigation return flow for each month of the year is a specified proportion of the total annual return flow, which itself is a fixed proportion of the annual target output for irrigation. Pertinent values are shown for our system in Table 7.4. Decreases in return flow in any month, because of irrigation deficits in preceding months, are calculated by applying return-flow reduction factors to the deficits. The

factors assumed for our simplified river-basin system are given in Table 7.5.

Denote the total releases from reservoirs A and B by RLSTA(IMONTH) and RLSTB(IMONTH), and the flow into reservoir C by FLOWC(IMONTH). Denote (1) the reduction in irrigation return flow for the present and next six months by the array DFCRIR(J), where J varies from 1 to 7; (2) the factors by which the current deficit, if any, must be multiplied to provide later reductions in irrigation return flow by the array DFRIFC(J); and (3) the irrigation return flow for the current month, if there have been no irrigation deficits in the previous seven months, by RIRNML(MONTH). Then

$$\text{RLSTA(IMONTH)} = \text{RLS1A} + \text{RLS2A},$$

$$\text{RLSTB(IMONTH)} = \text{RLS1B} + \text{RLS2B},$$

and

$$\text{FLOWC(IMONTH)} = \text{OUTFLE(IMONTH)} + \text{RIRNML(MONTH)} - \text{DFCRIR(1)}.$$

(12) Compute the initial contents of reservoirs A and B for the succeeding month, after the basic and supplementary releases have been made. Note that the contents of A and B may be altered later in the procedure if additional energy is required.

Denote the contents of reservoirs A and B at the beginning of the succeeding month by CONTSA(IMONTH + 1) and CONTSB(IMONTH + 1), respectively. Then

$$\text{CONTSA(IMONTH} + 1) = \text{CONTSA(IMONTH)} + \text{FLOWA(IMONTH)} - \text{RLSTA(IMONTH)}$$

and

$$\text{CONTSB(IMONTH} + 1) = \text{CONTSB(IMONTH)} + \text{FLOWB(IMONTH)} - \text{RLSTB(IMONTH)}.$$

(13) Compute the average contents of reservoir B for the month and the associated average head. Use as the head–capacity function the seventh-order polynomial shown for the simplified system in Table 9.1 and presented graphically in Fig. 7.8.

Define the following quantities for reservoir B: average contents for the month, AVCONB; average head associated with the average contents, AVHEAD; dead storage, DEADST; and head–capacity function, HEADFN. Then

$$\text{AVCONB} = \tfrac{1}{2}[\text{CONTSB(IMONTH)} + \text{CONTSB(IMONTH} + 1)],$$

and

$$\text{AVHEAD} = \text{HEADFN(AVCONB, DEADST)}.$$

(The symbols AVCONB and DEADST in parentheses are arguments or independent variables of the head–capacity function, rather than subscripts.)

(14) Compute the water capacity of the turbines in power plant B

associated with the average head. As shown in Table 9.1, water capacity is a function of the rated power-plant capacity and the maximum and average power heads.

Let the rated power capacity of power plant *B*, given as input data, be denoted by PPB. For any given period the maximum water capacity, FLCPB, of the variable-head power plant at *B* is directly proportional to the power capacity and the square root of the average head AVHEAD and inversely proportional to the $\frac{3}{2}$ power of the maximum head HDMX. Or formally,

$$\text{BETA} = 8387.6\ \text{PPB}/\text{HDMX}^{3/2},$$

$$\text{FLCPB} = \text{BETA} \times \text{AVHEAD}^{1/2}$$

when the units of power and head are megawatts and feet, respectively, and BETA is an intermediate quantity denoting the constant monthly energy factor.

(15) Compute the effective monthly flow through the turbines at reservoir *B*.

Denote the effective monthly flow through the turbines at reservoir *B* as EFPOFL. Then, if RLSTB(IMONTH) $\leq$ FLCPB,

$$\text{EFPOFL} = \text{RLSTB(IMONTH)};$$

if not,

$$\text{EFPOFL} = \text{FLCPB}.$$

(16) Compute the spill at reservoir *B*.

Denote the spill at reservoir *B* during any one month by SPILB(IMONTH). Then, if RLSTB(IMONTH) $-$ FLCPB ≥ 0,

$$\text{SPILB(IMONTH)} = \text{RLSTB(IMONTH)} - \text{FLCPB};$$

if not,

$$\text{SPILB(IMONTH)} = 0.$$

(17) Calculate the energy generated at power plant *B* and the additional energy needed to meet the target output for the month.

Denote the energy generated by power plant *B* during the month by ENERB(IMONTH) and the power units and efficiency conversion factor by PREFFB. Then

$$\text{ENERB(IMONTH)} = \text{PREFFB} \times \text{AVHEAD} \times \text{EFPOFL}.$$

Denote the monthly target energy output by REQEN(MONTH) and the energy deficit for the month by DEFEN(IMONTH). Then

$$\text{DEFEN(IMONTH)} = \text{REQEN(MONTH)} - \text{ENERB(IMONTH)}.$$

(18) Calculate the flow through power plant *G* that will generate the further energy needed to meet the deficit. This flow cannot exceed the

water capacity of the turbines, which is constant for a fixed-head power plant.

Let REQFLG(IMONTH) denote the required flow through power plant G, WCAPG the water capacity of its turbines, and PRFCTG the power unit and efficiency conversion factor, namely, 14.4 $Mw/10^2$ acre ft. If WCAPG $\leq$ DEFEN(IMONTH)/PRFCTG, then

$$\text{REQFLG} = \text{WCAPG};$$

if not, $$\text{REQFLG(IMONTH)} = \text{DEFEN(IMONTH)/PRFCTG}.$$

(19) Compute the basic release from reservoir *C* that will prevent spill and provide the flood storage specified in step (1), and set the supplementary release (for additional energy) to zero.

Denote the initial contents, the basic and supplementary releases for the current month, and the capacity of reservoir *C* by CONTSC(IMONTH), RLS1C, RLS2C, and CAPC, respectively. Let FRACC be the fraction of the total flood storage allocated to reservoir *C*. Denote any spill at reservoir *C* by EXCESC.

Let EXCESC = CONTSC(IMONTH) + FLOWC(IMONTH) + FRACC × FLDALW − CAPC. If EXCESC > 0,

$$\text{RLS1C} = \text{EXCESC};$$

if not, $$\text{RLS1C} = 0$$

and $$\text{RLS2C} = 0.$$

(20) Compute the basic release from reservoir *D* that will prevent spill and provide the specified flood storage, and set the supplementary release (for additional energy) to zero.

Denote the initial contents, the basic and supplementary releases for the current month, and the capacity of reservoir *D* by CONTSD(IMONTH), RLS1D, RLS2D, and CAPD, respectively. Let FRACD be the fraction of the total flood storage allocated to reservoir *D*. Denote any spill at reservoir *D* by EXCESD.

Let EXCESD = CONTSD(IMONTH) + FLOWD(IMONTH) + FRACD × FLDALW − CAPD. If EXCESD > 0,

$$\text{RLS1D} = \text{EXCESD};$$

if not, $$\text{RLS1D} = 0$$

and $$\text{RLS2D} = 0.$$

(21) Compute the difference between (a) the flow through power plant *G* resulting from the basic releases from reservoirs *C* and *D*, as determined in steps (19) and (20), plus the natural runoff entering between points *F* and *G*, and (b) the required flow at *G* as computed in step (18).

If the difference is negative or zero, go to step (27); otherwise, go to step (22).

Denote the natural flow entering between points F and G by FLOWFG(IMONTH), and the difference between required and available flows at G by DFHYFL. Then

$$\text{DFHYFL} = \text{REQFLG(IMONTH)} - \text{RLS1C} - \text{RLS1D} - \text{FLOWFG(IMONTH)}.$$

If DFHYFL is $\left.\begin{matrix}\text{negative or zero} \\ \text{positive}\end{matrix}\right\}$ go to step $\left\{\begin{matrix}(27). \\ (22).\end{matrix}\right.$

(22) Go to step (23) or (24) depending on the period of the year in which the month falls.

If MONTH lies in the period $\left.\begin{matrix}\text{September through January} \\ \text{February through August}\end{matrix}\right\}$ go to step $\left\{\begin{matrix}(23). \\ (24).\end{matrix}\right.$

(23) Reduce the deficit in flow through power plant G as much as is possible and necessary by making a supplementary release from reservoir D. Go to step (27) or (25) if the remaining deficit is zero or positive, respectively.

Denote the supplementary release from reservoir D by RLS2D. If DFHYFL < CONTSD(IMONTH) + FLOWD(IMONTH) − RLS1D, then

$$\text{RLS2D} = \text{DFHYFL};$$

if not, $\quad \text{RLS2D} = \text{CONTSD(IMONTH)} + \text{FLOWD(IMONTH)} - \text{RLS1D}$

and $\quad \text{DFHYFL} = \text{DFHYFL} - \text{RLS2D}.$

If DFHYFL is $\left.\begin{matrix}\text{zero} \\ \text{positive}\end{matrix}\right\}$ go to step $\left\{\begin{matrix}(27). \\ (25).\end{matrix}\right.$

(24) Reduce the deficit in flow through power plant G as much as is possible and necessary by drawing supplementary water from reservoir C. Go to step (27) or (25) if the remaining deficit is zero or positive, respectively.

Denote the supplementary release from reservoir C by RLS2C. If DFHYFL < CONTSC(IMONTH) + FLOWC(IMONTH) − RLS1C, then

$$\text{RLS2C} = \text{DFHYFL};$$

if not, $\quad \text{RLS2C} = \text{CONTSC(IMONTH)} + \text{FLOWC(IMONTH)} - \text{RLS1C}$

and $\quad \text{DFHYFL} = \text{DFHYFL} - \text{RLS2C}.$

If DFHYFL is $\left.\begin{matrix}\text{zero} \\ \text{positive}\end{matrix}\right\}$ go to step $\left\{\begin{matrix}(27). \\ (25).\end{matrix}\right.$

(25) Go to step (23), (24), or (26) if a prior supplementary release has been made at reservoir C only, at reservoir D only, or at both reservoirs C and D, respectively.

$$\text{If a prior release has been made} \begin{array}{l} \text{at } C \text{ only} \\ \text{at } D \text{ only} \\ \text{at both } C \text{ and } D \end{array} \Bigg\} \text{ go to step } \Bigg\{ \begin{array}{l} (23). \\ (24). \\ (26). \end{array}$$

(26) Reduce the deficit in the flow through power plant G as much as is possible or necessary by making a further supplementary release from reservoir A, and adjust the new contents of A, the total release from A, and the flow into reservoir C accordingly.

Denote a second supplementary release from reservoir A by RLS3A. If DFHYFL $<$ CONTSA(IMONTH $+$ 1), then

$$\text{RLS3A} = \text{DFHYFL};$$

if not,
$$\text{RLS3A} = \text{CONTSA}(\text{IMONTH} + 1).$$

Also,
$$\text{RLSTA}(\text{IMONTH}) = \text{RLSTA}(\text{IMONTH}) + \text{RLS3A},$$

$$\text{CONTSA}(\text{IMONTH} + 1) = \text{CONTSA}(\text{IMONTH} + 1) - \text{RLS3A},$$

$$\text{FLOWC}(\text{IMONTH}) = \text{FLOWC}(\text{IMONTH}) + \text{RLS3A},$$

$$\text{RLS2C} = \text{RLS3A} + \text{RLS2C},$$

and
$$\text{DFHYFL} = \text{DFHYFL} - \text{RLS3A}.$$

(27) Compute the total monthly releases from, and the new contents of, reservoirs C and D; the flow through, and energy produced at, power plant G; and the energy deficit, if any.

Denote the total releases for the month from reservoirs C and D by RLSTC(IMONTH) and RLSTD(IMONTH), the total flow through power plant G by FLOWG(IMONTH), and the energy produced there by ENERG(IMONTH). Then

$$\text{RLSTC}(\text{IMONTH}) = \text{RLS1C} + \text{RLS2C},$$

$$\text{RLSTD}(\text{IMONTH}) = \text{RLS1D} + \text{RLS2D},$$

$$\text{CONTSC}(\text{IMONTH} + 1) = \text{CONTSC}(\text{IMONTH}) + \text{FLOWC}(\text{IMONTH}) - \text{RLSTC}(\text{IMONTH}),$$

$$\text{CONTSD}(\text{IMONTH} + 1) = \text{CONTSD}(\text{IMONTH}) + \text{FLOWD}(\text{IMONTH}) - \text{RLSTD}(\text{IMONTH}),$$

and

$$\text{FLOWG}(\text{IMONTH}) = \text{RLSTC}(\text{IMONTH}) + \text{RLSTD}(\text{IMONTH}) + \text{FLOWFG}(\text{IMONTH}).$$

If FLOWG(IMONTH) < WCAPG,

$$\text{ENERG(IMONTH)} = \text{FLOWG(IMONTH)} \times \text{PRFCTG};$$

if not,
$$\text{ENERG(IMONTH)} = \text{WCAPG} \times \text{PRFCTG}$$

and

$$\text{DEFEN(IMONTH)} = \text{REQEN(MONTH)} - \text{ENERB(IMONTH)} - \text{ENERG(IMONTH)}.$$

(28) Compute the spill at power plant *G* for the month.

Denote the spill at power plant *G* for the month by SPILG(IMONTH). If FLOWG(IMONTH) > WCAPG, then

$$\text{SPILG(IMONTH)} = \text{FLOWG(IMONTH)} - \text{WCAPG};$$

if not,
$$\text{SPILG(IMONTH)} = 0.$$

(29) Go to step (30) or (31) if the energy deficit is negative or zero, or positive, respectively.

If DEFEN(IMONTH) is $\left.\begin{matrix}\text{negative or zero}\\ \text{positive}\end{matrix}\right\}$ go to step $\left\{\begin{matrix}(30).\\ (31).\end{matrix}\right.$

(30) Compute the cumulative dump energy produced for the month, year, and study, and go to step (34).

Denote the dump energy produced for a month, a year, and the entire study by ENSURP(IMONTH), DUMPEN(IYEAR, and TOTDMP, respectively. Then

$$\text{ENSURP(IMONTH)} = -\text{DEFEN(IMONTH)},$$

$$\text{DUMPEN(IYEAR)} = \text{DUMPEN(IYEAR)} + \text{ENSURP(IMONTH)},$$

$$\text{TOTDMP} = \text{TOTDMP} + \text{ENSURP(IMONTH)},$$

and
$$\text{DEFEN(IMONTH)} = 0.$$

Go to step (34).

(31) Go to step (33) or (32) if the active contents of reservoir *B* are zero or positive, respectively.

If CONTSB(IMONTH + 1) is $\left.\begin{matrix}\text{zero}\\ \text{positive}\end{matrix}\right\}$ go to step $\left\{\begin{matrix}(33).\\ (32).\end{matrix}\right.$

(32) Call upon the compensating procedure, detailed later in this chapter, for a second supplementary release from reservoir *B* that will generate as much additional energy as is necessary and possible at power plants *B* and *G*, subject to the limitations specified for the residual contents of reservoir *B* and the water capacities of the turbines at the two

power plants. Then go to step (34) or (33) if the remaining energy deficit is zero or positive, respectively.

In the compensating procedure, the computed values of CONTSB(IMONTH) FLOWB(IMONTH), RLSTB(IMONTH), FLCPB, DEFEN(IMONTH), CONTSB(IMONTH, +1), ENERG(IMONTH), and ENERB(IMONTH) are used to determine the magnitude of the second supplementary release from reservoir B. The original quantities are then modified by this magnitude.

If, after the compensating procedure has been called, DEFEN(IMONTH) is

$$\left.\begin{matrix}\text{zero}\\ \text{positive}\end{matrix}\right\} \text{go to step} \left\{\begin{matrix}(34).\\ (33).\end{matrix}\right.$$

(33) Increase the number of monthly energy deficits by one, and classify the monthly deficit according to its magnitude.

For DEFEN(IMONTH): class I $\leq$ 0.1 REQEN, 0.1 REQEN $<$ class II $\leq$ 0.25 REQEN, and class III $>$ 0.25 REQEN.

Denote the cumulative annual energy deficit by CMANED(IYEAR). Then

$$\text{CMANED(IYEAR)} = \text{CMANED(IYEAR)} + \text{DEFEN(IMONTH)}.$$

(34) Go to step (36) or (35) if the monthly irrigation deficit is zero or positive, respectively.

If, after reservoirs A and B have been emptied, DEFIR(IMONTH) is

$$\left.\begin{matrix}\text{zero}\\ \text{positive}\end{matrix}\right\} \text{go to step} \left\{\begin{matrix}(36).\\ (35).\end{matrix}\right.$$

(35) Compute the cumulative irrigation deficits for the year and for the study as a whole.

Denote the cumulative irrigation deficit for a 50-yr study by CMRNID and the cumulative annual irrigation deficit by CMANID(IYEAR). Then

$$\text{CMANID(IYEAR)} = \text{CMANID(IYEAR)} + \text{DEFIR(IMONTH)}$$

and

$$\text{CMRNID} = \text{CMRNID} + \text{DEFIR(IMONTH)}.$$

(36) Test for the end of the simulation year and go to step (37) if the end has been reached; if not, return to step (1) and start the monthly cycle again.

If MONTH $=$ 12, then IYEAR $=$ IYEAR $+$ 1; set MONTH $=$ 0 and go to step (37). If MONTH $\neq$ 12, return to step (1).

(37) Classify the cumulative annual irrigation deficit by its magnitude.

Denote the annual target output for irrigation by ANIREQ.

For CMANID(IYEAR): class I $\leq$ 0.1 ANIREQ, 0.1 ANIREQ $<$ class II $\leq$ 0.7 ANIREQ, and class III $>$ 0.7 ANIREQ.

(38) Test for the end of the 50-yr study and call the benefit-evaluation procedure if the end has been reached; if not, return to step (1) and start the monthly cycle again.

If IYEAR = 50, call the benefit-evaluation procedure; if not, return to step (1).

FLOOD-ROUTING PROCEDURE

Basic flood-routing policy. When examination of the runoff data in step (2) of the monthly procedure shows that a particular month includes a flood, the flood-routing procedure is substituted for the monthly operating procedure. This is accomplished readily because the flood-routing and monthly procedures are identical in attempting to satisfy target irrigation and energy outputs by a sequence of releases from reservoirs. However, the flood-routing procedure differs from the monthly one in three respects: (1) its fundamental time interval is 1/120 of a month, or approximately 6 hours, hydrographic information as well as target outputs for irrigation and energy being referred to this foreshortened time scale; (2) it includes time lags of flow; and (3) it attempts to lower the peak flow to a prespecified nondamage level. Hence, the basic policy of the flood-routing procedure is expressed by the following instructions or commands.

If possible, meet target outputs for irrigation and energy by releases from reservoirs A, B, C, and D as in the monthly operating procedure. Make additional releases for flood control from reservoirs B, C, and D until either the reservoirs are lowered to the levels for flood storage specified for the month or the flow at point G reaches the prespecified limit of channel capacity. Store flows that exceed this value in reservoirs B, C, and D to the extent possible, storing first at reservoir B alone and then, if necessary, at both reservoirs C and D in proportion to the space available in each of them. Should there be an energy deficit in any time interval, do not call the compensating procedure; instead, make additional releases from reservoirs C, D, and A, as necessary.

Elaboration and formal statement of the flood-routing procedure. Unless otherwise stated, the notation for this procedure is the same as for the monthly procedure. A flow chart is presented in Fig. 9.4.

(1) Adjust the initial contents of reservoirs A, B, and C for a time

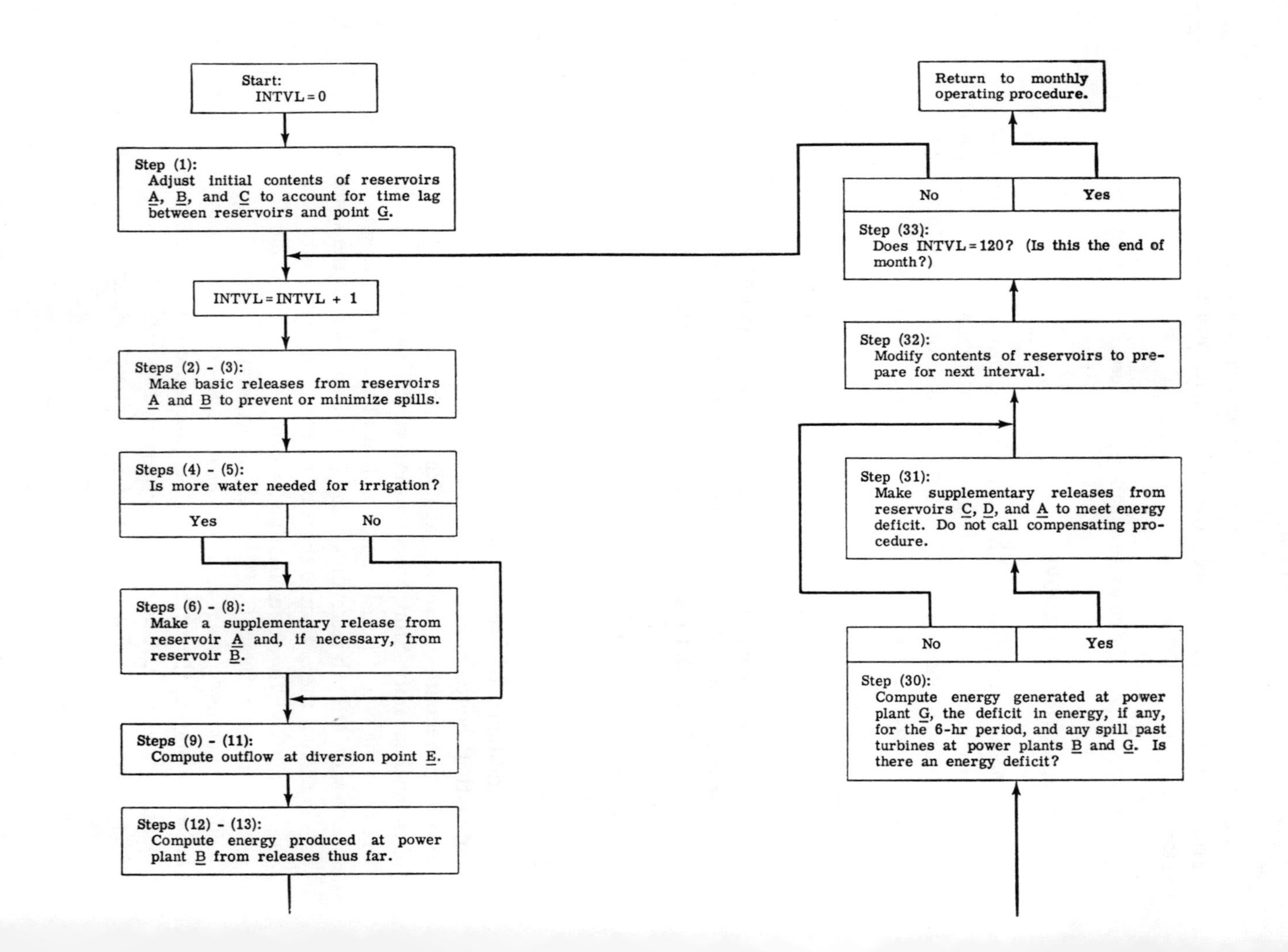

Start:
INTVL = 0
Step (1):
Adjust initial contents of reservoirs A, B, and C to account for time lag between reservoirs and point G.
INTVL = INTVL + 1
Steps (2) - (3):
Make basic releases from reservoirs A and B to prevent or minimize spills.
Steps (4) - (5):
Is more water needed for irrigation?
Yes
No
Steps (6) - (8):
Make a supplementary release from reservoir A and, if necessary, from reservoir B.
Steps (9) - (11):
Compute outflow at diversion point E.
Steps (12) - (13):
Compute energy produced at power plant B from releases thus far.
Return to monthly operating procedure.
No
Yes
Step (33):
Does INTVL = 120? (Is this the end of month?)
Step (32):
Modify contents of reservoirs to prepare for next interval.
Step (31):
Make supplementary releases from reservoirs C, D, and A to meet energy deficit. Do not call compensating procedure.
No
Yes
Step (30):
Compute energy generated at power plant G, the deficit in energy, if any, for the 6-hr period, and any spill past turbines at power plants B and G. Is there an energy deficit?

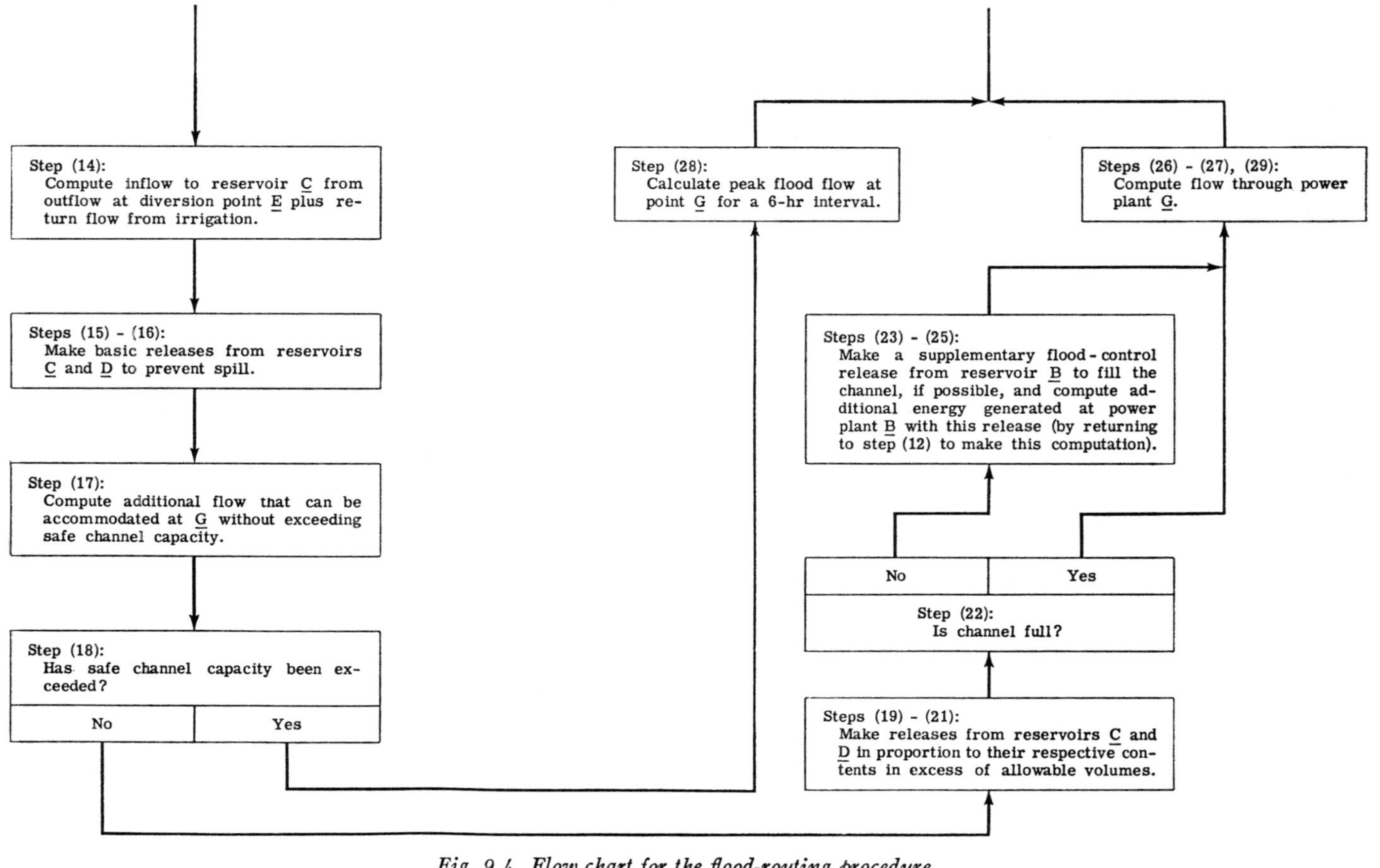

Step (14):
Compute inflow to reservoir C from outflow at diversion point E plus return flow from irrigation.
Steps (15) - (16):
Make basic releases from reservoirs C and D to prevent spill.
Step (17):
Compute additional flow that can be accommodated at G without exceeding safe channel capacity.
Step (18):
Has safe channel capacity been exceeded?
No
Yes
Step (28):
Calculate peak flood flow at point G for a 6-hr interval.
Steps (26) - (27), (29):
Compute flow through power plant G.
Steps (23) - (25):
Make a supplementary flood-control release from reservoir B to fill the channel, if possible, and compute additional energy generated at power plant B with this release (by returning to step (12) to make this computation).
No
Yes
Step (22):
Is channel full?
Steps (19) - (21):
Make releases from reservoirs C and D in proportion to their respective contents in excess of allowable volumes.

Fig. 9.4. Flow chart for the flood-routing procedure.

lag of 12 hours between point *G* and reservoirs *A* and *B*, and of 6 hours between *G* and reservoir *C*.

Denote by CONA6, CONB6, CONC6, and COND6 the respective contents of reservoirs *A*, *B*, *C*, and *D* associated, in the case of reservoirs *A* and *B* with a 12-hr time lag, in that of reservoir *C* with a 6-hr time lag, and in that of reservoir *D* with no time lag. Then the initial contents of reservoir *A*, for example, equal those at the beginning of the month, as calculated in the monthly procedure, minus $\frac{1}{60}$ of the flow into reservoir *A* during the previous month plus $\frac{1}{60}$ of the release from reservoir *A* during that same month, where 12 hours equal $\frac{1}{60}$ of a month. The fractions for reservoirs *B* and *C* are $\frac{1}{60}$ and $\frac{1}{120}$, respectively. Therefore,

$$\text{CONA6} = \text{CONTSA(IMONTH)} + \tfrac{1}{60}[\text{RLSTA(IMONTH} - 1) - \text{FLOWA(IMONTH} - 1)],$$

$$\text{CONB6} = \text{CONTSB(IMONTH)} + \tfrac{1}{60}[\text{RLSTB(IMONTH} - 1) - \text{FLOWB(IMONTH} - 1)],$$

$$\text{CONC6} = \text{CONTSC(IMONTH)} + \tfrac{1}{120}[\text{RLSTC(IMONTH} - 1) - \text{FLOWC(IMONTH} - 1)],$$

and

$$\text{COND6} = \text{CONTSD(IMONTH)}.$$

(2) Compute the basic release from reservoir *A* that will prevent or minimize spill, and set the two supplementary releases (for irrigation and for energy) to zero.

Successive 6-hr intervals of the month are enumerated by the subscript INTVL which assumes values from 1 to 120. Denote the natural flow into reservoir *A* for any particular 6-hr interval by FLOWA6(INTVL) and the basic and supplementary 6-hr releases by RLS1A6, RLS2A6, and RLS3A6. Then, since initially INTVL = 0, begin the procedure by setting INTVL = INTVL + 1.

Let EXCESA = CONA6 + FLOWA6(INTVL) − CAPA. If EXCESA > 0,

$$\text{RLS1A6} = \text{EXCESA};$$

if not,

$$\text{RLS1A6} = 0.$$

Also,

$$\text{RLS2A6} = 0$$

and

$$\text{RLS3A6} = 0.$$

(3) Compute the basic release from reservoir *B* that will prevent or minimize spill, and set the two supplementary releases (for irrigation and for flood storage) to zero.

Let FLOWB6(INTVL), RLS1B6, RLS2B6, and RLS3B6 be the natural flow into reservoir *B* for the interval, and the basic and supplementary releases from reservoir *B*, respectively.

Let EXCESB = CONB6 + FLOWB6(INTVL) − CAPB. If EXCESB > 0,

$$\text{RLS1B6} = \text{EXCESB};$$

if not, RLS1B6 = 0.

Also, RLS2B6 = 0

and RLS3B6 = 0.

(4) Determine the deficit in irrigation incurred by holding releases from reservoirs A and B down to their basic values, as computed in steps (2) and (3).

Denote the 6-hr target output for irrigation by REQIR6, the natural flow into E for this interval by FLOWE6(INTVL), and the associated irrigation deficit by DEFIR6. Then

DEFIR6 = REQIR6 − FLOWE6(INTVL) − RLS1A6 − RLS1B6.

(5) Go to step (10), (9), or (6) if the irrigation deficit is negative, zero, or positive, respectively.

If DEFIR6 is negative / zero / positive go to step (10). / (9). / (6).

(6) Reduce the irrigation deficit as much as is possible and necessary by drawing supplementary water from reservoir A.

If DEFIR6 ≤ CONA6 + FLOWA6(INTVL) − RLS1A6, then

RLS2A6 = DEFIR6;

if not, RLS2A6 = CONA6 + FLOWA6(INTVL) − RLS1A6.

Then DEFIR6 = DEFIR6 − RLS2A6.

(7) Go to step (9) or (8) if the remaining irrigation deficit is zero or positive, respectively.

If DEFIR6 is zero / positive go to step (9). / (8).

(8) Reduce the irrigation deficit as much as is possible and necessary by making a supplementary release from reservoir B. Record any irrigation deficit.

If DEFIR6 ≤ CONB6 + FLOWB6(INTVL) − RLS1B6, then

RLS2B6 = DEFIR6;

if not, RLS2B6 = CONB6 + FLOWB6(INTVL) − RLS1B6.

Then DEFIR6 = DEFIR6 − RLS2B6.

(9) Set the flow past diversion point E to reservoir C at zero and go to step (11).

Denote the flow past diversion point E for the interval by OTFLE6.

Then
$$\text{OTFLE6} = 0.$$

Go to step (11).

(10) Set the flow past diversion point E to reservoir C equal to the difference between (a) the releases from reservoirs A and B plus the runoff entering between A and B and E, and (b) the target output for irrigation. Set the irrigation deficit for the interval to zero.

$$\text{OTFLE6} = -\text{DEFIR6}$$

and
$$\text{DEFIR6} = 0.$$

(11) Record the total releases from reservoirs A and B, and the remaining volume of water in reservoir B.

Denote the total releases for the interval from reservoirs A and B and the new contents of reservoir B by RLSTA6, RLSTB6, and CONB6N, respectively. Then

$$\text{RLSTA6} = \text{RLS1A6} + \text{RLS2A6},$$

$$\text{RLSTB6} = \text{RLS1B6} + \text{RLS2B6},$$

and
$$\text{CONB6N} = \text{CONB6} + \text{FLOWB6(INTVL)} - \text{RLSTB6}.$$

(12) Compute the average contents of reservoir B during the period of release and, by the same method as in the monthly procedure, the head associated with this average.

Denote the average contents of reservoir B during the interval by AVCNB6 and the head by HEADB. Then

$$\text{AVCNB6} = \tfrac{1}{2}[\text{CONB6N} + \text{CONB6}]$$

and
$$\text{AVHEAD} = \text{HEADFN(AVCNB6, DEADST)}.$$

(As previously mentioned, the symbols AVCNB6 and DEADST are the arguments of the head–capacity function HEADFN.)

(13) Similarly, compute the maximum water capacity of the turbines in power plant B, the effective flow through them, and the energy generated by them. Go to step (14) or (29) if the supplementary release for flood storage is zero or positive, respectively.

Denote the flow capacity associated with the release from reservoir B by FLCPB6, the cumulative energy generated during the interval by ENERB6, the effective power flow through the turbines during the interval by EFPOF6, and a constant energy factor by BETA6. Then

$$\text{BETA6} = 8387.6\ \text{PPB}/\{120[\text{HDMX}]^{3/2}\}$$

If this is the first time that energy has been generated during this interval,

$$\text{PRLSB} = 0$$

and $$\text{FLCPB6} = \text{BETA6} \times \text{AVHEAD}^{1/2};$$

if not, $$\text{FLCPB6} = \text{BETA6} \times \text{AVHEAD}^{1/2} - \text{PRLSB},$$

where PRLSB denotes any previous effective flow at B.

Also, if RLSTB6 < FLCPB6, then

$$\text{EFPOF6} = \text{RLSTB6};$$

if not, $$\text{EFPOF6} = \text{FLCPB6}$$

and $$\text{ENERB6} = 0.0871\ \text{AVHEAD} \times \text{EFPOF6} + \text{ENERB6}.$$

Set the original contents of reservoir B equal to the new contents, or

$$\text{CONB6} = \text{CONB6N}.$$

If RLS3B6 is $\left.\begin{matrix}\text{zero}\\ \text{positive}\end{matrix}\right\}$ go to step $\left\{\begin{matrix}(14).\\ (29).\end{matrix}\right.$

(14) Compute, as in step (10), the flow into reservoir C that derives from the flow past diversion point E. Also compute, as defined and calculated in step (11) of the monthly procedure, the return flow from irrigation minus any decrease in return flow because of irrigation deficits in preceding months.

Denote the normal irrigation return flow, the reduction in return flow, and the flow into reservoir C for the interval by RIRNM6, DFCRI6, and FLOWC6, respectively. Then

$$\text{FLOWC6} = \text{OTFLE6} + \text{RIRNM6} - \text{DFCRI6}.$$

(15) Compute the basic release from reservoir C that will prevent spill, and set the two supplementary releases (for flood storage and additional energy) to zero.

Let the basic and supplementary releases from reservoir C for the interval be RLS1C6, RLS2C6, and RLS3C6, respectively.

Let EXCESC = CONC6 + FLOWC6 − CAPC. If EXCESC > 0, then

$$\text{RLS1C6} = \text{EXCESC};$$

if not, $$\text{RLS1C6} = 0,$$

$$\text{RLS2C6} = 0,$$

and

$$\text{RLS3C6} = 0.$$

(16) Compute the basic release from reservoir D that will prevent spill, and set the two supplementary releases (for flood storage and additional energy) to zero.

Let the basic and supplementary releases from reservoir D for the interval be RLS1D6, RLS2D6, and RLS3D6, respectively, and denote the natural flow into reservoir D by FLOWD6(INTVL).

Let EXCESD = COND6 + FLOWD6(INTVL) − CAPD. If EXCESD > 0, then

$$\text{RLS1D6} = \text{EXCESD};$$

if not,

$$\text{RLS1D6} = 0,$$

$$\text{RLS2D6} = 0,$$

and

$$\text{RLS3D6} = 0.$$

(17) Compute the additional flow that can be received at G without exceeding the safe channel capacity.

Denote by FLOOD the safe channel capacity at G, by FLOFG6(INTVL) the natural flow to G for the interval, and by REMFLG the channel capacity at G remaining after the total flow through G to this point has been subtracted. Then

$$\text{REMFLG} = \text{FLOOD} - \text{FLOFG6(INTVL)} - \text{RLS1C6} - \text{RLS1D6}.$$

(18) Go to step (28), (26), or (19) if the safe channel capacity remaining at G is negative, zero, or positive, respectively.

$$\text{If REMFLG is } \left.\begin{array}{r}\text{negative}\\ \text{zero}\\ \text{positive}\end{array}\right\} \text{ go to step } \left\{\begin{array}{l}(28).\\ (26).\\ (19).\end{array}\right.$$

(19) Compute the allowable contents of reservoirs B, C, and D as the total active capacity of each reservoir reduced by the proportion of total flood-control space of the system that is allocated to the reservoir, as specified in the input data.

Denote the allowable contents of reservoirs B, C, and D after monthly flood storage releases by ALCONB, ALCONC, and ALCOND, respectively. Then

$$\text{ALCONB} = \text{CAPB} - \text{FRACB} \times \text{FLDALW},$$

$$\text{ALCONC} = \text{CAPC} - \text{FRACC} \times \text{FLDALW},$$

and

$$\text{ALCOND} = \text{CAPD} - \text{FRACD} \times \text{FLDALW}.$$

(20) Draw down the reservoirs to their allowable levels in order to afford maximum flood storage during the flood month, whenever this is possible. If, as computed in step (19), the volumes of water stored in either reservoir C or D are larger than allowable, make supplementary releases from reservoirs C and D in proportion to the amounts by which the allowable storage in each reservoir is exceeded. The total water to be released then equals the smaller of the following two amounts: the total contents of the two reservoirs above their allowable levels, or the additional flow that can be received at G as computed in step (17). Reduce the residual free channel capacity at G by the amounts of these supplementary releases.

Denote by X the water in reservoir C above the level of ALCONC and by Y the water in reservoir D above the level of ALCOND. Then

$$\text{X} = \text{CONC6} - \text{ALCONC} + \text{FLOWC6(INTVL)} > 0;$$

if not, $$\text{X} = 0.$$

Furthermore, $$\text{Y} = \text{COND6} - \text{ALCOND} + \text{FLOWD6(INTVL)} > 0;$$

if not, $$\text{Y} = 0.$$

In order to draw down both reservoirs in proportion to their surplus of water, form a proportion denoted by PROPO, or PROPO = X/[X + Y], and make a supplementary release from reservoir C, namely

$$\text{RLS2C6} = \text{PROPO} \times \text{REMFLG} < \text{X};$$

if not, $$\text{RLS2C6} = \text{X}.$$

Make a supplementary release from reservoir D, namely

$$\text{RLS2D6} = [1 - \text{PROPO}] \times \text{REMFLG} < \text{Y};$$

if not, $$\text{RLS2D6} = \text{Y}.$$

Reduce the remaining free capacity of the channel at G, or

$$\text{REMFLG} = \text{REMFLG} - \text{RLS2C6} - \text{RLS2D6}.$$

(21) Compute the total releases from reservoirs C and D as the respective sums of the basic releases computed in steps (15) and (16) and the supplementary releases computed in step (20).

Denote the total releases for the interval from reservoirs C and D by RLSTC6 and RLSTD6, respectively. Then

$$\text{RLSTC6} = \text{RLS1C6} + \text{RLS2C6}$$

and $$\text{RLSTD6} = \text{RLS1D6} + \text{RLS2D6}.$$

(22) Go to step (27) or (23) if the free channel capacity at G is zero or positive, respectively.

If REMFLG is $\left.\begin{matrix}\text{zero}\\\text{positive}\end{matrix}\right\}$ go to step $\left\{\begin{matrix}(27).\\(23).\end{matrix}\right.$

(23) Go to step (25) if the volume of water in reservoir B is less than or equal to the allowable value; go to step (24) if it exceeds the allowable value.

If CONB6 − ALCONB is $\left.\begin{matrix}\text{negative or zero}\\\text{positive}\end{matrix}\right\}$ go to step $\left\{\begin{matrix}(25).\\(24).\end{matrix}\right.$

(24) Make a supplementary release from reservoir B equaling the smaller of the following two amounts: the excess water stored in reservoir B, or the additional flow that can be received at G as computed in step (20). Recompute the total releases from reservoirs B and C, the flow into reservoir C, and the contents of reservoir B, and return to step (12) to compute the additional energy produced by this supplementary release.

If REMFLG < CONB6 − ALCONB, then

$$\text{RLS3B6} = \text{REMFLG};$$

if not,

$$\text{RLS3B6} = \text{CONB6} - \text{ALCONB},$$

$$\text{RLSTB6} = \text{RLSTB6} + \text{RLS3B6},$$

$$\text{FLOWC6} = \text{FLOWC6} + \text{RLS3B6},$$

$$\text{RLSTC6} = \text{RLSTC6} + \text{RLS3B6},$$

and

$$\text{CONB6N} = \text{CONB6} - \text{RLS3B6}.$$

Return to step (12) to compute the additional energy generated by RLS3B6.

(25) Set the supplementary release from reservoir B to zero and go to step (29).

$$\text{RLS3B6} = 0.$$

Go to step (29).

(26) Compute the total releases from reservoirs C and D, which in this case are equal to the basic releases from them.

$$\text{RLSTC6} = \text{RLS1C6}$$

and

$$\text{RLSTD6} = \text{RLS1D6}.$$

(27) Compute the flow through G that just equals the maximum safe channel capacity. Go to step (30).

Denote the total flow at G for the interval by FLOWG6. Then

$$\text{FLOWG6} = \text{FLOOD}.$$

Go to step (30).

(28) Calculate the peak flood flow at G that equals the sum of the basic releases from reservoirs C and D and the natural flow through G. Compute the total releases from reservoirs C and D, which once again just equal the basic releases. Set the flow at G equal to the peak flood flow. Proceed to step (30).

Denote the peak of the flood at G during the interval by PEAKN. Then

$$\text{PEAKN} = \text{FLOFG6(INTVL)} + \text{RLS1C6} + \text{RLS1D6},$$

$$\text{RLSTC6} = \text{RLS1C6},$$

$$\text{RLSTD6} = \text{RLS1D6},$$

and

$$\text{FLOWG6} = \text{PEAKN}.$$

Go to step (30).

(29) Set the flow through G equal to the sum of all the releases from reservoirs C and D and the natural flow through G during the interval.

$$\text{FLOWG6} = \text{RLSTC6} + \text{RLSTD6} + \text{FLOWFG6(INTVL)}.$$

(30) Compute the spills at power plants B and G, the energy generated at G, and the energy deficit which equals the difference between the target output for energy and the total energy generated at power plants B and G. Ordinarily there will be no deficit during a flood month because flows will be excessive and all flows will be effective under our assumption of constant tailwater elevation. Go to step (32) or (31) if the energy deficit is zero or positive, respectively.

Denote the spills at power plants B and G during the interval by SPILB6 and SPILG6, the effective energy-generating flow through power plant G by EFPOW, the energy produced at G by ENERG6, the energy deficit by DEFEN6, and the water capacity of the turbines at power plant G by WCAPG6. Then if $\text{RLSTB6} - \text{FLCPB6} > 0$,

$$\text{SPILB6} = \text{RLSTB6} - \text{FLCPB6};$$

if not,

$$\text{SPILB6} = 0.$$

If FLOWG6 − WCAPG6 > 0, then

$$\text{SPILG6} = \text{FLOWG6} - \text{WCAPG6};$$

if not,
$$\text{SPILG6} = 0.$$

If FLOWG6 < WCAPG6, then

$$\text{EFPOW} = \text{FLOWG6};$$

if not,
$$\text{EFPOW} = \text{WCAPG6},$$

$$\text{ENERG6} = \text{EFPOW} \times \text{PRFCTG},$$

and
$$\text{DEFEN6} = \text{REQEN6} - \text{ENERB6} - \text{ENERG6}.$$

If DEFEN6 is $\left.\begin{matrix}\text{zero}\\ \text{positive}\end{matrix}\right\}$ go to step $\left\{\begin{matrix}(32).\\ (31).\end{matrix}\right.$

(31) Make supplementary releases from reservoirs *C*, *D*, and *A* to erase the energy deficit if possible, as in steps (22) through (26) of the monthly procedure. Unlike the monthly procedure, the flood-routing procedure never calls the compensating procedure to erase an energy deficit.

(32) Modify all relevant quantities, such as the contents of reservoirs, to prepare for the next interval.

$$\text{CONA6} = \text{CONA6} + \text{FLOWA6(INTVL)} - \text{RLSTA6},$$

$$\text{CONB6} = \text{CONB6} + \text{FLOWB6(INTVL)} - \text{RLSTB6},$$

$$\text{CONC6} = \text{CONC6} + \text{FLOWC6(INTVL)} - \text{RLSTC6},$$

and
$$\text{COND6} = \text{COND6} + \text{FLOWD6(INTVL)} - \text{RLSTD6}.$$

(33) Test for the end of the month. If INTVL = 120, return to the monthly operating procedure. Otherwise return to step (2) and repeat the flood-routing procedure for the next interval.

COMPENSATING PROCEDURE

Basic compensating policy. The purpose of the compensating procedure is to eliminate or reduce, by means of further releases from reservoir *B*, any deficit in firm energy that may remain after reservoirs *A*, *C*, and *D* have been emptied. Drafts from reservoir *B* can generate energy at power plants *B* and *G* as long as the full water capacity of the turbines at *B* and *G* has not been pre-empted. The exact release that will do away with the energy deficit cannot be determined directly, because the energy generated at power plant B is a function of the power head, which in turn is a function of the release. The compensating procedure determines the required release step by step, according to the following instructions.

Calculate the remaining effective flow at power plant G and make a trial release, of perhaps 10,000 acre ft, from reservoir B. Compute the additional energy generated at both power plants. If the energy deficit has not been erased to within a preset tolerance, make further trial releases of 10,000 acre ft until the deficit has been erased or a surplus has been created, or until no more water can be drawn from reservoir B, or until the full water capacity of the turbines has been reached. If a surplus of energy is generated, reduce the last trial release fractionally, step by step, until the surplus is wiped out or falls within the tolerance limits.

Elaboration and formal statement of the compensating procedure. A flow chart for the compensating procedure is depicted in Fig. 9.5.

(1) Compute for power plant G the remaining effective flow, that is, the additional flow that can be put to use in generating further energy. In magnitude this is the smaller of the following two amounts: (a) the total water capacity of the turbines at power plant G minus the flow through them that has been considered thus far, or (b) the remaining active contents of reservoir B.

Denote by FURFLG the remaining effective flow at power plant G, by WCAPG the maximum flow capacity at power plant G, by FLOWG(IMONTH) the previous monthly flows through power plant G (counting releases from reservoirs C and D as well as natural flows entering between F and G), and by CONTSB(IMONTH + 1) the contents of reservoir B after initial releases have been made to prevent spills and to provide irrigation and flood storage. Then if CONTSB(IMONTH + 1) $\geq$ WCAPG $-$ FLOWG(IMONTH),

$$\text{FURFLG} = \text{WCAPG} - \text{FLOWG(IMONTH)};$$

if not,

$$\text{FURFLG} = \text{CONTSB(IMONTH} + 1).$$

(2) Compute the maximum water capacity of the turbines in power plant B. This is a function of the initial contents, inflow, and release, and is obtained by an iterative procedure. The maximum possible release from reservoir B that can be utilized by power plant B for the month in question is calculated by the iterative process of making a trial release, computing the associated average head, and finding the resulting water capacity of the turbines. When the trial release equals the perti-

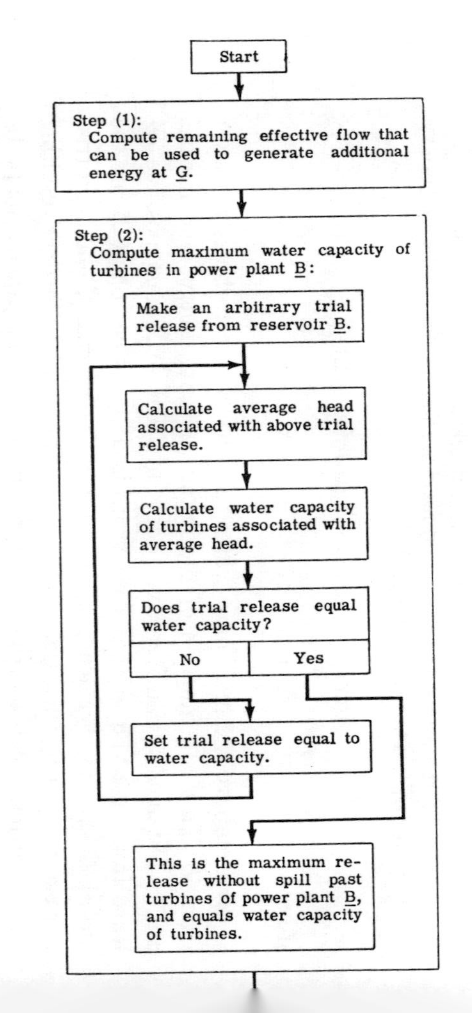
Start
Step (1):
Compute remaining effective flow that can be used to generate additional energy at G.
Step (2):
Compute maximum water capacity of turbines in power plant B:
Make an arbitrary trial release from reservoir B.
Calculate average head associated with above trial release.
Calculate water capacity of turbines associated with average head.
Does trial release equal water capacity?
No
Yes
Set trial release equal to water capacity.
This is the maximum release without spill past turbines of power plant B, and equals water capacity of turbines.

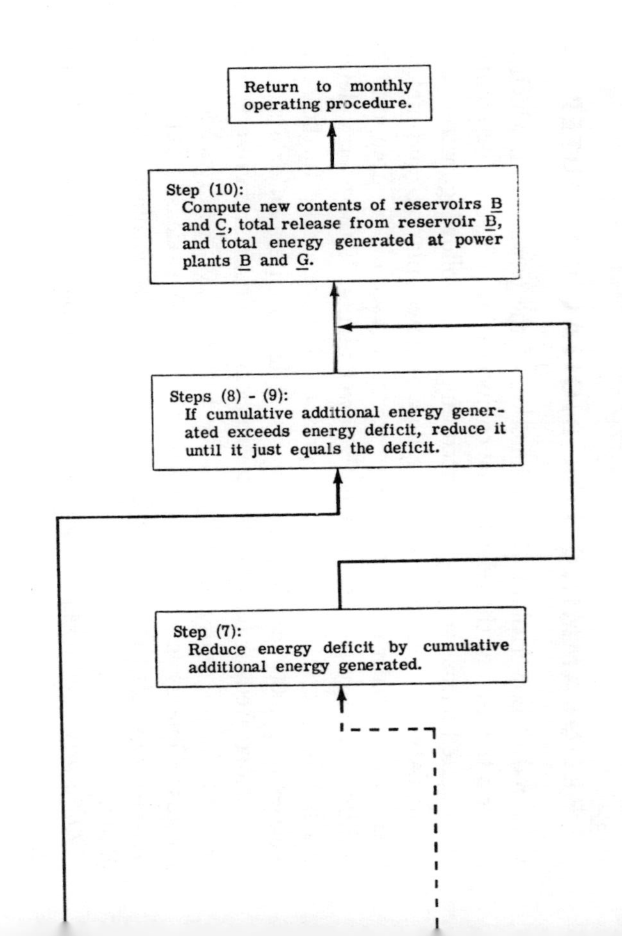
Return to monthly operating procedure.
Step (10):
Compute new contents of reservoirs B and C, total release from reservoir B, and total energy generated at power plants B and G.
Steps (8) - (9):
If cumulative additional energy generated exceeds energy deficit, reduce it until it just equals the deficit.
Step (7):
Reduce energy deficit by cumulative additional energy generated.

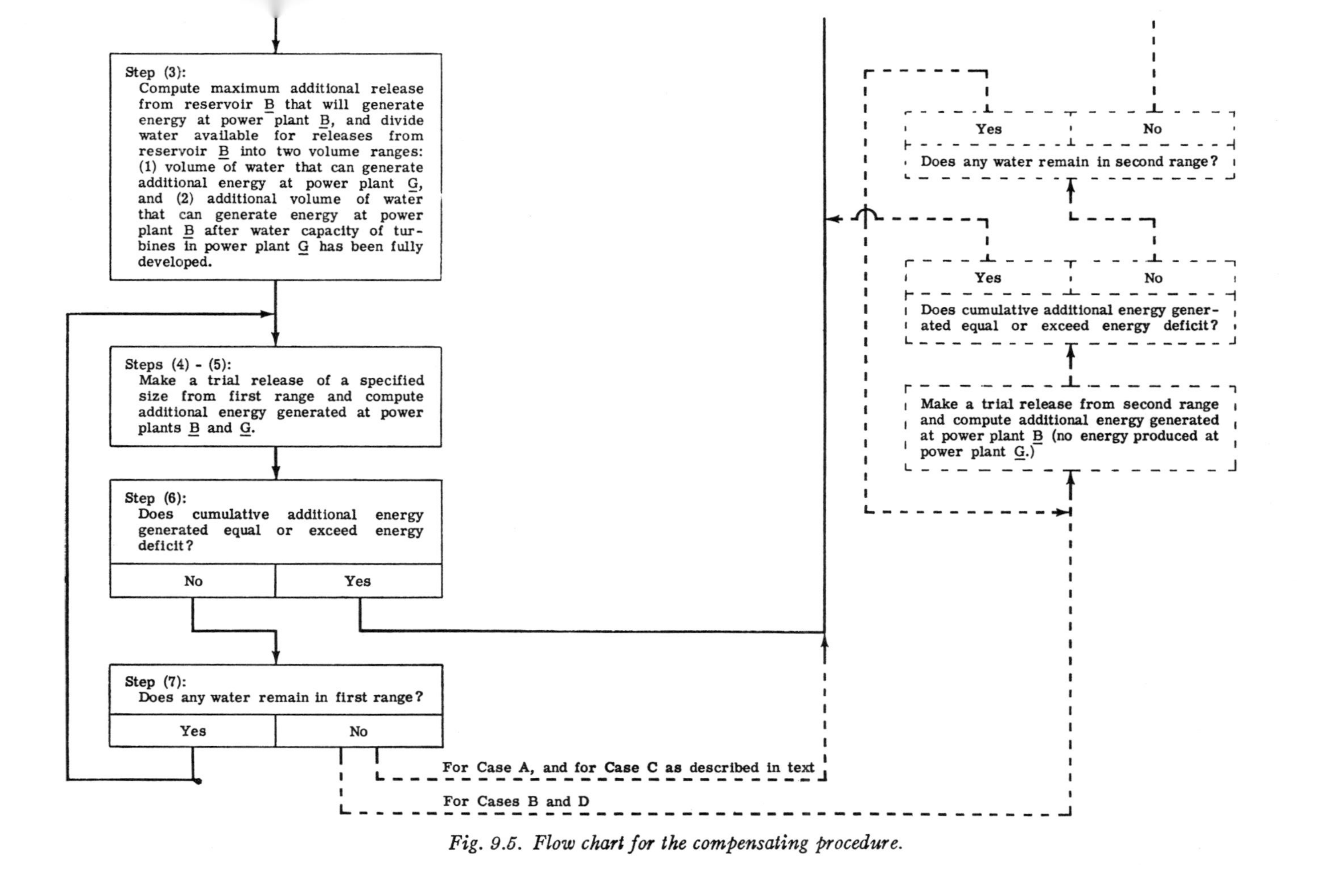

Step (3):
Compute maximum additional release from reservoir B that will generate energy at power plant B, and divide water available for releases from reservoir B into two volume ranges: (1) volume of water that can generate additional energy at power plant G, and (2) additional volume of water that can generate energy at power plant B after water capacity of turbines in power plant G has been fully developed.
Steps (4) - (5):
Make a trial release of a specified size from first range and compute additional energy generated at power plants B and G.
Step (6):
Does cumulative additional energy generated equal or exceed energy deficit?
No
Yes
Step (7):
Does any water remain in first range?
Yes
No
For Case A, and for Case C as described in text
For Cases B and D
Yes
No
Does any water remain in second range?
Yes
No
Does cumulative additional energy generated equal or exceed energy deficit?
Make a trial release from second range and compute additional energy generated at power plant B (no energy produced at power plant G.)

Fig. 9.5. Flow chart for the compensating procedure.

nent water capacity of the turbines, the maximum possible release that will generate energy at power plant B has been found.

(*a*) Let CONTSB(IMONTH), FLOWB(IMONTH), and RLSTRL be the contents of reservoir B at the beginning of the month, the natural monthly flow to reservoir B, and the trial release, respectively. Then the first trial release is arbitrarily

$$\text{RLSTRL} = \tfrac{1}{2}[\text{CONTSB(IMONTH)} + \text{FLOWB(IMONTH)}].$$

(*b*) Let the average head associated with the trial release be AVHEAD and the dead storage in reservoir B be DEADST. Then

$$\text{AVHEAD} = \text{HEADFN}[\{\text{CONTSB(IMONTH)} + \tfrac{1}{2}[\text{FLOWB(IMONTH)} - \text{RLSTRL}]\}, \text{DEADST}]$$

where HEADFN is the head–capacity function of reservoir B as given in Table 9.1.

(*c*) Denote the maximum head possible at reservoir B by HDMX. If the computed value AVHEAD > HDMX, because the capacity of reservoir B is small and the flow into reservoir B is large, then set AVHEAD = HDMX.

(*d*) Denote by PPB the rated power capacity of power plant B, and by FLOCAP the effective flow that can be used at power plant B and that is associated with the average head at power plant B. As in step (14) of the monthly procedure, if FLOCAP < CONTSB(IMONTH) + FLOWB(IMONTH), then

$$\text{FLOCAP} = \text{BETA} \times \text{AVHEAD}^{1/2};$$

if not,
$$\text{FLOCAP} = \text{CONTSB(IMONTH)} + \text{FLOWB(IMONTH)}.$$

(*e*) Let an arbitrary error tolerance be ERRTOL. If

$$\text{RLSTRL} - \text{FLOCAP} \begin{matrix} > \text{ERRTOL} \\ \leq \text{ERRTOL} \end{matrix} \Bigg\} \text{ go to step } \Bigg\{ \begin{matrix} (2f). \\ (2g). \end{matrix}$$

(*f*) Begin another iteration to determine RBTLMX, the maximum possible release from reservoir B that can be utilized by power plant B for the month in question, by setting the new value of the trial release equal to the computed value of the flow capacity associated with the last trial release, or RLSTRL = FLOCAP. Return to (2*b*) and continue iterating.

(*g*) Since the trial release differs from its associated flow capacity at most by a specified tolerance, the maximum release from reservoir B without spill is equal to this flow capacity, or RBTLMX = FLOCAP.

(3) By means of (a) the maximum additional release from reservoir B that will generate energy at G, as computed in step (1); (b) the maximum total release from reservoir B that will generate energy at power plant B, as computed in step (2); and (c) the release from reservoir B that has been made during the month, as computed in step (11) of the monthly operating procedure; determine whether or not power plant B has excess

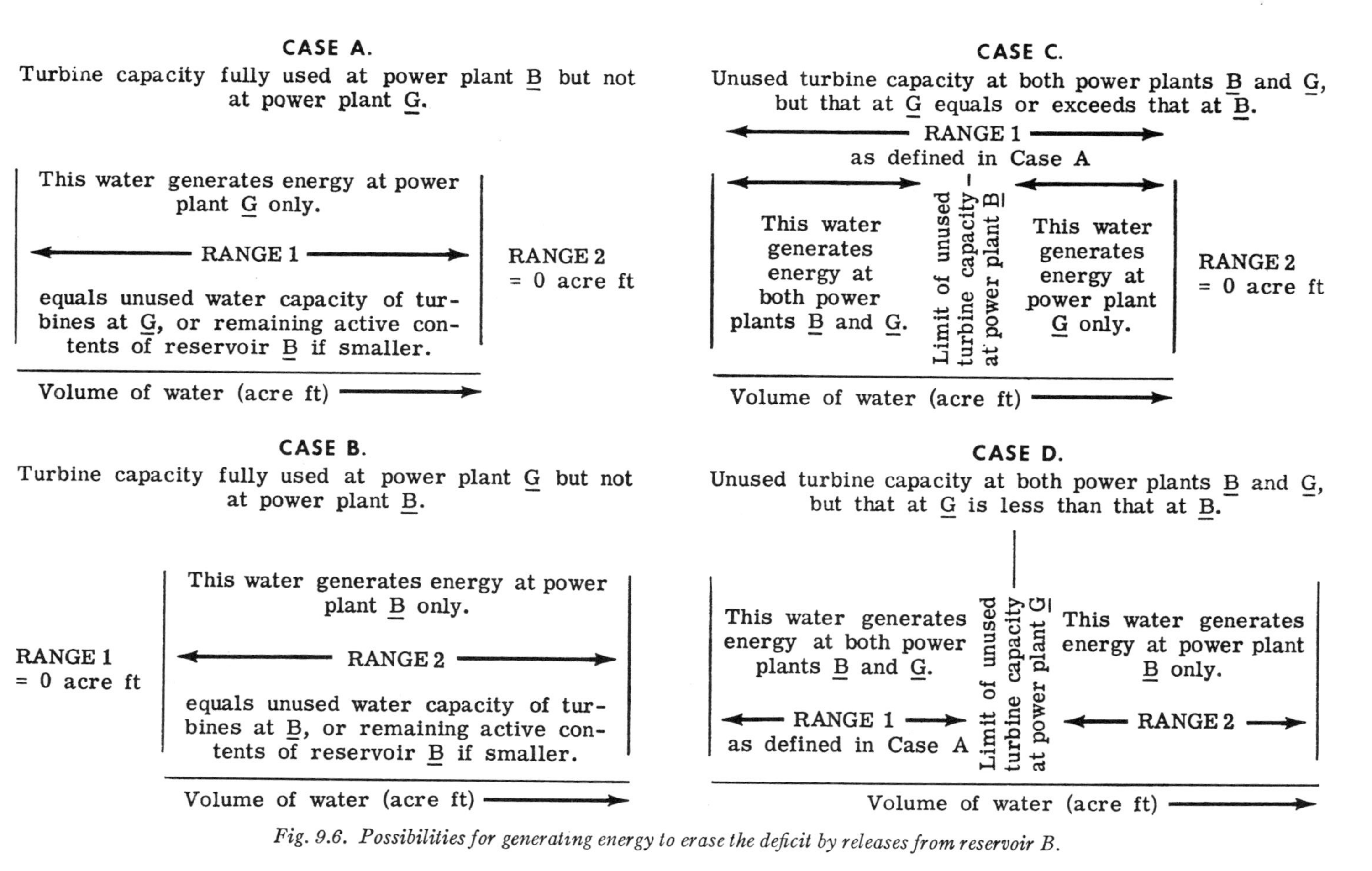

Fig. 9.6. Possibilities for generating energy to erase the deficit by releases from reservoir B.

capacity. Compute the maximum additional release from reservoir *B* that will generate energy at this power plant.

In making releases from reservoir *B* to erase the energy deficit, power plants *B* and *G* can be exploited in one of the four ways illustrated in Fig. 9.6:

Case A. Turbine capacity fully used at power plant *B*, but not at power plant *G*.

Case B. Turbine capacity fully used at power plant *G*, but not at power plant *B*.

Case C. Unused turbine capacity at both power plants *B* and *G*, but that at *G* equals or exceeds that at *B*.

Case D. Unused turbine capacity at both power plants *B* and *G*, but that at *G* is less than that at *B*.

Divide the active storage in reservoir *B* into two volume ranges: RANGE1 equal to the maximum additional release from *B* that will generate energy at power plant *G*, and RANGE2 equal to the difference between the maximum additional releases that will generate energy at both power plants *B* and *G*. The difference (that is, RANGE2) will be zero when the power plant at *G* is left with more unused capacity than that at *B*.

Denote the two ranges in volume by RANGE1 and RANGE2, and the total monthly release that has been made at reservoir *B* by RLSTB(IMONTH). Then

$$\text{RANGE1} = \text{FURFLG}$$

and

$$\text{RANGE2} = \text{RBTLMX} - \text{RLSTB(IMONTH)} - \text{FURFLG}.$$

(4) To generate additional energy at power plants *B* and *G*, make successive trial releases of approximately 10,000 acre ft each, first in RANGE1 insofar as possible and next in RANGE2 if possible. Calculate in a similar way for each case, as illustrated here for case C only, the number and exact size of the trial releases in RANGE1. Set their cumulative value and the additional energy generated by them to zero.

Denote by WDTHMX, which equals 10,000 acre ft, the maximum size of the trial releases, by WIDTH1 the actual size of the trial releases in RANGE1, by NOTRIL their number. Denote further by RLSBAH the cumulative sum of the successive trial releases equal to WIDTH1 that have been made, and by ADNLNH the additional energy generated at both power plants from RLSBAH. Then

$$\text{NOTRIL} = \text{FURFLG/WDTHMX} + 1$$

and

$$\text{WIDTH1} = \text{FURFLG/NOTRIL}.$$

At the beginning of the procedure, furthermore,

$$\text{RLSBAH} = 0$$

and

$$\text{ADNLNH} = 0.$$

(5) Knowing the limits of the excess turbine capacity in power plants *B* and *G*, make a trial release of 10,000 acre ft and calculate the additional energy generated at these plants.

(*a*) Make a RANGE1 release, or

$$\text{RLSBAH} = \text{RLSBAH} + \text{WIDTH1}.$$

(*b*) Denote the average contents of reservoir *B* before the compensating procedure was called by AVCONB. Then

$$\text{AVCONB} = \text{CONTSB(IMONTH)} + \tfrac{1}{2}[\text{FLOWB(IMONTH)} - \text{RLSTB(IMONTH)}]$$

and

$$\text{AVHEAD} = \text{HEADFN}([\text{AVCONB} - \tfrac{1}{2}\text{RLSBAH}], \text{DEADST})$$

where [AVCONB − ½RLSBAH] is a measure of the average contents of reservoir *B* after the cumulative release RLSBAH has been made.

(*c*) If RLSTB(IMONTH) + RLSBAH − RBTLMX is $\left.\begin{matrix}\text{negative or zero}\\ \text{positive}\end{matrix}\right\}$, go to step $\left\{\begin{matrix}(5d).\\ (5e).\end{matrix}\right.$

(*d*) Denote the total energy generated at *B* during the month by TOTNB and the units and efficiency conversion factor by PREFFB. Then

$$\text{TOTNB} = \text{PREFFB} \times \text{AVHEAD} \times [\text{RLSTB(IMONTH)} + \text{RLSBAH}].$$

Go to step (5*f*).

(*e*) Since RBTLMX = FLOCAP,

$$\text{TOTNB} = \text{PREFFB} \times \text{AVHEAD} \times \text{FLOCAP}.$$

(*f*) Denote by ADENG the additional energy generated by power plant *G* from the release RLSBAH, and by PRFCTG the factor that converts water flow to energy. Then

$$\text{ADENG} = \text{PRFCTG} \times \text{RLSBAH}.$$

(*g*) Let ENERB(IMONTH) be the energy generated at *B* during the month before the compensating procedure was called. Then

$$\text{ADNLNH} = \text{TOTNB} - \text{ENERB(IMONTH)} + \text{ADENG},$$

where ADNLNH is the total additional energy generated from RLSBAH.

(6) Compare the energy deficit with the additional energy generated. If the difference between the additional energy generated and the energy

deficit is negative, zero, or positive, go to step (7), (9), or (8), respectively.

(*a*) Let DEFEN(IMONTH) be the monthly energy deficit recorded before the compensating procedure was called.

$$\text{If ADNLNH} - \text{DEFEN(IMONTH) is } \left.\begin{matrix}\text{negative}\\ \text{zero}\\ \text{positive}\end{matrix}\right\} \text{go to step} \left\{\begin{matrix}(7).\\ (6b).\\ (8).\end{matrix}\right.$$

(*b*) Denote by RLSBAP the release from reservoir *B* that will exactly meet the deficit. Then

$$\text{RLSBAP} = \text{RLSBAH}$$

and go to step (9).

(7) Compare the number of trial releases that have been made with the number that are possible, as computed in step (4). If additional releases can be made, return to step (5). If all possible trial releases have been made, decrease the energy deficit and go to step (10).

If the number of RANGE1 releases is less than NOTRIL, make another release and return to step (5). If not, denote the cumulative release from reservoir *B* by RLSBAP. Then

$$\text{RLSBAP} = \text{RLSBAH}$$

and
$$\text{DEFEN(IMONTH)} = \text{DEFEN(IMONTH)} - \text{ADNLNH}.$$

Go to step (10).

(8) Compute by iteration the fractional part of the last trial release that will meet the deficit. Recompute the additional energy generated.

Since more energy has been generated than is necessary, divide into fractions the last release equal to WIDTH1 and make these fractional releases until the energy generated comes close to erasing the deficit.

$$\text{RLSBAP} = \text{RLSBAH} - \text{WIDTH1} + \text{some fraction of WIDTH1}$$

and
$$\text{AVHEAD} = \text{HEADFN}([\text{AVCONB} + \tfrac{1}{2}\text{RLSBAP}], \text{DEADST}).$$

The revised total energy at power plant *B* is then

$$\text{TOTNB} = \text{PREFFB} \times \text{AVHEAD} \times \text{RLSBAP};$$

the revised additional energy at power plant *G* is

$$\text{ADENG} = \text{PRFCTG} \times \text{RLSBAP}$$

and
$$\text{ADNLNH} = \text{TOTNB} - \text{ENERB(IMONTH)} + \text{ADENG}.$$

(9) Set the monthly energy deficit to zero.

$$\text{DEFEN(IMONTH)} = 0.$$

(10) Calculate the new or next-month contents of reservoirs B and C and the total energy generated at power plants B and G for the month. Then modify the total release from reservoir B and return to the monthly procedure.

$$\text{CONTSB(IMONTH)} = \text{CONTSB(IMONTH)} - \text{RLSBAP}$$

and

$$\text{RLSTB(IMONTH)} = \text{RLSTB(IMONTH)} + \text{RLSBAP}.$$

Denote the initial contents of reservoir C for the following month by CONTSC(IMONTH). Since the compensating procedure has been called, reservoir C presumably has been drained. Consequently, if RLSBAP $\leq$ FURFLG, then

$$\text{CONTSC(IMONTH)} = 0;$$

if not,

$$\text{CONTSC(IMONTH)} = \text{RLSBAP} - \text{FURFLG},$$

so that at reservoir C is stored the amount that would otherwise be spilled at G.

The total energy generated at power plant G for the month ENERG(IMONTH) is then

$$\text{ENERG(IMONTH)} = \text{ENERG(IMONTH)} + \text{ADENG}$$

and

$$\text{ENERB(IMONTH)} = \text{TOTNB}.$$

Return to the monthly operating procedure.

BENEFIT-EVALUATION PROCEDURE

Basic benefit-evaluation policy. The present value of net benefits extending over the economic life of the system at a given rate of interest is calculated as the sum, over the economic life of the system, of the total gross benefits from irrigation, energy, and flood control for each year discounted to the present, minus (1) the sum, over the same period, of the discounted total annual operation, maintenance, and replacement (OMR) costs, and (2) the total capital costs, which are not discounted. Table 9.2 indicates the functions and constants used in making these computations for our simplified river-basin system.

For each year, the total gross irrigation benefits equal the nominal benefits minus the loss in benefits resulting from failures to meet the target output for irrigation. The nominal benefits for each year are obtained by applying a unit irrigation-benefit function, such as that shown in item 14 of Table 9.2, to the target output for irrigation. The loss in benefits is that associated with an irrigation deficit. It is obtained by applying to this deficit an irrigation-deficit loss function such as that listed in item 15 of Table 9.2.

Table 9.2. Functions and constants used in the benefit-evaluation procedure.

Function	Independent variable (x)	Dependent variable (y)	Element of system	*Functions* Relevant figure in Chapter 7	Relationship assumed for simplified river-basin system[a]
1	Streamflow (10^2 acre ft/6 hr)	Flood damage (10^5 dollars)	Point G	Fig. 7.9	Four segments: if $x < 655$, $y = 5.36 \times 10^{-6}x^3 - 0.02405x^2 + 24.55x - 6883.89$; if $655 \leq x < 760$, $y = 383.2$; if $760 \leq x < 1065$, $y = -5.75 \times 10^{-33}x^9 - 4.56 \times 10^{-29}x^8 + 1.87 \times 10^{-23}x^7 + 1.11 \times 10^{-19}x^6 + 3.38 \times 10^{-17}x^5 + 4.97 \times 10^{-14}x^4 - 2.37 \times 10^{-5}x^3 + 0.0598x^2 - 46.8x + 11801.5$; if $x \geq 1065$, $y = 1176.7$
2	Reservoir capacity (10^2 acre ft)	Capital cost (10^3 dollars)	Reservoir A	Fig. 7.10	$y = -8.488 \times 10^{-13}x^4 + 3.069 \times 10^{-8}x^3 - 4.155 \times 10^{-4}x^2 + 3.00x + 594.1$
3	Reservoir capacity (10^6 acre ft)	Capital cost (10^6 dollars)	Reservoir B	Fig. 7.10	$y = -14.459 \times 10^{-4}x^5 + 0.01269x^4 + 0.0343x^3 - 3.299x^2 + 35.995x + 0.078$
4	Reservoir capacity (10^6 acre ft)	Capital cost (10^6 dollars)	Reservoir C	Fig. 7.10	$y = -16.426 \times 10^{-4}x^5 - 0.0281x^4 + 0.5651x^3 - 6.2099x^2 + 39.24x - 1.184$
5	Reservoir capacity (10^6 acre ft)	Capital cost (10^6 dollars)	Reservoir D	Fig. 7.10	$y = -1.511 \times 10^{-7}x^7 + 5.199 \times 10^{-4}x^6 - 0.006499x^5 + 0.02665x^4 + 0.4050x^3 - 7.240x^2 + 46.70x - 0.763$
6	Reservoir capacity (10^2 acre ft)	OMR cost (10^3 dollars)	Reservoir A	Fig. 7.11	$y = 1.431 \times 10^{-11}x^3 - 4.128 \times 10^{-7}x^2 + 4.418 \times 10^{-3}x + 1.554$
7	Reservoir capacity (10^2 acre ft)	OMR cost (10^3 dollars)	Reservoir B, C, or D	Fig. 7.11	$y = 1.222 \times 10^{-14}x^3 - 4.109 \times 10^{-9}x^2 + 9.027 \times 10^{-4}x + 7.375$
8	Power-plant capacity (Mw)	Capital cost (10^3 dollars)	Power plant B	Fig. 7.6	Two segments: if $x < 600$, $y = 5.021 \times 10^{-7}x^4 + 8.486 \times 10^{-4}x^3 - 0.523x^2 + 244.1x + 106.34$; if $x \geq 600$, $y = 130x + 9500$
9	Power-plant capacity (Mw)	Capital cost (10^6 dollars)	Power plant G	Fig. 7.6	$y = -3.1476 \times 10^{-10}x^4 + 6.068 \times 10^{-7}x^3 - 4.187 \times 10^{-4}x^2 + 0.2333x + 1.55$
10	Power-plant capacity (Mw)	OMR cost (10^3 dollars)	Power plant B	Fig. 7.7	$y = 2.00 \times 10^{-9}x^4 + 3.812 \times 10^{-6}x^3 - 2.56 \times 10^{-3}x^2 + 1.972x + 6.67$

11	Power-plant capacity (Mw)	OMR cost (10^3 dollars)	Power plant *G*	Fig. 7.7	$y = -8.338 \times 10^{-9}x^4 + 7.886 \times 10^{-6}x^3 - 4.12 \times 10^{-3}x^2 + 2.73x - 5.70$
12	Annual target output for irrigation (10^2 acre ft)	Capital cost (10^3 dollars)	Irrigation-diversion, distribution, and pumping works	Fig. 7.4	Two segments: if $x < 42{,}000$, $y = 3.2x - 5 \times 10^{-6}(x - 17{,}500)^2 + 8300$; if $x \geq 42{,}000$, $y = 4.8x - 57{,}750$
13	Annual target output for irrigation (10^2 acre ft)	OMR cost (10^3 dollars)	Irrigation-diversion, distribution, and pumping works	Fig. 7.5	Two segments: if $x \leq 42{,}000$, $y = 0.25x$; if $x > 42{,}000$, $y = 0.15x + 4200$
14	Annual target output for irrigation (10^5 acre ft)	Unit irrigation benefits [($1/($10^2$ acre ft)]	Target output for irrigation	Fig. 7.2	Two segments: if $x \leq 55$, $y = 0.05278x^2 - 6.412x + 644.924$; if $x > 55$, $y = 450$
15	Annual irrigation deficit (10^2 acre ft)	Loss of irrigation benefits (10^3 dollars)	Irrigation deficit	Fig. 7.3	Three segments: if $x \leq 0.1$ ANIREQ,[b] $y = 0.7x$; if 0.1 ANIREQ $< x \leq 0.7$ ANIREQ, $y = [(2\ \text{BENNIR}^{b} - 0.07)/0.6](x - 0.1\ \text{ANIREQ}) + 0.07$ ANIREQ; if $x > 0.7$ ANIREQ, $y = 0.2(x - 0.7\ \text{ANIREQ}) + 0.07\ \text{ANIREQ}\ (2\ \text{BENNIR} - 0.07)$ ANIREQ

Constants

Unit firm-energy benefits: 7.0 mills/kw hr
Unit dump-energy benefits: 1.5 mills/kw hr
Unit loss for energy deficits: 2.0 mills/kw hr

[a] All functions except 12, 13, and 15 are polynomials fitted to the curves shown in Chapter 7. The polynomials were computed by means of a least-squares curve-fitting routine (SCUF1) used on the IBM 704 computer. The expressions for the three remaining functions were computed manually.

[b] ANIREQ is annual target output for irrigation in 10^2 acre ft; BENNIR is unit irrigation benefits in 10^3 dollars associated with ANIREQ.

For each year, similarly, the total gross energy benefits equal the nominal firm-energy benefits plus the dump-energy benefits, minus the loss in benefits resulting from failures to meet the target output for energy. The nominal firm-energy benefits are obtained by applying a fixed unit-benefit factor to the target output for energy. The loss in benefits is that associated with energy deficits. Dump-energy benefits and losses from energy deficits are calculated by applying a fixed unit benefit or penalty to the quantity in each case. The energy benefit and penalty constants are listed in Table 9.2.

Gross flood-control benefits for each year are computed by determining from a flood-damage function, such as that shown in item 1 of Table 9.2 for our river-basin system, the flood damages prevented by the reduction in peak flood flow.

Total capital costs and total annual OMR costs are calculated also from the functions listed in items 2 to 13 of Table 9.2.

Elaboration and formal statement of benefit-evaluation procedure. Figure 9.7 presents a flow chart for the benefit-evaluation procedure.

(1) Go to step (2) or (3) if the annual target output for irrigation is zero or positive, respectively.

Denote the annual target output for irrigation by ANIREQ. If ANIREQ is

$$\left.\begin{array}{l}\text{zero}\\\text{positive}\end{array}\right\}\text{ go to step }\left\{\begin{array}{l}(2).\\(3).\end{array}\right.$$

(2) Set the gross annual irrigation benefits to zero and go to step (4).

Denote the gross annual irrigation benefits by CASHIR. Set CASHIR = 0 and go to step (4).

(3) Compute the gross annual irrigation benefits without deficits, using for this purpose the quadratic function, shown in item 14 of Table 9.2 for our river-basin system, relating unit irrigation benefits to the annual target output for irrigation.

Denote the unit irrigation-benefit function by BENFN and the unit irrigation benefits by BENNIR. Then

$$\text{BENNIR} = \text{BENFN}(\text{ANIREQ})$$

and

$$\text{CASHIR} = \text{BENNIR} \times \text{ANIREQ}.$$

(4) Compute the gross annual firm-energy benefits without deficits by applying to the annual target output for energy the constant unit value for firm-energy benefits given in Table 9.2.

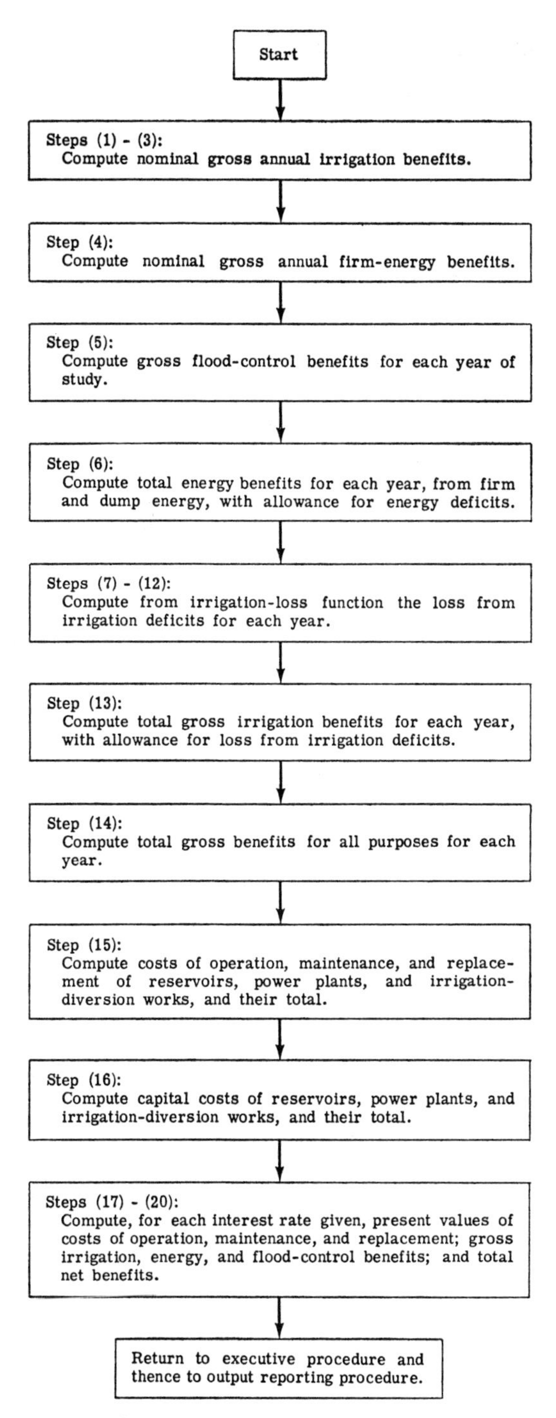

Fig. 9.7. Flow chart for the benefit-evaluation procedure.

Denote the annual target output for firm energy by ANENRQ, the gross annual firm-energy benefits by CASHEN, and the constant unit firm-energy benefits by FEBEN. Then

$$\text{CASHEN} = \text{FEBEN} \times \text{ANENRQ}.$$

(5) For each year of the study compute (a) the flood damages that would have been experienced under natural or uncontrolled conditions by introducing into the flood-damage function given in Table 9.2 the peak natural flood flows given as input data, and (b) the flood damages that would remain, in the presence of flood-control works, after the flood-routing procedure has been completed. Then calculate the flood-control benefits, or flood damages prevented by the assumed system, as the difference between the two kinds of damages for each year.

For each year of the study denote the natural peak flood flows in the input data by PEKFLG(IYEAR), the peak flows in the presence of the flood-routing procedure by PEAKO(IYEAR), the flood damage caused by the natural peak flood flows for the year by DAMAGE, and the gross flood-control benefits for the year by FLOBEN(IYEAR). Then

$$\text{DAMAGE} = \text{FLDAMF}[\text{PEKFLG}(\text{IYEAR})],$$

where FLDAMF is the flood-damage function

and $$\text{FLOBEN}(\text{IYEAR}) = \text{DAMAGE} - \text{FLDAMF}[\text{PEAKO}(\text{IYEAR})].$$

(6) For each year compute the total energy benefits by (a) adding to the firm-energy benefits calculated in step (4) the dump-energy benefits, obtained by multiplying the unit value of dump energy by the total dump energy generated, and (b) subtracting the losses from energy deficits, computed by multiplying the unit energy-deficit loss by the sum of the energy deficits for the year. These constants are shown in Table 9.2.

Denote by BENEGY(IYEAR) the total energy benefits for the year, with shortages and surpluses counted, by DPRICE the unit dump-energy benefits, by DUMPEN(IYEAR) the dump energy generated for the year, by CMANED(IYEAR) the cumulative annual energy deficit, and by ENLOSS the unit loss from energy benefits. Then

$$\text{BENEGY}(\text{IYEAR}) = \text{DPRICE} \times \text{DUMPEN} - \text{ENLOSS} \times \text{CMANED}(\text{IYEAR}) + \text{CASHEN}!$$

(7) Beginning with this step and continuing through step (12), compute the loss created by the irrigation deficit. To do so, introduce the total irrigation deficit for the year into the irrigation loss function, which, as shown in item 15 of Table 9.2 for our river-basin system, is composed

of three linear segments. Deficits up to 10 percent of the annual target output for irrigation are assessed a moderate penalty, those between 10 and 70 percent a severe penalty, and those over 70 percent a nominal additional penalty.

Denote the three classes of irrigation deficits by CLASS(1), CLASS(2), and CLASS(3). Then

$$\text{CLASS}(1) = 0.1 \times \text{ANIREQ},$$

$$\text{CLASS}(2) = 0.7 \times \text{ANIREQ},$$

and

$$\text{CLASS}(3) = \text{ANIREQ}.$$

(8) Compute the unit losses of irrigation benefits for each of the three classes of deficits.

Denote the unit loss of irrigation benefits for CLASS(1) deficits by SLOPIL(1), for CLASS(2) deficits by SLOPIL(2), and for CLASS(3) deficits by SLOPIL(3). Then

$$\text{SLOPIL}(1) = 0.7 \times 10^3 \text{ dollars}/10^2 \text{ acre ft},$$

$$\text{SLOPIL}(2) = [(2\ \text{BENNIR} - 0.07)/0.6] \times 10^3 \text{ dollars}/10^2 \text{ acre ft},$$

and

$$\text{SLOPIL}(3) = 0.2 \times 10^3 \text{ dollars}/10^2 \text{ acre ft}.$$

(9) Define and calculate six intermediate quantities that simplify the computation of irrigation losses.

Denote these quantities by BASIL(1), BASIL(2), BASIL(3), SUBTRA(1), SUBTRA(2), and SUBTRA(3). Then

$$\text{BASIL}(1) = 0,$$

$$\text{BASIL}(2) = 0.7 \times \text{CLASS}(1),$$

$$\text{BASIL}(3) = \text{BASIL}(2) + \text{SLOPIL}(2) \times [\text{CLASS}(2) - \text{CLASS}(1)],$$

$$\text{SUBTRA}(1) = 0,$$

$$\text{SUBTRA}(2) = \text{CLASS}(1),$$

and

$$\text{SUBTRA}(3) = \text{CLASS}(2).$$

(10) Repeat this and subsequent steps through step (14) for each year of the study. Go to step (13) or (11) if the annual irrigation deficit is zero or positive, respectively.

Denote the cumulative annual irrigation deficit by CMANID(IYEAR). If CMANID(IYEAR) is $\left.\begin{matrix}\text{zero}\\ \text{positive}\end{matrix}\right\}$ go to step $\left\{\begin{matrix}(13).\\ (11).\end{matrix}\right.$

(11) Determine the class of the irrigation deficit, for use in step (12).

(*a*) Set the subscript I = 1.

(*b*) If CMANID(IYEAR) − CLASS(1) is $\left.\begin{array}{l}\text{negative or zero}\\ \text{positive}\end{array}\right\}$ go to step $\left\{\begin{array}{l}(12).\\ (11c).\end{array}\right.$

(*c*) Set I = I + 1 and return to step (11*b*).

(12) Compute the loss in irrigation benefits for the year.

Denote the loss in irrigation benefits for the year by DEBITI(IYEAR). Then

$$\text{DEBITI(IYEAR)} = \text{SLOPIL}(1) \times [\text{CMANID(IYEAR)} - \text{SUBTRA}(1)] + \text{BASIL}(1).$$

(13) Compute the total gross irrigation benefits for each year by subtracting from the gross annual irrigation benefits computed in step (3) the loss caused by the irrigation deficit, and set it to zero.

Denote the irrigation benefits for the year, after the deduction of losses, by BENIRR(IYEAR). Then

$$\text{BENIRR(IYEAR)} = \text{CASHIR} - \text{DEBITI(IYEAR)}$$

and

$$\text{DEBITI(IYEAR)} = 0.$$

(14) Compute the total gross benefits by summing the gross flood-control, energy, and irrigation benefits for the year.

Denote the total gross benefits for the year by ANNBEN(IYEAR). Then

$$\text{ANNBEN(IYEAR)} = \text{FLOBEN(IYEAR)} + \text{BENIRR(IYEAR)} + \text{BENEGY(IYEAR)}.$$

(15) Compute from the cost functions the OMR costs of the reservoirs, power plants, and irrigation-diversion works, and find the resultant total for the system. The cost functions for our simplified river-basin system are listed in items 6, 7, 10, 11, and 13 in Table 9.2.

Denote OMR costs for reservoirs *A*, *B*, *C*, and *D* by OMRA, OMRB, OMRC, and OMRD, respectively. Furthermore, let OMRPPB, OMRPPG, and OMRIRW be the OMR costs for power plants *B* and *G* and the irrigation-diversion works, respectively. Finally, denote their total cost by TOTOMR. Then

$$\text{TOTOMR} = \text{OMRA} + \text{OMRB} + \text{OMRC} + \text{OMRD} + \text{OMRPPB} + \text{OMRPPG} + \text{OMRIRW}.$$

(16) Compute from the cost functions in items 2 to 5, 8, 9, and 12 of Table 9.2 the capital costs of reservoirs, power plants, and irrigation-diversion works, and find the resultant total cost of the system.

Denote capital costs for reservoirs *A*, *B*, *C*, and *D* by COSTA, COSTB, COSTC, and COSTD, respectively. Furthermore, let COSPPB, COSPPG, and DVIRWK be the respective capital costs of the power plants at B and G and the irrigation, diversion works. Finally, denote the total capital costs by TCCOST. Then

$$\text{COSTA} = \text{CSTFNA(CAPA)}$$

where CSTFNA denotes the cost function for reservoir A and similarly for the remaining units of the system. Also,

$$\text{TCCOST} = \text{COSTA} + \text{COSTB} + \text{COSTC} + \text{COSTD} + \text{COSPPB} + \text{COSPPG} + \text{DVIRWK}.$$

(17) Set the cumulative present values of gross irrigation, energy, and flood-control benefits and total gross benefits serially at each of ten specified interest rates (provided as input data) to zero. This repetitive procedure is followed also in steps (18), (19), and (20).

Denote the cumulative present value of irrigation benefits for a given interest rate by PVAIRR(NRATE) where NRATE varies from 1 to 10, the cumulative present value of energy benefits for that interest rate by PVAENY(NRATE), the cumulative present value of flood-control benefits by PVAFLO(NRATE), and the cumulative present value of total gross benefits by PVABEN(NRATE). Then

$$\text{PVAIRR(NRATE)} = 0,$$

$$\text{PVAENY(NRATE)} = 0,$$

$$\text{PVAFLO(NRATE)} = 0,$$

and

$$\text{PVABEN(NRATE)} = 0.$$

(18) For each year of the study compute the cumulative present values of gross irrigation, energy, flood-control, and total benefits at the specified interest rates, by applying the discount operator to these benefits for each year and adding the result to the cumulated totals for the previous year. The discount operator is $[1 + i]^{-t}$ where i is the interest rate and t is the number of the year.

Denote the interest rate in percent by RATE and apply the discount operator. Then

$$\text{PVAIRR(NRATE)} = \text{PVAIRR(NRATE)} + \text{BENIRR(IYEAR)} \times [1 + 0.01\ \text{RATE}]^{-\text{IYEAR}},$$

$$\text{PVAENY(NRATE)} = \text{PVAENY(NRATE)} + \text{BENEGY(IYEAR)} \times [1 + 0.01\ \text{RATE}]^{-\text{IYEAR}},$$

$$\text{PVAFLO(NRATE)} = \text{PVAFLO(NRATE)} + \text{FLOBEN(IYEAR)} \times [1 + 0.01\ \text{RATE}]^{-\text{IYEAR}},$$

and

$$\text{PVABEN(NRATE)} = \text{PVABEN(NRATE)} + \text{ANNBEN(IYEAR)} \times [1 + 0.01\ \text{RATE}]^{-\text{IYEAR}}.$$

(19) Compute the present value of the stream of annual OMR costs, which are assumed to be constant for each year, by applying to these costs, as computed in step (15), the following discount operator: $[1 - (1 + i)^{-T}]/i$, where i is the interest rate and T is the economic life of the system (set at 50 years).

Denote the present value of the stream of annual OMR costs for a given interest rate by PVAOMR(NRATE) and the number of years in the study by IYRMX. Normally, IYRMX is set at 50. Then if RATE $\neq$ 0,

$$\text{PVAOMR(NRATE)} = \{[1 - (1 + 0.01\ \text{RATE})^{-\text{IYRMX}}]/0.01\ \text{RATE}\} \times \text{TOTOMR};$$

if RATE = 0,

$$\text{PVAOMR(NRATE)} = \text{IYRMX} \times \text{TOTOMR}.$$

(20) Compute the present value of net benefits by subtracting from the present value of total irrigation, energy, and flood-control benefits, as computed in step (18), (a) the present value of the annual OMR costs, as computed in step (19), and (b) the total capital costs of reservoirs, power plants, and irrigation-diversion works, as computed in step (16). When steps (17) through (20) have been performed for all ten interest rates, return to the executive or over-all management procedure.

Denote the present value of net benefits for a particular interest rate by BENNET(NRATE). Then

$$\text{BENNET(NRATE)} = \text{PVABEN(NRATE)} - \text{PVAOMR(NRATE)} - \text{TCCOST}.$$

Return to the executive procedure after all interest rates have been considered.

OUTPUT-REPORTING PROCEDURE

The output-reporting procedure assembles in concise and readable form for printing as computer output certain physical and economic data that are either provided as input or generated by the simulation. The detailed steps of this procedure follow well-established technical programming conventions and accordingly are not included in this book. From this procedure we obtain a print-out by the computer of information such as that shown for illustrative purposes only in Table 9.3.

CODING THE SIMULATION PROGRAM

Before the procedures elaborated above were fed to the computer for compiling, certain modifications in form were introduced, and an execu-

Table 9.3. Information printed by the computer as output for each simulation study of the simplified river-basin system.

Capacities and costs of structures

	Capacity (10^6 acre ft)	Capital cost (10^6 dollars)
Reservoir		
A	0.0	0.0
B, total storage	2.8	76.5
B, dead storage	2.0	
C	0.0	0.0
D	1.5	54.5
Total, active storage	2.3	
Total, active and dead storage	4.3	131.0
Power plant	(Mw)	
B	200	34.0
G	170	31.8
Total, power plants	370	65.8
Irrigation-diversion and distribution works		56.3
Total capital costs		253.1
Total annual OMR costs		1.5

Output specifications

Target output for irrigation (10^6 acre ft)	1.5
Target output for energy (10^9 kw hr)	1.0
Flood-control storage in reservoir (10^6 acre ft)	
A	0.0
B	0.5
C	0.0
D	0.0
Total	0.5

Analysis of reservoir contents

	Maximum		Minimum			No. of times reservoir was —	
Reservoir	Contents (10^6 acre ft)	Date of first occurrence	Contents (10^6 acre ft)	Date of first occurrence	Final contents (10^6 acre ft)	full	empty
A	0.0		0.0		0.0		
B	0.8	Jan. yr 1	0.0	Oct. yr 17	0.8	178	8
C	0.0		0.0		0.0		
D	1.5	Feb. yr 1	0.0	Feb. yr 2	1.5	189	50

Greatest spill (10^6 acre ft) and date of occurrence	
Reservoir *B*	1.703 (May yr 21)
Point *E*	3.830 (May yr 33)
Power plant *G*	1.884 (May yr 23)

Table 9.3. (cont.)

Physical data on delivered output

Irrigation deficits (10^6 acre ft)	
Total deficit	0.494
Maximum in one season	0.312 (yr 29)
Number of deficits in:	
Class 1	5
2	2
3	0
Energy (10^9 kw hr)	
Total energy output	82.22
Total dump-energy output	32.35
Total energy deficits	0.13
Maximum deficit in one month	0.05 (Mar. yr 48)
Number of deficits in	
Class 1	1
2	2
3	3
Flood control	
Total number of damaging floods	5
Greatest 6-hr flow at G (10^3 acre ft) which occurred at 18 hours, Dec. 23, yr 40	61.6

Benefit data

	Present value (10^6 dollars)					
		Gross benefits for —				
Interest rate	OMR costs	Irrigation	Energy	Flood control	Gross benefits less OMR costs	Net benefits
0	75.4	415.3	397.3	367.5	1104.7	851.6
2.5	42.8	236.1	225.8	213.6	632.7	379.6
3.0	38.8	214.3	204.9	195.4	575.7	322.6
3.5	35.4	195.4	186.8	179.7	526.5	273.4
4.0	32.4	179.1	171.1	166.3	484.0	230.9
4.5	29.8	164.8	157.4	154.6	447.0	193.9
5.0	27.5	152.3	145.4	144.5	414.7	161.6
5.5	25.5	141.3	134.8	135.7	386.4	133.3
6.0	23.8	131.6	125.5	128.1	361.4	108.3
7.0	20.8	115.4	109.9	115.3	319.8	66.7

tive or control program was written to tie the procedures together. Only two of the modifications need be mentioned here: division of the compensating procedure into four subroutines to simplify the checking of its many complicated loops, and division of the benefit-evaluation procedure into two subroutines to separate the calculation of system costs from that of system benefits. In addition, a part of the bulky output-reporting procedure was placed for convenience in the executive program, to keep the output subroutine to manageable size for ease of checking.

Because standard service subroutines of the Fortran system handle many tedious details, the programmer need encode in Fortran language only the following instructions required for the executive program: (1) from cards and magnetic tape (which contains the large volume of hydrologic information) read in necessary input data; (2) based on input data for the study, compute necessary constants (such as target outputs for irrigation and energy for each month, and parameters relating water flow to power capacity for power plants *B* and *G*); (3) specify initial values for certain quantities (such as the number of the month, and numbers and values of irrigation and energy deficits) that cumulate throughout the study; and (4) call the operating subroutines at proper times.

As shown diagrammatically in Fig. 9.8, the full simulation program consists of nine major subroutines, controlled by an executive program. The following notation was adopted for the purpose of instructing the computer to follow a specific routine or procedure. OPROCW is the monthly operating procedure; FLDRUT is the flood-routing procedure; subroutines COMPEN, RLSBFN, ADNLEN, and BRKTFN, taken together, constitute the compensating procedure; subroutines PRICER and BENFIT comprise the benefit-evaluation procedure; and OUTPUT is the major part of the output-reporting procedure. Figure 9.8 outlines the way in which operating control passes from the executive program to the nine subroutines.

Appendix

Notational symbols used for the simulation program and their definitions.

Symbol	*Definition*
ADENG	additional energy generated at power plant *G* from release RLSBAH.
ADNLEN	additional energy subroutine, a part of the compensating procedure of the simulation program.
ADNLNH	additional energy, high, or total additional energy produced at both power plants *B* and *G* from release RLSBAH.
ALCONB	allowable contents of reservoir *B* after monthly releases for flood storage.
ALCONC	allowable contents of reservoir *C* after monthly releases for flood storage.

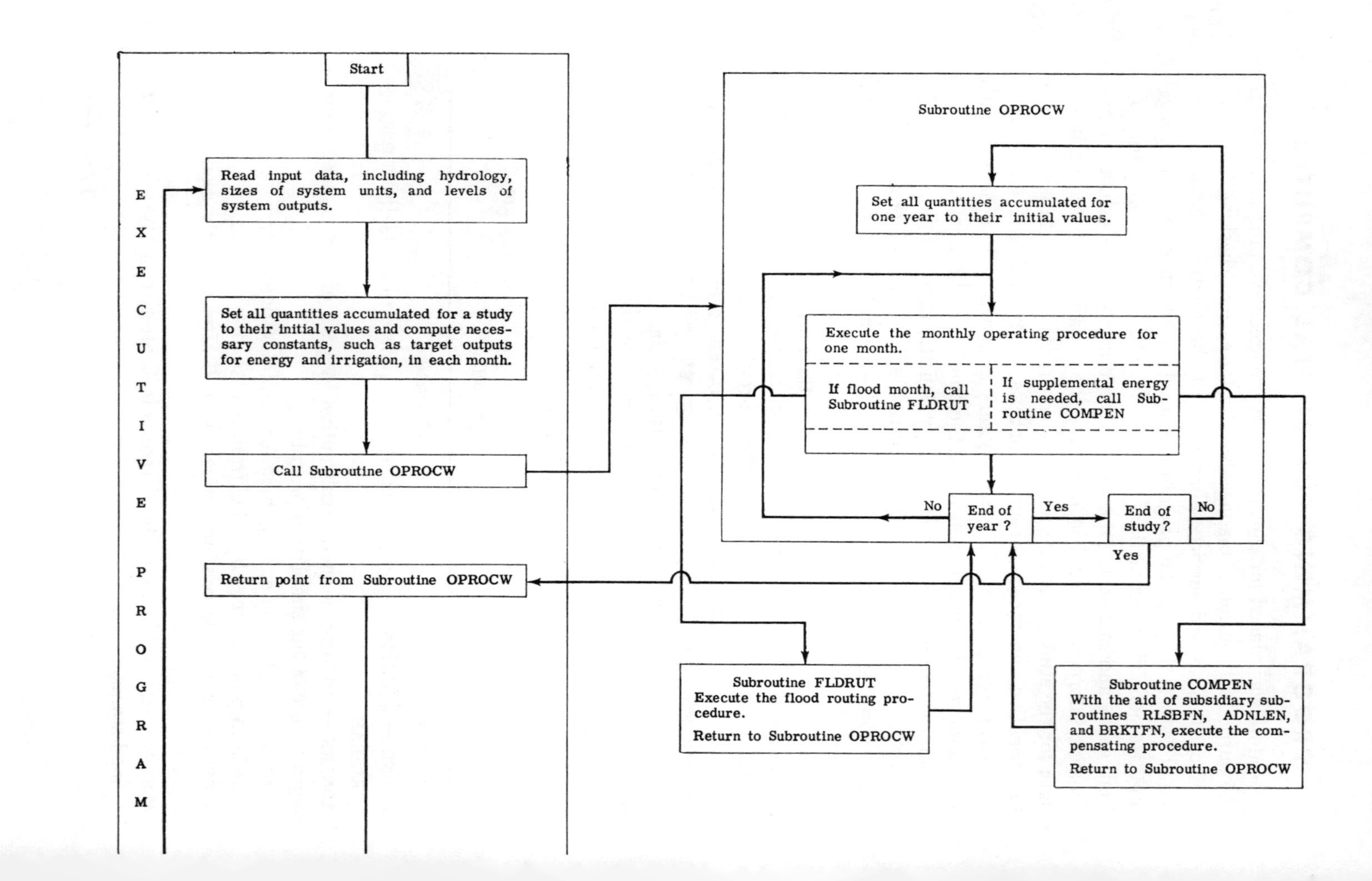
Start
EXECUTIVE PROGRAM
Read input data, including hydrology, sizes of system units, and levels of system outputs.
Set all quantities accumulated for a study to their initial values and compute necessary constants, such as target outputs for energy and irrigation, in each month.
Call Subroutine OPROCW
Return point from Subroutine OPROCW
Subroutine OPROCW
Set all quantities accumulated for one year to their initial values.
Execute the monthly operating procedure for one month.
If flood month, call Subroutine FLDRUT
If supplemental energy is needed, call Subroutine COMPEN
No
End of year ?
Yes
End of study?
No
Yes
Subroutine FLDRUT
Execute the flood routing procedure.
Return to Subroutine OPROCW
Subroutine COMPEN
With the aid of subsidiary subroutines RLSBFN, ADNLEN, and BRKTFN, execute the compensating procedure.
Return to Subroutine OPROCW

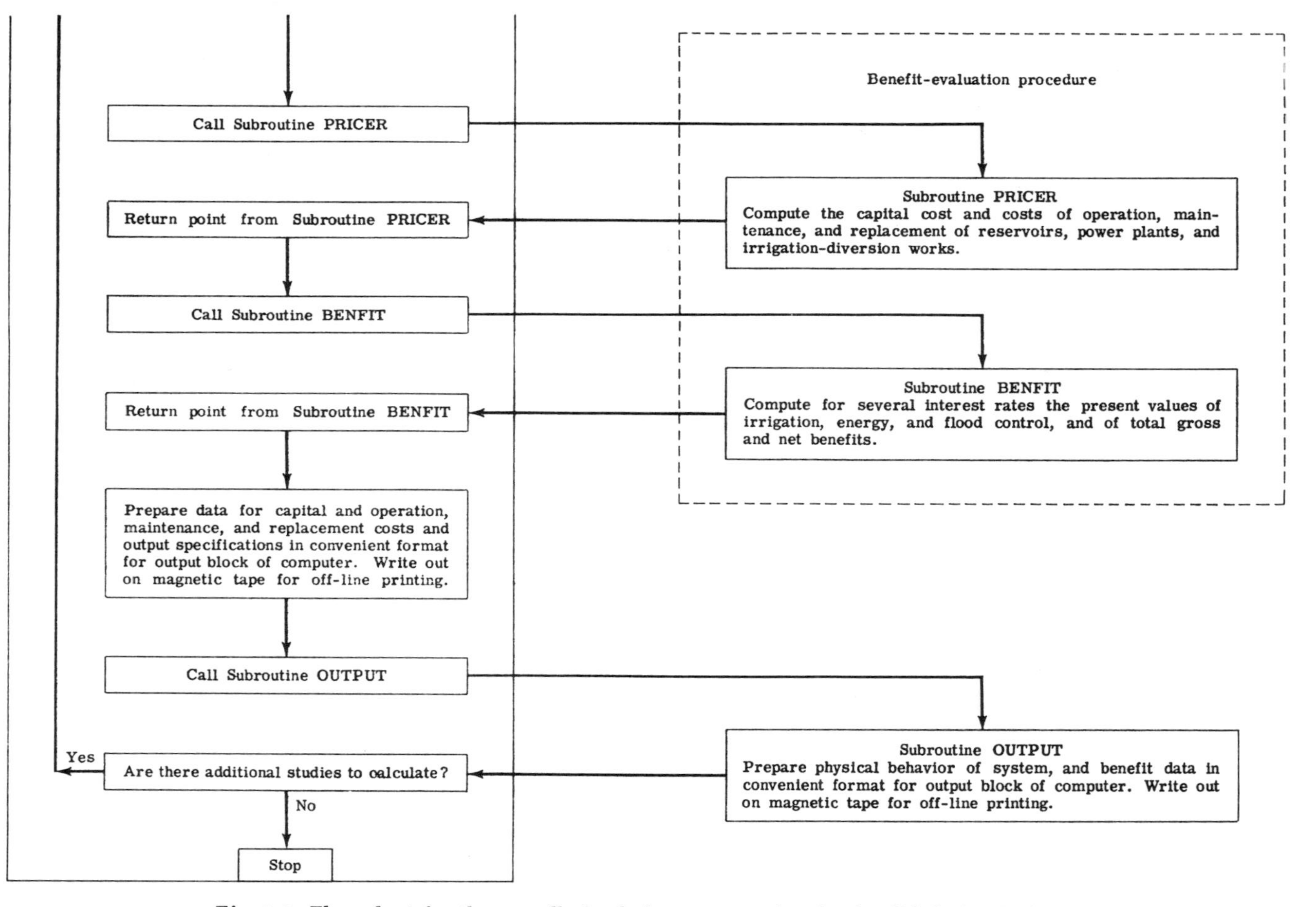

Fig. 9.8. Flow chart for the over-all simulation program for the simplified river-basin system.

Symbol *Definition*

ALCOND — allowable contents of reservoir *D* after monthly releases for flood storage.

ANENRQ — annual energy requirement, or annual target output for energy.

ANIREQ — annual irrigation requirement, or annual target output for irrigation.

ANNBEN(IYEAR) — annual benefits, or total gross benefits for the year.

AVCNB6 — average contents of reservoir *B* for the 6-hr interval.

AVCONB — average contents of reservoir *B* for the month.

AVHEAD — average head at reservoir *B*, or head associated with average contents at reservoir *B*.

BASIL(1), BASIL(2), BASIL(3) — intermediate quantities that simplify computation of irrigation losses.

BENEGY(IYEAR) — benefit from energy, or total energy benefits for the year.

BENFIT — benefit subroutine, a part of the benefit-evaluation procedure of the simulation program.

BENFN — unit irrigation-benefit function.

BENIRR(IYEAR) — benefit from irrigation, or total gross irrigation benefit less losses incurred from irrigation deficits, for the year.

BENNET(NRATE) — benefits, net, or present value of net benefits for the particular interest rate.

BENNIR — benefits from irrigation, or unit irrigation benefits.

BETA — constant monthly energy factor for power plant *B*.

BETA6 — constant energy factor for power plant *B* for 6-hr interval.

BRKTFN — bracket-function subroutine, a part of the compensating procedure of the simulation program.

CAPA — capacity of reservoir *A*.

CAPB — capacity of reservoir *B*, active.

CAPC — capacity of reservoir *C*.

CAPD — capacity of reservoir *D*.

CASHEN — cash from energy, or gross annual firm-energy benefits.

CASHIR — cash from irrigation, or gross annual irrigation benefits.

CLASS(1) — class of irrigation deficits up to and including 10 percent of annual target output.

CLASS(2) — class of irrigation deficits between 10 and 70 percent of annual target output.

Symbol *Definition*

CLASS(3) — class of irrigation deficits above 70 percent of annual target output.

CMANED(IYEAR) — cumulative annual energy deficit for the year.

CMANID(IYEAR) — cumulative annual irrigation deficit for the year.

CMRNID — cumulative-run irrigation deficit, or cumulative irrigation deficit for the 50-yr study.

COMPEN — compensating subroutine, a part of the over-all compensating procedure of the simulation program.

CONA6 — contents of reservoir *A* at start of flood-routing procedure, with 12-hr time lag.

CONB6 — contents of reservoir *B* at start of flood-routing procedure, with 12-hr time lag.

CONB6N — contents of reservoir *B*, 6-hr period, new, or contents of reservoir *B* after releases for irrigation.

CONC6 — contents of reservoir *C* at start of flood-routing procedure, with 6-hr time lag.

COND6 — contents of reservoir *D* at start of flood-routing procedure, with no time lag.

CONTSA(IMONTH) — contents of reservoir *A*, initial, for the month.

CONTSB(IMONTH) — contents of reservoir *B*, initial, for the month.

CONTSC(IMONTH) — contents of reservoir *C*, initial, for the month.

CONTSD(IMONTH) — contents of reservoir *D*, initial, for the month.

CONTSB(IMONTH + 1) — contents of reservoir *B* after initial releases have been made in the compensating procedure to prevent spills and provide irrigation and flood storage.

COSPPB — cost of power plant *B*, capital.

COSPPG — cost of power plant *G*, capital.

COSTA — cost of reservoir *A*, capital.

COSTB — cost of reservoir *B*, capital.

COSTC — cost of reservoir *C*, capital.

COSTD — cost of reservoir *D*, capital.

CSTFNA — cost function of reservoir *A*, capital.

DAMAGE — damage from floods for the year.

DEADST — dead storage in reservoir *B*.

DEBITI(IYEAR) — debit from irrigation deficits, or loss in irrigation benefits for the year.

DEFEN(IMONTH) — deficit in energy for the month.

DEFEN6 — deficit in energy for the 6-hr interval.

Symbol *Definition*

DEFIR(IMONTH) — deficit in irrigation for the month.

DEFIR6 — deficit in irrigation for the 6-hr interval.

DFCRIR(J), J = 1 to 7 — deficits in return irrigation flow for the present month and six successive months.

DFCRI6 — deficit in return irrigation flow for the 6-hr interval.

DFHYFL — deficit in hydroelectric flow, or difference between required and available flows at *G*.

DFRIFC(J), J = 1 to 7 — deficits in return-irrigation-flow fractions, an array of fractions for seven months used in computing deficits in irrigation return flow.

DPRICE — dump price for energy, or unit dump-energy benefits.

DUMPEN(IYEAR) — dump energy generated for the year.

DVIRWK — diversion- (irrigation-) works cost, or capital cost of irrigation-diversion works.

EFPOFL — effective power flow, or effective water flow through turbines at power plant *B* for the month.

EFPOF6 — effective power flow for 6 hours, or effective water flow through turbines at power plant *B* for the 6-hr interval.

EFPOW — effective power water, or effective water flow through turbines at power plant *G*.

ENERB(IMONTH) — energy generated at power plant *B* for the month.

ENERB6 — energy generated at power plant *B* for the 6-hr interval.

ENERG(IMONTH) — energy generated at power plant *G* for the month.

ENERG6 — energy generated at power plant *G* for the 6-hr interval.

ENLOSS — energy loss, or unit loss from energy deficits.

ENSURP(IMONTH) — energy surplus, or dump energy generated for the month.

ERRTOL — error tolerance value used in subroutine RLSBFN.

EXCESA — excess water, or spill at reservoir *A*.

EXCESB — excess water, or spill at reservoir *B*.

EXCESC — excess water, or spill at reservoir *C*.

EXCESD — excess water, or spill at reservoir *D*.

FDSTAL — flood-storage allowance, or maximum system flood-storage allocation for a month.

FEBEN — firm-energy benefit, or unit firm-energy benefits.

FLCPB — flow capacity of power plant *B*, or maximum water capacity of turbines at power plant *B* for a month.

Symbol *Definition*

FLCPB6 — flow capacity of power plant *B* for 6-hr interval, associated with the release from reservoir *B*.

FLDALW — flood allowance, or total system flood-storage allocation for next consecutive month.

FLDAMF — flood-damage function below point *G*.

FLOAPR — flow in April, or combined flows into reservoirs *A*, *B*, and *D* for April.

FLOBEN(IYEAR) — flood-control benefit, or gross flood-control benefits for the year.

FLOCAP — flow capacity, or effective flow associated with average head at reservoir *B* which can be used at power plant *B*.

FLOFG6(INTVL), INTVL = 1 to 120 — flow between points *F* and *G*, or natural inflow below point *F* and above point *G*, array for all 6-hr intervals of the flood month.

FLOJUN — flow in June, or combined flows into reservoirs *A*, *B*, and *D* for June.

FLOMAY — flow in May, or combined flows into reservoirs *A*, *B*, and *D* for May.

FLOOD — flood-free capacity of channel below point *G*, or maximum safe channel capacity expressed in 10^2 acre ft.

FLOWA(J), J = 1 to IMNMX — flow into reservoir *A*, array for all months of the study.

FLOWA6(INTVL), INTVL = 1 to 120 — flow into reservoir *A*, array for all 6-hr intervals of flood month.

FLOWB(J), J = 1 to IMNMX — flow into reservoir *B*, array for all months of the study.

FLOWB6(INTVL), INTVL = 1 to 120 — flow into reservoir *B*, array for all 6-hr intervals of flood month.

FLOWC(IMONTH) — flow into reservoir *C* for the month.

FLOWC6 — flow into reservoir *C* for the 6-hr interval.

FLOWD(J), J = 1 to IMNMX — flow into reservoir *D*, array for all months of the study.

FLOWD6(INTVL), INTVL = 1 to 120 — flow into reservoir *D*, array for all 6-hr intervals of flood month.

FLOWE(J), J = 1 to IMNMX — flow into point *E*, or natural inflow between reservoirs *A* and *B* and point *E*, array for all months of the study.

FLOWE6(INTVL), INTVL = 1 to 120 — flow into point *E*, or natural inflow

Symbol *Definition*

between reservoirs A and B and point E, array for all 6-hr intervals of flood month.

FLOWFG(J), J = 1 to IMNMX — flow between points F and G, or natural inflow below point F and above point G, array for all months of the study.

FLOWG(IMONTH) — flow through power plant G for the month.

FLOWG6 — flow through power plant G for the 6-hr interval.

FRACB — fraction for B, or fraction of total flood storage allocated to reservoir B.

FRACC — fraction for C, or fraction of total flood storage allocated to reservoir C.

FRACD — fraction for D, or fraction of total flood storage allocated to reservoir D.

FURFLG — further flow at G, or the remaining effective flow at power plant G.

HDMX — head, maximum, or maximum possible head at reservoir B.

HEADB — head at reservoir B, or head associated with average contents of reservoir B during the 6-hr interval.

HEADFN — head–capacity function, shown in Table 9.1.

IMNMX — subscript denoting number of months in the study.

IMONTH — subscript enumerating successive months of the study.

INTVL — subscript enumerating successive 6-hr intervals of the month.

IYEAR — subscript enumerating successive years of the study.

IYRMX — subscript denoting number of years in the study.

MONTH — subscript enumerating successive months of the year.

NOTRIL — number of trial releases in RANGE1.

NRATE — subscript denoting number of interest rates used in the study.

NUMFLD — subscript denoting number of floods occurring during the study.

OMRA — OMR costs of reservoir A.

OMRB — OMR costs of reservoir B.

OMRC — OMR costs of reservoir C.

OMRD — OMR costs of reservoir D.

OMRIRW — OMR costs of irrigation-diversion works.

OMRPPB — OMR costs of power plant B.

OMRPPG — OMR costs of power plant G.

OPROCW — monthly operating procedure of the simulation program,

Symbol *Definition*

OUTFLE(IMONTH) — outflow from point *E*, or flow past diversion point *E* to reservoir *C* for the month.

OTFLE6 — outflow from point *E*, or flow past diversion point *E* to reservoir *C* for the 6-hr interval.

OUTPUT — output-reporting subroutine in the simulation program.

PEAKN — peak, new, or peak flood flow at point *G* during an interval.

PEAKO(IYEAR) — peak flow incurred for the year under flood-routing procedure.

PEKFLG(I), I = 1 to NUMFLD — peak natural flows, array for all floods of the study.

PPB — power plant *B*, rated power capacity.

PPG — power plant *G*, rated power capacity.

PREFFB — power efficiency at power plant *B*, or units and efficiency conversion factor at power plant *B*.

PRFCTG — power factor for power plant *G*, or units and efficiency conversion factor at power plant *G*.

PRICER — price-of-inputs subroutine, a part of the benefit-evaluation procedure of the simulation program.

PRLSB — previous release from reservoir *B*, or any previous effective flow at reservoir *B* during an interval.

PROPO — proportion, or ratio of contents of reservoir *C* exceeding allowable flood storage to sum of such excess contents for both reservoirs *C* and *D*.

PVABEN(I), I = 1 to NRATE — present value of total gross benefits for the 50-yr study, array for all interest rates.

PVAENY(I), I = 1 to NRATE — present value of energy benefits for the 50-yr study, array for all interest rates.

PVAFLO(I), I = 1 to NRATE — present value of flood-control benefits for the 50-yr study, array for all interest rates.

PVAIRR(I), I = 1 to NRATE — present value of irrigation benefits for the 50-yr study, array for all interest rates.

PVAOMR(I), I = 1 to NRATE — present value of annual OMR costs for the 50-yr study, array for all interest rates.

RANGE1 — first range of trial releases made from reservoir *B*.

RANGE2 — second range of trial releases made from reservoir *B*.

RATE(I) — array of interest rates expressed as percentages.

Symbol *Definition*

RBTLMX — release from reservoir *B*, total maximum, or maximum possible monthly release from reservoir *B* that can be used.

REMFLG — remaining flow at point *G* that can be accommodated without exceeding safe channel capacity.

REQEN(MONTH) — required energy, or monthly target output for energy.

REQEN6 — required energy for 6 hours, or target output for energy for the 6-hr interval.

REQFLG(IMONTH) — required flow through power plant *G* to fulfill energy deficit for the month.

REQIR(MONTH) — required irrigation, or monthly target output for irrigation.

REQIR6 — required irrigation for 6 hours, or target output for irrigation for the 6-hr interval.

RIRNML(MONTH) — return flow, irrigation normal, or return flow from irrigation for the month in the absence of irrigation deficits over the preceding seven months.

RIRNM6 — return flow, irrigation normal for 6 hours, or normal return flow from irrigation for the 6-hr interval.

RLSBAH — release from reservoir *B*, additional, high, or cumulative value of successive trial releases in RANGE1.

RLSBAP — release from reservoir *B*, appropriate, or release from reservoir *B* that will exactly satisfy the irrigation deficit.

RLSBFN — release-from-reservoir-*B* function, a part of the compensating procedure of the simulation program.

RLSTRL — release, trial, or trial release from reservoir *B*.

RLS1A — release from reservoir *A*, first or basic for monthly interval.

RLS1B — release from reservoir *B*, first or basic for monthly interval.

RLS1C — release from reservoir *C*, first or basic for monthly interval.

RLS1D — release from reservoir *D*, first or basic for monthly interval.

RLS1A6 — release from reservoir *A*, first or basic for 6-hr interval.

RLS1B6 — release from reservoir *B*, first or basic for 6-hr interval.

RLS1C6 — release from reservoir *C*, first or basic for 6-hr interval.

RLS1D6 — release from reservoir *D*, first or basic for 6-hr interval.

RLS2A — release from reservoir *A* for irrigation, second or supplementary for monthly interval.

RLS2B — release from reservoir *B* for irrigation, second or supplementary for monthly interval.

Symbol *Definition*

RLS2C — release from reservoir *C* for generation of firm energy, second or supplementary for monthly interval.

RLS2D — release from reservoir *D* for generation of firm energy, second or supplementary for monthly interval.

RLS2A6 — release from reservoir *A* for irrigation, second or supplementary for 6-hr interval.

RLS2B6 — release from reservoir *B* for irrigation, second or supplementary for 6-hr interval.

RLS2C6 — release from reservoir *C* to provide flood-storage space, second or supplementary for 6-hr interval.

RLS2D6 — release from reservoir *D* to provide flood-storage space, second or supplementary for 6-hr interval.

RLS3A — release from reservoir *A* for generation of firm energy, third or additional supplementary for monthly interval.

RLS3A6 — release from reservoir *A* for generation of firm energy, third or additional supplementary for 6-hr interval.

RLS3B6 — release from reservoir *B* to provide flood-storage space, third or additional supplementary for 6-hr interval.

RLS3C6 — release from reservoir *C* to generate firm energy, third or additional supplementary for 6-hr interval.

RLS3D6 — release from reservoir *D* for generation of firm energy, third or additional supplementary for 6-hr interval.

RLSTA(IMONTH) — release from reservoir *A*, total for the month.

RLSTA6 — release from reservoir *A*, total for the 6-hr interval.

RLSTB(IMONTH) — release from reservoir *B*, total for the month.

RLSTB6 — release from reservoir *B*, total for the 6-hr interval.

RLSTC6 — release from reservoir *C*, total for the 6-hr interval.

RLSTD6 — release from reservoir *D*, total for the 6-hr interval.

SCUF1 — least-squares curve-fitting routine used on IBM 704 computer.

SLOPIL(1) — slope of irrigation loss function 1, or unit loss of irrigation benefits for CLASS(1) deficits.

SLOPIL(2) — slope of irrigation loss function 2, or unit loss of irrigation benefits for CLASS(2) deficits.

SLOPIL(3) — slope of irrigation loss function 3, or unit loss of irrigation benefits for CLASS(3) deficits.

SPILB(IMONTH) — spill from power plant *B* for the month.

SPILB6 — spill from power plant *B* for the 6-hr interval.

Symbol	*Definition*
SPILG(IMONTH)	spill from power plant *G* for the month.
SPILG6	spill from power plant *G* for the 6-hr interval.
SUBTRA(1), SUBTRA(2), SUBTRA(3)	intermediate quantities that simplify computation of irrigation losses.
TCCOST	total capital costs.
TODFIR	total deficit for irrigation for the study.
TOTDMP	total dump energy generated for the study.
TOTNB	total energy generated at power plant *B* during the month.
TOTOMR	total OMR costs.
WCAPG	water capacity of turbines at power plant *G* for a month.
WCAPG6	water capacity of turbines at power plant *G* for a 6-hr interval.
WDTHMX	width, maximum, or maximum size of trial releases made under compensating procedure.
WIDTH1	width one, or size of release increments in RANGE1.
X	contents of reservoir *C* in excess of allowable flood-storage contents.
Y	contents of reservoir *D* in excess of allowable flood-storage contents.

10 Analysis by Simulation: Examination of Response Surface

MAYNARD M. HUFSCHMIDT

To refresh the reader's memory as to the nature of the simplified river-basin system being analyzed by simulation, it may be well to recapitulate its salient points.

(1) Our simplified system comprises 12 design components or variables: the storage capacities of reservoirs *A*, *C*, and *D*; the capacities of power plants *B* and *G*; the allocations of active and dead storage within reservoir *B*; the allocations of flood-control storage to reservoirs *B*, *C*, and *D*; and the annual target outputs for irrigation and energy. These variables we are free to test by simulation in as many different combinations and magnitudes as we wish.

(2) A fixed operating procedure is accepted for all trials. Chapter 7 describes the procedure developed specifically for our system.

(3) Each simulation run is based upon the same 60 years of streamflow information, the first 10 years of which constitute a warm-up period preparatory to the run proper. Monthly flows are employed except when the river rises to flood stage; at such times 6-hr flows are interpolated. See Chapters 6 and 12 for a discussion of synthesized streamflow records and their use in simulation studies.

(4) The computer evaluates the physical and related economic results. Table 9.3 shows an example of the computer output for a typical simulation run.

(5) The response of the system is measured as the present value of net benefits at a specific interest rate of $2\frac{1}{2}$ percent, with the assumption that there are no constraints on capital and operation, maintenance, and replacement (OMR) funds. See Chapter 4 for a discussion of the effects on optimal design of introducing budgetary constraints.

Before proceeding to the analysis proper, let us review also what we know or can infer from certain *a priori* considerations of the response of the simplified system to changes in its components. Chapter 3 presents basic concepts of this nature.

The response of the system, or "net-benefit response surface," can be represented by the function

$$B = f(x_1, \ldots, x_{12}),$$

where B denotes net benefits and $x_1, \ldots, x_{12}$ are the sizes of the 12 design components. The two components that are target outputs are not necessarily the same as those actually obtained in the simulation over the 50-yr period; there are deficits in irrigation and energy on the one hand and surpluses in energy, or dump energy, on the other. Actual outputs for irrigation, energy, or flood control are therefore implicit functions of the 12 design variables, including the target outputs.

It is our task to select for the 12 design components magnitudes that will maximize net benefits. Unfortunately, the resulting net-benefit function is too complicated to be described explicitly, unless we consider the entire computer program for simulation a definition of this function. Because the net-benefit function is so complicated, the only practical means of finding the optimal combination of the 12 variable design components is suitable sampling. In each observation of a given sample a specified value is assigned to the individual components. The observation as a whole then becomes a design alternative, which responds to a simulation run by yielding a certain value of net benefits.

Each response is the result of several complex operations by the computer. Given the specified combination of components, the computer follows the operating procedure of making releases from the reservoirs so as to satisfy the target outputs, if this can be done. By simulation the computer derives deficits from failures to meet target outputs for irrigation and energy, dump energy produced, and maximum flood flows. From basic data the computer attaches benefits to these target outputs, to surplus energy, and to reductions in peak flood flows, and losses to deficits. To obtain net benefits, therefore, (1) capital costs associated with the inputs, OMR costs, and penalties assessed against deficits are deducted from benefits; and (2) all cost and benefit streams are discounted back to the present on a 50-yr, $2\frac{1}{2}$ percent interest basis.

A particular design is "nonoptimal" if the same target outputs — or, indeed, actual outputs — can be attained when the capacities of the input units (reservoirs and power plants) are dropped below the design level. Designs of this kind have excess capacity and are classed as technologically inefficient. They are off the production function, as defined in Chapter 3. On the other hand, designs may be technologically efficient, yet unable to achieve the target outputs without frequent and severe shortages. This is so, for example, when the specified reservoir capacities or the volumes of water in the system are too small. When a powerful economic loss function penalizes deficits, or when a tolerance limit is set for the permissible number of deficits of given magnitude, such combinations also prove to be either nonoptimal or inadmissible.

It may help to clarify these concepts if we consider the relationships that exist between only two of the system variables, one input and one target output, all other components being held constant. (This implies that the target outputs held constant must actually be met.) If we take the relationship between the active capacity of reservoir *B* and the target output for irrigation as our first example, it is obvious that for each active capacity of reservoir *B* there must be some maximum irrigation output that the system can just meet over the 50-yr period without incurring deficits.[1] Conversely, for each target output for irrigation there must be some minimum active capacity of reservoir *B* that will provide the required output without deficits or excess capacity.[2]

The balanced combinations of active reservoir size without excess capacity and target output for irrigation achievable without deficit can be represented by a curve such as AA′ in Fig. 10.1. Combinations that lie below AA′ are not in balance; they are seen to be wasteful, because the input could be smaller and still meet the target output without deficits. As long as there is excess reservoir capacity, a combination cannot be optimal.

Conversely, for unbalanced combinations that lie above AA′, we encounter deficits which are penalized according to the irrigation loss function. Combinations that lie only slightly above AA′ have only small to moderate deficits, but as we move away from AA′ deficits increase

[1] This is the case for a fixed operating procedure and a given number of years of hydrologic data. The fact that the target outputs were achieved with a particular pattern of hydrology does not imply that they could be attained also with any other pattern. Only a single observation was made available to us, so to speak. In principle, the probability of meeting the target outputs with the assigned inputs should be estimated from the results obtained with a substantial number of streamflow patterns of equal length. Since the probability of satisfying our requirements must lie somewhere between zero and one, we should be forced to face up to the question of how low a probability we could accept.

[2] Of course, the target output may be set so high that the available water becomes the limiting factor; then, no matter how large the reservoir, the output cannot be realized without deficits.

rapidly. Hence, deficits are penalized more and more severely until it no longer pays to incur them. We can conceive of the existence of a curve such as BB′ in Fig. 10.1, above which deficits become so great

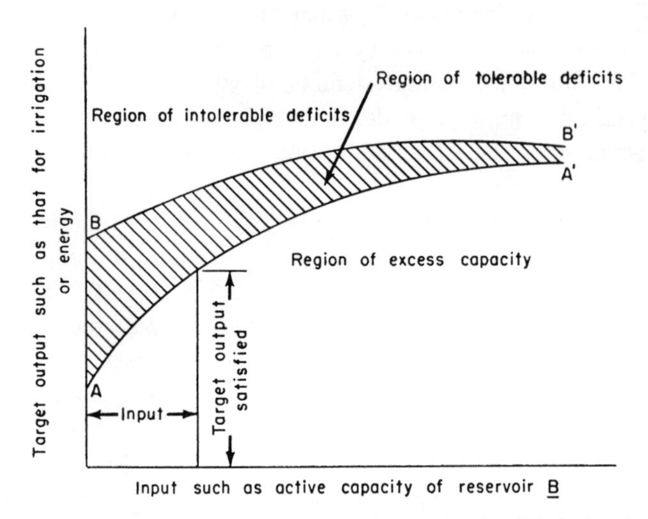

Fig. 10.1. Input-output function for two variables of the simplified river-basin system, assuming (1) fixed values of all other variables, and (2) a fixed operating procedure and a single 50-yr trace of hydrology.

that points in this region too are obviously nonoptimal. A nonoptimal region in which large increases in reservoir capacity cause little gain in irrigation output is entered, moreover, as curve AA′ approaches its upper limit of output.

For our second example, let us substitute energy for irrigation as the variable output. If, as in Chapter 7, designs with monthly energy deficits equaling 25 percent or more of the target output (designated class 3 deficits) are ruled intolerable, curve BB′ can be drawn to delimit the minimum active capacity of reservoir *B* that will generate the target output for energy without incurring an intolerable shortage. The significance of curve AA′ remains basically the same. Related in this instance are the minimum active capacities of reservoir *B* that generate the output for energy with no deficits and no excess capacity. Once again, combinations of system variables in the region of excess capacity below AA′ are obviously inefficient. Although combinations above BB′ are not technologically inefficient, this portion of the response surface fails to be of ultimate interest because of the rule that class 3 energy shortages are intolerable.

These considerations lead us to the conclusion that large regions of the response surface cannot possibly comprehend the optimal system design. If such regions could be isolated quickly and rejected, we could profit in time and effort. Nonetheless, because one of the purposes of our research was to learn something about the configuration of the response surface as a whole and, in doing so, to experiment with different sampling techniques, we deliberately paid little attention to *a priori* information on the nature of our response surface.

SAMPLING METHODS AND STRATEGY

The sampling methods available to us for analysis of our system fall into two broad categories: systematic sampling and random sampling. In systematic sampling, we select the values of our system variables in accordance with some ordering principle; in random sampling, we do so purely by chance from an appropriate population of values of the variables. In deciding upon a sampling strategy, however, we remain free to use the two types of sampling methods either separately or together.[3]

Systematic sampling may follow a number of different courses that vary within themselves. Several of them are of interest to us, in particular: (1) the uniform-grid or factorial method; (2) the single-factor method; (3) the method of incremental or marginal analysis; and (4) the method of steepest ascent.[4] These are discussed individually below, with their application to our river-basin system detailed later in the chapter.

The selection of sampling methods suited to the nature of our experiment is an important concern. We have two contrasting situations: on the one hand, a large number of designs to be examined in one big experiment; on the other hand, one or at most a small number of designs to be tested in a few tight experiments, in which the results of one are used to determine the nature of the next. Making these distinctions, we conclude that random sampling, and also the uniform-grid or factorial method of systematic sampling, are suitable for the first situation, and that the remaining three methods which, as we shall see, are iterative, are applied best to the second situation.

[3] William G. Cochran and Gertrude M. Cox, *Experimental designs* (John Wiley, New York, ed. 2, 1957), p. 335.

[4] Cochran and Cox, reference 3, except that the method of marginal analysis is not discussed there.

To each of these methods, finally, append certain advantages and disadvantages that vary in kind and degree with the nature of the response surface under investigation. Together they constitute a collection of techniques that can be combined in strategic ways for the effective examination of response surfaces such as those created by the evaluation of our river-basin system.[5]

SYSTEMATIC SAMPLING

Uniform-grid or factorial sampling. The essence of this method is coverage of the relevant range of each variable by a series of uniformly spaced values. For n variables, each with m values, the size of the sample is m^n. If $n = 2$ and $m = 4$, for example, $m^n = 16$, as in the illustration given in Fig. 10.2. The intensity of coverage of a response surface with a given number of variables, or the fineness of the superimposed grid, depends therefore on the number of possible combinations of the items in the sample. As the number of variables increases, the number of combinations required for the desired intensity of coverage rises exponentially. To cover our own system with the coarsest possible grid, namely two values for each of our 12 variables, a sample of 2^{12} or 4096 combinations is needed. Obviously, if sample sizes are not to become unreasonable, the number of variables to be examined by uniform-grid sampling must be kept down.

The great advantage of uniform-grid sampling is its ability to map the entire response surface of systems with a small number of variables. The effectiveness of the method depends to a significant degree on the nature of the response surface. The gentler the slopes and the rounder the peaks and ridges, the more exactly does a uniform-grid sample of given size portray the surface and approximate its highest points. If the topography is jagged, the range of each variable subtending high values on the response surface is small in relation to the total range of the variable, and the likelihood of any one combination of variables or any one design falling into the region subtended by high points will not be great.

Similarly, the likelihood of a uniform-grid sample approximating the highest points on the surface becomes better as the area of high elevation becomes proportionately larger.

[5] For a discussion of the merits of these sampling techniques in exploring different types of response surfaces, see S. H. Brooks, "A comparison of maximum-seeking methods," *Operations Research 3*, 430 (1959).

These relationships are illustrated graphically in Fig. 10.2. A comparison of the two component surfaces shows that, where sharp peaks with high values of 40 predominate, not one of the 16 combinations composing the grid is as high as 20, whereas in the presence of rounded ridges of the same maximum elevation, at least one of the 16 combinations is higher than 30 and six are higher than 20.

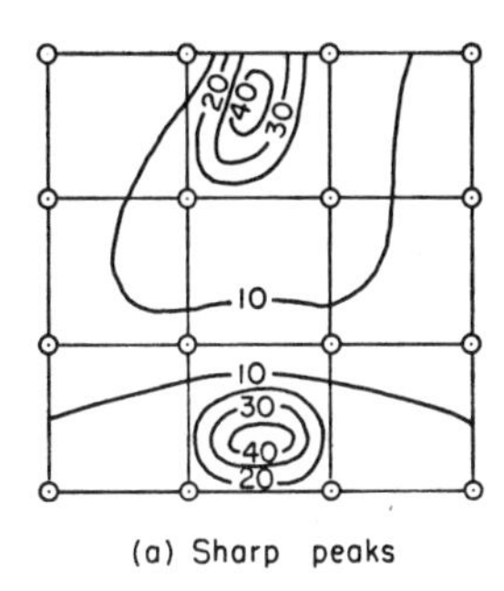

(a) Sharp peaks

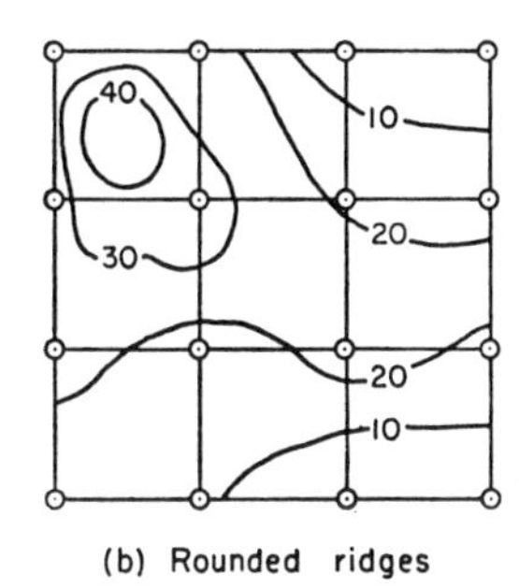

(b) Rounded ridges

Fig. 10.2. Uniform-grid sampling of two-variable response surfaces with high values of about 40.

Single-factor sampling. In this method the values of all but a single variable of the system are held constant while the values of the chosen variable or component are altered step by step until the greatest response of the system to this variable has been registered. The process is then repeated with each component in succession until there is no further improvement in the design as a whole. When this sampling technique is applied to the simplified system, the most promising initial values of its 12 variables are chosen from information at hand. When the twelfth variable has been adjusted for greatest response, the previously determined best values of one or more of the other 11 variables may no longer be optimal. Successive readjustment may then be necessary.

Single-factor sampling works best when the variables are independent of one another. When they are not, the process of varying a single factor at a time leads to the optimal combination by a roundabout path at best. In certain instances, indeed, the optimum may be missed completely.[6]

The component variables of water-resource systems are closely interrelated. A change in the target output for irrigation, for example, may well alter both the minimum volume of reservoir storage needed and the amount of energy that the system can generate. It follows that single-

[6] Cochran and Cox, reference 3.

factor sampling is inherently unsuited to our purpose. Nevertheless, it can be useful in taking up the slack when an input variable, such as a reservoir or a power plant, has been given more than the required capacity. Under conditions such as this, the variable assumes an independent status within the range of its excess capacity. No final judgment can be passed, however, until the nature of the response surface of the simplified system has been fully explored.

Sampling by incremental or marginal analysis. Instead of successively altering a single variable only, marginal analysis alters two variables at a time, with the other system variables held constant. The interdependence of system variables is thereby acknowledged. The relationship between them is isolated and provides information on the direction in which the change should be made. The technique clearly derives from the principles of marginal analysis set forth in Chapter 3.

In our system, for example, we might select target output for irrigation and active capacity of reservoir B as the variables to be changed. Starting with a combination that includes no excess reservoir capacity, we should test the effects of an arbitrary small trial increase in target output for irrigation in combination with a suitable trial increase in reservoir capacity. Although the former increase must be large enough clearly to show its effect on gross irrigation benefits, it must not be so large that the response in the interval between the two values becomes markedly nonlinear. Computed in succession would then be (1) from irrigation-benefit data, the resulting increase in gross benefits; and (2) from available cost data, the increase in reservoir capacity that would cost just as much as the gain in gross benefits. If the trial increases should combine to create excess reservoir capacity, it should be possible to lower the initial trial increase in capacity and still produce both the higher irrigation output and a concomitant gain in net benefits. If, on the other hand, the trial increases result in substantial irrigation deficits, gross irrigation benefits will drop below their estimated magnitudes and net benefits will become less than their starting values. In these circumstances we should proceed to decrease the two variables. When the optimal combination of irrigation output and active capacity has been found, another pair of variables is tested. The system design becomes optimal when it is no longer profitable and feasible to change any pair of variables upward or downward.

Sampling by marginal analysis is best suited to the study of systems in which there are only a few variables or in which important pairs are readily identified by prior analysis. The method should, therefore, be a good final check when apparently optimal combinations have been determined. Information would probably be available at that time on the major variables and the degree of their interrelatedness.

Sampling by steepest ascent. This iterative technique derives its name from the fact that sampling moves sequentially from lower to higher elevations on the response surface and that, as shown in Fig. 10.3, it

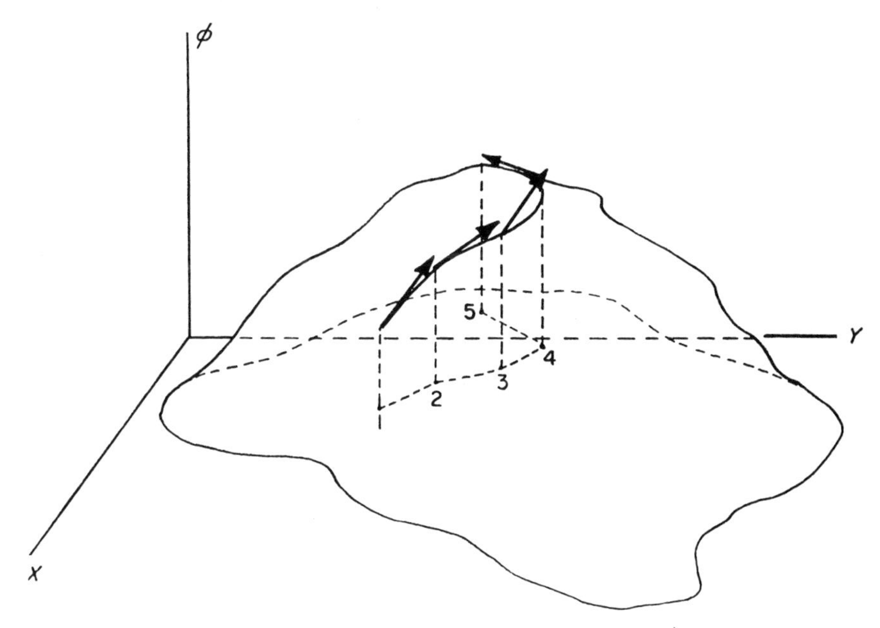

Fig. 10.3. Graphic example of finding the maximum of a function of two variables by a discrete-step, steepest-ascent procedure.

does this along the steepest and hence the shortest path up the slope.[7] Once again, all available and pertinent information is adduced to choose a starting point — or in this case a starting base, because the plane of reference is the magnitude of all system components. The magnitude of each variable is then changed separately by an arbitrary preselected amount, while the values of the other components are held constant. Again the change is made large enough to register a measurable net-benefit response, but not so large as to make the response manifestly

[7] For a description of the method, see G. E. P. Box and K. B. Wilson, "On the experimental attainment of optimum conditions," *J. Roy. Stat. Soc.*, *B*, *13*, 1 (1951). See also Cochran and Cox, reference 3, pp. 357–362.

nonlinear. Parenthetically, these changes are empirical approximations to the partial derivatives of the net-benefit function with respect to each variable. When the response to each change has been computed, the method seeks to revise each of the changed variables in such manner that a new base demarking the maximum response to the collective change is reached. This implies that both distance and direction of motion have been correctly chosen.

Two considerations enter into the choice of distance: the greater it is, the faster the approach to the summit, whether point or ridge; but the greater, too, the risk of overshooting the mark and the poorer the assumption of linear responses implicit in the method. A poor decision on selection of the distance to be moved will slow down the rate of approach to the top but will not defeat the method.

Choice of direction as well as distance can be justified most clearly by expressing the sampling procedure in mathematical language as follows.

If we denote the first-choice, or starting, series of the design variables by $x_1^0, \ldots, x_n^0$, the starting value of net benefits B^0 can be expressed as a function of these system components:

$$B^0 = f(x_1^0, \ldots, x_n^0).$$

After a small increment Δx_i, where the subscript i denotes any one component of the system, has been added to each variable in turn, the associated change in net benefits, ΔB_i, and its rate of change, $\Delta B_i/\Delta x_i$, are computed in a simulation run.

Moving from the collection of starting values x_i^0 to the first-revised-base values x_i^1 requires that the change in each variable be made proportional to the associated rate of improvement $\Delta B_i/\Delta x_i$. This ratio may be negative or zero as well as positive for a particular component of the system. The move to the new base is symbolized, therefore, by the equation

$$x_i^1 - x_i^0 = C(\Delta B_i/\Delta x_i), \tag{10.1}$$

where C is a constant of proportionality. The relationship between the distance d from the original base to the revised base and the proportionality constant C is given by the Pythagorean equation

$$d = \left[\sum_1^n (x_i^1 - x_i^0)^2\right]^{\frac{1}{2}} = C\left[\sum_1^n (\Delta B_i/\Delta x_i)^2\right]^{\frac{1}{2}},$$

or

$$C = d\Big/\left[\sum_1^n (\Delta B_i/\Delta x_i)^2\right]^{\frac{1}{2}}. \tag{10.2}$$

Alternatively, C can be chosen of such magnitude that the maximum absolute value of change for any variable will equal a specified magnitude, k. Or,

$$\max_i |x_i^1 - x_i^0| = k,$$

and it follows from Eq. (10.1) that

$$C \max |\Delta B_i/\Delta x_i| = k.$$

Since $$k/\max |\Delta B_i/\Delta x_i| \leq k/|\Delta B_i/\Delta x_i|,$$

$$C \leq k|\Delta x_i/\Delta B_i|,$$

or $$C = k \min |\Delta x_i/\Delta B_i|. \tag{10.3}$$

Proceeding from the base combination $(x_1^0, \ldots, x_n^0)$ and the selected jumps $(\Delta x_1, \ldots, \Delta x_n)$, therefore, we find by simulation the values of $(\Delta B_1, \ldots, \Delta B_n)$. After deciding on a magnitude for d, or, alternatively, for k, we compute C from Eq. (10.2) or (10.3) and calculate the terms $[(x_1^1 - x_1^0), \ldots, (x_n^1 - x_n^0)]$ from Eq. (10.1). Subtraction gives the values of the variables for the new base $(x_1^1, \ldots, x_n^1)$.

Upon completion of the first lift the routine of calculations is repeated until there is no further improvement. If necessary or expedient, the values of x_i, d, or k can be redefined at the beginning of each successive lifting operation.

Application of the steepest-ascent method to our simplified system runs into complications. Most important is the matter of units. Equation (10.1) assumes that all increments Δx_i are stated in the same units. Our increments Δx_i, however, are acre feet of irrigation output and reservoir storage, and megawatt hours of energy, for example. To overcome this difficulty, we can make the units of the system components, as well as the sampling parameters d and k, dimensionless by expressing them as ratios of their permissible range. If we let R_i be the permissible range of values for the variable x_i,

$$\begin{aligned}&\Delta x_i \text{ (as a percentage-range ratio)}\\&\qquad = 100\, \Delta x_i \text{ (in natural units)}/R_i \text{ (in natural units)}.\end{aligned}$$

The major advantage of steepest-ascent sampling over single-factor and incremental-analysis sampling is that, by measuring in each step or lift the combined influence of the variables on the net benefits of the system, it progresses quickly to an optimal value. Its major disadvantage is that it may miss the true summit on a response surface if there is more than one enclosed or boundary high elevation. This follows readily from an examination of Fig. 10.2. The starting point is seen to be the important determinant. If it lies on a lesser slope, the procedure will lead to the top of that slope only, and the true summit will not be reached.

If, as is usual for water-resource systems, the general nature of the response surface is not known in advance, the method is uncertain unless preliminary analyses and supplementary checks are introduced. It is uncertain too when there are discontinuities, either in the net-benefit function or in its derivatives with respect to one or more of the variables. Examples in the simplified river-basin system are (1) points at which class 3 energy deficits are recorded and, because the design is therefore inadmissible, net benefits drop to zero; (2) the boundary AA′ in Fig. 10.1, which lies between designs with excess capacity and output deficits; (3) the two sharp changes in the slope of the irrigation loss function shown in Fig. 7.3; and (4) the discontinuity in costs of the irrigation-diversion, distribution, and pumping works shown in Figs. 7.4 and 7.5.

RANDOM SAMPLING

Our river-basin system is analyzed by random sampling as follows. System designs are created for analysis by assigning to our 12 components numerical values selected completely at random from the entire population of values within reasonable ranges of the individual components.[8] The method spots the topography of the response surface of the simplified river-basin system in much the same way as a stadia survey spots the natural topographic features of a given land area. The information obtained is of prime importance in developing a strategy for identifying the region of optimal combination of system components by subsequent sampling techniques or refinements.

As in uniform-grid sampling, the effectiveness of random sampling varies with the size of the sample and the nature of the response surface. The larger the sample, the better the information. With a response surface of smooth topography the probability of a random combination of the variables falling into the high net-benefit region is considerably better

[8] The total population of relevant values of the 12 system variables is very great, even if the size of increments is made as large as 100×10^3 acre ft for reservoir capacities and target output for irrigation, and as large as 100×10^6 kw hr for target output for energy. It should be noted that the capacities of power plants *B* and *G* are not chosen directly, but derive from the randomly selected target output for energy.

than with jagged topography. Furthermore, as the proportion of high elevation in a response surface becomes larger, the probability of a random sample approximating the highest points on the surface becomes better.

The mathematical probabilities in a given random sample of including a combination of large value can be explored as follows. For 12 or, more generally, n design variables, the values of which are variable, the net-benefit response surface is represented, as before, by the equation

$$B = f(x_1, \ldots, x_n),$$

where $x_1, \ldots, x_n$ have now been chosen at random from the full range of conceivable values and the resulting magnitude of the net benefits B has been calculated by simulation. The value of B is therefore itself a random variable with unknown probability distribution. However, if we let θ_a be the probability that B will be less than or equal to some quantity a, or

$$\theta_a = \text{Prob } \{B \leq a\},$$

it follows that for m observations drawn at random, the probability is θ_a^m that none will exceed a, and $1 - \theta_a^m$ that at least one will.

These boundary values enable us to make some useful statements, even if we do not know the probability distribution of B itself. Let us suppose, for example, that a is the value of net benefits exceeded by only 10 percent of the possible random observations. Then $\theta_a = 0.9$. Stated in more general terms,

$$\theta_a = 1 - p/100,$$

where p is the percentage of times that a value of a is equaled or exceeded in random sampling. The probability that the highest value in m observations will fall into the upper p percent of the population then becomes $1 - (1 - p/100)^m$. For instance, if the size of the sample is 30, the probability of at least one of the observations falling into the upper 10 percent of the population is $1 - (1 - 0.10)^{30} = 0.957$, or 95.7 percent. Moreover, if the response surface slopes gently and includes a preponderance of high-altitude topography, all observations greater than the upper 5- or 10-percentile value will yield net benefits close to the optimum. If the response surface is jagged and predominantly low in altitude, the 10-percentile value of the population may be a net benefit far below the optimum. In that case probability statements no longer retain their significance.

SAMPLING STRATEGY

It is clear from what has been said that sampling strategy may be most effective when it employs a combination of methods. The nature of the response surface is the most important factor in arriving at a decision. Since the general nature of the surface is largely unknown until sampling is begun, the first samples should be chosen to reveal salient topographic features. Both random and uniform-grid samples are informative. As interesting regions of the surface are traversed and optimal values are approached, the surface can be examined more closely by intensifying random or uniform-grid samples, or by shifting to the iterative techniques of the single-factor, marginal-analysis, or steepest-ascent methods. The particular combination of methods chosen depends on their properties and on two additional considerations: the nature of the experiments (such as computer simulation) by which individual responses are found, and the availability of manpower, computers, money, and time for experimentation and for analysis of results.

As we learn more about the response surfaces generated by typical water-resource systems, and as planners and designers become more experienced samplers, it should become possible to identify an effective sampling strategy for a given system in quick order and with high promise of success.[9]

APPLYING SAMPLING METHODS TO THE SIMPLIFIED RIVER-BASIN SYSTEM

An important aim in applying different sampling methods to the simplified river-basin system was to learn something about their relative usefulness in the analysis of water-resource developments in general. Because a simulation run of our system on the IBM 704 computer took

[9] Depending upon circumstances, a strategy may be judged efficient if it discovers the optimal design in the least number of trials, or if it minimizes, on the one hand, the time required for analysis at the expense of computer time or, on the other hand, the elapsed time. The degree of certainty with which the optimal design can be found is another criterion of effectiveness. Random sampling, for example, will surely lead to the optimal region; some systematic sampling methods may not.

Brooks has proposed as a measure of effectiveness the magnitude of the response achieved in a fixed number of trials, or the average achievement of the sampling method. He notes among other possible criteria of effectiveness (a) the number of trials required to achieve a fixed response, say 90 percent of the maximum response, and (b) the distance of the estimated from the true optimal combination, given a fixed number of trials. See Brooks, reference 5.

approximately two minutes of time, about 200 trials appeared to be in balance with our resources.[10] Accordingly, we adopted the following strategy.

(1) A preliminary analysis of the system told us which variables we should choose for initial study.

(2) The entire range of these variables was sampled by a uniform grid of moderate size.

(3) The results of this sampling identified the most promising region of the response surface, which was then explored by systematic sampling of the most important system variables, starting with the method of steepest ascent.

(4) After a seemingly optimal region had been isolated, other regions that appeared to promise high net benefits were sampled, including boundaries along which large discontinuities could exist. The general contours of the optimal region were thus confirmed.

(5) Once we were reasonably certain of the optimal region, we examined the remaining variables by iterative techniques, starting with the most important variables and continuing with the lesser ones until no further improvement in net benefits was obtained.

(6) The apparent optimal design was then checked by marginal analysis with successive pairs of variables.

(7) The large region enclosed in the uniform grid and the smaller one in the vicinity of the apparent optimum were sampled at random to give still more detailed information about the response surface.

Our choice of uniform-grid sampling, rather than random sampling, for the first computer run was made arbitrarily, with the intent of covering the range of each variable systematically. Selection of the steepest-ascent method as the first iterative technique, too, was arbitrary, although we did know that the single-factor method is not well suited to the analysis of water-resource systems. The order in which the variables were examined was also somewhat arbitrary, as little was known beforehand about the relative influence of the variables on net benefits.

DECIDING ON THE MAJOR VARIABLES

As indicated previously, the operating procedure and the hydrology were fixed for all trials discussed in this chapter. Among the system

[10] Limitations of this order would not necessarily apply to the design of actual systems, where computer time might not be so severely curtailed.

components allied to reservoir size and operation, active storage in reservoir B was judged a key variable because it controls over 80 percent of the streamflow for irrigation, contributes its waters for the generation of power, and affords space for the retardation of flood flows. Believed to occupy strategic positions also were dead storage in reservoir B, which maintains the head at power plant B; and reservoir D, which controls about 40 percent of the entire runoff for power generation in power plant G and for the reduction of floods in the downstream zone. Although reservoir C can control about half the streamflow of the system, it was considered a less significant component because it lies below reservoirs A and B and the irrigation system which to a large degree regulates streamflow at C. Least significant, it was thought, was reservoir A, because the volume of streamflow at this site is but a small fraction of the whole.

System outputs of energy and irrigation were classed as important; however, there was no advance information as to their relative contribution to net benefits. A preliminary analysis suggested that less than 1.0×10^6 acre ft of combined flood-control capacity was needed in reservoirs B, C, and D to exercise a very high degree of flood protection. Because this was so small a proportion of the total reservoir capacity of the system, space allocated to flood control was not considered strategically important. The capacities of the power plants at B and G, it was judged, were dependent variables deriving their proper magnitude primarily from the target output for energy.

Accordingly, we started our studies with an examination of the active- and dead-storage capacities of reservoir B, the capacity of reservoir D, and the target outputs for irrigation and energy — or a total of five variables — by the uniform-grid or factorial method of sampling.

APPLYING THE UNIFORM-GRID METHOD

Selecting the sample. Because our computer time was limited, we had to be satisfied with two values for each of the five free or multiple-valued variables of the system. At that, about an hour of computer time was consumed in the analysis of the $2^5 = 32$ possible combinations of variables. As representative values of the free system components we chose the one-third and two-thirds points of their range in magnitude for all but the target output for irrigation. The high value of that variable was shifted from the two-thirds point to the point of discontinuity in the irrigation-cost curve shown in Fig. 7.4, because net irrigation benefits were expected to drop sharply at the discontinuity.

The five pairs of selected values are listed in Table 10.1, together with

Table 10.1. Uniform-grid sample of components of the simplified river-basin system.

Variable	Range	Starting values	
Free or multiple-value variables			
Reservoir capacities (10^6 acre ft)			
B, active storage	0–12.0[a]	4.0	8.0
B, dead storage	0– 6.0[a]	2.0	4.0
D	0–12.0	4.0	8.0
Target outputs			
Irrigation (10^6 acre ft)	0– 6.0	2.0	4.2
Energy (10^9 kw hr)	0– 4.0	1.35	2.7
Fixed or single-value variables			
Reservoir capacities (10^6 acre ft)			
A	0– 1.5	0.5	
C	0–15.0	2.0	
Power-plant capacities (Mw)			
B	0–800	500	
G	0–800	500	
Flood-control storage (10^6 acre ft)			
Reservoir *B*	0–2.0[b]	0.2	
Reservoir *C*	0–2.0[b]	0.3	
Reservoir *D*	0–2.0[b]	0.5	

[a] Subject to the constraint that the sum of the active- and dead-storage capacities of reservoir *B* cannot exceed 15 × 10^6 acre ft.

[b] Subject to the constraint that flood-control storage in any reservoir cannot exceed its own active-storage capacity.

the single values that were assigned to the remaining or fixed variables in accordance with preliminary information about them. It is seen that the capacity of reservoir *C* was made less than one-sixth of its maximum value. This was done because the primary task of reservoir *C* is to re-regulate flows that are already controlled by reservoirs *A* and *B* and by the irrigation works. Reservoir *A* was given a capacity equal to one-third its maximum possible value. At this level the reservoir can hold about 83 percent of the mean annual tributary flow and should be large enough to exercise adequate streamflow control under the adopted operating procedure, which calls first on reservoir *A* for irrigation releases. The capacities of power plants *B* and *G* were made large enough for either plant to meet the target output for energy of the system as a whole.

Any energy deficits in the simulation could not then be ascribed to lack of plant capacity. Total flood-control storage was made 1.0×10^6 acre ft, which appeared adequate for complete protection against the greatest flood of record. Space was allocated to the three large reservoirs in such a way as to place each of the component catchment areas under some control.

It was recognized that the values assigned to the fixed variables did not have to be strictly correct, because each of them could be examined in detail later on in the analysis. The net benefits associated with the 32 possible combinations of values for the 12 system variables were determined by simulation runs on the computer.

Analyzing the results. To exemplify the many detailed considerations that entered into our analysis of the results of individual sampling procedures, a full account is given below of one analysis of the uniform-grid results. Discussions of subsequent samplings are in much less detail.

Table 10.2. Net benefits identified by uniform-grid sample of 32 combinations of 5 system variables.

Selected combination of target outputs (10^6 acre ft for irrigation and 10^9 kw hr for energy)		Selected combination of reservoir capacities (10^6 acre ft)							
		Res. *B* active, 4.0 Res. *B* dead, 2.0 Res. *D*, 4.0	Res. *B* active, 4.0 Res. *B* dead, 2.0 Res. *D*, 8.0	Res. *B* active, 4.0 Res. *B* dead, 4.0 Res. *D*, 4.0	Res. *B* active, 4.0 Res. *B* dead, 4.0 Res. *D*, 8.0	Res. *B* active, 8.0 Res. *B* dead, 2.0 Res. *D*, 4.0	Res. *B* active, 8.0 Res. *B* dead, 2.0 Res. *D*, 8.0	Res. *B* active, 8.0 Res. *B* dead, 4.0 Res. *D*, 4.0	Res. *B* active, 8.0 Res. *B* dead, 4.0 Res. *D*, 8.0
I. Irrigation, 2.0; energy, 1.35	(a)	341	290	328	277	316	264	295	244
	(b)	0	0	0	0	0	0	0	0
	(c)	3.5	3.5	3.5	3.5	7.5	7.5	7.5	7.5
II. Irrigation, 4.2; energy, 1.35	(a)	521	452	493	442	481	430	461	410
	(b)	0	0	0	0	0	0	0	0
	(c)	1.3	1.1	1.1	1.1	5.1	5.1	5.1	5.1
III. Irrigation, 2.0; energy, 2.7	(a)	561	511	580	530	583	533	582	532
	(b)	92	90	61	59	36	35	0	0
	(c)	0	0	0	0	0	0	1.1	1.1
IV. Irrigation, 4.2; energy, 2.7	(a)	628	577	671	623	637	634	709	660
	(b)	120	120	77	76	85	58	34	30
	(c)	0	0	0	0	0	0	0	0

Note: Line (a) for each combination gives resulting net benefits in 10^6 dollars;
line (b) for each combination gives the number of class 3 energy deficits;
line (c) for each combination gives the excess capacity of reservoir *B* in 10^6 acre ft.

Results of the simulation runs for the uniform-grid sample indicated how net benefits varied with the magnitude of each of the five free variables when the remaining four were held constant. Starting with the target-output variables, we related the results of the 32 studies to the four possible combinations of output values identified by roman numerals in the left-hand column of Table 10.2. For each output combination there are eight possible combinations of the three reservoir-storage variables, with the sum of the capacities ranging from (10 to 20) $\times$ 10^6 acre ft. These combinations are shown in the eight columns of Table 10.2 and constitute eight different designs for providing the target outputs.

The broad range in the magnitude of reservoir capacities suggested the possibility of excess reservoir capacities as well as of irrigation and energy deficits, and many such instances were indeed discovered. For the eight pertinent storage combinations, output combinations I and II each showed excess reservoir capacity for all eight; output combination III, for two of the eight; and only output combination IV, none. Conversely, associated with all output combinations for which there were no excess reservoir capacities were substantial irrigation and energy deficits, including some class 3 energy deficits in each instance.

The sizable spread of net benefits shown in Table 10.2 resulted from the wide range in magnitude of the variables for the 32 system combinations. Extending from a low of $244,000,000 for output combination I, with greatest excess reservoir capacity, to a high of $709,000,000 for output combination IV, with no excess reservoir capacity but with measurable irrigation and energy deficits, net benefits rose as target outputs were made larger. They were greatest when reservoir storage was least for output combinations I and II. With net benefits of $341,000,000 there was still some excess reservoir capacity. When the cost of this overage was deducted from the total, revised net benefits of $430,000,000 reached their highest mark for output combination I.[11]

For output combination II maximum net benefits rose from $521,000,000 to $596,000,000 when excess storage was removed from the system.

Quite different were the benefits for output combinations III and IV

[11] To save on computer time, we made the revision without a rerun of the adjusted combination. Instead, we used information on excess reservoir capacity (given in the output block of Table 9.3) and on reservoir costs (Figs. 7.10 and 7.11) to make a rough estimate of the excess costs. The adjustment is only approximate because when the capacity of reservoir *B* is lowered, the average head of (and hence energy output from) power plant *B* is also lowered. A rerun of the adjusted combination would be necessary to establish its correct energy output and hence net benefits.

For these, the smallest reservoir combinations could not meet the target outputs in question without heavy deficits. As shown in Table 10.2, there were 92 class 3 or intolerable energy deficits for output combination III and the first reservoir combination. This was inadmissible. Either the reservoir capacities were too small or the outputs were too large. Accordingly, we studied the results of the remaining seven possible reservoir combinations. As is evident from Table 10.2, increasing the capacity of reservoir D alone was ineffective. Class 3 energy deficits remained large in number, while net benefits fell by almost 10 percent. Augmenting the active and dead storage in reservoir B singly increased net benefits by about 5 percent but failed to eliminate class 3 energy deficits. When both active and dead storage were enlarged, however, intolerable energy deficits vanished, net benefits rose over $20,000,000, and an excess capacity of 1.1×10^6 acre ft was uncovered. Eliminating this excess raised net benefits to about $595,000,000.

Class 3 energy deficits were encountered for all combinations of reservoir capacity with output combination IV. Highest net benefits accrued when the active- and dead-storage capacities of reservoir B were increased. However, 34 class 3 energy deficits were associated with the net-benefits maximum of $709,000,000. Whether a further increase in active and dead storage in reservoir B might achieve our target outputs, either in full or with tolerable deficits, remained unknown.

From this analysis we reached the following conclusions.

(1) The highest target outputs for irrigation and energy are accompanied by the greatest net benefits but also by intolerable energy deficits. Whether the deficits can be held in check by added storage requires further study.

(2) Within the accepted magnitudes of reservoir capacities all capacities of reservoir D above 4.0×10^6 acre ft are ineffective. Conversely, both the active and dead storages in reservoir B are beneficial up to their maximum sampling values of 8.0×10^6 and 4.0×10^6 acre ft, respectively. The effects of even higher values of these variables should be examined.

Accordingly, the ranges in the values of the free variables shown in Table 10.1 with which we had started were modified for subsequent examination by assigning capacities of $(0 \text{ to } 4.0) \times 10^6$ acre ft to reservoir D; active-storage capacities of $(1.5 \text{ to } 10.0) \times 10^6$ acre ft and dead-storage capacities of $(4.0 \text{ to } 10.0) \times 10^6$ acre ft to reservoir B; and target outputs of $(2.0 \text{ to } 5.5) \times 10^6$ acre ft to irrigation and of $(1.35 \text{ to } 2.70) \times$

Table 10.3. Application of steepest-ascent procedure to a first set of four free variables.

	Variables				Results			
	Target output for —		Capacity, reservoir B (10^6 acre ft)		Net benefits (10^6 dollars)			
Combination	Irrigation (10^6 acre ft)	Energy (10^9 kw hr)	Active storage	Dead storage	Actual	Change[b]	No. of class 3 energy deficits	Excess capacity at reservoir B (10^6 acre ft)
(1)	(2)	(3)	(4)	(5)	(6)	(7)	(8)	(9)
First sample:								
1. Starting base	4.2	2.0	5.5	7.0	611	—	0	2.6
Change in —								
2. Irrigation = +0.5	4.7	2.0	5.5	7.0	621	+10	0	0.8
3. Energy = +0.2	4.2	2.2	5.5	7.0	642	+31	0	2.3
4. Reservoir B, active storage = +2.0	4.2	2.0	7.5	7.0	580	−31	0	4.6
5. Reservoir B, dead storage = +1.0	4.2	2.0	5.5	8.0	596	−15	0	2.6
Second sample:								
1. Second base[a]	4.45	2.27	4.43	6.48	671	—	0	0
Change in —								
2. Irrigation = +0.25	4.70	2.27	4.43	6.48	652	−19	4	0
3. Energy = +0.13	4.45	2.40	4.43	6.48	663	−8	12	0
4. Reservoir B, active storage = +1.0	4.45	2.27	5.43	6.48	667	−4	0	0.73
5. Reservoir B, dead storage = +0.5	4.45	2.27	4.43	6.98	667	−4	0	0
Third sample:								
1. Third base	4.200	2.22	4.32	6.37	667	—	0	0.7
Change in —								
2. (a) Irrigation = +0.175	4.375	2.22	4.32	6.37	671	+4	0	0.1
(b) Irrigation = −0.175	4.025	2.22	4.32	6.37	661	−6	0	1.2
3. Energy = +0.14	4.200	2.36	4.32	6.37	676	+9	6	0
4. Reservoir B, active storage = +1.0	4.200	2.22	5.32	6.37	658	−9	0	1.8
5. Reservoir B, dead storage = +0.5	4.200	2.22	4.32	6.87	663	−4	0	0.7

[a] From column (10) of Table 10.4.
[b] The change in net benefits equals their actual value for the combination minus their value for the base within each sample.

10^9 kw hr to energy. The reasons for these ranges are obvious from Table 10.2 and will not be elaborated here.

APPLYING THE ITERATIVE TECHNIQUE OF STEEPEST ASCENT

Selecting the starting base. To reduce computations, we decided to start the steepest-ascent procedure with only four free variables. The capacity of reservoir D was transferred arbitrarily to the list of fixed variables shown in Table 10.1, its assigned value being 4.0×10^6 acre ft. Each of the four free variables was then given a value near the midpoint of its range, the discontinuity in the cost function for irrigation being allowed for by making the target output for irrigation 4.2×10^6 rather than 3.75×10^6 acre ft.

By assigning a new value to one of the four free variables, we added four combinations to the sample of basic variables. The size of the change or jump, Δx_i, for each variable was set arbitrarily at about one-fifth its range, R_i. Although the jump could be up or down, we decided arbitrarily to deal at this time with increases only. The results obtained in simulation runs of the five combinations are included in the first sample shown in Table 10.3. It is evident that the starting series of values was inefficient both in net benefits and in excess reservoir capacity. The latter was associated, too, with all four variant combinations, indicating that the boundary AA′ of Fig. 10.1 between the region of excess capacity and the region of output deficits was not breached. As pointed out earlier, the discontinuity in slope of the partial derivatives along this boundary may obstruct effective sampling by steepest ascent.

Finding the second base. The second base was found from Eqs. (10.1) and (10.3) through the intermediary use of dimensionless quantities or percentage-range ratios.

To k, the magnitude of change for each variable, we assigned a percentage-range ratio of 20 range units, or one-fifth of the range. Necessary calculations are shown and fully explained in Table 10.4. Column 10 of this table lists the values of variables x_i^1 for the second base.

Further iterations. As shown in the second sample, Table 10.3, the results obtained in simulation runs with the second base combination of variables were impressive. Net benefits rose from $611,000,000 to $671,-000,000 and excess reservoir capacity vanished. However, a moderate number of irrigation deficits did appear. A single iteration thus had carried us from an inefficient to an efficient region near the boundary

Table 10.4. Application of steepest-ascent procedure to a first set of four free variables. Finding the second base.

Variable and its natural units	Change in net benefits, ΔB_i (10^6 dollars)	Range, R_i (natural units)	Jump, Δx_i (natural units)	Jump, Δx_i (range units)	$\frac{\Delta x_i}{\Delta B_i}$	$x_i^1 - x_i^0$ (range units)	$x_i^1 - x_i^0$ (natural units)	Original base, x_i^0 (natural units)	Second base, x_i^1 (natural units)
(1)	(2)	(3)	(4)	(5)	(6)	(7)	(8)	(9)	(10)
Target output for —									
Irrigation (10^6 acre ft)	+10	3.50	0.5	14.29	1.43	6.71	0.24	4.20	4.45
Energy (10^9 kw hr)	+31	1.35	0.2	14.82	0.48	20.00	0.27	2.00	2.27
Reservoir *B* storage:									
Active (10^6 acre ft)	−31	8.50	2.0	23.53	−0.76	−12.63	−1.07	5.50	4.43
Dead (10^6 acre ft)	−15	6.00	1.0	16.67	−1.11	−8.65	−0.52	7.00	6.48

Note: In accordance with Eqs. (10.1) and (10.3), $x_i^1 - x_i^0 = C(\Delta B_i/\Delta x_i)$; $C = k \min \Delta x_i/\Delta B_i$; and for $k = 20$, $C = 20$ (min of column 6) $= 20(0.48) = 9.6$.

Column (2) from column (7) of Table 10.3.
Column (3) = ranges modified as described in text.
Column (4) from column (1) of Table 10.3.
Column (5) = 100 × column (4)/column (3).
Column (6) = column (5)/column (2).
Column (7) = C/column (6).
Column (8) = column (7) × column (3)/100.
Column (9) from starting base of first sample, Table 10.3.
Column (10) = column (8) + column (9).

along which target outputs can be just fulfilled without deficits and excess capacity.

For the next iteration of the second sample, sizes of the jumps were cut in half with the hope of avoiding discontinuities either in the net-benefit function itself through the creation of class 3 energy deficits or in the slope of the net-benefit function. As shown in Table 10.3, this objective was not realized. None of the variants improved the net benefits, two of them produced class 3 energy deficits, and one crossed the boundary between regions of output deficit and excess reservoir capacity.

Repeating the steepest-ascent procedure with these data established the third base magnitudes shown in Table 10.3. A computer run recorded a decline in net benefits from $671,000,000 to $667,000,000 and in addition some excess capacity at reservoir *B*. It was evident that we had crossed the boundary between the regions of output deficit and of excess reservoir capacity. The limitations of the steepest-ascent procedure in the presence of discontinuities in the derivative of the net-benefit function were clear.

The results of a third and final series of changes are also shown in Table 10.3. Because the third base combination was close to the boundary between output deficits and excess capacity, it was evident that the boundary would again be crossed in any move to a fourth base, and good results could not be expected.

Applying a modification of uniform-grid sampling — a diversion. At this point we temporarily abandoned the steepest-ascent procedure in favor of a modification of uniform-grid sampling whereby three variables were changed simultaneously. As shown in Table 10.5, the target output for energy was increased in three steps above its third base value of 2.22×10^9 kw hr, to the upper limit of its range at 2.7×10^9 kw hr. Two combinations of active and dead storage in reservoir *B* were tried for each target output for energy, the target output for irrigation being held constant at 4.2×10^6 acre ft. This value was assumed to be in the nature of a local optimum because of the coincident discontinuity in the cost of irrigation works.

The results were encouraging, particularly for the conditions in combination 2 of Table 10.5, for which there were no class 3 energy deficits. Elimination of the small excess in reservoir capacity placed the net benefits, without a computer rerun, at $693,000,000, the highest level attained thus far.

Table 10.5. Application of modified uniform-grid procedure to three free variables in the first set.

Combination	Fixed value	Variables			Results			
	Target output for irrigation (10^6 acre ft)	Target output for energy (10^9 kw hr)	Capacity, reservoir B (10^6 acre ft)		Net benefits (10^6 dollars)		No. of class 3 energy deficits	Excess capacity at reservoir B (10^6 acre ft)
			Active storage	Dead storage	Actual	Change[b]		
Low target output for energy								
1. Low active storage in res. B	4.2	2.435	5.14	6.86	682	+11	4	0
2. High active storage in res. B	4.2	2.435	6.00	6.00	687[a]	+16	0	0.4
Medium target output for energy								
3. Low active storage in res. B	4.2	2.57	5.57	7.43	681	+10	14	0
4. High active storage in res. B	4.2	2.57	6.50	6.50	694	+23	4	0
High target output for energy								
5. Low active storage in res. B	4.2	2.7	6.00	8.00	672	+1	22	0
6. High active storage in res. B	4.2	2.7	7.00	7.00	689	+18	18	0

[a] 693×10^6 dollars, or an increase of 22×10^6 dollars over previous admissible high value, when excess capacity at reservoir B is eliminated.

[b] The change in net benefits equals their actual value for the combination minus 671×10^6 dollars, net benefits for combination 2(a), third sample, Table 10.3. (previous admissible high value).

SPOT-CHECKING OTHER REGIONS OF THE RESPONSE SURFACE

At this juncture in our studies other likely regions of the net-benefit response surface were spot-checked for possible high values. On the one hand, these might lie along a boundary demarked by the absence of one or the other of the output variables of the simplified river-basin system.

Table 10.6. Application of modified uniform-grid procedure to a two-purpose system of irrigation and flood control, with two reservoirs (*A* and *B*) and a fixed target output for irrigation of 4.2 × 10⁶ acre ft.

	Reservoir capacity (10^6 acre ft)			
		B		
Combination	*A*	Total	Allocated to flood control	Net benefits (10^6 dollars)
1	0.4	2.2	0.8	560
2	0.4	2.0	0.6	563
3	0	2.6	0.8	565
4	0	2.4	0.6	567
5	0.4	2.0	0.8	557
6	0.4	1.8	0.6	557
7	0	2.4	0.8	565
8	0	2.2	0.6	564

Note: Reservoir *B* showed no excess capacity.

Table 10.7. Application of systematic sampling to a two-purpose system of irrigation and flood control, with a single reservoir *B* including a fixed flood-control allocation of 0.8 × 10⁶ acre ft.

Combination	Target output for irrigation (10^6 acre ft)	Capacity of reservoir *B* (10^6 acre ft)	Net benefits (10^6 dollars)
1	1.50	0.8	411
2	2.25	1.0	462
3	3.00	1.5	524
4	3.60	2.0	553
5	4.50	4.0	546
6	4.75	5.5	535
7	5.00	7.5	531

Note: Reservoir *B* showed no excess capacity.

On the other hand, high target outputs of one kind might combine with low outputs of another kind to create high interior net benefits.

Our spot checks, three in number, are summarized in Tables 10.6, 10.7, and 10.8. In the first check, the system was reduced to the two purposes

Table 10.8. Application of steepest-ascent procedure to the first set of four free variables, including high energy and low irrigation target outputs.

Combination	Variables: Target output for — Irrigation (10^6 acre ft)	Energy (10^9 kw hr)	Capacity, reservoir *B* (10^6 acre ft): Active storage	Dead storage	Results: Net benefits (10^6 dollars)	No. of class 3 energy deficits
1. Base	2.5	3.0	4.0	7.5	626	73
Change in —						
2. Irrigation = +0.5	3.0	3.0	4.0	7.5	651	78
3. Energy = +0.2	2.5	3.2	4.0	7.5	627	103
4. Reservoir *B*, active storage = +2.0	2.5	3.0	6.0	7.5	610	57
5. Reservoir *B*, dead storage = +1.0	2.5	3.0	4.0	8.5	619	69

Note: Reservoir *B* showed no excess capacity.

of irrigation and flood control and two component structures, reservoirs *A* and *B*, with a fixed target output for irrigation. The second check used the same two purposes, but reservoir *A* was removed, flood-control storage in reservoir *B* was given a fixed value, and the target output for irrigation was allowed to vary. In the third check, the four free variables of the system previously examined were restudied, by means of the steepest-ascent procedure, for the effects of high target outputs for energy and low ones for irrigation.

As shown in the three tables, the net benefits ranged from $557,000,000 to $567,000,000, from $411,000,000 to $553,000,000, and from $610,000,000 to $651,000,000, respectively. Optimal values, obviously, did not lie along the boundaries examined and high target outputs for energy could not be achieved without numerous class 3 energy deficits. Of note is the fact that the highest net benefits in Table 10.8 point in the direction of the previously observed optimum and not to a local interior high value.

These spot checks left us reasonably certain of where the highest net benefits for the four free variables must be situated. The results obtained thus far are brought together in Fig. 10.4. Further experimentation might

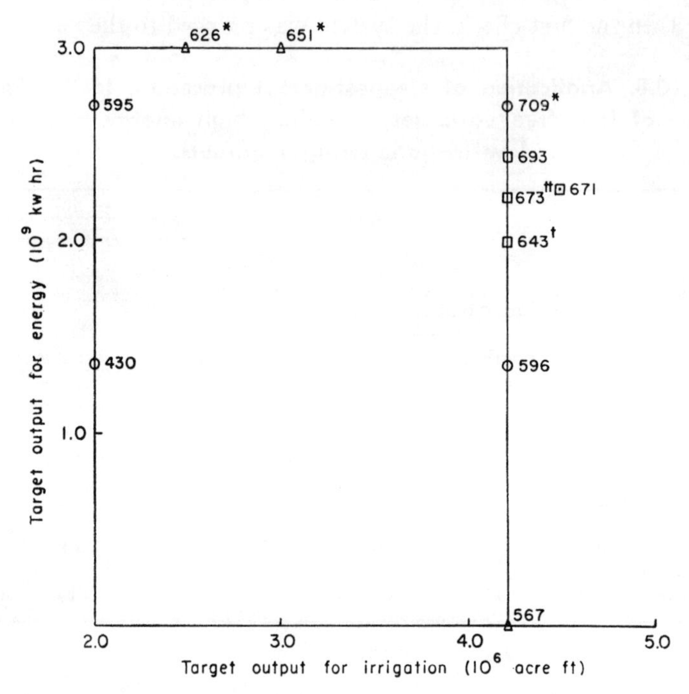

Fig. 10.4. Net benefits obtained in the first 74 simulation runs and associated with high-value combinations of target outputs for irrigation and energy. Net-benefit figures are adjusted to minimum required reservoir sizes and recorded in 10^6 dollars. Circles denote the first sample of 5 free variables in 32 combinations; squares, the subsequent samples of 4 free variables; triangles, the spot checks; asterisks, combinations inadmissible because of class 3 energy deficits. The single dagger after value 643 indicates that the unadjusted value from Table 10.3 is 611; similarly, the double dagger after value 673 indicates an unadjusted value of 667.

have isolated somewhat better combinations, but freeing some of the remaining variables was considered more promising.

EXAMINING THE REMAINING VARIABLES

A study of reservoirs A, C, and D by the steepest-ascent procedure. Of the seven fixed variables shown in Table 10.1, we selected the capacities of reservoirs *A*, *C*, and *D* to be varied next, along with energy output and hence power-plant capacities. Attention was focused on the downstream portion of the simplified river-basin system, which contributes nothing to irrigation output. As a starting base for the new steepest-ascent procedure, the variables were assigned the best feasible values thus far attained (those listed for combination 2 of Table 10.5), but eliminating the excess capacity at reservoir *B*.

The composition of the starting base is shown in Table 10.9. In the

Table 10.9. Steepest-ascent sample of a second set of components of the simplified river-basin system.

Variable	Range	Starting base
Free or multiple-value variables		
Reservoir capacities (10^6 acre ft)		
A	0–0.5	0.5
C	0.3–4.0	2.0
D	0.5–5.0	4.0
Target output		
Energy (10^9 kw hr)	2.2–2.5	2.435
Power-plant capacities (Mw)[a]		
B		470
G		470
Fixed or single-value variables		
Target output		
Irrigation (10^6 acre ft)		4.2
Reservoir capacities (10^6 acre ft)		
B, active storage		5.6
B, dead storage		6.0
Flood-control storage (10^6 acre ft)		
Reservoir *B*		0.2
Reservoir *C*		0.3
Reservoir *D*		0.5

[a] Power-plant capacities are dependent on the selected target output for energy.

absence of advance information the ranges for the free variables were chosen somewhat arbitrarily. Storage in reservoir *A* ranges well above its mean annual inflow, and storage in reservoir *C* is up to about 70 percent of its mean annual inflow. Although capacities above 4.0×10^6 acre ft for reservoir *D* had not proved useful, storage space as high as 5.0×10^6 acre ft was allocated. The lower limits of storage in reservoirs *C* and *D* equal the flood-storage capacities previously assigned to them. The span of the target output for energy is narrowed to $(2.2 \text{ to } 2.5) \times 10^9$ kw hr, the upper limit being such that class 3 energy deficits appear even when the reservoirs are of maximum size. The capacities of power plants *B* and *G*

Table 10.10. Application of steepest-ascent procedure to a second set of four free variables.

Combination	Variables				Results			
	Target output for energy (10^9 kw hr)	Reservoir capacity (10^6 acre ft)			Net benefits (10^6 dollars)		No. of class 3 energy deficits	Excess capacity at reservoir B (10^6 acre ft)
		C	D	A	Actual	Change[a]		
First sample:								
1. Starting base	2.435	2.000	4.00	0.5	700	—	0	0
Change in —								
2. Energy = +0.065	2.500	2.000	4.00	0.5	697	−3	12	0
3. Reservoir C = −0.725	2.435	1.275	4.00	0.5	715	+15	0	0
4. Reservoir D = −0.9	2.435	2.000	3.10	0.5	710	+10	0	0
Second sample:								
1. Second base	2.425	1.26	3.45	0.5	721	—	0	0.2
Change in —								
2. Energy = −0.07	2.355	1.26	3.45	0.5	714	−7	0	1.0
3. (a) Reservoir C = −0.48	2.425	0.78	3.45	0.5	734	+13	2	0
(b) Reservoir C = −0.96	2.425	0.30	3.45	0.5	738	+17	6	0
4. (a) Reservoir D = −0.55	2.425	1.26	2.90	0.5	730	+9	0	0.1
(b) Reservoir C = −0.48 and reservoir D = −0.55	2.425	0.78	2.90	0.5	742	+21	2	0
Third sample:								
1. Third base	2.425	1.26	2.90	0.5	730	—	0	0.1
Change in —								
2. Energy = −0.04	2.385	1.26	2.90	0.5	728	−2	0	0.6
3. Reservoir C = −0.2	2.425	1.06	2.90	0.5	735	+5	0	0.1
4. Reservoir D = −0.25	2.425	1.26	2.65	0.5	734	+4	0	0
5. Reservoir A = −0.5	2.425	1.26	2.90	0	738	+8	0	0.1
Fourth base	2.400	1.06	2.70	0	747	—	0	0.3

[a] The change in net benefits equals their actual value for the combination minus their value for the base within each sample.

were computed from the selected target output for energy, each plant being able to meet the peak energy load of the entire system.

Jumps, up or down, for the free variables were set at about one-fourth their range. Net benefits and other pertinent results of computer runs are shown in Table 10.10 for three sequential samplings. Three variants were included in the first procedure, the second base being found without considering class 3 energy deficits as a constraint. In the second procedure the number of variants was raised to five, and class 3 energy deficits were penalized arbitrarily by $5,000,000 for each recorded deficit. Four variants were studied in the third procedure, including the elimination of reservoir *A*. The resulting fourth base recorded net benefits of $747,-000,000 and an excess capacity at reservoir *B* of 0.3×10^6 acre ft.

We may conclude that the steepest-ascent procedure worked well with these variables. Large reductions were possible in the reservoir-capacity variables without significantly reducing actual outputs and hence gross benefits, and there was a substantial improvement in the design of the system in only 16 trials. However, it is recognized that the introduction of heavy penalties for class 3 energy deficits and the occasional crossing of the boundary of excess capacity and output deficits brings into play discontinuities in the benefit function or its derivative which divert us from the steepest gradient.

Table 10.11. Systematic sampling of power-plant capacities for a constant target output of 2.4×10^9 kw hr.

Combination	Power-plant capacities (Mw)		Net benefits (10^6 dollars)		No. of class 3 energy deficits
	B	*G*	Actual	Change	
Starting base	460	460	747	—	0
Change in capacity at power plant—					
1. $B = +40$	500	460	742	−5	0
2. $B = -60$	400	460	756	+9	0
3. $B = -85$	375	460	757	+10	0
4. $B = -85$; $G = -60$	375	400	765	+18	0
5. $B = -135$; $G = -110$	325	350	776	+29	0
6. $B = -185$; $G = -110$	275	350	769	+22	10
7. $B = -135$; $G = -160$	325	300	783	+36	0
8. $B = -185$; $G = -160$	275	300	776	+29	10

Further analysis of these four variables was suspended in order to examine power-plant capacities as a group.

A systematic study of power-plant capacities. Thus far in our analysis power plants had either been assigned fixed capacities or fixed relationships to the target output for energy, which had to be satisfied in either case. This requirement was now relaxed.

Once again, the best design thus far obtained — in this instance, the fourth base combination of Table 10.10 — was made the starting base. For this base power-plant capacities were found to be 460 Mw at each of the two plants. The eight variants chosen for analysis in groups of four are shown in Table 10.11, together with their effects upon the system. For the first group of variants, net benefits reached $765,000,000 when both power plants were reduced in size. Evidently, the target output for energy could be met, while the loss of dump energy was more than offset by savings in capital costs. For the second group of variants, best results were associated with capacities of 325 Mw at *B* and 300 Mw at *G*. The systematic procedure employed in this analysis of the two system variables, it should be noted, advanced the base to an approximately optimal combination of these variables in only eight jumps.

A systematic study of reservoir capacities for flood storage. The last remaining variables, namely, the allocations of reservoir capacity for

Table 10.12. Systematic sampling of flood-storage capacities and allocations.

	Flood-storage capacity (10^6 acre ft)				Gross flood-control benefits (10^6 dollars)
Combination	Total	Reservoir *B*	Reservoir *C*	Reservoir *D*	
1	0.8	0.16	0.24	0.4	352
2	0.8	0.32	0.48	0	352
3	0.8	0.8	0	0	352
4	0.8	0	0.8	0	352
5	0.6	0.12	0.18	0.3	352
6	0.6	0.24	0.36	0	352
7	0.6	0.6	0	0	352
8	0.6	0	0.6	0	352
9	0.4	0.08	0.12	0.2	352
10	0.4	0.16	0.24	0	352
11	0.4	0.4	0	0	352
12	0.4	0	0.4	0	352
13	0.2	0	0.2	0	352
14	0.1	0	0.1	0	352
15	0	0	0	0	317

flood control, were now studied by the systematic sampling of a uniform grid with the 15 variants shown in Table 10.12. In all prior sampling with high outputs for irrigation and energy 1.0×10^6 acre ft of flood storage, allocated to reservoirs *B*, *C*, and *D* in the ratio 2:3:5, had eliminated all flood damages. These damages were still prevented when as little as 0.1×10^6 acre ft of space in reservoir *C* was allocated to flood control with no flood-control storage in reservoirs *B* and *D*. Flood-storage capacity in *C* as high as 0.4×10^6 acre ft does not diminish the amount of dump energy or the net benefits. Net benefits for this combination (no. 12), when rerun on the computer with a slightly lower capacity of reservoir *B* were $784,000,000.

CHECKING THE APPARENT OPTIMAL DESIGN BY MARGINAL ANALYSIS

As summarized in Table 10.13, systematic analysis of all 12 system variables had identified the values of the variables that would yield net benefits of $784,000,000 while meeting target outputs without class 3 energy shortages and with only a negligible amount of excess space in

Table 10.13. Best combination of system variables after sampling all 12 variables.

Variable	*Value*
Reservoir capacities (10^6 acre ft)	
A	0
B, active storage	5.6
B, dead storage	6.0
C	1.06
D	2.7
Target outputs	
Irrigation (10^6 acre ft)	4.2
Energy (10^9 kw hr)	2.4
Power-plant capacities (Mw)	
B	325
G	300
Flood-control storage (10^6 acre ft)	
Reservoir *B*	0
Reservoir *C*	0.4
Reservoir *D*	0
Results	
Net benefits (10^6 dollars)	784
Class 3 energy deficits	0
Irrigation deficits	0
Excess capacity, reservoir *B* (10^6 acre ft)	0.05

reservoir *B*. Some testing of the variables for changes that might further improve net benefits seemed to be in order. Accordingly, all variables were freed for sampling except the flood-storage allowances which had only just been determined. Selected for this testing was a simple version of marginal analysis by which pairs of system variables were changed until the marginal benefit of the last pair of increments equaled their marginal costs. More specifically, the value of one variable was changed by a small amount, after which the amount by which some other variable had to be altered so as to increase net benefits was calculated from the available cost and benefit data. The new combination of variables was then run through the computer and the results examined. To give an example: an arbitrary small reduction of the target output for irrigation was followed by calculating (from irrigation-benefit data) the associated reduction in gross irrigation benefits, and estimating (from cost data) the amount by which the capacity of reservoir *B* must be decreased to offset the decline in benefits.

A check of system outputs. A check of the target outputs for irrigation and energy is shown in Table 10.14. It is seen that no gains were achieved by varying either irrigation or energy outputs.

Table 10.14. Checking the apparent optimal design by marginal analysis of system outputs.

	Variables			Results		
	Target output for —		Active-storage capacity, reservoir *B*	Net benefits[a] (10^6 dollars)		No. of class 3 energy deficits
Combination	Irrigation (10^6 acre ft)	Energy (10^9 kw hr)	(10^6 acre ft)	Actual	Change	
Starting base	4.2	2.4	5.6	784	—	0
Change in —						
1. Irrigation = −0.4; reservoir *B*, active storage = −1.6	3.8	2.4	4.0	780	−4	0
2. Irrigation = +0.2; reservoir *B*, active storage = +0.9	4.4	2.4	6.5	772	−12	4
3. Energy = −0.1; reservoir *B*, active storage = −1.0	4.2	2.3	4.6	782	−2	0
4. Energy = +0.1; reservoir *B*, active storage = +1.4	4.2	2.5	7.0	784	0	0

[a] Excess capacity at reservoir *B* has been eliminated where necessary.

Table 10.15. Checking the apparent optimal design by marginal analysis of reservoir capacities. First study.

	Variables					Results		
	Reservoir capacity (10^6 acre ft)					Net benefits[a] (10^6 dollars)		
Combination	A	B, active storage	B, dead storage	C	D	Actual	Change	No. of class 3 energy deficits
1. Starting base	0	5.6	6.0	1.06	2.7	784	—	0
Change in capacity of —								
2. Reservoir $A = +0.2$; reservoir B, active storage $= -0.3$	0.2	5.3	6.0	1.06	2.7	782	−2	0
3. Reservoir B, dead storage $= +0.5$; reservoir B, active storage $= -0.3$	0	5.3	6.5	1.06	2.7	782	−2	0
4. Reservoir B, dead storage $= -0.5$; reservoir B, active storage $= +0.5$	0	6.1	5.5	1.06	2.7	792	+8	0
5. Reservoir $C = +0.5$; reservoir B, active storage $= -0.3$	0	5.3	6.0	1.56	2.7	775	−9	0
6. Reservoir $C = -0.46$; reservoir B, active storage $= +0.5$	0	6.1	6.0	0.60	2.7	790	+6	2
7. Reservoir $D = +0.5$; reservoir B, active storage $= -0.4$	0	5.2	6.0	1.06	3.2	781	−3	0
8. Reservoir $D = -0.2$; reservoir B, active storage $= +0.1$	0	5.7	6.0	1.06	2.5	788	+4	2

[a] Excess capacity at reservoir B has been eliminated where necessary.

A check of reservoir capacities. Checking our reservoir capacities by incremental analysis consisted first of varying the capacities of reservoirs

Table 10.16. Checking the apparent optimal design by further marginal analysis of reservoir capacities. Second study.

	Variables				Results		
	Reservoir capacity (10^6 acre ft)				Net benefits[a] (10^6 dollars)		
Combination	*B*, active storage	*B*, dead storage	*C*	*D*	Actual	Change[b]	No. of class 3 energy deficits
First check:							
1. Starting base	5.6	6.0	1.0	2.7	784	—	0
Change in capacity of —							
2. Reservoir *B*, dead storage = −1.0; reservoir *B*, active storage = +0.3	5.9	5.0	1.0	2.7	793	+9	0
3. Reservoir *C* = −1.0; reservoir *B*, active storage = +1.8	7.4	6.0	0	2.7	791	+7	8
4. Reservoir *D* = −0.4; reservoir *B*, active storage = +0.2	5.8	6.0	1.0	2.3	792	+8	0
Second check:							
1. Second base	7.4	6.0	0	2.7	791	—	8
Change in capacity of —							
2. Reservoir *B*, dead storage = −1.0; reservoir *B*, active storage = +0.9	8.3	5.0	0	2.7	798	+7	0
Third check:							
1. Third base	8.3	5.0	0	2.7	798	—	0
Change in capacity of —							
2. Reservoir *B*, dead storage = −0.5; reservoir *B*, active storage = 0	8.3	4.5	0	2.7	805	+7	0
3. Reservoir *D* = −0.4; reservoir *B*, active storage = +0.3	8.6	5.0	0	2.3	799	+1	2
Fourth check:							
1. Fourth base	8.3	4.5	0	2.7	805	—	0
Change in capacity of —							
2. Reservoir *B*, dead storage = −0.5; reservoir *B*, active storage = +0.5	8.8	4.0	0	2.7	804	−1	0
3. Reservoir *B*, dead storage = −1.0; reservoir *B*, active storage = +0.5	8.8	3.5	0	2.7	808	+3	8
4. Reservoir *B*, dead storage = −1.5; reservoir *B*, active storage = +1.1	9.4	3.0	0	2.7	810	+5	0
5. Reservoir *B*, dead storage = −2.0; reservoir *B*, active storage = +1.4	9.7	2.5	0	2.7	811	+6	2
Fifth check:							
1. Fifth base	9.4	3.0	0	2.7	810	—	0
Change in capacity of —							
2. Reservoir *D* = −0.5; reservoir *B*, active storage = +0.6	10.0	3.0	0	2.2	811	+1	0

[a] Excess capacity at reservoir *B* has been eliminated where necessary.

[b] The change in net benefits equals their actual value for the combination minus their value for the base within each sample.

A, C, and D and the dead storage in reservoir B, up and down, and, in each case, making a compensating change in the active capacity of reservoir B. The results for this first study are shown in Table 10.15. They suggested that it might pay to give still smaller sizes to reservoirs C and D and less dead storage to reservoir B.

How this was done in a second study comprising five successive checks is shown in Table 10.16. It is seen that reservoir C now drops out of the system. Also, the fifth check established a feasible combination of reservoir capacities of 3.0×10^6 acre ft of dead storage and 10.0×10^6 acre ft of active storage in reservoir B and 2.2×10^6 acre ft of storage in reservoir D, and produced net benefits of $811,000,000. Any further decrease in dead storage in reservoir B gave class 3 energy deficits.

Other checks. At this point a final check was made of power plants B and G. Their capacities were each reduced by 25 Mw. Although net benefits rose by $4,000,000, they did so only at the price of two class 3 energy deficits. In addition, final marginal-analysis checks of the target outputs for irrigation and energy showed that no further gain in net benefits was possible when their magnitudes were moved up or down. Accordingly, the combination of variables that yielded net benefits of $811,000,000

Table 10.17. Optimal combination of the 12 variables constituting the simplified river-basin system.

Variable	*Value*
Reservoir capacities (10^6 acre ft)	
A	0
B, active storage	10.0
B, dead storage	3.0
C	0
D	2.2
Target outputs	
Irrigation (10^6 acre ft)	4.2
Energy (10^9 kw hr)	2.4
Power-plant capacities (Mw)	
B	325
G	300
Flood-control storage (10^6 acre ft)	
Reservoir B	0.4
Reservoir C	0
Reservoir D	0
Results	
Net benefits (10^6 dollars)	811
Class 3 energy deficits	0
Irrigation deficits	0
Excess capacity, reservoir B (10^6 acre ft)	0

was judged to be optimal. The values of the 12 design components included in this combination are brought together in Table 10.17.

EXPLORING THE RESPONSE SURFACE BY RANDOM SAMPLING

The first systematic sample of 32 combinations of the variables, in which only 5 of the 12 components were free to change, had provided only a rough measure of the topography of the response surface. We therefore turned to random sampling for more detailed mapping of this surface. Random sampling was also used for exploration of the area surrounding the apparent optimal combination.

First random sample. All variables were considered free except the capacities of the power plants, which depended implicitly on the target outputs for energy. With the results of the first systematic sample at hand we selected ranges for the variables and imposed constraints that would rule out combinations obviously inadmissible or resulting in low net benefits. The accepted values are shown in Table 10.18.

Thirty combinations of the 12 system variables were sampled at random. This was done by assigning values to the variables in each com-

Table 10.18. Ranges of free and dependent variables and constraints accepted for first random sample.

Variable	Range	Constraint
	Free variables	
Reservoir capacities (10^6 acre ft)		
A	0–1.0	
B, total storage		$\nless 8.0$ and $\ngtr 15.0$
B, active storage	0–11.0	
B, dead storage	4.0–9.0	
C	0–5.0	
D	0–5.0	
Target outputs		
Irrigation (10^6 acre ft)	2.0–5.5	$\nleq 2.0$ with energy ≤ 1.35 $\ngeq 4.2$ with energy ≥ 2.7
Energy (10^9 kw hr)	1.0–3.0	$\nleq 1.35$ with irrigation ≤ 2.0 $\ngeq 2.7$ with irrigation ≥ 4.2
Flood-control storage (10^6 acre ft)		
Reservoir *B*	0–0.5	
Reservoir *C*	0–0.5	$\ngtr$ total capacity
Reservoir *D*	0–0.5	$\ngtr$ total capacity
	Dependent variables	
Power-plant capacities (Mw)		
B	140–435	
G	125–375	

Table 10.19. Values of variables assigned for first random sample of 30 combinations.

	Reservoir capacity (10^6 acre ft)					Target output for —		Power-plant capacity (Mw)		Flood-control storage (10^6 acre ft)		
		B										
Combi-nation	*A*	Active storage	Dead storage	*C*	*D*	Irrigation (10^6 acre ft)	Energy (10^9 kw hr)	*B*	*G*	Reservoir *B*	Reservoir *C*	Reservoir *D*
1	0.332	3.55	6.95	2.16	2.52	4.60	1.38	185	175	0.407	0.296	0.392
2	0.250	2.84	7.72	4.61	1.03	2.95	1.63	220	205	0.419	0.269	0.202
3	0.049	4.62	7.00	1.75	3.11	4.93	1.02	140	130	0.044	0.281	0.225
4	0.747	1.86	8.85	0.92	0.50	4.95	1.25	170	155	0.367	0.450	0.059
5	0.659	8.38	5.77	2.73	1.76	2.04	2.00	270	250	0.435	0.220	0.110
6	0.486	5.15	4.91	1.06	1.35	5.09	2.41	325	300	0.223	0.244	0.117
7	0.943	5.64	6.81	4.97	1.63	3.74	2.34	315	295	0.221	0.449	0.314
8	0.676	7.74	5.15	4.81	2.72	4.14	2.76	375	345	0.324	0.320	0.238
9	0.688	6.36	4.17	2.19	4.60	2.09	1.49	200	185	0.337	0.067	0.042
10	0.692	3.25	7.82	0.79	2.28	2.93	2.88	390	360	0.150	0.318	0.244
11	0.654	4.24	7.24	0.52	3.55	5.00	1.75	235	220	0.239	0.190	0.368
12	0.780	2.79	8.70	2.37	2.39	4.88	1.17	160	145	0.292	0.126	0.399
13	0.500	1.93	8.29	4.49	4.78	2.29	2.13	290	265	0.007	0.188	0.044
14	0.748	5.94	6.78	2.86	0.43	4.36	2.29	310	285	0.278	0.467	0.254
15	0.878	4.26	5.61	1.58	1.79	3.12	2.17	295	270	0.159	0.115	0.496
16	0.432	6.11	7.49	2.56	1.00	5.34	1.26	170	160	0.437	0.312	0.089
17	0.796	5.83	7.03	1.98	1.83	3.03	2.24	305	280	0.331	0.455	0.043
18	0.002	4.57	5.57	2.41	1.17	4.55	1.05	140	130	0.205	0.072	0.203
19	0.104	6.72	7.08	1.24	0.51	4.95	1.91	260	240	0.306	0.175	0.465
20	0.668	6.54	5.12	1.24	4.78	2.97	2.52	340	315	0.080	0.373	0.365
21	0.356	2.20	7.74	0.48	4.78	4.96	1.12	150	140	0.126	0.408	0.496
22	0.367	6.09	4.38	4.35	2.67	3.98	2.55	345	320	0.379	0.174	0.251
23	0.393	7.49	4.38	4.33	4.19	2.43	1.83	250	230	0.002	0.432	0.169
24	0.984	4.72	5.92	1.56	4.34	3.50	1.46	200	185	0.174	0.184	0.261
25	0.522	4.27	5.88	0.77	4.89	2.21	2.73	370	340	0.470	0.293	0.219
26	0.004	6.69	7.13	0.93	3.32	2.08	2.78	375	350	0.139	0.394	0.404
27	0.571	3.32	5.15	2.76	1.58	5.36	2.52	340	315	0.394	0.459	0.428
28	0.655	3.94	5.19	1.92	2.01	4.34	1.16	155	145	0.274	0.194	0.063
29	0.561	6.00	6.37	4.52	2.52	2.66	2.17	295	270	0.154	0.023	0.287
30	0.861	7.13	6.99	2.54	2.76	4.80	2.51	340	315	0.256	0.470	0.203

bination through the medium of random number tables. Ours were taken from the RAND series, and are shown in Table 10.19. Table 10.20 gives

Table 10.20. Summary of results, first random sample of 30 combinations.

	No. of deficits		Minimum active reservoir contents (10^6 acre ft)[a]			Reservoir sizes in relation to output[b]			Net benefits (10^6 dollars)	
Combination	Irrigation, class 2	Energy, class 3	*B*	*C*	*D*	*B*	*C*	*D*	Unadjusted	Adjusted[c]
1	6	0	0	1.03]	0.89	−	+	+	557	603
2	0	0	1.43	0	0	+	...	...	584	597
3	12	0	0	1.01	2.13	−	+	+	487	564
4	41	5	0	0	0	−	−	−	498	498
5	0	0	7.02	0	0	+	...	...	517	593
6	45	78	0	0	0	−	−	−	643	643
7	0	0	1.44	0	0	+	...	...	696	713
8	4	35	0	0	0	−	−	−	714	714
9	0	0	5.79	0	0	+	...	...	409	456
10	4	77	0	0	0	−	−	−	731	731
11	16	0	0	0	0	−	...	...	650	650
12	16	0	0	1.47	0.38	−	+	+	477	519
13	0	0	0.50	0	0	+	...	...	540	545
14	8	20	0	0	0	−	−	−	738	738
15	0	0	1.94	0	0	+	...	...	697	712
16	16	0	0	0.92	0	...	+	...	500	519
17	0	0	3.47	0	0	+	...	...	657	696
18	0	0	0	1.64	0.14	...	+	+	532	576
19	30	22	0	0	0	−	−	−	652	652
20	0	0	2.24	0	0	+	...	...	685	707
21	10	0	0	0	3.55	−	...	+	509	569
22	4	28	0	0	0	−	−	−	722	722
23	0	0	6.74	0.45	0	+	+	...	499	561
24	0	0	3.06	0.30	2.18	+	+	+	519	580
25	0	43	0	0	0	−	−	−	664	664
26	0	4	0	0	0	−	−	−	656	656
27	70	101	0	0	0	−	−	−	533	533
28	0	0	0.64	1.01	0.75	+	+	+	561	611
29	0	0	4.47	0	0	+	...	...	578	625
30	12	20	0	0	0	−	−	−	685	685

[a] Considered to be an approximate measure of excess reservoir capacity.
[b] + = too large; − = too small.
[c] Adjustment consists of deducting capital costs of excess reservoir capacity and increasing net benefits accordingly.

the results of the simulation runs of the 30 randomized combinations of the design components.

Not unexpectedly, a number of combinations proved to be either inefficient or inadmissible. According to Table 10.20, excess reservoir capacities rendered 18 combinations inefficient, and class 3 energy deficits made 11 inadmissible. Indeed, only one of the 30 designs (combination 11) was both efficient and admissible.

When the net benefits of the inefficient combinations were adjusted roughly by deducting the costs of excess reservoir capacities, and net benefits of all combinations were plotted as functions of the target outputs for irrigation and energy, the net-benefit ridge or hill shown in Fig. 10.5 was outlined by interpolated lines of equal net benefits. Starting

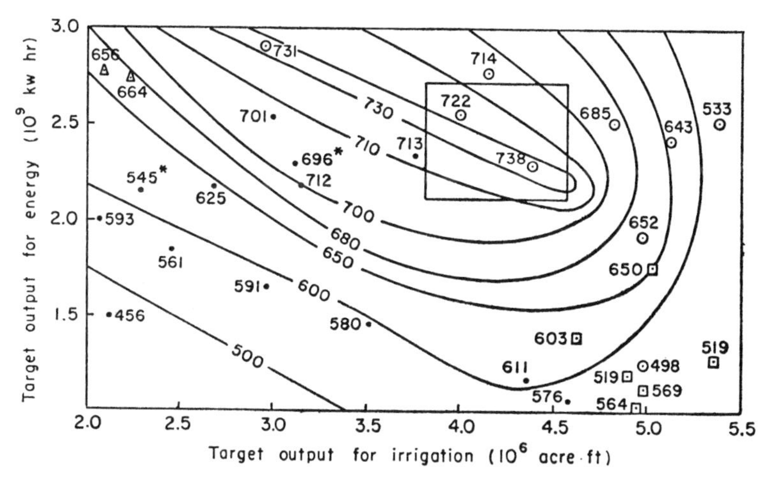

Fig. 10.5. Net-benefit response surface for first random sample of 30 combinations. Net-benefit figures are adjusted to minimum required reservoir sizes and recorded in 10^6 dollars. Circles denote combinations with class 3 energy deficits plus irrigation deficits; triangles, combinations with energy deficits only; squares, combinations with irrigation deficits only; and asterisks, those with obviously nonoptimal combinations of reservoirs. Box indicates region covered by second random sample of 20 combinations.

from the lower left-hand corner of the plot, where target outputs of both kinds possess low values, we see that net benefits increase in magnitude moving diagonally upward to higher values of the two target outputs, until the region of highest net benefits is reached when target outputs for irrigation lie between (3.0 and 4.5) $\times 10^6$ acre ft and target outputs for energy between (2.0 and 3.0) $\times 10^9$ kw hr. At still higher target outputs of both kinds the irrigation and energy loss functions become depressing factors, net benefits drop sharply, and class 3 energy deficits appear.

The net-benefit response surface that emerges in Fig. 10.5 is obviously only approximately correct because the sample was so small. Some combinations could not be fitted into the first contour map of the response surface. Examples are the two combinations (13 and 17) marked with asterisks. These comprise uneconomically high total storages for reservoirs *C* and *D* and dead storage for reservoir *B*. The combinations in the lower right-hand corner of Fig. 10.5, too, fail to be consistent in their arrangement, because the sizing of their reservoirs varies widely in effectiveness. These are the exceptional results within the general pattern of response. A better combination of reservoirs within each design should do away with these minor variations and permit lines of equal net benefits, not unlike those shown in Fig. 10.5, to emerge in more definitive form.

The reader will recall the discussion, earlier in this chapter, of useful probability statements that can be made about results obtained from random sampling. If, in fact, net-benefit values are left unadjusted and

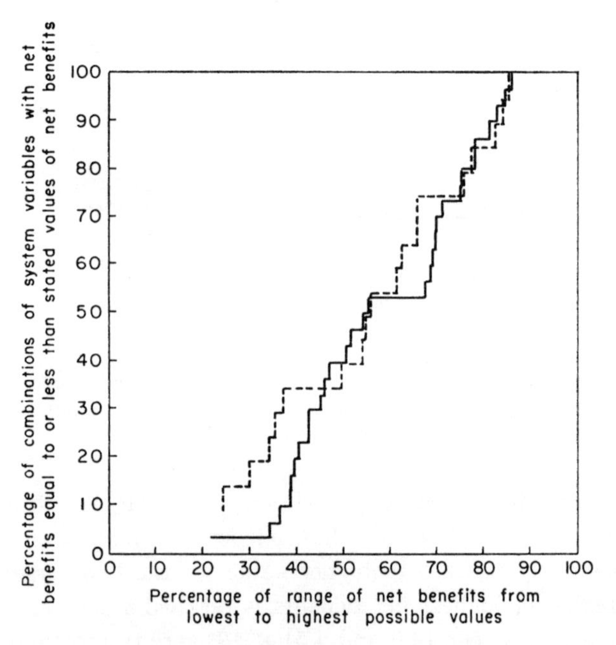

Fig. 10.6. Cumulative frequency distribution of net benefits obtained in first random sample of 30 and second of 20 combinations of selected system variables. Net benefits are expressed as percentages of the range in net benefits.

——— *First random sample, range in net benefits taken from 300 × 10⁶ dollars at 0 percent to 811 × 10⁶ dollars at 100 percent.*

- - - - *Second random sample, range in net benefits taken from 720 × 10⁶ dollars at 0 percent to 790 × 10⁶ dollars at 100 percent.*

no combinations, not even those with class 3 energy deficits, are left out, from page 403 we see that the probability of at least one of our 30 combinations of variables recording net benefits within the upper 10 percent of the entire population is 95.7 percent. Our ten free variables (excluding the dependent power-plant capacities), with increments ranging in number from 200 for some variables to 1000 for others, constitute a large population of the order of 5×10^{26} possible combinations. The probability statement for our first random sample thus becomes: The best design obtained by random sampling, namely $738,000,000 in unadjusted net benefits for combination 14 of Table 10.20, has a 95.7 percent chance of being at least as good as the best 5×10^{25} combinations in the total population.

The meaning of this statement is illustrated in Fig. 10.6, which contains a plot of the cumulative frequency distribution of the net benefits recorded in the first random sample. The range of this frequency distribution was constructed in the following way. The highest value achieved in our systematic sampling, $811,000,000, was made the ultimate maximum, or 100 percent magnitude, and the value of the least profitable combination lying within the assumed ranges of variables, $300,000,000, was made the minimum or 0 percent magnitude.

It is seen that the highest observed net benefits in the random sample, $738,000,000, lie at approximately the 86th percentile of the range between highest and lowest net benefits. The calculation is $100 \times (738 - 300)/(811 - 300) = 86$ percent. According to combination 14 in Table 10.20, however, this design has 20 class 3 energy deficits and so is inadmissible by our criterion. Net benefits of the best admissible combination, $713,000,000 in combination 7, equal about 81 percent of the possible maximum.

We may conclude that, had we chosen a strategy that began with a random sampling, our first sample would have identified the shape of the net-benefit response surface sufficiently closely to establish a good starting point for further analysis.

Second random sample. In our second random sample the ranges of the system variables were narrowed down to values within apparently optimal combinations.[12] Only the six clearly most influential variables were kept free. The remainder were fixed — the capacities of the two power

[12] As it turned out, the values of active and dead storage of reservoir B and total storage of reservoir C in the optimal combination (shown in Table 10.17) lay beyond the ranges for these variables established for the second random sample.

plants being based, as before, upon the target output for energy. All combinations that had proved inadmissible in prior studies were ruled out, and it became possible to cover the variables in a sample of only 20 combinations of system components. The region covered is indicated on the plot of the first sample in Fig. 10.5. Table 10.21 lists the accepted

Table 10.21. Ranges of free and dependent variables, values of fixed variables, and constraints accepted for second random sample: region close to apparent optimal combination.

Variable	*Range or value*						
Free variables							
Reservoir capacities (10^6 acre ft)							
B, active storage	5.0–6.5						
B, dead storage	5.5–6.5						
C	0.4–1.5						
D	2.2–3.2						
Target outputs							
Irrigation (10^6 acre ft)	3.8–4.6						
Energy (10^9 kw hr)	2.1–2.7						
Dependent variables							
Power-plant capacities (Mw)							
B	300–400						
G	250–340						
Fixed variables							
Reservoir capacity (10^6 acre ft)							
A	0						
Flood-control storage (10^6 acre ft)							
Reservoir *B*	0						
Reservoir *C*	0.4						
Reservoir *D*	0						
Constraints							
When reservoir capacity (10^6 acre ft):							
B, active storage	≤6.2	5.6	5.6	5.1	5.6	5.6	5.6
B, dead storage	≤6.0	6.0	6.0	6.0	5.5	6.0	6.0
C	≤1.06	8.0	1.06	1.06	1.06	1.06	2.0
D	≤2.7	2.7	2.35	2.7	2.7	2.7	4.0
Then:							
Target output for irrigation (10^6 acre ft)	<4.2	4.2	4.2	4.2	4.2	4.6	4.2
Target output for energy (10^9 kw hr)	<2.5	2.4	2.4	2.4	2.4	2.4	2.5

values and constraints for the second sample.

Table 10.22 indicates that the simulation runs of the 20 randomized combinations of system variables once again proved a substantial number of system designs to be inefficient or inadmissible. In fact, eleven designs were inadmissible because of class 3 energy deficits, eight had excess reservoir capacity, and only two (combinations 10 and 19) were both efficient and admissible.

Table 10.22. Summary of results, second random sample: region close to apparent optimal combination.

	Variables									Results		
	Reservoir capacity (10^6 acre ft)					Target output for —		Power-plant capacity (Mw)		Flood-control storage, reservoir C (10^6 acre ft)	No. of class 3 energy deficits	Net benefits (10^6 dollars)	
Combination	A	B, active storage	B, dead storage	C	D	Irrigation (10^6 acre ft)	Energy (10^9 kw hr)	B	G			Unadjusted	Adjusted[a]
1	0	5.75	6.34	0.833	2.67	4.17	2.60	380	325	0.40	31	780	780
2	0	6.11	6.08	0.886	3.01	4.46	2.15	315	270	0.40	0	745	757
3	0	5.19	5.61	1.100	2.36	3.89	2.68	390	335	0.40	54	774	774
4	0	5.19	6.08	1.280	2.70	4.20	2.24	330	280	0.40	0	758	771
5	0	5.44	6.00	1.190	2.77	4.56	2.31	340	290	0.40	8	744	744
6	0	5.98	6.30	1.000	2.92	4.31	2.20	320	275	0.40	0	746	766
7	0	5.34	6.13	1.170	2.63	4.17	2.50	365	310	0.40	18	778	778
8	0	5.92	5.67	1.400	2.79	4.37	2.50	365	310	0.40	22	758	758
9	0	6.11	5.92	0.745	2.92	4.28	2.68	390	335	0.40	47	764	764
10	0	6.02	6.05	0.611	2.80	4.48	2.15	315	270	0.40	0	759	759
11	0	5.25	6.15	1.430	2.94	4.51	2.48	365	310	0.40	26	737	737
12	0	6.35	6.40	0.712	3.20	3.96	2.24	330	280	0.40	0	737	771
13	0	6.15	5.90	1.420	3.04	3.94	2.58	375	320	0.40	6	773	773
14	0	5.68	5.69	1.220	2.83	4.52	2.19	320	275	0.40	0	755	757
15	0	5.14	6.23	1.200	3.10	4.52	2.59	375	320	0.40	46	726	726
16	0	6.38	5.95	1.450	2.25	3.88	2.45	360	305	0.40	0	763	783
17	0	5.25	6.06	1.320	3.04	4.13	2.58	375	320	0.40	27	766	766
18	0	5.15	6.15	0.738	2.33	4.00	2.19	320	275	0.40	0	766	780
19	0	5.64	6.49	1.158	2.42	4.21	2.40	350	300	0.40	0	779	779
20	0	6.60	6.34	0.648	3.08	4.59	2.40	315	270	0.40	18	741	741

[a] Adjustment consists of deducting capital costs of excess reservoir capacity and increasing net benefits accordingly.

As shown in Fig. 10.7, a plot of the net benefits, adjusted as before for excess reservoir capacity, in relation to target outputs for irrigation and energy identified a plateau of high net benefits between target outputs for irrigation of (3.8 and 4.2) $\times 10^6$ acre ft and between target outputs for energy of (2.1 and 2.6) $\times 10^9$ kw hr. The sizing of reservoirs *B*, *C*, and *D* is poor for combinations 12, 13, and 17, which have significantly lower net benefits in the plateau region. Between combinations of record there is an undefined band, or no man's land, beyond which lie inadmissible designs. As sketched into Fig. 10.7, the band stretches

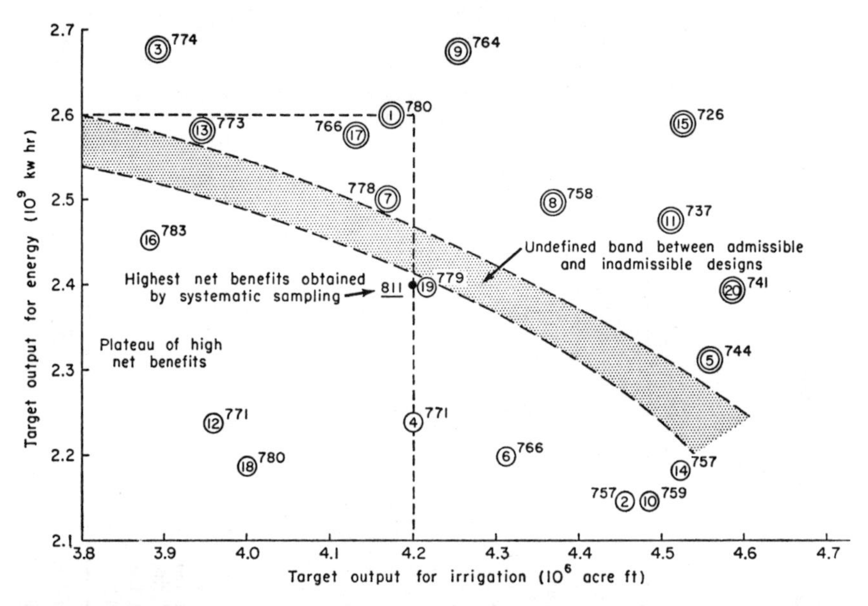

Fig. 10.7. Net-benefit response surface for second random sample of 20 combinations. Net-benefit figures are adjusted to minimum required reservoir sizes and recorded in 10^6 dollars. The combination numbers of Table 10.22 are enclosed in he circles. A concentric second circle is added when the combination is inadmissible because of class 3 energy deficits.

diagonally across the plot. Net benefits drop sharply beyond it where output combinations are highest. Only a small part of the high net-benefit plateau extended into the inadmissible region, but the random sample was so small that this observation is inconclusive.

Based on unadjusted figures, there is an 87.8 percent probability that the high net benefits of $780,000,000 in combination 1 of Table 10.22 are at least as good as the best 3×10^{11} combinations in a total population of 3×10^{12} combinations. The cumulative frequency distribution of the net benefits recorded in the second random sample is plotted in Fig. 10.6 with the distribution of net benefits in the first sample. If our ranges of variables and the results of systematic sampling tell us that the 100 percent magnitude of net benefits is $790,000,000 and the 0 percent magnitude is $720,000,000, the best observed net-benefit value of $780,000,000 lies at the 86th percentile of the range between these values. However, this combination has 31 class 3 energy deficits. Combination 19 in Table 10.22, which is the best admissible combination, produces net benefits of $779,000,000 or the 84th percentile of the range.

SUMMARY AND CONCLUSIONS

The reader will recall that the aim of our analysis of the simplified river-basin system has been to derive the optimal design, and incidentally to explore the net-benefit response surface. Having indicated the optimal design, we can now summarize our findings and compare the usefulness of the different sampling methods employed.

NATURE OF THE RESPONSE SURFACE

The most striking feature of the net-benefit response surface is its complexity. Much of the topography is low, a fact that reflects the response of net benefits to combinations of system components that are either inefficient or include target outputs too large for the reservoir capacities or streamflows. These parts of the surface are of little importance, and our sampling techniques tend to carry us away from them. Of interest is the fact that the response surface in the region of inefficient combinations — the n-dimensional counterpart of the region below curve AA′ in Fig. 10.1 — is largely smooth, although it may slope steeply. The reason for this is that the input cost functions, except for power plant *B* and the irrigation works, are smooth curves (see Figs. 7.4, 7.6, and 7.10). The steepest-ascent procedure is most effective in this part of the surface.

The region of efficient combinations contains but a single major enclosed peak or interior ridge. This is shown graphically in Figs. 10.5 and 10.7 for the response of net benefits to the two target output variables. In more generalized form, and assuming optimal combinations of all other input variables, Fig. 10.8 shows that the response surface for combinations of target outputs for irrigation and energy, along with complete flood protection, rises smoothly as we move to higher outputs for both purposes. The top is reached in the region of 4.2×10^6 acre ft for irrigation and 2.4×10^9 kw hr for energy; beyond this area the irrigation and energy loss functions come into play and the surface drops steeply.

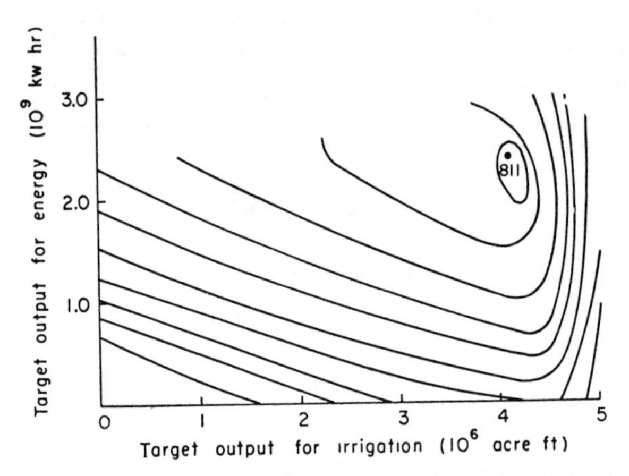

Fig. 10.8. Generalized trace of net-benefit response surface for irrigation and energy target-output variables, assuming optimal combinations of all other variables. The point labeled 811 defines the optimal combination obtained by sampling, with net benefits of $811,000,000.

Optimal combinations of reservoirs and power plants were not included in Figs. 10.5 and 10.7, which are based on the results of one systematic and two random samples. Optimal mixes of these structures may well pull the net-benefit response of all combinations to the configuration of values shown in Fig. 10.8.

USE OF SAMPLING METHODS

Our conclusions about sampling methods, based on our experience in simulating the simplified river-basin system, are given below.

(1) Before simulation runs are started, major system variables should be identified and discontinuities and constraints spotted. This will save time in subsequent detailed analyses. The five key system variables that we started to sample all turned out to be important factors in moving from low to high values of net benefits, and in narrowing the search for the optimum to a relatively small part of the response surface.

(2) Random sampling can be used effectively early in the analysis. It will indicate the general configuration of the response surface and provide starting points for a more detailed examination of parts of the surface with high net benefits. Thus Figs. 10.5 and 10.7 show how well samples of only 30 and 20 combinations, respectively, mapped portions of the response surface.

When the number of components is large, random sampling is particularly effective since each sample combination permits the selection of a different value for each component. Rarely in a large sample would the same value be chosen more than once. In contrast, uniform-grid sampling retains the same value of a given component for many different combinations.

Random sampling is indifferent to the number of variables. If some of them prove to be unimportant, nothing is lost but the little extra effort involved in randomly selecting new values for these variables rather than keeping them fixed. Random sampling, however, does not permit easy identification of the relative importance of individual variables.

(3) Uniform-grid sampling is most effective when the number of variables and their respective values are small; otherwise, the size of sample required for systematic coverage of the surface grows rapidly. If one or more variables prove to be unimportant, many of the combinations sampled are redundant, and there are several identical sets of combinations of the important variables that differ only in the values assigned to the unimportant variables. In the 32-combination uniform-grid sample, for instance, half of the combinations differed from the other half solely in the storage assigned to reservoir *D*. Only four studies were actually needed to establish the influence of reservoir *D*; the remaining twelve merely repeated this information.

Uniform-grid sampling covers the response surface uniformly and shows the influence of each variable clearly. These advantages are achieved at the cost of large samples when the number of variables is great, and wasted effort when some of the variables are unimportant. We conclude that, for problems no more complex than the simplified river-basin system, uniform-grid samples are most suitable for examining relatively small regions of the response surface, where the effects of two or three variables are to be tested by assigning two to four values to each of them. In these instances the number of combinations in a sample would range from 4 to 64.

(4) Generally speaking, the response surface proved much too complex

for us to rely solely on the steepest-ascent procedure to arrive at the optimal combination of components. The procedure is effective in regions of the response surface that are free from discontinuities. In our case good results were obtained when the starting-base combination contained substantial amounts of excess reservoir capacity. Here the procedure speedily wiped out the excess capacity and moved to an efficient region of the response surface. It was ineffective, however, in regions contiguous to the net-benefit ridge that separates combinations with excess capacity and combinations with heavy deficits in target outputs. The procedure also became cumbersome when it encountered the constraint on class 3 energy deficits or the sharp change in slope of the irrigation loss function.

We found that it is unnecessary to limit to a very few the number of variables in a single steepest-ascent analysis. We could easily have handled all 12 variables. As shown in Table 10.4, the calculations involved are few and uncomplicated.

(5) The single-factor method is of only limited usefulness. In our studies it did eliminate the excess capacity of a reservoir or power plant. Where more than one such input unit had excess capacity, however, the steepest-ascent method was superior in that it could adjust several units in one operation, instead of one at a time.

(6) The marginal-analysis method is effective in testing combinations that are on the production function and near the optimum. As shown in Tables 10.15 and 10.16, the method enabled us to adjust reservoir capacities and allocations to obtain the best combination and to check whether or not the target-output levels were optimal. The method is slow and cumbersome, however, as only two variables are changed at a time; it would not be suitable for exploring large regions of the response surface.

A REVISED STRATEGY

The questions naturally arise as to how we should examine the response surface if we were to repeat the analysis with foreknowledge of our own experiences, and what changes, if any, we should make in our strategy and in the manner in which the analysis was carried out. Assuming that we still have a single 50-yr trace of hydrology and a single operating procedure, certain major changes and revisions suggest themselves.

In setting up the problem, we should substitute a steeply sloping, continuous loss function for the constraint on class 3 energy deficits. The end result sought by introducing the constraint would be achieved just as well; designs with class 3 deficits would still be rejected as clearly nonoptimal, and application of the steepest-ascent procedure would be simplified immeasurably. We should also replace the three-piece, linear irrigation loss function with a continuous, nonlinear loss function.

In place of the systematic sample of 32 combinations of only five active variables, we should take a random sample of about the same size but including all variables. Of these, the two power-plant capacities would remain dependent on the value of the target output for energy selected for each combination, but the values of all other variables would be randomly chosen.

From the results obtained with the random sample, modified by approximately eliminating excess reservoir capacity, we should decide whether to apply a second random sample to a more restricted part of the response surface or to begin an iterative procedure. A second random sample might encompass an area about as large as that covered by the first random sample described earlier in the chapter and summarized in Tables 10.19 and 10.20.

After delineating the interesting part of the response surface — hopefully by one or two random samples — we should turn to the steepest-ascent procedure. All but obviously unimportant variables would be allowed to change. We should probably free the capacities of reservoirs *A* and *C*, for instance, and perhaps freeze only the allocations of flood storage. Power-plant capacities would remain dependent on the target outputs for energy. In effect, we should be combining into one the two steepest-ascent applications summarized in Tables 10.3 and 10.10.

Small uniform-grid samples would be applied to the best combination obtained by steepest ascent, to find the best combination of the two power-plant sizes and the best allocation of flood-storage capacity to reservoirs *B*, *C*, and *D*. This is essentially what was done in the procedure actually applied.

As in our example, boundary conditions would be examined for regions of high net benefits. Because boundary cases involve typically fewer variables than do interior cases, some variant of the uniform-grid method might well be suitable. This was so in our examination of the irrigation–flood-control boundary.

The best combination yielded by the procedure outlined above would be tested by marginal analysis to determine whether or not net benefits could be increased by changing any two of the variables. As long as improvements resulted, further trial changes would be made.

If the response surface in the vicinity of the optimal combination appeared to be flat, as in our problem, we might study a systematic sample of the region around the optimum to determine the approximate boundaries of the flat region. The range in which different values for the variables could be selected without significantly changing net benefits would thereby be indicated.

It stands to reason that as more experience is gained with exploring response surfaces, improvements and refinements in computer programming will become evident, and the entire process will become more automatic. From the revised strategy given above, we would recommend at least two changes in programming the simulation for the computer.

First, the computer program should be revised so that, if desired, designs that show excess reservoir capacity would be automatically rerun with the capacity reduced by some proportion of the excess. The process would be repeated until the excess capacity was either eliminated or reduced to or below tolerance. Second, a subroutine should be added to the computer program to evaluate the results of a steepest-ascent analysis, calculate a new base, and continue with steepest ascent until there is no further improvement in net benefits.

11 Operating Procedures: Their Role in the Design of Water-Resource Systems by Simulation Analyses

BLAIR T. BOWER, MAYNARD M. HUFSCHMIDT, AND WILLIAM W. REEDY

UNDER proper management the structures and nonstructural elements of a water-resource system mobilize the latent usefulness of natural water resources while allaying their potential destructiveness. The rules under which this is done constitute the operating procedure of the system. The importance of this procedure in the performance analysis of water-resource systems at the design stage and the desirability of devising operating procedures that are optimal or near-optimal over a wide range of system designs are discussed in other chapters of this volume. In the present chapter the formulation of effective operating procedures is presented in greater detail in connection with simulation analyses.

In simplest terms an operating procedure is a set of rules for storing and releasing water from surface and ground-water reservoirs in a given water-resource system.[1] Three major kinds of decisions must be made about the apportioning of storage and release of water: apportionment among reservoirs, among purposes, and among time periods.[2] Certain basic considerations that we shall call determinants govern the making of these detailed decisions. In the context of a simulation analysis they are the system-design variables and parameters, the basic hydrologic data, and the internal conditions of the system.

[1] In the context of Chapters 3 and 14, the line between water-resource systems and operating procedures is arbitrary. Together, they transform naturally available water into desired outputs. The process is symbolized by a magician's hat into which he pours water and out of which emerge irrigation water, energy, recreational services, and the like. Within the hat are the transforming mechanisms: the structures, facilities, nonstructural measures, and operating rules.

[2] A possible fourth decision is the method of release — for example, whether water is drawn from upper or lower levels of the reservoir to provide water of suitable temperature or quality in downstream flows.

The system-design variables and parameters are associated with system inputs and outputs. Examples of governing inputs are active- and dead-storage capacities of surface reservoirs; capacities of ground-water reservoirs, power plants, and stream channels; reservoir head-capacity functions; levee heights; and areas of flood plains to be zoned. Governing outputs are the magnitudes of monthly target outputs for irrigation, energy, water supply, and the like (including desired quality characteristics); and the economic benefit and loss functions associated with them.

The basic hydrologic data include primarily information on inflows over the simulation period for monthly intervals, or for shorter periods for some purposes such as flood control and power, and are obtained from the observed or a synthesized record of the stream system. Also included are data on rainfall and snow cover that may be used in forecasts of inflows formulated specifically for simulation purposes.

The internal conditions of the system are created jointly by past inflows and by the operating procedure. For any one time interval they show up as initial surface- and ground-water reservoir contents, quality of contents, and total amounts of energy and irrigation water developed in preceding time intervals, with their associated surpluses or deficits in relation to target outputs.

The system-design determinants are fixed for any one design and either remain unchanged throughout the period of simulation or change in a manner specified in advance. Hydrologic determinants change throughout the period, but in a manner specified in advance. Internal conditions of the system, which together with the hydrology identify its state, fluctuate throughout the simulation period in response to operating decisions taken in preceding time intervals. In any given interval during simulation, any or all determinants may be relevant to the operating procedure, that is, their inclusion and proper application will increase the value of the objective function significantly.

OPERATING PROCEDURES IN SYSTEM DESIGN AND OPERATION [3]

There should be close correspondence between the performance of a water-resource system as estimated at the design stage and that attain-

[3] For discussions of the operation of multiunit, multipurpose systems, see:

(*a*) N. W. Bowden, "Multiple purpose reservoir operation," *Civil Eng. 11*, 291 and 337 (1941).

(*b*) R. M. Morris, "Operation experiences: Tygart multipurpose reservoir," *Trans. Am. Soc. Civil Engrs. 107*, 1389 (1942).

able after the system is built. Accordingly, we must use at the design stage an operating procedure that is consistent with feasible management of the system-in-being. Crucial factors in attaining consistency are the length of the time interval and the degree of knowledge of future hydrologic events. As to the time interval, most simulation studies are usefully based on monthly flows except when flood control and certain detailed water-supply and power studies call for shorter intervals. Whatever the length of time, however, total target outputs (of water and energy, for example) are commonly specified without reference to the way in which they are to be distributed within the assumed interval. Releases are related to total outputs, the assumptions being that the system can be operated so as to meet a specified demand pattern within the selected interval and that the natural pattern of flow itself can be estimated accurately enough for releases to be in accord with the internal demand pattern during the time interval. Actual operating experience has shown that these assumptions are often reasonable for monthly intervals. In any event, however, system designers are free to select a basic time interval that validates these assumptions.

As to future flows, operation studies for purposes of design are often based on historical sequences of streamflow as if they represented perfect foreknowledge. Obviously, this is not so.[4] In our own simulation of the simplified river-basin system we have, with some exceptions, assumed that little or nothing is known about inflows beyond the current month. This is an unnecessarily confining assumption. If flow forecasts are to be made, they can be based either on the past history of flows or on runoff data modified by information on current and antecedent rainfall, snow cover, soil moisture, and temperature.[5] For runoff from snowmelt, fore-

(*c*) C. E. Blee, "Multiple-purpose reservoir operation of Tennessee River system," *Civil Eng. 15*, 220 (1945).

(*d*) R. K. Linsley, M. A. Kohler, and J. L. H. Paulhus, *Applied hydrology* (McGraw-Hill, New York, 1949), pp. 613–614.

(*e*) Francis P. Blanchard, "Operational economy through applied hydrology," *Proc. Western Snow Conf. 23*, 35 (1955).

(*f*) Victor A. Koelzer, "The use of statistics in reservoir operations," *Proc. Am. Soc. Civil Engrs.*, Paper 1008 (1956).

(*g*) R. J. Pafford, "Operation of Missouri River main stem reservoirs," *Proc. Am. Soc. Civil Engrs.*, Paper 1370 (1957).

[4] Linsley, Kohler, and Paulhus, reference 3(*d*), p. 611.

[5] For attempts to make long-range forecasts, see L. A. Dean, "The use of long range streamflow forecasts in the control of the Columbia River power system," paper presented at the Fifth Annual Meeting of the Pacific Northwest Region of the American Geophysical Union (1959); and I. P. Krick, "Long range weather forecasting as a water supply tool," *J. Am. Water Works Assoc. 51*, 1366 (1959).

casts can be made with reasonable accuracy for several months in advance; for other types of runoff, this is not generally the case. Whatever the quality of forecasts, however, we are entitled to take account of forecasting possibilities at the design stage, to the extent that we would do so in operating a system-in-being.

CLASSIFICATION OF OPERATING PROCEDURES

Determinants may be omitted from operating procedures or included to the fullest extent of their relevancy. Their absence makes a procedure rigid; their presence gives it flexibility.[6]

Let us use as an example a river-basin system with a single flood-control reservoir discharging through adjustable gates into a stream channel of restricted capacity. The objective of the system is to minimize expected flood damages. If the reservoir is large and the streamflow such that the probability of reservoir spill and consequent inundation is small when a simple rule of making releases from the reservoir up to the limit of the channel capacity is followed, the only relevant determinant is the downstream channel capacity. However, if the reservoir is small and the probability of reservoir spill and consequent inundation is high when the same simple rule is followed, reservoir capacity, initial reservoir contents, and forecasts of inflow may become additional relevant determinants. A more complex operating procedure is then required, for example, one which allows moderate, purposeful overbank flooding to reduce the chances of severe and relatively more damaging floods. Even for very simple systems, therefore, rigid procedures may give optimal answers for only a few of the many possible designs. Procedures with a high degree of flexibility are required if we are to be certain of optimal answers over a wide range of system designs.

Because such a highly flexible procedure may unnecessarily complicate performance analysis at the design stage, a less flexible procedure with a smaller number of determinants may be preferred, especially if the chosen

[6] Because nature is so complex, the determinants are many. Even hourly streamflows, for example, are only gross averages in terms of power-canal surges or slug flows in flumes. There is no theoretical upper limit to the number of determinants, only a practical limit. Furthermore, we distinguish here between the concept of the operating procedure involving choices and determinants and the actual manner of executing an operating procedure. A system can be designed with a "built-in" operating procedure: for example, a system of fixed-orifice flood-control reservoirs, the operation of which is fully determined once the structures have been built. Such systems react automatically to determinants such as reservoir contents and associated head.

determinants introduce in an effective way important information on the state of the system.[7] In this event, if (as in the simulation of our simplified system) many combinations must be tested in order to discover the optimal system design, a large number of discrete operating procedures that are relatively rigid must be developed.

A single master procedure (which includes within it a multiplicity of such relatively rigid procedures) can probably be formulated by the introduction of alternative rules for storing or releasing water. Each such rule might contain a number of different ways of allocating storage and releases to reservoirs, purposes, and time intervals, and each set of allocations might constitute a discrete operating procedure. Although such a master procedure could presumably be programmed without difficulty for simulation on a computer, any one of the resulting discrete procedures would probably be far from optimal for a wide range of designs. A search for more flexible procedures is therefore in order.

INTRODUCTION OF FLEXIBILITY

Although the operating procedure for the simplified river-basin system (described in Chapter 7) is fundamentally rigid, it contains two rules that illustrate how some flexibility can be introduced without much increase in complexity. The provision of flood-storage space in reservoirs at the proper times is one; the other, which is part of the flood-routing procedure, is the allocation of flood releases from two reservoirs on separate tributaries in proportion to the flood waters stored in them. With respect to the first rule, the determinants are the recorded inflows for the flood months of April, May, and June (the general magnitudes of which are assumed to be known as of April 1). With respect to the second rule, the determinants consist of the flood-storage space that is occupied in each reservoir.

These rules (described in detail in Chapter 7) form only a small part of the procedure. In a simulation run they deal with only the one purpose of flood control, and with relatively few time intervals. More flexible procedures for more than one purpose are established by what we shall call a "space rule," and by a "pack rule" and its counterpart a "hedging rule." As we shall see, releases are apportioned among reservoirs by the

[7] The "rule curve" is a basically rigid procedure often used in actual system operation and has the advantages of simplicity of formulation and operation. Where, however, a series of rule curves are developed, each to be applied under specified conditions, a degree of flexibility is introduced. Examples of rule curves will be found in the sources listed in reference 3.

space rule, and among time intervals by the pack and hedging rules. The space and pack rules will be discussed in some detail, because they help to explain the internal workings of simulation operating procedures — about which there is little information in the literature, especially in relation to the design of river-basin systems.

THE SPACE RULE

When parallel reservoirs serve the same purposes, the apportionment of required releases among the reservoirs becomes a major decision. A rigid procedure specifies fixed proportions or sequences of releases for all reservoir conditions and time intervals. A flexible procedure, by contrast, may elect to change these proportions for sequent intervals of simulation. If we assume that spills leaving the system as unused water are undesirable for withdrawal uses such as irrigation and water supply, or for a nonwithdrawal use such as power, we can choose minimization of spill as the physical objective of a flexible operating rule for apportioning needed releases. In terms of probability it is then the objective of the operating procedure, at a given time, to equalize the probability that the parallel reservoirs (or the space within them utilized for active storage) will have refilled at the end of the drawdown-refill cycle.[8] Actually, all reservoirs will be full and spilling, full and not spilling, or partly full, the unoccupied space being proportioned to inflows during the drawdown-refill cycle. The inefficient condition created by some reservoirs being full and spilling while others remain unfilled can be avoided by apportioning required releases[9] among parallel reservoirs in such a way that, after the water has been withdrawn, the ratio of space available in each reservoir to that in all reservoirs equals, insofar as possible, the ratio of the predicted flow into each reservoir during the remainder of the drawdown-refill cycle to that into all reservoirs. This is the space rule.[10]

[8] The drawdown-refill cycle begins when reservoirs are full and ends when they have been refilled by inflows after withdrawals. In well-watered regions the cycle is usually 12 months long except when the reservoir is large in relation to streamflow. In semiarid and arid regions the cycle often extends over several years — as much as 10 to 20 years for relatively large reservoirs. For a given stream hydrology the length of the cycle depends on reservoir size and level of development of system outputs.

[9] Required releases are defined as those necessary to meet output requirements after all unregulated flows entering the stream below the reservoirs but above the point of diversion have been put to use.

[10] The concept of the space rule was proposed by Loren A. Shoemaker, refinements were suggested by Ralph M. Peterson and William B. Shaw, and the rule was programmed for the IBM 704 computer by Hugh Blair-Smith.

In mathematical terms, it is

$$\frac{S_{\max j} - S_{Ijk} - Q_{jk} + R_{jk}}{\sum_{j}^{m} (S_{\max j} - S_{Ijk} - Q_{jk}) + R_T} = \frac{Q_{j,n-k}}{\sum_{j}^{m} Q_{j,n-k}}, \tag{11.1}$$

where $S_{\max j}$ is the full capacity of the jth in a series of m reservoirs; S_{Ijk} is the initial contents of the jth reservoir in the kth of a series of n months; Q_{jk} is the flow into the jth reservoir in the kth month; R_{jk} is the release from the jth reservoir in the kth month; R_T is the sum total of the releases required to fulfill the target outputs; and $Q_{j,n-k}$ is the predicted flow into the jth reservoir for the remaining $n - k$ months of the drawdown-refill cycle.

We can solve Eq. (11.1) for R_{jk}, the release from the jth reservoir in the kth month:

$$R_{jk} = S_{Ijk} + Q_{jk} + \left[\sum_{j}^{m} (S_{\max j} - S_{Ijk} - Q_{kj}) + R_T\right] \times (Q_{j,n-k}/\sum_{j}^{m} Q_{j,n-k}) - S_{\max j}, \tag{11.2}$$

subject to the constraint $0 \leq R_{jk} \leq (S_{Ijk} + Q_{jk})$.

Application of the space rule is exemplified in Table 11.1 for a single month and two parallel reservoirs, A and B, both supplying water to irrigated lands. We assume that there is no unregulated inflow available for irrigation use below the reservoirs and above the point of irrigation diversion. Given current streamflows and required releases as well as the maximum capacity and initial contents of the reservoirs, and predicting flows for the remainder of the drawdown-refill cycle, we can calculate the storage space available at the beginning, item (3), and end, item (6), of the interval for each reservoir and for the system, as

$$\sum_{j}^{2} (S_{\max j} - S_{Ij})$$

and

$$\sum_{j}^{2} (S_{\max j} - S_{Ij} - Q_j) + R_T.$$

Here the total space available at the end of the month cannot be a negative quantity. After deriving the ratios of predicted flows into each reservoir for the remainder of the cycle to those into the system as a whole, then multiplying by the total end-of-the-month space, we can establish the space allocation for each reservoir, items (8) and (9). End-of-the-month reservoir contents and, from Eq. (11.2), required re-

Table 11.1. Application of the space rule to two reservoirs in parallel. The reservoirs are operated for a typical month, during which the target output for irrigation is 250 × 10³ acre ft. No unregulated inflow below the reservoirs and above the point of diversion is available.

	Capacities, contents, inflows and releases (10^3 acre ft)		
Item	Reservoir A	Reservoir B	Total, reservoirs A and B
(1) Maximum reservoir capacity	150.0	2500.0	2650.0
(2) Reservoir contents at beginning of month	50.0	1500.0	1550.0
(3) Available storage space at beginning of month, (1) − (2)	100.0	1000.0	1100.0
(4) Inflow during month	90.0	400.0	490.0
(5) Total target output or release required during month	—	—	250.0
(6) Total space available at end of month, (3) − (4) + (5)	—	—	860.0
(7) Predicted inflow between end of month and end of refill cycle	110.0	1100.0	1210.0
(8) Proportion of required space at end of month	$\frac{110.0}{1210.0}$	$\frac{1100.0}{1210.0}$	$\frac{1210.0}{1210.0}$
(9) Allocation of space at end of month, (6) × (8)	78.2	781.8	860.0
(10) Reservoir contents at end of month, (1) − (9)	71.8	1718.2	1790.0
(11) Release during month, (2) + (4) − (10)	68.2	181.8	250.0

leases from each reservoir are then calculated, items (10) and (11). In accordance with the constraints, releases cannot be negative, nor can they exceed the sum of the initial reservoir contents and inflows for the time interval.

We have programmed this space rule for the IBM 704 computer and applied it to reservoirs A and B and the irrigation area in the upper portion of our simplified river-basin system (see Fig. 7.1), and also to a modification of this system in which flows into reservoir D enter reservoir A instead. In a series of 50-yr simulation runs the effects of a rigid rule calling for withdrawals from reservoir A first until it is emptied were compared with the effects of the space rule. Different combinations of reservoir sizes and target outputs for irrigation were studied, and mean monthly flows, calculated from the observed record, were used as the predicted flows for the remainder of the drawdown-refill cycle. In each instance the space rule performed as well or better than the rigid rule, performance being measured not only by the quantity of spill but also by the number and magnitudes of irrigation deficits. Both of these quantities were to be minimized by the operating procedure.

Our tests of the space rule were so few in number that no firm conclusions can be drawn from them. For the simple cases illustrated, however, programming was relatively easy. Since for these cases we were dealing with snowmelt hydrology, the pattern of mean flows seemed to approximate the pattern of future inflows quite well. Furthermore, the opportunity for recalculating releases each month allowed us to adjust for sharp deviations of actual from predicted flows. Finally, a high degree of accuracy in predicted inflows did not seem to be required in the early months of the drawdown-refill cycle. Toward its end, however, predictions had to be more accurate if unnecessary spill was to be avoided. Since the end of the 12-month drawdown-refill cycle was close to the period of active snowmelt, and forecasting techniques (based on snowpack measurements) are well developed, predicted flows should match actual flows reasonably well.

For other types of streamflow the effectiveness of the space rule would be a function of the coefficient of variation of the mean monthly flows, the correlation between flows on adjacent streams, and the reliability of flow forecasts that a system-designer can justifiably assume.

In modified form the space rule can also be used to apportion releases among reservoirs for flood control, based on 6-hr or other short intervals of time. It is valid, too, when each unit of water is of equal value in a given reservoir but not of equal value in different reservoirs. However, the space rule must be modified in form to deal with this situation. Unequal values of water are created in our system, for instance, when a fixed-head power plant downstream from reservoir *B* generates firm energy from irrigation releases, but no such plant exists below reservoir *A*. In an application of the modified space rule, the economic value of system output was maximized by preventing or minimizing spills of the higher-valued water from reservoir *B* at the expense of spills of lower-valued water from reservoir *A*. Had the water within either of the two reservoirs not possessed a fixed value, however (as in the case of a variable-head power plant), application of the rule would have become much more difficult.

THE PACK RULE

Releases from reservoirs for different purposes need not necessarily be allocated to current time intervals by rigid rules that confine them to target-output requirements. Flexible rules may specify useful releases in excess of target outputs instead. For example, it may be advantageous to generate dump energy by stepping up releases toward the end of the drawdown-refill cycle in order to free reservoir space for predicted inflows that would otherwise spill. The pack rule is a flexible rule of this kind.[11]

Rigid and flexible operation are compared in Fig. 11.1. In accordance with the pack rule, inflows expected during the last four months of the drawdown-refill cycle of the reservoir are estimated, and spills are avoided by releasing excess water to the fullest possible extent for the generation of dump energy. For application of the rule, the sum of the available reservoir space and the amount of water that can usefully be run through the turbines during the entire period must be determined, assuming no further releases during the current month. The pack rule is so named because the possible future spill is packed as tightly as possible into future spare turbine capacity.

The rule is written in mathematical terms as follows:

$$R_d = Q_{n-k} - (S_{\max} - S_{Tk}) - P_{n-k}, \tag{11.3}$$

where R_d are the additional releases in the current month, k, for the generation of dump energy; Q_{n-k} is the predicted flow into the reservoir for the remaining $n - k$ months of the drawdown-refill cycle; $S_{\max}$ is the full reservoir capacity; S_{Tk} is the reservoir contents in the current month after current inflows have been added and releases made to meet the target output for energy; and P_{n-k} is the useful water capacity of the turbines for the remaining $n - k$ months of the drawdown-refill cycle. If the right-hand side of the equation is not positive, $R_d = 0$. Equation (11.3) is also subject to the constraint $P_c \geq R_d \leq S_{Tk}$, where P_c is the useful water capacity of the turbines in the current month after releases have been made through the turbines to meet the target output for energy.

Since predictions are seldom perfect, we must expect either that the reservoir will not refill because releases have been too large or that there

[11] The pack rule as here stated was devised and tested by Hugh Blair-Smith.

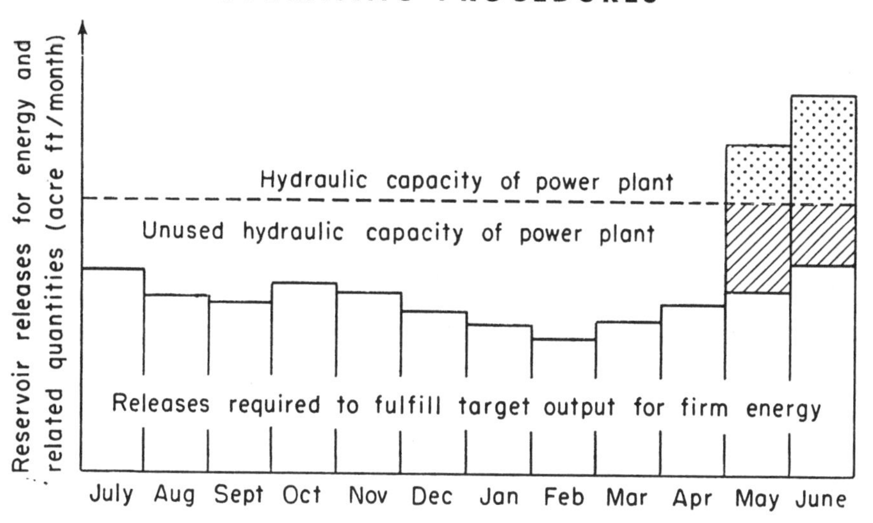

(a) Pack rule is not applied. Spills occur because dump energy is generated only in May and June when the reservoir is overflowing. Shaded area represents dump-energy releases and dotted area represents spills.

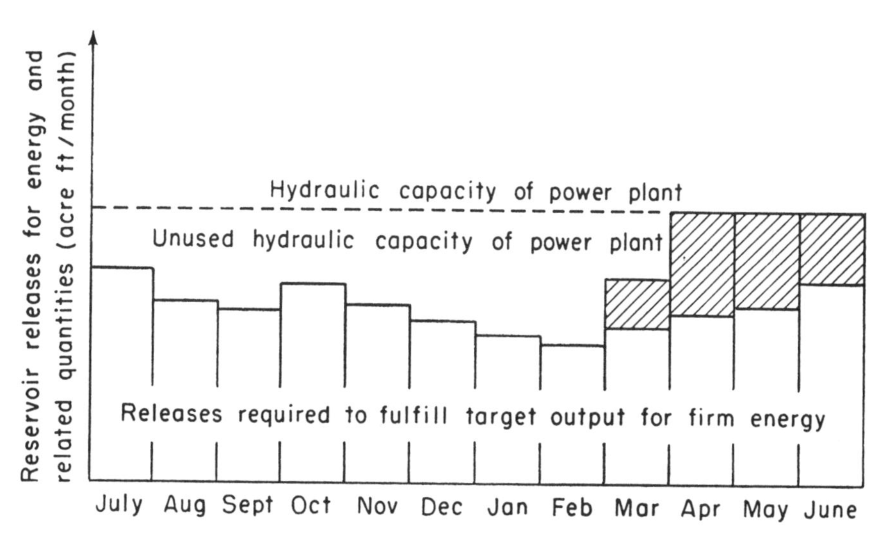

(b) Pack rule is applied through explicit releases in March and April for generation of dump energy in anticipation of end-of-cycle surpluses in May and June. Shaded area represents packed dump-energy releases during these months.

Fig. 11.1. Comparison of the pack rule with a rigid rule for a single-reservoir, fixed-head power system.

will be some spill toward the end of the cycle because releases have been too small. Fortunately, flow forecasts become increasingly reliable toward the end of the reservoir cycle when the pack rule is normally brought into use.

The pack rule can be applied whenever releases beyond specified output requirements arc of value. Examples, in addition to the operation of variable- and fixed-head power plants, are releases for water-pollution control and groundwater recharge. While the rule can be of assistance in making decisions involving time for simple systems or for parts of

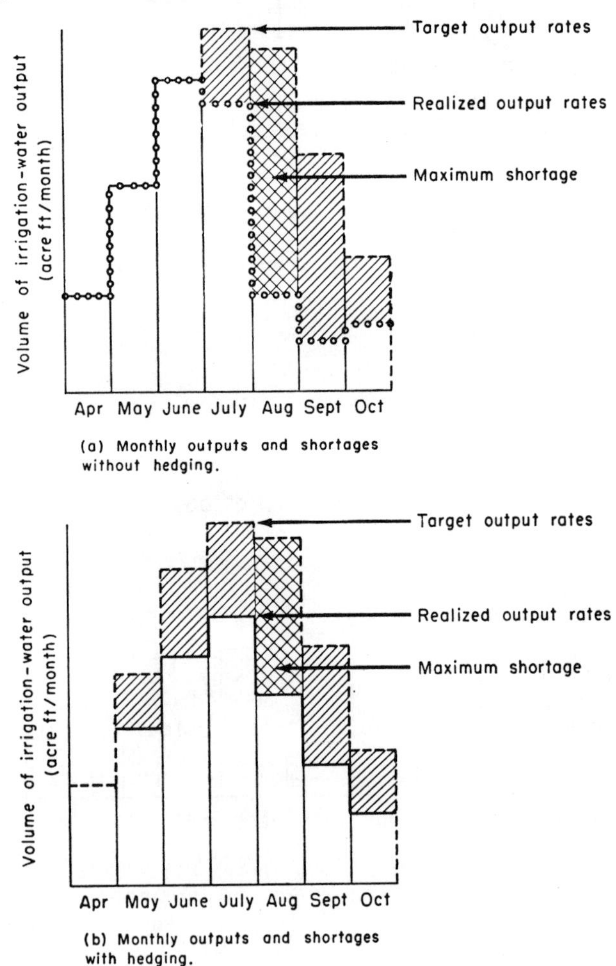

Fig. 11.2. Effect of the hedging rule on irrigation output. Realized outputs are the same, but shortages differ in intensity.

complex systems, it cannot be used to apportion water among purposes or reservoirs.

THE HEDGING RULE

It is sometimes economical to accept a small current deficit in output so as to decrease the probability of a more severe water or energy shortage later in the drawdown-refill cycle. The resulting hedging procedure is illustrated in Fig. 11.2 for an irrigation use. Hedging is introduced early in the cycle, with small shortages accepted early and excessively large ones thereby avoided later.

Economically, a hedging rule can be justified only if the proposed uses of water have nonlinear loss functions. If the marginal values of water for specific uses are constant, the economic losses from shortages must be linear and, because streamflows are stochastic, it follows that it is optimal to postpone shortages for as long as possible. In spite of a high probability of a severe deficit later, therefore, an assured full supply now is preferable to a definite deficit now with a lesser probability of a heavy deficit later.[12] If severe deficits are penalized proportionately more than mild deficits, however, it may pay to reduce the probabilities of suffering heavy deficits.

Hedging within a single annual drawdown-refill cycle, which is common in actual operations, may be unimportant at the design stage. Since the irrigation loss function of our simplified river-basin system is based solely on the annual irrigation shortage, hedging is of no value. However, it is feasible, by hedging, to reduce possible monthly energy shortages in our system to levels below the intolerable shortage level of 25 percent of the monthly output. Difficulties of flow-forecasting interfere with hedging for more than a single 12-month cycle in the design phase.

A FLEXIBLE OPERATING PROCEDURE

A highly flexible operating procedure that incorporates all important determinants must include rules for allocating releases among purposes as well as among reservoirs and time intervals. A procedure is optimal when, in a given interval, the economic values of releases and storage are marginally equal. This implies, for example, that the value of the last

[12] Preference for present over future supply is often expressed in economic terms by a discount or interest rate.

acre foot of water released from each reservoir is the same and equal to the expected value of the last acre foot of water stored. The goal remains maximization of the value of an economic objective function rather than maximization of physical quantities, such as the quantity of electrical energy or irrigation water.

The values of an economic objective function to be maximized for each interval equal the sum of the benefits during the current interval and the expected benefits during the remainder of the drawdown-refill cycle. The benefits of single-purpose energy generation, for instance, comprise the value of firm and dump energy generated by releases in the current interval plus the value of firm and dump energy expected to be generated during the remainder of the cycle, including the value of expected reservoir storage at the end of the cycle.[13] For irrigation, the benefits equal the value of the current irrigation releases plus the value of irrigation water expected to be released during the remainder of the cycle, including the value of expected reservoir storage at the end of the cycle. For flood control, benefits consist of the value of current storage space for withholding flood waters in terms of expected future damages prevented.

One way of obtaining a highly flexible procedure for simulation might be by letting the flexible release rules for operating the system for each relevant purpose become the basis for formulating a single multiple-purpose procedure. This technique is described below.

A flexible single-purpose procedure. A flexible single-purpose procedure would specify an economic objective function and release rules that would maximize its expected value in sequent time intervals. Space, pack, and hedging rules would be introduced where applicable. For single-purpose power, all three can be applied in connection with economic benefit and loss functions to determine, sequentially in time, the optimal volume of releases and their optimal allocation among reservoirs. For single-purpose irrigation and for flood control, only the hedging and space rules are applicable.

We have applied these rules with good results to relatively simple situations, such as two reservoirs in parallel and fixed-head power plants. For more complex systems with several reservoirs in series and parallel, or with variable-head power plants, more sophisticated basic rules would

[13] As discussed in Chapter 3 under "Uncertainty," maximization of expected value is only one of many possible ways of formulating an objective function that deals with uncertain future benefits.

have to be established. Programming for the computer would become more difficult, requiring much iteration within each time interval. To combine the rules into a single procedure for optimizing physical output for multireservoir, single-purpose power systems, a general framework has been constructed within which marginal releases from reservoirs in any one interval can be evaluated in terms of their contribution to energy output during the entire drawdown-refill cycle.[14] Insofar as possible, releases are then made to meet the target output for energy in the current interval and to maximize the expected generation of energy during the remainder of the cycle.

Further elaboration of this framework may make it possible to maximize the value of an economic rather than a physical objective function. Efforts thus far on the multiunit power procedure and on the space and pack rules give promise of the eventual development and programming of flexible procedures for any single purpose such as power, irrigation, and flood control.[15]

A flexible multipurpose procedure. We would specify for a multipurpose procedure, just as for the single-purpose procedure, an economic objective function and release rules that would maximize its expected value in sequent time intervals. But the multipurpose procedure must take account of benefits for all purposes. The difficult task of maximizing the value of an objective function of this form could be greatly simplified by specifying overriding economic priorities for individual purposes, such priorities to be valid for the entire period of simulation, and by applying rules for apportioning water among system units and time intervals with the constraints of these priorities in mind. Although these priorities would be set on economic grounds, it would probably be necessary to try several combinations of priorities before the best combination was found. While we made no tests of this method, it appears to be a straightforward extension of the single-purpose experiments mentioned above.

Much more difficult of attainment is maximizing the value of the ob-

[14] The conceptual and detailed work on this framework was done by Loren A. Shoemaker and others at the North Pacific Division, Corps of Engineers, Department of the Army, Portland, Oregon.

[15] Considerable research has been done during the past ten years on optimizing the operation of water-power and combined water-power and steam-power systems. A pioneer study is that of R. J. Cypser, "Computer search for economical operation of a hydrothermal electric system," *Trans. Am. Inst. Elec. Engrs. 73*, 1260 (1954). For examples of further research, see reference 3 of Chapter 9, and the discussion of previous research on mathematical models in Chapter 14.

jective function during each interval by the apportionment of releases among purposes or combinations of purposes in sequential time periods free from any constraints imposed by priorities.

In common with the examples that have been discussed, a highly flexible operating procedure requires estimates or predictions of future inflows. Although such predictions can be based on actual streamflow records, better results could be obtained if some more definitive method of forecasting were possible. Because flexible rules, such as the space, pack, and hedging rules, can be applied for each month, or for shorter time intervals if necessary, good results can still be obtained even where the accuracy of forecasting is not high.

Such operating procedures also require an assumption concerning the length of the drawdown-refill cycle. Our experiments with the space and pack rules seem to indicate that, in very many cases, an assumed cycle length of 12 months is adequate to give good results. In some cases, where reservoir capacity is small in relation to inflow, shorter cycles may be appropriate; conversely, where reservoir capacity is large in relation to inflow, longer cycles may be required. Unfortunately, no rigorous method exists for determining the correct length of the cycle for use in simulation studies.

CONCLUSION

The value of design techniques is measured ultimately by the difference between their cost of development and application on the one hand and the benefits obtained on the other hand. The costs of simple operating procedures, such as the space and pack rules, are relatively small and seem to be justified by potential benefits. Development and implementation of a highly flexible operating procedure, however, may well be too expensive both at the design stage and in system operation itself. Capital and operating costs may be increased by the necessity for installing additional measuring and control devices, such as gates, by-passes, and gaging stations, as well as systems for collecting and processing basic data.

Still to be determined is the justifiable degree of refinement in the operating procedure on which a system design is based. Refinement may indeed imply sharply increasing costs. The need is for more experience in operating complex systems, and in developing and programming operating procedures for system design.

12 Mathematical Synthesis of Streamflow Sequences for the Analysis of River Basins by Simulation

HAROLD A. THOMAS, JR.,
AND MYRON B FIERING

A portion of our simulation study of the simplified river-basin system is based on a 500-yr streamflow sequence, or streamflow hydrograph, that we computed by mathematical synthesis from 32 years of observed monthly runoffs and 6-hr flood flows of the Clearwater River and its tributaries, in Idaho. As explained in Chapter 6, the relative brevity of existing streamflow records almost always impairs the precision of the final designs of river-basin developments that are analyzed by simulation techniques. Rarely does one have as many as 50 years of recorded observations for which the hydrologic system has been stable, and even that long a record may lack a critical sequence of years of low (or high) regional runoff. If the most severe droughts or floods on record are not representative of the statistical population, it is obvious that the design will become distorted.

Even though a short record may not identify the true frequency of years or seasons of unusually low or high flows, unless the record is very short indeed it will give fairly precise estimates of mean annual and mean seasonal flows and their variances. Together with a few assumptions about the population of flows (which may be based on experience with nearby rivers), these statistical parameters make it possible to construct a stochastic model that will generate synthetic flow sequences for as long a period of time as desired. Since our own chief interest was confined to an economic period of 50 or 100 years, our synthetic hydrograph of 500 years was assumed to represent ten replicate streamflow sequences of 50 years each. That many alternative hydrographs enabled us to test a given system design more exhaustively than would have been possible with the observed flows alone. Had it been desirable, we could have generated and tested a still greater number of synthetic flow sequences, but only at the cost of added computations.

It is the function of mathematical synthesis to create the critical patterns of low and high runoff that are probably not included in brief records of streamflow but that, based on statistical considerations, would be expected to be part of an actual record of sufficient length. If a stochastic model of flow sequences is suitable in all respects, it becomes impossible to distinguish between real and synthesized hydrographs by the usual statistical tests of significance.

The concept of synthesis in hydrology is not a new one. Several hydraulic engineers have used it in one form or another in deriving design criteria for storage reservoirs. Allen Hazen, for example, synthesized a runoff sequence of 300 years by combining the annual-mean-flow series for 14 streams in which the flows of each were expressed in terms of the individual mean flows.[1] This method in effect combines samples from different populations and thus is not precisely applicable to any particular stream. It was useful, however, for deriving generalized storage tables that give a first approximation to the desired reservoir size.

Charles E. Sudler employed a deck of 50 cards, on each of which was printed a "representative" annual streamflow.[2] By dealing this deck 20 times, he obtained an artificial record of 1000 years. The adequacy of this method depends on how the values printed on the deck are determined. Furthermore, the method has the unrealistic limitation that the largest flow in the first 50 years is also the largest flow for the entire record. In real hydrologic records the maximum value tends to increase as the length of the record is increased. Also, in real streams the mean flow in one 50-yr period differs from that in any other 50-yr period. In the Sudler method all such periods have the same mean, the same standard deviation, and the same range. This defect would tend to cause underdesign of reservoirs.

F. B. Barnes used a method similar to that of Sudler, except that the synthetic flows were (approximately) made normal variates with the same mean and standard deviation as the flows of the historical record (for the Upper Yarra River).[3] Barnes introduced the improvement of using a table of random numbers in synthesizing a 1000-yr sequence of

[1] Allen Hazen, "Storage to be provided in impounding reservoirs for municipal water supply," *Trans. Am. Soc. Civil Engrs. 77*, 1539 (1914).

[2] Charles E. Sudler, "Storage required for the regulation of stream flow," *Trans. Am. Soc. Civil Engrs. 91*, 622 (1927).

[3] F. B. Barnes, "Storage required for a city water supply," *J. Inst. Engrs., Australia 26*, 198 (1954).

streamflows. His method does not have the defects of the Sudler method, but its use is limited to the representative annual flows of a single stream that are approximately normally distributed and that do not exhibit serial correlation.

The method of synthesizing streamflow sequences to be described in this chapter has the following advantages over the earlier methods that have been discussed: it may be used for weekly, monthly, or seasonal flows as well as for annual flows; it does not require that the flow data be normally distributed, and may be used with skewed distributions as well; it incorporates serial correlation between successive flows so as to accord with observed streamflows; and, by means of the cross-correlation technique, the method is adapted to the design of multireservoir projects.

AN ILLUSTRATIVE ANALOGUE

A pack of playing cards can be used to elucidate the synthesis of streamflow sequences. To this purpose let us generate a sequential series of 52 numbers analogous to a 52-yr record of streamflow by calling the pips volume units (assigning values of 11, 12, and 13 to the jack, queen, and king, respectively), shuffling the pack, and drawing cards in succession until the pack is exhausted. We can then consider the 52 values as a random set of yearly flows into a single reservoir, their mean value being 7.0 volume units. If the reservoir is to provide a constant annual draft equal to the mean inflow, the deviations of successive inflows from the mean equal the change in storage during each time unit, and the cumulative sum of the deviations evaluates the storage required for the period under consideration. The range of the cumulative deviation from the mean then defines the reservoir capacity required to provide exactly the specified draft for the period of 52 years. Each time the cards are shuffled and dealt, another series of 52 inflows is created, and another, usually different, required storage is found.

We played 20 games in this way to exemplify the results. Expressed in volume units, the recorded mean storage requirement was 31.1, the standard deviation 7.5, and the range in storage values from 22 to 45. Because each of the 20 records has an equal chance of occurring, the wide spread of required reservoir sizes inherent in this model suggests that there may be large errors in some of the sizes and that the trace of any single sequence cannot be considered a reliable datum for design. To obtain a more reliable value, we must take into account the probability distribution of further inflows through a statistical analysis of the recorded sequence. The necessary distribution cannot be deduced from any single trace alone, because such a trace does not indicate the full potentialities of the population; however, a random sample of a sufficiently large number of traces reveals the characteristics of the population.

We might think that a different storage requirement would be generated by varying the sequence through cutting the cards without reshuffling. However, we would conclude after a little consideration that cutting the cards leaves the storage requirement unchanged for any cyclical hydrograph of flow versus time, when the draft equals the mean annual runoff. Expressed in another way, the integral of inflow minus draft over one period is independent of the time-phase angle. For any cyclic function of fixed period, the time-phase angle is any time of starting during the period. It is usually expressed as an angle between zero and 360 degrees.

The original procedure can be modified conceptually in two ways. The first modification corrects for the fact that once a card is drawn the probability of drawing another of the same rank in any successive try becomes smaller. Accordingly, to represent more closely a natural streamflow system, we must replace each card before drawing another. Testing this modification in 12 games, we obtained a mean storage requirement of 39.5, a standard deviation of 13.8, and a range in values from 16 to 58 volume units. The large increase in the standard deviation and the wider spread in range are particularly noteworthy and result from the greater variety of possible outcomes.

The second modification of the original procedure introduces the concept of a "population deck." This modification has the important property of providing a realistic distribution of inflows and ensuring a fixed probability of drawing any rank at any draw. This is done by labeling the individual cards so as to accord with the probability distribution of streamflows. By means of successive games with appropriate population decks, the statistical distribution of the storage requirement (regarded as a stochastic variable) is determined. Such a deck need not be a physical pack of cards. In fact, the stochastic model for our investigation was contrived to utilize published tables of random-sampling numbers instead.

SYNTHETIC STREAMFLOW SEQUENCES

MATHEMATICAL SYNTHESIS OF MONTHLY STREAMFLOW SEQUENCES

To simulate the response and evaluate the benefits of proposed designs for our simplified river-basin system, we synthesized two types of flow sequences: monthly flow sequences for all benefits except flood control (derived, as previously stated, from the records of the U. S. Geological Survey for the Clearwater River and several of its tributaries); and 6-hr flood-flow sequences for flood-control benefits, because flood damages, and hence flood benefits, are functions not only of monthly mean flows but also of the magnitude and duration of specific flood flows and the general shape of flood hydrographs during flood months. We chose intervals of 6 hours because our system was to be operated on a 6-hr basis during flood periods, and 6-hr readings provided sufficient data (120 values) to define adequately the peak and shape of the flood hydrographs of the Clearwater River.[4]

Our synthesis of monthly streamflow sequences made use of two types of mathematical analysis. The first employed serial correlation of monthly flows at a given station, 12 sets of coefficients being computed from the record by the least-squares method of linear-regression analysis for the purpose of relating the discharge during any month to that in the month immediately preceding it. Monthly records could then be extended by the appropriate recursion formula. As used here, the term "serial correlation" connotes a month-to-month relationship associated with seasonal fluctuations of discharge. This relationship in turn induces a small amount of year-to-year serial correlation in the synthesized flows, which accords with that found in the observed flows.

The second type of mathematical analysis employed cross correlation between two or more streams or two or more gaging stations on different reaches of the same stream. In its simplest form, this procedure is equivalent to filling in gaps in a runoff record when gagings exist at one or more nearby stations for the missing flow periods.[5] In our analysis, we calculated least-squares coefficients for overlapping records at stations within our river system and made estimates of missing information for

[4] This decision was based on the time lag from the upstream stations to point *G*, the area of potential flood damage.

[5] The term "nearby" is used to imply not so much physical proximity as similarity of climate and geology.

interpolation and extrapolation so as to create uninterrupted sequences of flow of like duration.

Within the river system diversion dam E was made a pivotal gaging station for all correlation studies, since it offered the longest observed record of streamflow and was centrally located within the river basin. It acquired special status in the stochastic model, because it appeared best to extend this record to a 500-yr synthetic sequence and subsequently to develop from it the concurrent synthesized flow sequences at all other stations.

For a comprehensive study of flood damages at point G, we converted large synthetic monthly flows within the synthesized hydrograph into sequences of 6-hr flows for flood months.[6] The method by which this was done was subsequently refined and generalized. In its final form it covered the five structural sites, or stations, of the simplified river-basin system and included floods of various magnitudes that occurred with frequencies and durations compatible with the recorded floods.

Ultimately, 510 years of monthly flows were synthesized for each of the five inflow sites of our simplified river-basin system by combining serial and cross correlation. The extra ten years were generated to provide data for a preliminary warm-up period in the simulation of any given design, so as to incorporate a random start when the benefit functions were turned on at the beginning of the first 50-yr synthetic sequence.

OUTLINE OF CORRELATION ANALYSIS

To assist the reader, correlation analysis is outlined at this point as an introduction to the application of serial and cross correlation to the streamflow of our simplified river basin.

In order to identify mathematical correlation, let X_i and Y_i represent numerical values corresponding to concurrent measurements of two variables (flows or transformed flows), and let there be n such pairs. Assume that there are additional X_i values and that we wish to estimate the most probable value of the associated Y_i values. Further, let $\overline{X}$ and $\overline{Y}$ be the observed arithmetic means of the respective concurrent measurements. If we assume that a linear-regression model is representative of the relationships, the method of least squares asserts that the best estimate, $\hat{Y}$, of the Y_i value corresponding to a given X_i value is

[6] In our study the definition of a flood is any discharge at point G exceeding 105×10^3 cu ft/sec for 6 hours. In this chapter the term "flood" connotes this only, and has no reference to actual floods that have occurred on the Clearwater River or its tributaries.

$$\hat{Y} = \bar{Y} + b(X_i - \bar{X}), \tag{12.1}$$

where

$$b = \left(\sum_{i=1}^{n} X_i Y_i - n\bar{X}\bar{Y}\right) \Big/ \left(\sum_{i=1}^{n} X_i^2 - n\bar{X}^2\right).$$

Dropping the limits of the summations, we then define the coefficient of correlation, r, by

$$r = (\Sigma X_i Y_i - n\bar{X}\bar{Y})/[(\Sigma X_i^2 - n\bar{X}^2)^{\frac{1}{2}}(\Sigma Y_i^2 - n\bar{Y}^2)^{\frac{1}{2}}],$$

where $-1.0 \leq r \leq 1.0$.

If the observed variances of the X_i and Y_i values are denoted by $\sigma_X{}^2 = (1/n)\,\Sigma(X_i - \bar{X})^2$ and $\sigma_Y{}^2 = (1/n)\,\Sigma(Y_i - \bar{Y})^2$, respectively, the standard error of estimate of the Y_i values is defined by $\sigma_Y(1 - r^2)^{\frac{1}{2}}$, its magnitude[7] being a measure of the random or unexplained variation of the Y_i. The proportion of the total variance of Y, namely $\sigma_Y{}^2$, that can be attributed to variation of the X_i values is equal to r^2. Thus for $r = 0$, $\sigma_Y(1 - r^2)^{\frac{1}{2}} = \sigma_Y$, and there is no explained variance; and for $r = \pm 1$, $\sigma_Y(1 - r^2)^{\frac{1}{2}} = 0$, and there is no unexplained variance. For intermediate values of r, the observed Y_i values are distributed about Y, with a closeness of grouping about the regression line related to r^2.

Another way to visualize the standard error is to imagine a band of width $\sigma_Y(1 - r^2)^{\frac{1}{2}}$ on both sides of the trend line defined by Eq. (12.1), within which observations will fall with a definite frequency, and of which the width is a function of the random influences or inaccuracies in the estimate.

INTRODUCTION OF A RANDOM COMPONENT

If we assume that the Y_i values for a given X_i value have a probability distribution about $\hat{Y}$, Eq. (12.1) may be rewritten to include a random component, as follows:

$$\hat{Y}_R = \bar{Y} + b(X_i - \bar{X}) + t_i\sigma_Y(1 - r^2)^{\frac{1}{2}}, \tag{12.2}$$

in which $\hat{Y}_R$ denotes $\hat{Y}$ with a random component added, and t_i is a standardized random, normal, and independently distributed variate with zero mean and unit variance. This has the effect of adding to $\hat{Y}$ in Eq. (12.1) a positive or negative component that exceeds in magnitude the band width of one standard error 32 percent of the time and the

[7] See any standard textbook on elementary statistics for the proof of this statement, for example, J. F. Kenney and E. S. Keeping, *Mathematics of statistics* (Van Nostrand, New York, ed. 3, 1954), pt. 1, pp. 267 ff.

band width of two standard errors 5 percent of the time. The nature of the distribution of the t_i values is such that parameters computed from a sample of many estimates of the type in Eq. (12.2) will not differ significantly from those in Eq. (12.1).

METHOD OF GENERATING SYNTHETIC MONTHLY STREAMFLOW SEQUENCES

In the stochastic model of basin hydrology described in this chapter, it was assumed that the population of monthly flows was statistically normal for each stream or river branch and for each month of the year, and that there was a bivariate normal population for the flows at two stations. Before normality of distribution was assumed, the probability distribution of the observed flows themselves and of two types of flow transformation, logarithmic and square root, were tested for the Clearwater River Basin record. Of the three, the logarithmic transformation best fitted a normal probability distribution and had the further advantage of eliminating the negative flows that occur occasionally when untransformed flows are used in the model. However, both these advantages were so slight for the Clearwater Basin that no transformation was used in our analysis.

Serial correlation at diversion dam E. For serial correlation of monthly flows at point E, we assumed that the statistical population of streamflows (or their transforms) at point E was normally distributed and that the 12 population correlation coefficients between successive pairs of months were the same as those calculated from the sample of historical flows.

The following recursion equation represents this bivariate model for unit time intervals of months:

$$Q_{i+1} = \overline{Q}_{j+1} + b_j(Q_i - \overline{Q}_j) + t_i\sigma_{j+1}(1 - r_j{}^2)^{\frac{1}{2}}, \qquad (12.3)$$

where Q_i and Q_{i+1} (10^2 acre ft) are the discharges during the ith and $(i + 1)$st month, respectively, reckoned from the start of the synthesized sequence; $\overline{Q}_j$ and $\overline{Q}_{j+1}$ (10^2 acre ft) are the mean monthly discharges during the jth and $(j + 1)$st month, respectively, within a repetitive annual cycle of 12 months; b_j is the regression coefficient for estimating flow in the $(j + 1)$st from the jth month; t_i is a random normal deviate with zero mean and unit variance, as before; σ_{j+1} is the standard deviation of flows in the $(j + 1)$st month; and r_j is the correlation coefficient between the flows of the jth and $(j + 1)$st month. In our computations the index j ran cyclically from 1 to 12 (12 values, or 1 year of 12 monthly flows) and the index i sequentially from 0 to 6119 (6120 values, or 510 years of 12 monthly flows).

Equation (12.3) characterizes a circular random walk, a model in which the discharge in the $(i + 1)$st month is comprised of a component linearly related to that in the ith month and a random additive component. Given a table of normal random deviates and the 36 calculated statistical parameters of monthly flows ($\bar{Q}_j$ and σ_j for each month and r_j for each pair of consecutive months), the computation of Q_{i+1} is a straightforward matter of arithmetic.

The sequence of flows obtained from this recursion equation possesses the requisites of the population deck discussed previously. The variation in sign and magnitude of the random additive component makes for a continuous, unbounded, and serially correlated sequence of data for simulation studies.

An examination by N. C. Matalas[8] of correlograms for serial correlation of the Clearwater River discharges, and of those of similar streams, to ascertain the effects of lag on the regression analysis validated the use

Table 12.1. Serial-correlation parameters of monthly flows at diversion dam *E* obtained from 32 years of observed record.

Month	Regression coefficient b_j $(10^2$ acre ft$)^{-1}$	Correlation coefficient r_j	Standard deviation σ_j $(10^2$ acre ft$)$	Mean flow $\bar{Q}_j$ $(10^2$ acre ft$)$
January	0.5470	0.7411	1219	1,698
February	1.0231	0.6367	899	1,778
March	1.4901	0.6388	1445	3,066
April	0.3700	0.2347	3372	8,939
May	0.1825	0.1515	5317	18,100
June	0.2281	0.7804	6404	12,618
July	0.1637	0.8755	1873	3,469
August	0.3876	0.4668	350	1,079
September	2.0764	0.8511	291	845
October	2.1884	0.8328	710	1,264
November	0.6815	0.7842	1868	1,813
December	0.6464	0.8612	1623	1,989
			Average	4,722

[8] N. C. Matalas, "Memorandum to the Harvard Water Resources Program" (unpublished report, September 1958), and "Statistical analysis of droughts" (Ph. D. thesis, Harvard University, 1958).

of the simple regression model of lag one (one month) incorporated in Eq. (12.3). He found that more elaborate models involving lags of two or more months did not give significantly better results.

Table 12.1 gives the pertinent parameters of serial correlation for each month at point E of the simplified river-basin system. The correlation coefficients in this table are statistically significant for a sample of $n = 32$ years, the length of the runoff record at point E. This is indicated by the fact that except for the April–May and the May–June correlation coefficients of $r_j = 0.235$ and 0.152 respectively, the values of r_j exceed the conventionally accepted minimum of 0.349 given in R. A. Fisher's table for testing the statistical significance of sample product-moment correlation coefficients at the 95 percent level of probability.[9] The change in time and duration of the spring thaw from year to year accounts for the weak correlations between the monthly spring runoffs. The resulting large random component corresponding to the last term in Eq. (12.3), apart from the random normal deviate t_i, is a reflection of the large random fluctuations in the observed data for the spring months.

Table 12.2. Number and magnitude of negative values generated in the sequence of monthly flows at diversion dam *E* synthesized by serial correlation.

	Magnitude of mean monthly flows (10^2 acre ft)		Negative values for stated month (50 yr)		
Month	Observed (32 yr)	Synthesized (510 yr)	Number	Magnitude (10^2 acre ft)	Percentage of total discharge
January	1,698	1,720	39	22,046	2.52
February	1,778	1,780	10	4,635	0.51
March	3,066	3,127	8	6,344	0.40
April	8,939	9,044	3	5,340	0.06
May	18,100	17,794	0	0	0.00
June	12,618	11,992	16	39,238	0.64
July	3,469	3,348	15	11,043	0.65
August	1,079	1,059	1	217	0.00
September	845	843	1	41	0.00
October	1,264	1,250	25	8,514	1.34
November	1,813	1,783	89	92,588	10.20
December	1,989	2,007	59	44,928	4.39
Total	56,658	55,747	266	234,935	0.83
Average	4,722	4,645			

[9] R. A. Fisher, *Statistical methods for research workers* (Oliver and Boyd, London, ed. 13, 1958).

The mean monthly flow of 0.4645×10^6 acre ft obtained for 510 years of synthesized streamflow at point E compares favorably with the observed monthly mean of 0.4722×10^6 acre ft, as indicated in Table 12.2. The standard deviation of monthly flows about the mean, 0.5935×10^6 acre ft for the synthesized sequence, also shows good agreement with the corresponding parameter of 0.5948×10^6 acre ft for the observed record. The distribution function, or cumulative frequency of the observed and synthesized runoffs at point E, is plotted on log-probability paper in

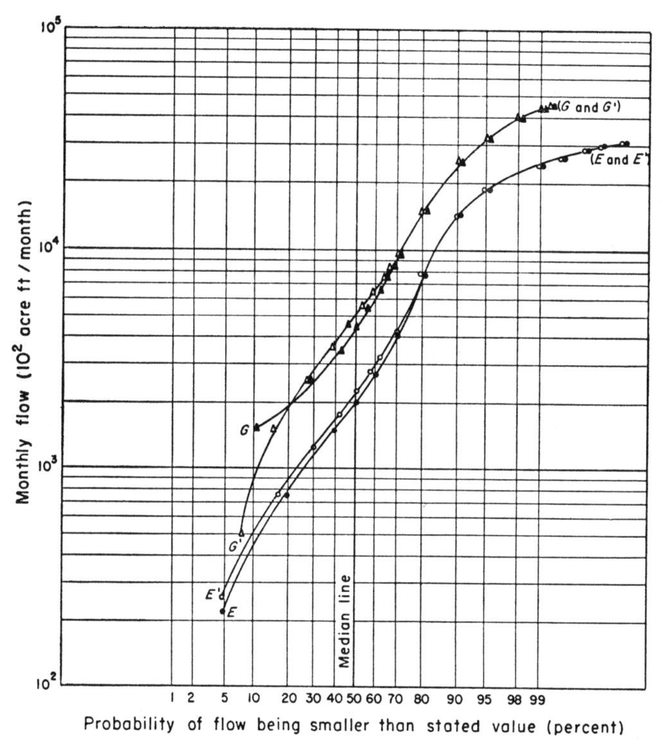

Fig. 12.1. Distribution functions for observed and synthesized monthly flows at stations E and G. Letters E and G denote lines of observed flows; E′ and G′, lines of synthesized flows.

Fig. 12.1 (along with point G) by the "center-of-strip" method, in which the midpoint of each increment is made the abscissa and the sum of the areas up to the midpoint in question the ordinate.[10] The median value

[10] Because of the absence of 0 and 100 percent ordinates on probability paper, ordinary cumulative-plotting techniques necessarily sacrifice one point of the data.

is seen to be 0.20×10^6 acre ft per month for the actual record and 0.22×10^6 acre ft per month for the synthetic record, again a close agreement.

The moments included in these results were computed by using negative values whenever they occurred in the synthesized sequence; they were subsequently set to zero. The significance of negative values is shown in Table 12.2, which suggests that negative values tend to cluster in groups of five or six successive months (as months of low flow are observed to do in actual records) and that the percentage of the total 510-yr flow affected by setting negative values to zero is small. Of the 6120 months of synthesized runoff, 266 or 4.35 percent had negative values. However, the total amount of water added to the system at point E by increasing negative values to zero was only 23.49×10^6 acre ft, or 0.83 percent of the total discharge during the 510 years. Since this was well within the 95 percent confidence limits for the population mean flow as estimated from the record,[11] elimination of negative values did not distort the population significantly; nor did negative values appreciably affect net benefits, as will be shown later in this chapter.

Required calculations were performed on a Univac I computer by writing separate programs for synthesizing 510 years of serially correlated monthly flows at point E, computing their mean and standard deviation, evaluating their distribution or cumulative frequency function, and extracting negative values and replacing them with zeros.

Cross-correlations between flows at different pairs of gaging stations. For cross-correlations between flows at pairs of gaging stations in our simplified river-basin system, we used another bivariate linear-regression model, flows at station X (independent) and station Y (dependent) being related by the equation:

$$Q_{(Y,i)} = \bar{Q}_{(Y,j)} + b_j(Q_{(X,i)} - \bar{Q}_{(X,j)}) + t_i\sigma_{(Y,j)}(1 - r_j^2)^{\frac{1}{2}},$$

where $Q_{(Y,i)}$ is the discharge at station Y during the ith month, reckoned from the start of the synthesized sequence; $\bar{Q}_{(Y,j)}$ is the mean discharge at station Y during the jth month within a repetitive annual cycle of 12 months; b_j is the regression coefficient for Y on X during the jth month; $Q_{(X,i)}$ is the discharge at station X during the ith month; $\bar{Q}_{(X,j)}$ is the mean discharge at station X during the jth month; $\sigma_{(Y,j)}$ is the standard devia-

[11] The approximate values of the 95 percent confidence limits for the population mean annual flow at point E expressed as a percentage of the observed mean flow (32 years) are 100 + 9.2 percent.

tion of discharges at station Y during the jth month; t_i is a random normal and independently distributed variate with zero mean and unit variance, as before; and r_j is the correlation coefficient between flows at stations X and Y during the jth month. In our computations the index j once again ran cyclically from 1 to 12, and the index i from 0 to 6119.

Simple bivariate correlation coefficients were computed for flows at combinations of the several stations. From the synthesized monthly record at point E it was our aim to synthesize monthly flows into reservoirs A, B, and D and for the river stretches between reservoirs A and B and diversion dam E, and between points F and G. Accordingly we examined carefully all of the possible intermediate regressions and settled eventually on the linked sequence of bivariate correlations that utilized the regressions with the highest coefficients of correlation. In this way the most reliable values were incorporated into our model. Although a multivariate correlation analysis of all stations would have extracted somewhat more information from the observed data, we felt that, for the Clearwater Basin, the gain in statistical efficiency would not be sufficient to warrant the increased work of computation.

Based on the synthetic record at point E, the bivariate cross-correlations (and subsidiary calculations) were computed in the following sequence: (1)* flows into reservoir A from those at diversion dam E; (2) flows at point G from those at diversion dam E; (3)* flows into reservoir D from those at point G; (4) flows in stretch E to A; (5) flows at B_1 from those in E to A;[12] (6) flows at B_2 from those at B_1; (7)* flows at B as the sum of those at B_1 and B_2; (8)* flows in stretch between reservoirs A and B and diversion dam E as the difference between the flow at E and the sum of the flows at A and B; and (9)* flows in stretch between points F and G as the difference between the flow at G and the sum of the flows at reservoir D and diversion dam E. Asterisks mark operations yielding final outputs, the unstarred values being of intermediate use in the computations.

The computed magnitudes of typical correlation parameters for the several populations are shown by months in Table 12.3. Once again, negative values appeared in the synthesized streamflow sequences. They were most common in computations of inflows between reservoirs A and B and diversion dam E and between stations F and G, because the inflows were often small and inherently erratic. For 510 years of synthesized

[12] The symbols B_1 and B_2 refer to flows at the Selway and Lochsa stations, respectively. As noted in Chapter 7, their sum constitutes the flow at point B.

Table 12.3. Cross-correlation parameters (regression coefficients b_j and correlation coefficients r_j) between monthly flows at different stations.[a]

Month	A from E		G from E		D from G		B_1 from E–A		B_2 from B_1	
	b_j	r_j	b_j	r_j	b_j	r_j	b_j	r_j	b_j	r_j
Jan.	0.09066	0.8701	2.71903	0.9789	0.43860	0.9836	0.41173	0.9936	1.11514	0.9799
Feb.	0.07874	0.9117	2.63516	0.9710	0.39131	0.9644	0.44928	0.9920	0.90525	0.9847
Mar.	0.10795	0.9151	2.48192	0.9734	0.39656	0.9545	0.49554	0.9826	0.80948	0.9794
Apr.	0.09336	0.8449	1.83716	0.9693	0.39121	0.9858	0.54894	0.9921	0.72113	0.9811
May	0.09771	0.7750	1.74181	0.9798	0.36543	0.9722	0.50747	0.9751	0.74211	0.9699
Jun.	0.09299	0.8945	1.57349	0.9915	0.31589	0.9856	0.50682	0.9908	0.76283	0.9905
Jul.	0.09473	0.8624	1.58063	0.9935	0.33371	0.9828	0.56895	0.9986	0.70185	0.9923
Aug.	0.12901	0.9329	1.80776	0.9868	0.38850	0.9863	0.51773	0.9930	0.74814	0.9722
Sept.	0.12767	0.9484	1.71726	0.9819	0.34674	0.9545	0.53001	0.9831	0.80289	0.9477
Oct.	0.08454	0.9088	1.82284	0.9843	0.37486	0.9639	0.55925	0.9939	0.80843	0.9668
Nov.	0.06883	0.9575	1.91494	0.9965	0.46371	0.9947	0.49932	0.9958	0.89700	0.9779
Dec.	0.07916	0.9163	2.58249	0.9582	0.47350	0.9756	0.44973	0.9960	1.04775	0.9700

[a] The stations are A, B_1, B_2, D, E, and G.

Table 12.4. Negative values generated in the several sequences of monthly flows synthesized by cross-correlation.

Gaging station or river stretch	No. of negative values	Greatest single magnitude (10^2 acre ft/month)
A	131	370
B	216	1049
A–B to *E*	1063	2872
D	218	2798
F to *G*	1025	2793

flows at diversion dam *E*, Table 12.4 gives the number of negative values in a total of 6120 monthly flows and the magnitude of the greatest single negative value at the five locations selected. Some 16 percent of the synthesized monthly values between reservoirs *A* and *B* and diversion dam *E* and between points *F* and *G* were negative, but their total effect on flow at point *G* was so small that they, too, were set to zero.

The observed and synthesized mean monthly flows and standard deviations of monthly flows are shown in Table 12.5 for the different gaging

Table 12.5. Flow-distribution parameters for the several cross-correlations of observed and synthesized monthly flows (10^2 acre ft/month).

Gaging station or river stretch	Mean monthly flow		Standard deviation of monthly flow		Median monthly flow	
	Observed	Synthesized	Observed	Synthesized	Observed	Synthesized
A	500	493	634	620	305	305
B	3888	3875	5045	4925	1750	1750
A–B to *E*	308	352	462	512	290	290
D	3345	3333	3628	3519	2050	2150
F to *G*	822	900	1046	1056	510	620

stations and river stretches. The distribution functions, or cumulative frequencies, of the observed and synthesized monthly runoffs derived by cross-correlation are shown in Figs. 12.1 and 12.2. The results, in good agreement, are represented in Table 12.5 by their median values. The medians of the synthesized distribution function for station *D* and river stretch *F* to *G* slightly exceeded those for the observed record, partially because of the bias introduced by setting negative values to zero.

As a final check on the validity of the model, the synthesized flow sequences at reservoirs *A* and *D* and diversion dam *E* were divided into ten 50-yr periods, and the mean flow, standard deviation, and extreme values were computed for four representative months of the year for each 50-yr set. As shown in Table 12.6, the results agree (within anticipated random fluctuations) with those of the actual record.

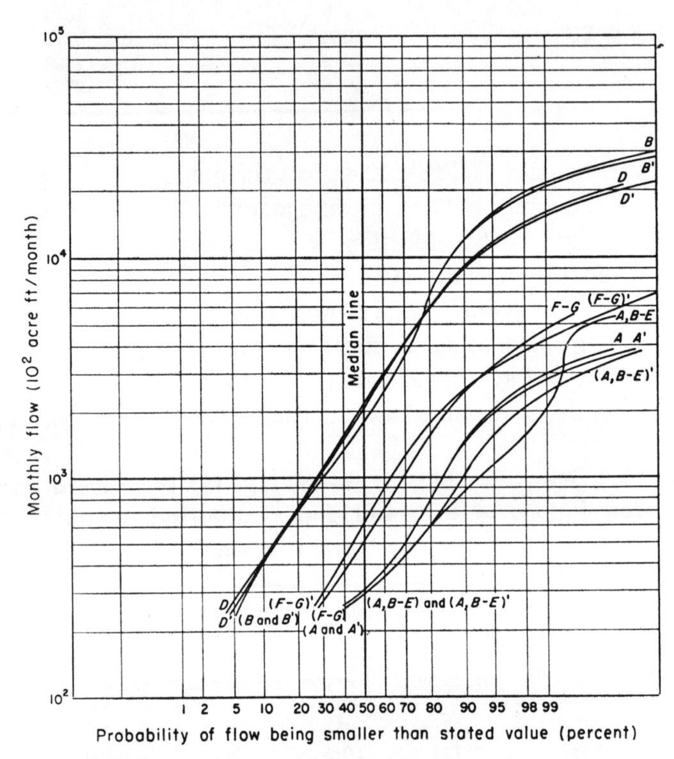

Fig. 12.2. Distribution functions for observed and synthesized monthly flows at stations A, B, D, A,B–E, and F–G. Letters with primes denote lines of synthesized flows; others denote lines of observed flows.

To perform the computations indicated above, separate Univac programs were used to do the following: synthesize the 510-yr flow records at five stations from the given synthetic record at diversion dam *E*, compute their means and standard deviations, evaluate their distribution or cumulative frequency functions, sum the five input stations and evaluate the distribution function at control point *G*, and determine the moments of monthly flows for a 50-yr period.

Discussion. The validity of basing river-basin analysis on synthesized

Table 12.6. Statistical parameters[a] describing ten replicate sets of 50-yr synthesized flow sequences for January, April, July, and October.

(a) At station A

Month	Set	50-yr mean flow	Std. dev.	50-yr max. flow	50-yr min. flow	Month	Set	50-yr mean flow	Std. dev.	50-yr max. flow	50-yr min. flow
Jan.	1	193	108	442	0 (4)[b]	Apr.	1	1142	460	2228	207
	2	193	112	426	19		2	1037	380	1760	355
	3	155[c]	113	392	0 (6)		3	1032	363	1837	174
	4	191	105	380	0 (4)		4	1046	340	1586	312
	5	176	102	455	0 (19)		5	1143	353	1818	333
	6	180	129	545	0 (4)		6	1096	327	1854	309
	7	204[d]	105	472	0 (45)		7	1157	406	2167	404
	8	190	125	430	0 (5)		8	1142	327	1807	138
	9	163	111	463	0 (5)		9	1018	394	1835	262
	10	199	129	463	0 (11)		10	1101	398	1886	381
	Obsd. value 32 yr	169	129	—	—		Obsd. value 32 yr	1113	379	—	—
July	1	347	215	1029	10	Oct.	1	135	79	360	0 (1)
	2	392	193	768	13		2	151	63	286	14
	3	426	212	857	0 (20)		3	150	71	237	0 (17)
	4	293	206	775	0 (22)		4	132	61	267	4
	5	338	187	696	0 (40)		5	140	67	268	0 (5)
	6	419	226	921	0 (34)		6	145	59	272	0 (19)
	7	401	194	818	0 (81)		7	150	61	292	20
	8	317	181	708	0 (10)		8	127	65	262	0 (12)
	9	365	215	769	0 (3)		9	132	57	272	24
	10	387	159	734	110		10	158	55	298	40
	Obsd. value 32 yr	374	206	—	—		Obsd. value 32 yr	143	65	—	—

(b) At station D

Month	Set	50-yr mean flow	Std. dev.	50-yr max. flow	50-yr min. flow	Month	Set	50-yr mean flow	Std. dev.	50-yr max. flow	50-yr min. flow
Jan.	1	1864	1436	6003	0 (187)	Apr.	1	7556	2964	13,495	1819
	2	1992	1429	5494	0 (78)		2	7340	2463	12,466	2653
	3	1627	1246	4616	0 (11)		3	7841	2373	12,581	2361
	4	1840	1297	4777	0 (177)		4	6996	2262	11,078	1723
	5	1851	1220	4287	0 (92)		5	7649	2826	13,382	1012
	6	2005	1434	5036	0 (165)		6	7697	2118	11,867	3004
	7	1972	1474	6210	0 (54)		7	8086	2505	13,843	3548
	8	2009	1429	5635	0 (51)		8	7625	2109	11,926	1929
	9	1691	1339	5332	0 (99)		9	7072	2763	13,721	1186
	10	2054	1368	4688	0 (49)		10	7922	2177	12,953	3079
	Obsd. value 32 yr	1845	1399	—	—		Obsd. value 32 yr	7321	2484	—	—

Table 12.6. (*cont.*)

Month	Set	50-yr mean flow	Std. dev.	50-yr max. flow	50-yr min. flow	Month	Set	50-yr mean flow	Std. dev.	50-yr max. flow	50-yr min. flow
July	1	2086	1093	4922	0 (165)	Oct.	1	1049	583	3004	216
	2	2150	932	3999	136		2	1133	438	2039	254
	3	2354	1010	4111	0 (641)		3	1013	549	2311	0 (50)
	4	1640	966	3991	212		4	1018	512	2221	95
	5	1951	935	3643	0 (202)		5	1033	495	2165	17
	6	2302	1119	5516	255		6	1037	448	1986	61
	7	2209	886	5106	365		7	1097	425	2143	167
	8	1809	977	3675	0 (164)		8	941	546	2045	0 (93)
	9	2001	1031	3910	0 (166)		9	995	456	2263	256
	10	2057	814	3948	305		10	1090	389	1950	280
	Obsd. value 32 yr	2165	1041	—	—		Obsd. value 32 yr	1067	536	—	—

(c) At station E

Month	Set	50-yr mean flow	Std. dev.	50-yr max. flow	50-yr min. flow	Month	Set	50-yr mean flow	Std. dev.	50-yr max. flow	50-yr min. flow
Jan.	1	1782	1202	5123	0 (66)	Apr.	1	9312	4146	17,087	0 (1059)
	2	1769	1205	4013	0 (112)		2	8817	3304	15,587	2026
	3	1478	1192	4611	0 (47)		3	9048	3223	15,098	2689
	4	1699	1119	3637	0 (7)		4	8375	2945	14,682	950
	5	1579	1121	4393	0 (285)		5	9417	3718	17,671	581
	6	1846	1284	4351	0 (68)		6	9017	2943	15,290	1619
	7	1785	1235	5048	0 (276)		7	9801	3239	17,250	3788
	8	1701	1345	4009	0 (81)		8	9336	2981	15,162	0 (939)
	9	1624	1194	4823	0 (45)		9	8268	3868	16,456	0 (1445)
	10	1828	1242	4299	0 (125)		10	9289	3025	16,069	2246
	Obsd. value 32 yr	1698	1238	—	—		Obsd. value 32 yr	8924	3441	—	—
July	1	3349	1958	8838	175	Oct.	1	1208	905	4078	0 (288)
	2	3521	1802	7064	0 (139)		2	1380	622	2540	0 (256)
	3	3958	1864	7334	0 (1212)		3	1242	817	2935	0 (8)
	4	2498	1952	7337	0 (255)		4	1178	744	2981	0 (127)
	5	3058	1941	6264	0 (93)		5	1214	821	2590	0 (18)
	6	3810	2065	9796	0 (335)		6	1250	605	2473	0 (105)
	7	3587	1805	8924	0 (191)		7	1351	626	2715	0 (87)
	8	2878	1848	6946	0 (41)		8	1084	785	2458	0 (96)
	9	3267	1968	7417	0 (59)		9	1154	680	3063	0 (142)
	10	3361	1499	7154	211		10	1394	495	2495	195
	Obsd. value 32 yr	3466	1877	—	—		Obsd. value 32 yr	1250	698	—	—

[a] All parameters are in 10^2 acre ft.
[b] Parenthetic values are the lowest positive values.
[c] Underscored values are minima.
[d] Overscored values are maxima.

flow sequences is limited by errors in measurement of observed flows and by random-sampling errors that occur when records are brief. The first limitation exists with equal force in designs based on operation studies that are performed with historical flow records alone. The second limitation is not so severe for decisions based on synthesized hydrographs as it is for their historical prototypes, because the numerous synthesized sequences will include possible low and high flow configurations more fully than the record of observed flows. However, the bias in the mean flow and in the variance of flows that is produced by random-sampling errors in the observed flows is not eliminated. Errors of this sort, nevertheless, are usually much less important than errors introduced by basing the design on a recorded critical-flow period when in fact this period is not fairly representative of the population of critical flows.

In the language of the statistician, the justification and importance of synthesized flow sequences derive from the fact that, from flow records of common length, the mean and variance of the streamflow population are estimated more precisely than is any quantile of the range of the cumulative departure from the mean, which is specified as the design reservoir capacity. With an appropriate stochastic model the required size can be estimated from the mean and variance of the population much more accurately than from conventional reservoir operation studies that utilize only the observed record.

The usefulness of synthesized streamflow sequences extends beyond the design of single-reservoir systems to that of multipurpose, multistructure systems. There they can test a proposed design or operating procedure by exposing the system to sets of equally likely inflows. The greater the number of replicate sets, the more thorough will be the test and the more reliable the evaluation of expected performance. The simulation of system operations with synthesized information does not merely provide a more precise estimate of the expected net benefits. It indicates also whether proposed structures or units are over- or underdeveloped and affords a thorough trial of proposed operating procedures.

MATHEMATICAL SYNTHESIS OF FLOOD FLOWS

ANALYSIS AND SYNTHESIS OF FLOOD-MONTH HYDROGRAPHS

As a means of simulating the operation of our river-basin system during periods of flood within the 510-yr synthesized monthly streamflow sequence, 6-hr flood hydrographs were synthesized in accordance with the following reasoning and procedure.

Our starting assumption was that the record of observed floods on the Clearwater River and the traces of the flood-month hydrographs constituted a representative sample of possible experiences. A general inspection of observed flood-month hydrographs showed great variation in their shape. Some rose to a well-defined sharp peak, others leveled off at a broad flat crest; some increased or decreased monotonically throughout the month, others were U- or trough-shaped; some were symmetric about the middle of the month, others were skewed; some were confined to a single month, others spread through several successive months. Obviously, the flood-month hydrographs recorded for the Clearwater Basin could not be expected to cover the complete range of possibilities any more than could the recorded monthly flows. To allow for the wider variations innate in the population of floods, we therefore varied the observed flood configurations still further by adding random components to a series of least-squares estimates of significant flood-month flow parameters.

Each flood month of record was characterized by two parameters: the mean monthly discharge and the coefficient of variation of discharge about the mean. To find these parameters, we replaced the hydrographs by a step diagram or histogram in which the time scale was divided, as shown in Fig. 12.3 for illustrative purposes only, into periods of convenient length. The associated rates of flow were averaged by making the volumes of discharge, represented by the areas of constituent rectangles of the histogram, equal in size to the associated areas under the observed hydrograph. The program was written so that any number of incremental time periods between 5 and 12 could be used. The mean monthly flow then equals the sum of the heights of the rectangles divided by their number, and the coefficient of variation equals the standard deviation of the heights, taken about their mean, divided by their mean. In Fig. 12.3 the sum of the heights of the rectangles is $2 + 4 + 5 + 4 + 2 + 1 = 18$ units of flow, and since their number is 6, the mean monthly flow is 3 units as shown; the deviations from this mean are -1, $+1$, $+2$, $+1$, -1, -2, the standard deviation consequently 1.5 units, and the coefficient of variation 0.5.

We assumed next that each flood month of record defined a type of flood hydrograph, and that the shapes (intensity–duration relationships) of synthesized flood-month hydrographs could be typed and subsequently constructed to resemble in over-all shape given floods of record. It was further assumed that histograms of synthesized flood-month flows could be generated by assigning to their constituent rectangles the proportionate shares of the total area under the histogram, or total monthly volume of discharge of specific floods of record. The synthesized histogram, therefore, would be an exact proportional replica of the histogram of a given flood month of record, differing from it only in scale, that is, in the ratio of the observed monthly river flow to the synthesized monthly mean flow.

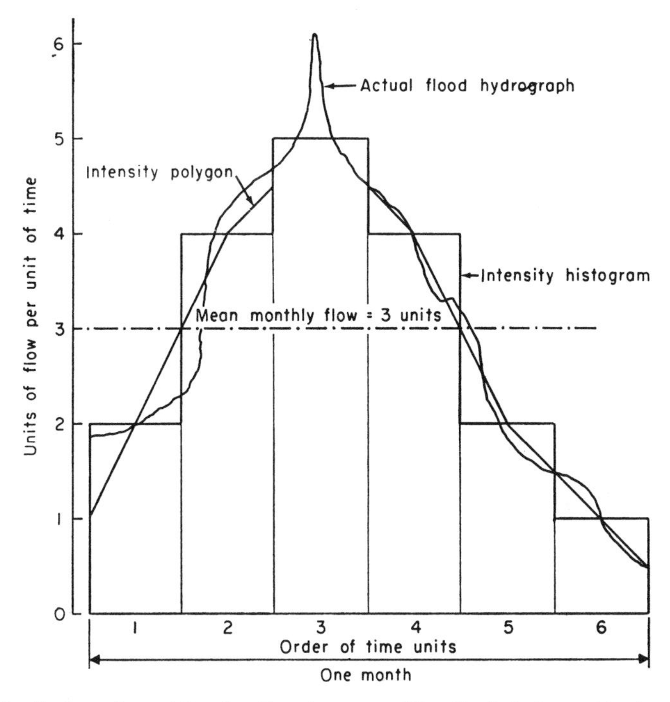

Fig. 12.3. Rectangular approximation to a monthly flood hydrograph, known as a histogram or step diagram.

Since floods synthesized in this way lack generality and flexibility and are too similar to their prototypes, our generating scheme was expanded by introducing linear-regression equations with an additive random component for computing histogram peaks, as dependent variates, from mean monthly flows. For flood months of record the correlation between the two variables was strongly positive. With the total volumes of monthly flows known from the synthesized flow sequences, the total volumes minus the flows during the peak periods, or the residual volumes, could be computed. The residual volumes, large enough to preserve the total flow, were then divided into proportional parts as outlined above for the flood flows. These modifications made it possible to produce any desired number of histograms resembling the observed flood-flow histograms in their over-all shape, but with greater variability in their peak intensities.

The 120 six-hourly readings required by the flood-routing subroutines were obtained by interpolation between the 5 to 12 rectangular values. Several intermediate values accordingly were added, the odd-numbered ones being the midpoints of the "risers" and the even-numbered the midpoints of the "treads" in the step diagram or histogram. Connecting successive midpoints by straight lines formed a flood-flow polygon or intensity polygon, as illustrated in Fig. 12.3. Subdividing its base into 120 divisions allowed average rates of discharge to be read from the intercepts of the vertical lines erected at the subdivisions with the intensity polygon. However, our task was not yet completed. Our concepts applied so far to only a single stream. They still had to be generalized and expanded into a model capable of defining floods at the five sites of our river basin.

In our simplified river-basin system the gaging station at reservoir *B*, from which the flow constitutes a substantial portion of the flow at control point *G*, was made the "pivot" or "deciding" point which determines the flood type for the other stations in the system. Accordingly, we allowed this flow to serve as the independent variate for the regression estimates needed to compute the flows at other stations. A justifiable alternative decision would have been to make *G* the pivot, because the flow at this station equals the sum of the system flows.

The type of the flood-month hydrograph was determined for all stations of the system by examination of the synthesized monthly flows at reservoir *B*, peak flows at *B* being calculated by regression on the monthly flows. The peak flow at *B*, in turn, became the independent variate from which estimates of the peak flows at each of the four other stations in the system were made. Each of the four new peak values included an expected value as well as a random component based on the appropriate standard error of estimate and a table of normal random-sample numbers. All of these parameters were obtained in a correlation analysis of the observed peak flows at the five stations of the system. After total and residual volumes had been calculated for each station, residual volumes were multiplied by appropriate fractional coefficients taken from tables of percentages derived for the flood types at each station. The final intensity polygon was then identified by linear interpolation.

The last phase of our flood-flow analysis and synthesis was concerned with the final combination and testing of a given synthetic flood peak against a stage-damage schedule. Since random components were included in the regression estimates, the peaks by chance could have been reduced sufficiently to render their total effect at point *G* harmless. Simultaneous inflows to *G* were summed, with due account taken of unequal natural time lags of flow from the several stations, and the largest sum was extracted for comparison against the safe channel capacity at point *G*. Where this sum exceeded the safe channel capacity, a bona fide flood had been generated. In this event, as described in Chapter 7, the 6-hr flood subroutine of the system operating procedure had to be invoked or triggered in the simulation run. If the sum did not exceed the safe channel capacity, the flood subroutine was not activated.

Trigger evaluation. The first decision to be reached was whether or not the current input data, consisting of the mean monthly flows at each of the five stations, foretold the possibility of a flood. Accordingly, the five input values were summed and compared against a predetermined quantity called a "trigger" value, which approximated the mean monthly flow at point *G* associated with a damaging flood. Clearly the trigger value could not be the lowest mean monthly flow that produced a flood at *G*. This would give far too many floods because of the seasonal distribution of inflows. Indeed, setting the trigger value low enough to include the rare winter flash floods associated with a low monthly mean would have been tantamount to classifying approximately half the April, May, and June inflows as floods.

Although low-peak floods could be expunged in due course by comparing the computed sum of peak flows with the channel capacity at point *G*, we found it uneconomical to invoke the long flood-generating routine many more times than necessary. Thus the decision as to whether or not a potential flood existed had to be based not only on the magnitude of the inflow but also on the month in question.

Among the possible methods for reaching a decision was a probability evaluation, in which a fixed probability for the occurrence of a flood, based on the floods of record, is associated with each range of discharge in each month at the pivot station. A random number is then compared against the appropriate probability level for the purpose of confirming or denying flood conditions. We did not use this method for a number of reasons; among them, aside from the difficulties of programming such a scheme, were the additional demands on computer memory and tape storage already heavily taxed by other parts of our studies, and also the fact that, because of the small amount of available data, the inclusion of synthesized flood flows in months without actual floods was not feasible, the flood probabilities for these months being indeterminate.

Instead we developed a simple and direct approach whereby two different trigger values were used. For a given potential flood within the synthesized sequence at pivot station *B*, its month and magnitude (monthly volume) together determined whether or not the flow was of flood proportions. In our model the months of November and December were assigned a low, or winter, trigger value because floods or near-floods of low magnitude had occurred in them during the period of record. The other ten months, including those with customarily low flows, had to show monthly inflows above a high, or spring, trigger value in order to activate the flood-generating routine. The magnitude of the spring trigger value was based on the seven observed, potentially damaging spring floods of record. These were characterized by high means and broad peaks. October, January, and February were essentially eliminated as flood-flow months by requiring a high mean, but this was not a drawback since experience with the type of hydrology associated with the Clearwater River showed that floods were generally rare during these months and were not part of the actual record. July, August, and September were also eliminated as flood-flow months by the magnitude of the trigger flow.

It could be argued that any reasonable operating procedure would leave reservoirs sufficiently empty in November and December to contain the low volumes that characterize winter floods, and that the generation of detailed monthly hydrographs for these months was therefore unnecessary. If the output were in some way limited to a maximum of one flood per year, this argument would be irrefutable; but there remains the possibility, however remote, of a flood in November followed by a flood in December, at which time reservoirs might not be sufficiently drawn down to contain the flood flows without damaging spills. In addition, winter reservoir drafts need not necessarily be large in proportion to reservoir size; thus little flood-storage volume may be left. Although slight, these possibilities were considered sufficiently important to warrant the generation of winter floods. Since they were few, the added amount of computer time was small.

With appropriate criteria our method of flood-flow synthesis chooses between flood and nonflood months. If a nonflood month is encountered, the computer program moves on to the next month and starts again, repeating this procedure until the input data are exhausted.

Type selection. When the trigger value indicated that the mean monthly flow was sufficiently high to include a possible flood, we introduced into the flow sequences synthesized for the simplified river-basin system one of the nine appropriate types of flood hydrograph that we had isolated from our study of the recorded floods. As we shall see shortly, flood types were assigned the probability-distribution function of the monthly flows that exceed a given trigger value and therefore represent potential floods. Ranges of mean monthly flow were identified by choosing percentiles appropriate to give each type of potential flood the same probability of being selected. Thus by a comparison of the particular input mean with these ranges, the flood type could be chosen with assurance that approximately the same number of each type of flood would occur in the over-all synthetic record.

APPLICATION TO THE SIMPLIFIED RIVER-BASIN SYSTEM

Since 9 floods occurred in our runoff record of 32 years, approximately $9 \times 510/32$, or 144, floods had to be introduced into the 510-yr runoff sequence prepared for our simplified river-basin system. Because of the effect of negative random components in the regression of peaks on means, more than this number of floods had to be generated. We estimated the required number at 200 and found that they could be obtained with a spring trigger value of 3.5×10^6 acre ft and a winter trigger value of 1.375×10^6 acre ft. Prior examination (but not processing) of the input data for several magnitudes of trigger values established the associated number of potential floods. Final values were selected from this information to create 197 floods. In addition, we determined the distribution function of those mean monthly flows that were larger than the trigger value at pivot point B and read from it the ranges of the mean monthly flow at point B shown in Table 12.7 for each type of flood hydrograph.[13]

With these values as the constants, the synthesized monthly flow sequences at five stations, and a table of normal random-sampling deviates as inputs, 197 floods were generated — of which 126 ultimately exceeded the channel capacity at point G.

[13] The ranges correspond to the noniles (11.1 percent quantiles) of the probability distribution of monthly flows associated with potential floods.

Table 12.7. Flood types and range of mean monthly flows in excess of seasonal trigger flows at station *B*.

Flood type	Season	Represented by observed flood no.	Range of mean monthly flows (10^2 acre ft)
1	Winter	4	0 to 17,000
2	Winter	6	17,000 to 18,600
3	Spring	2	18,600 to 19,200
4	Spring	1	19,200 to 20,000
5	Spring	5	20,000 to 20,700
6	Spring	7	20,700 to 21,400
7	Spring	3	21,400 to 22,000
8	Spring	9	22,000 to 23,100
9	Spring	8	23,100 to ∞

The flood-damage curve used in the simplified river-basin system (Fig. 7.9) was based on 6-hr peak flows at point *G*. Since the synthesized flood peaks were computed from rectangular approximations of the observed flows, the synthesized peak flows of each type of hydrograph were multiplied by the ratios of observed 6-hr peaks to associated rectangular peaks which are shown in Table 12.8. With this adjustment the synthesized peaks, measured against the same standard as the observed data, were used to evaluate flood-control benefits in our simulation studies. However,

Table 12.8. Flood types and number of floods synthesized and retained for simulation.

Flood type	Represented by observed flood no. —	No. of floods synthesized	No. of synthesized floods retained[a]	Ratio of observed 6-hr peak to associated rectangular peak
1	4	22	9	1.75
2	6	22	5	1.67
3	2	23	14	1.07
4	1	22	16	1.05
5	5	19	6	1.10
6	7	23	19	1.08
7	3	23	14	1.18
8	9	21	21	1.09
9	8	22	22	1.21
		197	126	

[a] Although 197 floods were synthesized, the peak sums of discharges for 71 were less than the safe channel capacity at point *G*; hence they were discarded and 126 values were retained.

in routing the synthetic floods through the simplified river-basin system, rectangular approximations to the peak flows rather than 6-hr peak values were employed for the sake of simplicity and also in order to keep the total volume of water passing through the reservoir unchanged.

The output of our flood-generating program included for each flood: the sequential number of the flood month (1 to 6120), the sequential number of the flood (1 to 126), the number of the month in the year during which the flood occurred (1 to 12), and the number of the flood type (1 to 9). For the trigger values selected, the results of our studies are summarized in Table 12.8.

Of the 126 floods, 14 were winter floods and 112 spring floods. Most of the spring floods occurred in May, and there were no floods in January, February, March, July, August, September, and October. Eight of the floods were twinned in four pairs of successive months. Three floods occurred in successive months in one instance. Multiple floods are the model's representation of the natural snowmelt phenomenon, which gives rise to multi-peaked floods, or to broad-peaked long hydrographs which may give rise to floods with peaks that occur at the beginning or end of a month and begin or end respectively in the preceding or following month.

The third and fourth columns in Table 12.8 show the distribution of flood types before and after it had been determined whether they would cause flood damage at point *G*. It is seen that a nearly rectangular distribution of types was generated. The floods are arrayed from the lowest to the highest according to their mean values. Type 1, for instance, is flood 4 in the observed 32-yr record. As shown in Table 12.7, it is the winter flood with the lowest mean. The number of type 1 and 2 floods discarded was relatively high because their means were so low. The degree of attrition among the different flood types depends on the relationship of the mean flow to the flood-damage threshold at point *G*, the sharpness of the hydrograph peak, and random-sampling effects. Attrition decreases as the means increase,[14] and flood types 8 and 9 have such high means that no flood had to be discarded because of the random-number effect. To be perfectly consistent with the observed flood record, we could have adjusted the ranges defining the types until the distribution of retained floods became uniform. This refinement, however, would not have been justified from the viewpoint of statistical compatibility of the synthesized series with the observed flood record. Several statistical tests of signifi-

[14] The law of chance took a heavy toll of type 5.

cance, including the chi-squared test, confirmed the compatibility of the observed and synthesized flood frequencies.

Thirteen synthesized flood crests exceeded the peak discharge of flood 8, the largest flood of record. This is in agreement with the theoretical value of $510/33 = 15$, which, according to probability theory, is the expected number of times that the largest flood on record in 32 past years will be exceeded during the next 510 years.[15]

Our programs of flood synthesis included replacing recorded floods by rectangular approximations, producing the distribution function of flows at reservoir *B* that exceeded the trigger values and printing out the number of floods to be obtained, synthesizing six-hourly readings for months in which the discharge exceeded the trigger value, and comparing all outputs with the channel capacity at point *G*.

The running time on the Univac I computer was about $2\frac{1}{4}$ hours for the entire computation, including both monthly flows and flood flows. All of the outputs were converted automatically from Univac tape to standard 80-column IBM cards for use in the several simulation studies. Clearly, most of the computational work required in our method was of a repetitive nature and ideally suited for programming on large-scale computers.

ADVANTAGES OF SYNTHESIZED STREAMFLOW SEQUENCES

Up to now the tables in this chapter have dealt primarily with the nature of synthesized flow sequences and their conformity to observed flow records. The remaining tables will show more particularly those properties of flow synthesis that make it useful in the planning process. As stated previously, the chief merit of the method is the expanded information it provides, beyond that of observed records, on the range and variety of flow shortages and the magnitude and frequency of floods that could occur during the future economic life of the system.

Listed in Table 12.6 are a number of descriptive statistical parameters for the ten replicate 50-yr synthesized flow sequences at reservoirs *A* and *D* and at diversion dam *E* for four representative months. These yardsticks measure the salient properties of the serial correlation at *E* and the cross correlation with the other gaging stations or river reaches.

[15] H. A. Thomas, Jr., "The frequency of minor floods," *J. Boston Soc. Civil Engrs.* *35*, 425 (1948).

(Similar results, not presented in this book, were obtained for the other stations of the system.) For ease of reference the maximum and minimum items of each column have been respectively overscored and underscored. The data indicate the intrinsic variability of inflows from one period to the next. Of particular significance is the greater variation, both relatively and absolutely, of the maximum and minimum monthly discharges in comparison with the variation of the mean discharges and the standard deviation of the discharges.

The extreme values within a record used for design are of signal importance in fixing the capacity and scale of the hydraulic structures of a system. To illustrate this fact, we calculated the cumulative departures of synthetic flows from the population mean flow at each station for the ten sets of 50-yr traces. In each 50-yr trace the range of the cumulative departure from the mean was determined. This statistical parameter is equivalent to the required or indicated reservoir size that would be obtained in a 50-yr operation study with 100-percent stream development, zero shortage, and uniform draft. The results are summarized in Table 12.9, which shows for each station the average of these ranges for the

Table 12.9. Range of cumulative departure from mean inflow for ten replicate sets of 50-yr synthesized inflow traces at the gaging stations and in the river stretches of the simplified river-basin system.[a]

Gaging station or river stretch	Average range	Maximum range of ten 50-yr traces	Minimum range of ten 50-yr traces
A	1.3365	1.7677	0.9795
B	10.6407	12.8268	8.8372
AB–E	1.3281	2.0137	0.9177
D	8.9204	11.7423	7.0049
F–G	3.1987	4.1863	1.8396
E[b]	12.6174	16.6316	6.1997

[a] All values are in 10^6 acre ft.
[b] The synthetic record at *E* was analyzed as a matter of interest, although it was not used as a system inflow.

ten 50-yr periods as well as their maximum and minimum. The data indicate the marked statistical instability of the range. For example, at reservoir *D* the maximum range (11.7423) is 66 percent larger than the minimum range (7.0049). These sets of 50-yr traces could have occurred with equal probability; yet each would have given quite different designs.

Table 12.10. Results of simulation run 1 using synthesized streamflow sequences: physical and economic data for the entire 500-yr simulation period.[a]

	Physical data						Benefits (10^6 dollars)			
		Energy output (10^9 kw hr)					Gross			
50-yr set	Total irrigation shortage (10^6 acre ft)	Total	Dump	Deficit	No. of class 3 energy shortages	Excess capacity, reservoir B (10^6 acre ft)	Irrigation	Energy	Flood control	Total, net[b]
1	2.495	129.954	11.626	1.672	12	0.000	546.2	479.2	246.8	660.3
2	0.000	134.777	14.778	0.000	0	0.482	558.4	490.0	240.4	676.9
3	0.000	134.998	14.998	0.000	0	0.592	558.4	487.8	295.8	730.1
4	0.619	125.999	6.525	0.526	4	0.000	556.7	480.4	226.9	652.1
5	0.788	130.234	10.981	0.747	5	0.000	555.6	482.2	271.3	697.2
6	0.000	136.628	16.628	0.000	0	0.812	558.4	491.6	409.7	847.7
7	0.000	135.388	15.388	0.000	0	2.714	558.4	489.5	256.5	692.4
8	0.748	129.919	10.050	0.130	1	0.000	556.3	485.3	240.1	669.8
9	0 000	130.008	10.009	0.000	0	0.084	558.4	484.9	470.7	902.1
10	0.929	130.947	11.890	0.943	9	0.000	555.0	483.7	338.0	764.8
Mean							556.2	485.5	299.6	729.3
Actual hydrology	0.000	131.354	11.354	0.000	0	0.048	558.4	485.9	352.0	784.3
Standard deviation of sample							3.7454	4.2027	82.099	84.681
95% confidence limits for population mean, μ							553.4 $<\mu<$ 559.0	482.3 $<\mu<$ 488.6	237.7 $<\mu<$ 361.5	665.5 $<\mu<$ 793.2

[a] There were no damaging floods and no excess capacity in reservoirs C and D for any of the 50-yr periods.
[b] Total gross benefits less capital and operation, maintenance, and replacement (OMR) costs.

Table 12.11. Results of simulation run 2 using synthesized streamflow sequences: physical and economic data for the entire 500-yr simulation period.[a]

50-yr set	Physical data					Benefits (10^6 dollars)			
	Total irrigation shortage (10^6 acre ft)	Energy output (10^9 kw hr)			No. of class 3 energy shortages	Gross			Total, net[b]
		Total	Dump	Deficit		Irrigation	Energy	Flood control	
1	0.797	133.838	6.821	2.981	20	421.9	506.8	219.8	592.9
2	0.000	138.936	8.494	0.008	0	425.5	524.6	240.4	635.0
3	0.000	139.215	9.283	0.068	1	425.5	522.6	295.8	688.3
4	0.000	131.039	2.301	1.262	9	425.5	514.6	226.9	611.5
5	0.159	134.991	6.685	1.694	0	424.7	513.8	271.3	654.2
6	0.000	140.497	10.566	0.068	1	425.5	525.7	353.9	749.5
7	0.000	139.201	9.200	0.000	0	425.5	523.8	256.5	650.2
8	0.266	134.793	5.713	0.920	7	424.7	517.8	240.1	627.1
9	0.000	135.136	5.481	0.346	3	425.5	518.6	452.9	841.4
10	0.286	135.686	7.292	1.606	11	424.5	516.2	338.0	723.2
Mean						424.8	518.5	289.6	677.3
Actual hydrology	0.000	135.784	7.071	1.285	14	424.5	515.3	352.0	737.2
Standard deviation of sample						1.1252	5.9057	73.249	75.571
95% confidence limits for population mean, μ						$424.0 < \mu < 425.7$	$514.0 < \mu < 522.9$	$234.3 < \mu < 344.8$	$620.4 < \mu < 734.3$

[a] There were no damaging floods and no excess reservoir capacity except for set 7, when excess capacity of 2.22×10^6 acre ft appeared at reservoir *B*.
[b] Total gross benefits less capital and OMR costs.

The instability of the cumulative departure from mean inflow during a design period is correlated positively with the coefficient of variation of annual inflow. In some western streams with large coefficients of variation, the results of computations analogous to those summarized in Table 12.9 would show still greater variability than the Clearwater River and its tributaries. In designs of high development (where target drafts approach the mean flow), the results of conventional operation studies may be erratic and unreliable as a consequence of hydrologic uncertainty, and an analysis with synthesized inflow sequences will afford the planner a more satisfactory insight into the potentialities of his system.

Tables 12.10 to 12.12 give the results of three simulation runs using

Table 12.12. Results of simulation run 3 using synthesized streamflow sequences: physical and economic data for the entire 500-yr simulation period.

	Physical data		Benefits (10^6 dollars)			
			Gross			
50-yr set	No. of natural floods	Greatest flood at point G (10^3 cu ft/sec)	Irrigation	Energy	Flood control	Total, net[a]
1	12	129.6	238	228	227	397
2	13	132.7	238	231	207	381
3	14	143.8	237	228	265	434
4	10	122.3	236	226	214	380
5	9	125.4	238	226	241	409
6	15	105.0	238	232	410	584
7	11	140.7	235	230	240	409
8	12	105.0	236	227	240	410
9	15	128.3	231	224	405	563
10	11	105.0	235	230	338	507
Mean			236.2	228.2	278.7	447.4
Standard deviation of sample					77.0	75.7
95% confidence limits for population mean, μ					$221 < \mu < 337$	$390 < \mu < 505$

[a] Total gross benefits less capital and OMR costs.

synthesized streamflow sequences of 500-yr duration (ten sets of 50-yr traces). Runs 1 and 2 were chosen to simulate designs in the region of high over-all net benefits based on simulation studies using 50-yr traces of actual flows, as described in Chapter 10. In run 1 the target output for irrigation was set at 4.2×10^6 acre ft and the target output for energy at 2.4×10^9 kw hr, while in run 2 irrigation output was lowered to 3.0×10^6

acre ft and energy output increased to 2.6×10^9 kw hr. In run 3 we set the reservoir capacities and target outputs for both irrigation and energy at extremely low levels in order to investigate the variability of flood-control benefits more precisely.

Two essentially different kinds of information were obtained. The first kind, similar to that presented in Table 12.9, indicates the extent to which a specific design of a system of hydraulic and related structures may be overloaded or underloaded by different sets of statistically compatible synthesized flow sequences. Looking at the results for run 1 in Table 12.10, for example, we find that half of the sets of synthesized flow sequences showed no excess capacity in reservoir *B*; in the other five sets, excess capacity was recorded. Moreover, the highest volume stored in reservoir *B* during the 50-yr period of set 7 was only a little more than one-half its design capacity of 5.6×10^6 acre ft ($5.6 - 2.7 = 2.9 \times 10^6$ acre ft). If the sequence of flows of this set had in fact been the historical record, reservoir *B* would have been considered grossly over-designed. Such a conclusion, however, is invalidated by the results for the other nine sets. These suggest a fairly well-balanced design.

The second kind of information pertains to the expected net benefits and the reliability with which this parameter is estimated. The simulations with recorded flows have no special status, and Table 12.10 shows that the net benefits obtained when these flows are used ($784,300,000) differ by no more than 8 percent from the mean net benefits of ten synthesized flow sequences ($729,300,000). Similar differences may be noted in Table 12.11. Since the population parameters of the stochastic model for synthesizing streamflow information were estimated from the sample of recorded flows, the results for gross and net benefits should properly be regarded as reflecting the extent of possible bias in the individual sets introduced by sampling errors rather than primarily as constituting a random sample of benefits drawn from the true population. Approximate 95 percent confidence limits for the population mean net benefits, shown in these tables, are indicative of this bias.

In simulation runs 1 and 2 the gross benefits for irrigation and energy, which have been discounted at $2\frac{1}{2}$ percent over a 50-yr economic life, are quite stable and close to the results based on simulation with observed runoffs. More generally, however, our studies of queuing theory indicate that the variation in benefits for these purposes from one simulation trace to another depend upon the range of fluctuation of inflows (as measured in particular by the coefficient of variation of annual inflows), and the characteristics of the economic loss function applicable to flow-shortage periods. In some designs on rivers of highly variable flow, much greater instability in gross benefits from irrigation and energy would be revealed by simulation runs with synthesized streamflow sequences. This statement should also hold true for irrigation and energy benefits where the floods intrude into a greater portion of the year and where there is marked competition between flood control and other uses.

In all these simulation runs flood-control benefits are much more variable and are largely responsible for the variability in net benefits among the ten sets of 50-yr traces. In run 3, for example, the maximum gross benefit from flood control was \$410,000,000 and the minimum was \$207,000,000 with a coefficient of variation of the entire distribution of $100 \times 77.0/278.7 = 28$ percent.

There is good reason to believe that for our simplified river-basin system the average net benefits of the ten sets of synthesized runoffs give a considerably more precise estimate of the population of expected net benefits than does the estimate resting on the historical sequence of flows, and that this average provides a more rational basis for evaluating alternative designs.

As stated previously, the assumption of a normal probability distribution for the statistical formulation of monthly flows led to the unrealistic inclusion of negative values in the synthesized sequences (Table 12.2).

Table 12.13. Frequency of negative flows at diversion dam *E* during June and July, with ranked benefits and irrigation shortages[a] for run 1.

No. of negative-flow months in 50 years in June and July	Set no.	Ranked benefits			Ranked irrigation shortages
		Gross irrigation	Gross energy	Total net benefits	
0	1	10	10	9	1
0	10	9	7	3	2
1	2	1	2	7	6
1	3	2	4	4	7
1	6	3	1	2	8
1	7	4	3	6	9
1	9	5	6	1	10
2	5	8	8	5	3
3	8	7	5	8	4
5	4	6	9	10	5

[a] Benefits and irrigation shortages are ranked in descending order on a scale of 1–10.

Although these could have been eliminated by employing a transformation, we did not think it worthwhile to do this for reasons that are supported by the comparison shown in Table 12.13. There the number of months with negative values for the flow at diversion dam *E* during June and July are compared for each 50-yr set with the associated benefit and irrigation-shortage data, which have been ranked in descending order on a scale of 1 to 10. The results show a complete absence of correlation between the number of months with negative flows in a given set and the benefits or irrigation shortages for that set. June and July mark the beginning of the irrigation season, and the runoff record shows that they are normally periods of high flow with minima well above zero. The fact that the presence of negative values for flows in the synthesized sequences did not appreciably alter the distribution of benefits seems to justify our avoiding the complication of transforming flows in our simplified river-basin system. The simple normal population for monthly flows served our purposes quite adequately.

This first large-scale use of synthesized streamflow sequences for the optimal design of a complex multipurpose, multiunit system has proved the value of this technique as an important tool of analysis. Although the suggested technique may well be elaborated in future applications — particularly in flood-damage analysis — the basic methodology should be satisfactory for a large number of river-basin investigations.

The variation in the results of the synthesized sets should be of help to the planner in identifying and developing his ultimate design. By following the technique, the planner can avoid the pitfall of overelaboration of design no matter whether he uses as an aid to judgment an objective function that incorporates an analytical uncertainty equivalent (as discussed in Chapter 3) or a more subjective criterion. At a higher level of the governmental process, the measure of variability in expected performance of a proposed design provided by synthesis of streamflow sequences should give the administrator a greater degree of insight into the evaluation and selection of projects and the setting of budgetary constraints. Indeed, it should provide a powerful and valuable means for the attainment of agency objectives such as those discussed in Chapter 2.

13 Mathematical Models: The Multistructure Approach

ROBERT DORFMAN

THE design of a water-resource system is so complex an undertaking that drastic and even illegitimate simplification is required, particularly in the early exploratory stages. The techniques of mathematical programming afford one group of useful simplifications, which will be discussed in terms of a succession of hypothetical problems of increasing difficulty.

The pertinence of mathematical-programming methods is suggested by the fact that many of the restrictions on the operation of a water-resource system are linear. If a system is fed by two reservoirs in parallel, for example, the following kind of constraint will arise:

Total water used in month $\leq$ releases from reservoir 1 + releases from reservoir 2 + natural runoff below the reservoirs.

Constraints of this type are precisely what mathematical programming is designed to analyze, while they present substantial difficulties to other modes of approach if they are at all numerous.

It thus seems clear that mathematical programming applies to the task of discovering optimal operating decisions for an extant water-resource system. It applies also to design problems by virtue of this simple fact: a design decision can be regarded as an operating decision that can be made once and thereafter cannot be changed. To illustrate this point, consider a single reservoir of capacity C^* that starts a month with contents C_t. Suppose that inflow during the month is F_t, not a decision variable, and denote releases by R_t, a decision variable. Then if we regard C^* as preassigned, the range of R_t is restricted by the requirements that the contents of the reservoir at the end of the month must be neither greater than C^*, at the one extreme, nor less than zero at the other. Algebraically stated, these are

$$C_t + F_t - R_t \leq C^*$$

and

$$C_t + F_t - R_t \geq 0,$$

assuming for simplicity that there is no requirement for dead storage. Then R_t is to be chosen, subject to these constraints, to maximize the value of some objective function. But the same two restraints are equally valid if C^* is not preassigned. In that case they become restrictions on our choice of R_t and C^*, and the two of them are to be chosen in the light of some objective function that involves them both. The sole formal difference between R_t and C^* is that R_t can be assigned different values in each month or year analyzed, whereas C^* must be the same in all months.

This point of view emphasizes the close relationship between design and operating decisions. The result of a mathematical-programming analysis will therefore be an optimal design together with an optimal procedure for operating it.

The way in which this method uses operating considerations to deduce optimal designs is illustrated by the first example below, which deals with a relatively simple configuration of uses and installations under conditions in which the pattern of inflows repeats itself each year with certainty so that overyear storage is not required. The second example will introduce methods applicable to situations in which overyear storage is needed. We shall see that this feature adds considerable complication to the analysis. Then we shall consider a third case in which uncertainty is taken into account. In the concluding section the role that can be played by mathematical programming in system design will be considered briefly.

EXAMPLES OF MATHEMATICAL-PROGRAMMING ANALYSES

THE SIMPLEST CASE: TWO SEASONS, PREDICTABLE HYDROLOGY

The mathematical-programming approach will be introduced by applying it to about the simplest problem that could be devised while retaining enough substance to present some challenge. In this spirit the configuration shown in Fig 13.1 has been abstracted from the simplified river-basin system discussed in Chapter 7. A glance at the figure shows that this configuration retains only two reservoir sites, an irrigation

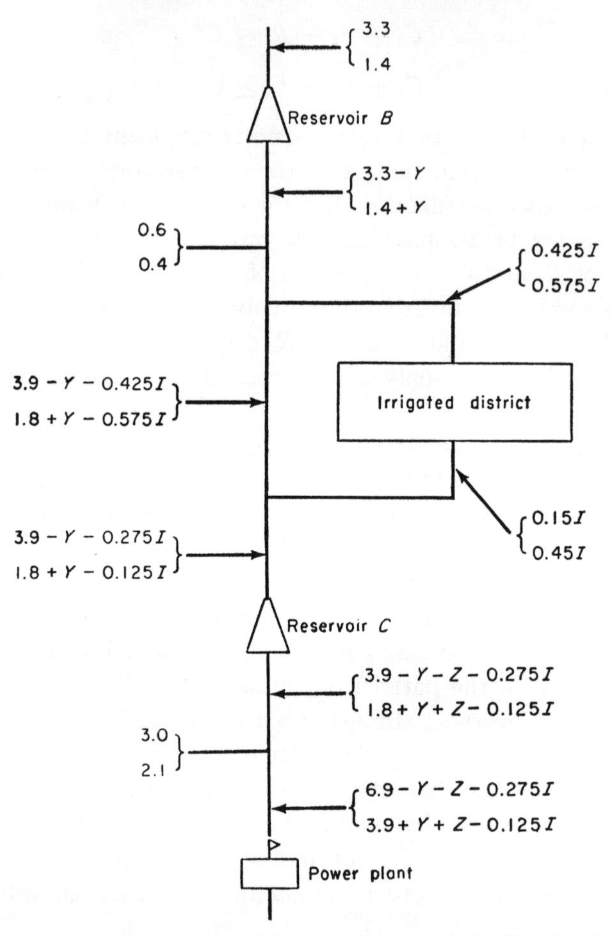

Fig. 13.1. Sketch of configuration used in two-season example. Top figure in each brace is wet-season flow, bottom figure is dry-season flow, both in 10^6 *acre feet. Y is active capacity of reservoir B, Z is active capacity of reservoir C, and I is annual irrigation supply, all in* 10^6 *acre feet.*

project, and a run-of-the-river power plant. An optimal plan for developing these opportunities will be evolved with the aid of mathematical programming.

Most of the data of the problem are contained in the figure, which may be regarded as a map of a reach of a river flowing from north to south. The numbers in braces give the mean flows in the river in two seasons of the year — the wet season (top figure) and the dry (bottom figure). Thus at the very top it is stated that the flow is 3.3×10^6 acre ft in the wet season and 1.4×10^6 acre ft in the dry.

The first topographic feature the river encounters is reservoir B; its tidal range, or active capacity, denoted by Y, is one of the unknowns of the problem. Since $Y \times 10^6$ acre ft will be retained in the wet season and released in the dry, the flows just below the reservoir are $3.3 - Y$ in the wet season and $1.4 + Y$ in the dry, as shown.

It should not be necessary to explain the remaining flows shown on the map, but a few more explanations of the hydrology are in order. Just below the confluence of the west branch with the main stem an irrigation-diversion canal takes off water to an irrigated area lying to the east. The total amount of irrigation water to be diverted, denoted by I, is the second unknown of the problem. The agricultural pattern of the region requires that, whatever I may be, 42.5 percent of the irrigation water must be provided in the wet season and 57.5 percent in the dry. The resulting return flows from the irrigated area are assumed to be 15 percent of I in the wet season and 45 percent in the dry. These assumptions explain the flows indicated for the irrigation segment and for the river as it passes by.

The third variable to be determined is the usable capacity of reservoir C, denoted by Z. Retaining $Z \times 10^6$ acre ft in the wet season and releasing them in the dry results in the flows shown just below reservoir C. The fourth and final variable is the energy output of the power plant, denoted by E.

In summary, the problem to be solved is: determine the capacities of reservoirs B and C, the supply of irrigation water, and the output of energy from the power plant so as to attain the maximum possible net benefits from the given system.

Economic and technical assumptions. Succinctly stated, we have to find values of four decision variables, Y, Z, I, and E, which (1) satisfy certain technologic constraints and (2) confer the maximum possible value on an "objective function" that will measure the net benefits derivable from the system.

The constraints can be divided into three groups. The first group requires simply that none of the four decision variables be negative. The second group of constraints states that the flows in all reaches of the system must be nonnegative. From the map four of these constraints, called ($F1$), ($F2$), ($F3$), and ($F4$) are

$$3.3 - Y \geq 0, \qquad (F1)$$

$$3.9 - Y - 0.425I \geq 0, \qquad (F2)$$

$$1.8 + Y - 0.575I \geq 0, \qquad (F3)$$

and $$3.9 - Y - Z - 0.275I \geq 0. \qquad (F4)$$

Although there are ten flow constraints that involve decision variables, only four are listed above, because if these four are satisfied all the others will be also For example, $1.4 + Y \geq 0$ is satisfied for all nonnegative values of Y, and $1.8 + Y + Z - 0.125I \geq 0$ is satisfied if ($F3$) holds.

The third group of constraints asserts that the flow at the power plant must be adequate in both the wet and the dry seasons to generate the amount of power that has been decided on.

The wet-season and dry-season power constraints, (W) and (D) respectively, follow from the technical relationship between flow and energy output, which is taken to be $E = 0.144\,F$, where E denotes the energy generated in any period in 10^9 kilowatt hours, and F denotes the flow through the turbines in 10^6 acre feet in the same period. Half the annual output of energy generated is assumed to be required in the wet season and half in the dry. Thus the two power constraints are

$$F_W = 6.9 - Y - Z - 0.275I \geq 0.5E/0.144 = 3.47E,$$

and

$$F_D = 3.9 + Y + Z - 0.125I \geq 0.5E/0.144 = 3.47E.$$

Rearranging terms, we can write:

$$Y + Z + 0.275I + 3.47E \leq 6.9 \qquad (W)$$

and $$- Y - Z + 0.125I + 3.47E \leq 3.9. \qquad (D)$$

The design sought is assumed to be the one which, while satisfying these constraints, yields the greatest possible present value of net benefits. Thus the objective function to be maximized is

$$\pi = B_1(E) + B_2(I) - K_1(Y) - K_2(Z) - K_3(E) - K_4(I) \qquad (13.1)$$

where π is the present value of net benefits in 10^6 dollars,

$B_1(E)$ is the present value of an output of $E \times 10^9$ kw hr per year in 10^6 dollars,

$B_2(I)$ is the present value of an irrigation supply of $I \times 10^6$ acre ft per year in 10^6 dollars,

$K_1(Y)$ is the capital cost of building reservoir B to capacity Y in 10^6 dollars,

$K_2(Z)$ is the capital cost of building reservoir C to capacity Z in 10^6 dollars,

$K_3(E)$ is the capital cost of building the power plant to capacity E per year in 10^6 dollars, and

$K_4(I)$ is the capital cost of building the irrigation system to capacity I per year in 10^6 dollars.

All six of these functions are given data. The first three capital-cost functions are simple, and are as follows:

$$K_1(Y) = 43Y/(1 + 0.2Y),$$

$$K_2(Z) = 47Z/(1 + 0.3Z),$$

and

$$K_3(E) = 20.6E - E^2.$$

Each of these, it will be seen, reflects the effects of increasing returns to scale. The function $K_4(I)$, however, is a bit more complicated because of the quirk that only 3×10^6 acre ft can be taken for irrigation without pumping, so that a pumping plant is required if more than this amount of irrigation water is to be supplied. The assumed data, then, are that the basic cost of the diversion works is $4,500,000 plus $44,000,000 per 10^6 acre ft of irrigation water. If more than 3×10^6 acre ft are to be supplied, a pumping plant must be constructed at a cost of $500,000 plus $20,000,000 per 10^6 acre ft of water to be pumped.

To express this capital-cost function, divide the irrigation water into two parts: I_1 denoting the nonpumped portion and I_2 the pumped portion. Then the capital-cost function for the irrigation works (in 10^6 dollars) is

$$K_4(I) = 44I_1 + 64I_2 + 4.5I_1^{+} + 0.5I_2^{+},$$

where $I_1 + I_2 = I$, $I_1 \leq 3$, and I^{+} is a function that takes on the value unity when $I > 0$ and otherwise equals zero.

The next task is to formulate the terms $B_1(E)$ and $B_2(I)$ of Eq. (13.1). The calculation in each case proceeds in four stages: first, the annual gross benefits are expressed as a function of scale (E or I as the case may be); second, annual operation, maintenance, and replacement (OMR) costs are expressed similarly; third, by subtraction, annual net benefits attributable to the output in question are computed;[1] and fourth, the present value of annual net benefits is computed by applying the appropriate present-value factor. For the last stage a planning period of 50 years and a discount rate of $2\frac{1}{2}$ percent are assumed, giving a present-value factor of 28.4. That is, under these assumptions the present value of a net benefit of $1 per year for 50 years is $28.40.

[1] In this context net benefits means gross benefits minus OMR costs.

Thus $B_1(E)$ is derived as follows. If we assume that energy is worth 9 mills per kilowatt hour, the gross benefits in 10^6 dollars are $9E$. If OMR costs are taken as $0.2E$, annual net benefits from electric power are $8.8E$. It follows that the present value of electric-power operations (gross of capital costs) is $B_1(E) = 250E$.

The calculation of the present value of the net benefits from irrigation, $B_2(I)$, is considerably more complicated, partly because of the discontinuity caused by the pumping plant and partly because the marginal value of irrigation water cannot be regarded as constant. The basic datum assumed for computing the gross benefit from irrigation (in dollars per acre foot) is

$$\text{Marginal gross benefit} = 2.1 + 3.2/(1 + 0.2I).$$

If we integrate this from 0 to I, total gross benefits in 10^6 dollars are found to be $2.1I + 36.8 \log (1 + 0.2I)$. The OMR costs are assumed to be $0.5I_1 + 1.56I_2$, so that annual net benefits from irrigation are $1.6I_1 + 0.54I_2 + 36.8 \log (1 + 0.2I)$. Finally, when we apply the present-value factor:

$$B_2(I) = 45.4I_1 + 15.3I_2 + 1045 \log (1 + 0.2I).$$

The objective function can now be computed by adding its six components. Hence:

$$\pi = 229.4E + E^2 + 1.4I_1 - 48.7I_2 + 1045 \log (1 + 0.2I) - 4.5I_1^{*} \\ -0.5I_2^{*} - 43Y/(1 + 0.2Y) - 47Z/(1 + 0.3Z). \qquad (13.2)$$

The problem has now been formulated in mathematical terms. For the most part, this discussion has consisted in corroborative detail intended to lend verisimilitude to an otherwise bald and unconvincing story. But it has served to indicate the kinds of data needed for the analysis of a design problem and to suggest the difficulties likely to be encountered in obtaining such data.

This problem requires finding the maximum of a nonlinear, and indeed discontinuous, function of some decision variables that are related by a number of linear constraints. It appears that this formalization is appropriate to the initial analysis of many water-resource design problems. Apart from the complexity of the objective function, a problem of this sort can be solved straightforwardly by the well-known computational techniques of linear programming. By refraining from further simplification of the objective function, we are testing the applicability of the mathematical-programming approach to problems in which nonlinearity of the objective function is a significant feature.

As a preliminary to the main work of solution, note that I_2 enters the objective function with a negative coefficient, which indicates that net benefits decrease as I_2 increases. Since pumped irrigation water is not a necessary input to any of the profitable variables in the objective function, it is obvious that I_2 should be chosen as small as possible, namely zero, and it will be held at this value from here on. Consequently $I = I_1$ and the subscripts are no longer required provided that the constraint $I \leq 3$ is imposed.

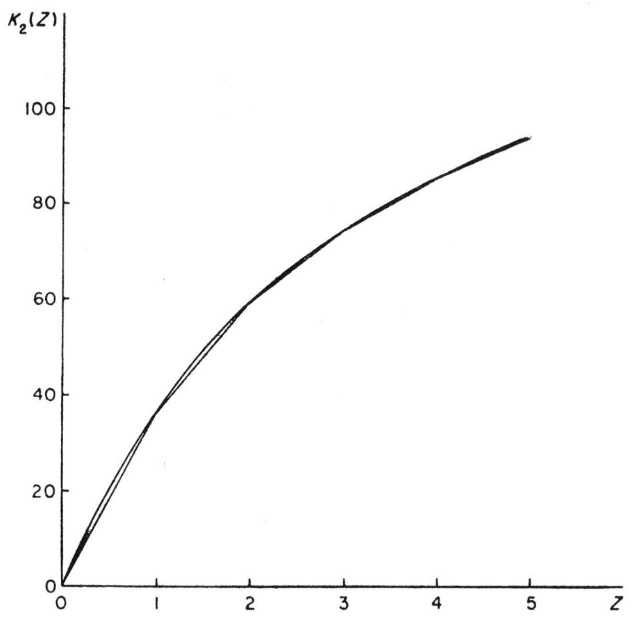

Fig. 13.2. The function $K_2(Z) = 47Z/(1 + 0.3Z)$ and approximation by five chords. The function $K_2(Z)$ gives the capital cost of building reservoir C to capacity Z.

Solution. This problem would be a straightforward case of linear programming if only the objective function were linear. An appropriate strategy, therefore, will be to replace the objective function, Eq. (13.2), by an approximation to it that is linear in some new variables introduced for the purpose.[2] To see how this is done, consider the term involving Z, namely $K_2(Z) = 47Z/(1 + 0.3Z)$. Suppose that a few values of Z are selected, chosen to cover the range of values deemed to be relevant, that

[2] This method of solution was suggested by Julius S. Aronofsky of the Socony Mobil Oil Company, who was also kind enough to perform many of the computations.

the value of $K_2(Z)$ is computed for each of these values, and that these computed points are connected by straight-line segments. The result will be a broken-line approximation to the original function as shown in Fig. 13.2, where the selected values of Z are 0, 1, 2, 3, 4, 5. To express this approximation algebraically, note that any value of Z in the range 0 to 5 can be expressed as a weighted sum of the numbers 0, 1, 2, 3, 4, 5. Thus

$$Z = 0\beta_0 + 1\beta_1 + 2\beta_2 + 3\beta_3 + 4\beta_4 + 5\beta_5,$$

where the weights, $\beta_0, \beta_1, \ldots, \beta_5$, are required to be nonnegative and to total unity. An infinite number of weighting schemes satisfy these restrictions, but only those in which no more than two of the weights are positive, and in which the two positive weights are adjacent, correspond to the chords in Fig. 13.2. For example, if $Z = 1.5$ the nonnegative weights are $\beta_1 = 0.5$ and $\beta_2 = 0.5$. If $Z = 2.8$ the weights are $\beta_2 = 0.2$ and $\beta_3 = 0.8$. These rules define a unique set of weights corresponding to each value of Z, as well as vice versa.

The virtue of this scheme is that the value of a chord at any point is a weighted average of the values of its endpoints where the weights obey the rules set forth above. Thus, given any value of Z and the weights corresponding to it, $K_2(Z)$ is given approximately by:

$$K_2(Z) \approx \beta_0 K_2(0) + \beta_1 K_2(1) + \beta_2 K_2(2) + \beta_3 K_2(3) + \beta_4 K_2(4) + \beta_5 K_2(5). \tag{13.3}$$

This approximation to $K_2(Z)$, the quality of which can be judged from Fig. 13.2, is a linear function of the weights. If this approximation is used in place of $K_2(Z)$ in the objective function, a nonlinear term will be eliminated in favor of a sum of six linear terms. Since the proliferation of variables presents only a minor computational problem, it is advantageous to do this.

To carry out this strategy, we express the original variables, Y, Z, E, and I, as functions of sets of weights, to be symbolized by Greek letters, according to the formulas

$$Y = \sum_{i=0}^{5} i\alpha_i,$$

$$Z = \sum_{j=0}^{5} j\beta_j, \tag{13.4}$$

$$E = \sum_{k=0}^{5} k\gamma_k,$$

and $$I = \sum_{l=0}^{5} l\delta_l,$$

where all weights are nonnegative, each set of weights totals unity, and no more than two weights in any set are permitted to be positive and those two must be adjacent. All the original variables must be replaced by their equivalents in the constraints and the objective function.

The replacement in the constraints is simple. For example, the first constraint in original form is $3.3 - Y \geq 0$, or $Y \leq 3.3$. If we substitute for Y in terms of α_i, the constraint becomes:

$$0\alpha_0 + 1\alpha_1 + 2\alpha_2 + 3\alpha_3 + 4\alpha_4 + 5\alpha_5 \leq 3.3.$$

From this formula, incidentally, it becomes clear that α_5 must equal zero, so that it can be dropped from further consideration. The substitutions in the other constraints are similar. Since the constraint $I \leq 3$ implies $\delta_4 = \delta_5 = 0$, these variables also disappear.

In order to replace $K_2(Z)$ in the objective function by the approximation given in Exp. (13.3), we must evaluate it at the values of Z corresponding to the ends of the chords, namely 0, 1, 2, . . . , 5. These values, presented in Table 13.1, yield the approximation:

$$K_2(Z) \approx 0\beta_0 + 36.154\beta_1 + 58.750\beta_2 + 74.211\beta_3 + 85.455\beta_4 + 94.000\beta_5.$$

Table 13.1. Chord-end values of components of the objective function, Eq. (13.1).

Y, Z, E, or I	$K_1(Y)$	$K_2(Z)$	$NB_1(E)$[a]	$NB_2(I)$[b]
0	0	0	0	0
1	35.833	36.154	230.4	84.643
2	61.429	58.750	462.8	151.006
3	80.625	74.211	697.2	207.159
4	95.556	85.455	933.6	—
5	—	94.000	1172.0	—

[a] $NB_1(E) = B_1(E) - K_3(E)$.
[b] $NB_2(I) = B_2(I) - K_4(I)$.

Table 13.1 includes also the chord-end values of $K_1(Y)$, of $B_1(E) - K_3(E)$ denoted by $NB_1(E)$, and of $B_2(I) - K_4(I)$ denoted by $NB_2(I)$. These values are used to obtain approximations to the components of the objective function, giving in each case a linear function of the Greek letters. Finally the four components of the objective function are added to obtain an objective function that is linear in all the Greek letters, which have now superseded the original variables.

The task of obtaining a linearized approximation to the original problem is complete. Clearly this approximation can be made as accurate as desired by using short enough chords; our choice was arbitrary. The only precaution necessary is to ensure that the range for which the approximation is constructed includes the final optimal solution. It pays, therefore, to make the range liberally wide. It is not even necessary for the endpoints of the chords to be equally spaced, although they are taken so in this example.

The result of these substitutions is displayed in detached-coefficient form[3] in Table 13.2. In displaying this table, we have further simplified the problem by suppressing α_0, β_0, γ_0, and δ_0 since they appear with zero coefficients in all the original constraints as well as in the objective function. The last four constraints in the revised problem, ($G1$), ($G2$), ($G3$), and ($G4$), are inserted to guarantee that each set of Greek symbols, inclusive of α_0, β_0, γ_0, and δ_0, will be nonnegative and will total unity.

The problem in this revised form now looks like a standard linear-programming problem but, in fact, is not quite one. Its peculiarity arises from two restrictions that are not expressed in the table: (1) no more than two Greek letters in any set can be positive, and (2) if two Greek letters in any set are positive, they must be adjacent. If the problem just formulated is solved by standard methods, there is no guarantee that these restrictions, which ensure that the solution will be on one of the chords illustrated in Fig. 13.2, will be met. In fact, there seems to be no automatic way to assure this result as a general rule.[4] The expedient chosen, therefore, is to start by solving the problem using standard linear-programming methods. If the side conditions are met, all is well; if not, it is necessary to determine by inspection of the unacceptable solution which of the Greek letters are likely to be zero in an acceptable solution. These symbols are then set equal to zero and deleted from the problem, which is solved again. This process of solving, inspecting, deleting, and solving again is repeated as many times as are necessary to obtain an acceptable solution. It will be demonstrated step by step below.

[3] A set of simultaneous equations or inequalities in detached-coefficient form is a table in which each equation or inequality occupies a row and each variable is assigned a column. The body of the table consists of the coefficients of the variables. Zeros are omitted when convenient.

[4] If the functions approximated are convex, these restrictions will be met automatically. In our problem we are not that fortunate.

Table 13.2. Linearized programming problem for the two-season example, in detached-coefficient form.

Constraint	α_1	α_2	α_3	α_4	β_1	β_2	β_3	β_4	β_5	γ_1	γ_2	γ_3	γ_4	γ_5	δ_1	δ_2	δ_3	$\leq$ constant
($F1$)	1	2	3	4														3.3
($F2$)	1	2	3	4											0.425	0.850	1.275	3.9
($F3$)	−1	−2	−3	−4											0.575	1.150	1.725	1.8
($F4$)	1	2	3	4	1	2	3	4	5						0.275	0.550	0.825	3.9
(W)	1	2	3	4	1	2	3	4	5	3.47	6.94	10.41	13.88	17.35	0.275	0.550	0.825	6.9
(D)	−1	−2	−3	−4	−1	−2	−3	−4	−5	3.47	6.94	10.41	13.88	17.35	0.125	0.250	0.375	3.9
($G1$)	1	1	1	1														1
($G2$)					1	1	1	1	1									1
($G3$)										1	1	1	1	1				1
($G4$)															1	1	1	1
Net benefits (10^6 dollars)	−35.83	−61.43	−80.62	−95.56	−36.15	−58.75	−74.21	−85.46	−94.00	230.40	462.30	697.20	933.60	1172.00	84.14	155.51	217.51	

The first step in the solution is to solve the linear-programming problem expressed in Table 13.2 as if it were an ordinary linear-programming problem.[5] The results of this preliminary computation are $\beta_5 = 0.255$, $\gamma_5 = 0.2767$, $\delta_3 = 1$, all other variables equal zero, and net benefits = \$517,800,000. These auxiliary variables can be converted to the original ones by using Eqs. (13.4), to obtain $Y = 0$, $Z = 1.275$, $E = 1.384$, and $I = 3$. This solution is, however, unacceptable. Consider, for example, the result $\beta_5 = 0.255$. This implies $\beta_0 = 1 - \beta_5 = 0.745$ (since all other betas in the linearized problem are zero). But this asserts that $K_2(Z)$ has been interpolated along the chord connecting the values for $Z = 0$ and $Z = 5$ (see Fig. 13.2) instead of along the closer approximation that was intended. The linear-programming solution will always select the best chord from the point of view of its objective function, whether or not it is a good approximation to the underlying problem. In this case it has chosen the chord that gives the smallest interpolated construction cost for a reservoir of given size, $Z = 1.275$. The result $\gamma_5 = 0.2767$ is similarly unacceptable. Here the linear-programming solution has chosen the interpolating chord that gives the highest apparent value of output for $E = 1.384$.

To eliminate these nonsensical results, we must solve the problem again after restrictions that forbid them have been imposed. Since Z was between 1 and 2, it is restricted to this range by setting $\beta_3 = \beta_4 = \beta_5 = 0$. Similarly, since E was between 1 and 2, γ_3, γ_4, and γ_5 are held equal to zero. The solution to the problem that results from the elimination of these variables is $\alpha_4 = 0.0687$, $\beta_1 = 1$, $\gamma_1 = 0.6167$, $\gamma_2 = 0.3833$, and $\delta_3 = 1$, corresponding to $Y = 0.275$, $Z = 1$, $E = 1.383$, $I = 3$, and net benefits = \$494,300,000. Note that the forced exclusions have caused a decrease in computed net benefits. Indeed, the preliminary solution provides at once an estimate of the upper limit to net benefits.

The presence of α_4 at a fractional level renders this solution unacceptable, since it implies that the cost of reservoir B has been estimated by interpolating along the chord connecting $K_1(0)$ and $K_1(4)$. To avoid it, since Y was found in the range 0 to 1, we must drop α_3 and α_4. Since two adjacent positive gammas appear in this solution, all the gammas are restored to free variation in the hope that they will now be well behaved.

[5] The standard methods of solving linear-programming problems will not be discussed here. Three standard references are (a) Robert Dorfman, Paul A. Samuelson, and Robert M. Solow, *Linear programming and economic analysis* (McGraw-Hill, New York, 1958); (b) Saul I. Gass, *Linear programming: methods and applications* (McGraw-Hill, New York, 1958); and (c) David Gale, *The theory of linear economic models* (McGraw-Hill, New York, 1960).

This permissiveness is motivated by the desire to avoid any more arbitrary restrictions than absolutely necessary. In this instance it proves ill advised.

Now the problem is solved for the third time with α_i, $i = 3, 4$, and β_j, $j = 3, 4, 5$, omitted, giving $\beta_2 = 0.6375$, $\gamma_5 = 0.2767$, $\delta_3 = 1$, and all other variables zero. In terms of the original variables this solution means $Y = 0$, $Z = 1.275$, $E = 1.384$, $I = 3$, and net benefits $=$ \$504,300,000. The relaxation of the constraints on the gammas has led to an increase in net benefits but otherwise was ill requited, because neither the betas nor the gammas in this solution have acceptable values. Thus the restrictions, on γ_3, γ_4, and γ_5 must be restored and the problem solved again.

The result of the fourth solution is $\beta_2 = 0.6375$, $\gamma_2 = 0.6916$, and $\delta_3 = 1$, corresponding to the same values of Y, Z, E, and I as in the third trial and net benefits of \$500,130,000. The additional restrictions, naturally, have decreased the computed value of net benefits. But this solution still will not do. Now β_2 and γ_2 are at fractional levels, implying β_0 and γ_0 at positive levels and hence an interpolation along an inadmissible chord. Since the optimal values of both Z and E seem to lie in the range 1 to 2, β_0 and γ_0 have to be forced out of the solution. This is done by regarding constraints ($G2$) and ($G3$) in Table 13.2 as strict equalities, rather than as inequalities as heretofore, since β_0 and γ_0 are, in linear-programming terminology, the slack variables in those constraints.

With this amendment the fifth solution produces $\beta_1 = 0.725$, $\beta_2 = 0.275$, $\gamma_1 = 0.6167$, $\gamma_2 = 0.3833$, and $\delta_3 = 1$. This solution is, finally, acceptable and corresponds to the previous values of the original variables, namely $Y = 0$, $Z = 1.275$, $E = 1.3834$, and $I = 3$. The net benefits are \$494,600,000. The problem is solved.

Some comments on this solution are in order. First, in the course of the experimenting to obtain a legitimate solution, the values of the underlying variables remained virtually unchanged. Their values at the end were the same as they were on the very first trial before any restrictions on the Greek variables were imposed. This circumstance was purely fortuitous; it cannot be relied on, and the successive revisions of the Greek symbols were necessary to assure that the first solution was indeed genuine. The computation would have been shortened greatly, however, if immediately after the first solution the permissible ranges of all variables had been restricted to the intervals in which the first approximations lay. The second trial would then have demonstrated that the optimal values lay in the interior of this permitted range. This would establish that a local, or relative, maximum had been found, but the possiblility would still remain that there was some higher maximum in some other range of the variables. The policy of being very chary of imposing restraints provides some, although incomplete, assurance on this point. This procedure for dealing with nonlinear objective functions, while effective, must be recognized as being somewhat delicate and dependent on the judgment of the user.

The convenience of the method depended to a large extent on the fact that the objective function, Eq. (13.2), although not linear, was "separable." By a separable function is meant one that consists of a sum of components, each of which involves only a single one of the variables in the problem. This characteristic of the objective function was exploited when each of its components was replaced by a chordal approximation employing linear interpolation of a single variable. If the objective function is not separable, the linearization procedure is substantially more involved. This case will be discussed briefly below.

No difficulties are added if some of the constraints are not linear, provided that they are separable. It would not be profitable to go through a complete problem merely to illustrate this point, but it can be illustrated by supposing that the power plant in the problem just solved was located at the dam forming reservoir B instead of downstream. Since the power plant is now located at the base of a dam, the effect of variations in the level of the reservoir on the head driving the turbines must be considered. The basic formula relating water flow to energy output is $E = 1.024\,Qhe$, where E = energy output in 10^6 kilowatt hours, Q = water flow in 10^6 acre feet, h = average head in feet, and e = efficiency of plant (assumed to be 0.85). If E denotes annual energy output, half to be supplied in the wet season and half in the dry, this relationship requires that the releases from reservoir B satisfy $Qh \geq 0.5E/(1.024 \times 0.85) = 0.574E$ in both the wet season and the dry. But from Fig. 13.1, in the wet season $Q = 3.3 - Y$, and in the dry season $Q = 1.4 + Y$. Thus the wet- and dry-season constraints are, respectively,

$$(3.3 - Y)h \geq 0.574E$$

and

$$(1.4 + Y)h \geq 0.574E.$$

The head, h, in these constraints is not a constant but depends upon the quantity of water in the reservoir, to be denoted by y, in a manner that reflects the topography of the reservoir. This relationship will be approximately linear if the surface area of the reservoir does not vary appreciably throughout the permitted range of its contents. Specifically, the head–contents relationship will be taken to be

$$h(y) = 335.65 + 14.3y.$$

The range of variation of y will be assumed to be from a maximum of Y, the capacity of the reservoir, to a minimum of 9×10^6 acre ft, the assumed size of the conservation pool. On these assumptions the average amount of water in the reservoir is $y = \frac{1}{2}(Y + 9)$, and the relationship between the average head h and the capacity of the reservoir is

$$\bar{h}(Y) = 335.65 + 14.3[\tfrac{1}{2}(Y + 9)] = 400 + 7.15Y. \qquad (13.5)$$

This equation applies in both seasons. When it is inserted, the constraints become:

$$(3.3 - Y)(400 + 7.15Y) \geq 0.574E$$

$$\text{and} \qquad (1.4 + Y)(400 + 7.15Y) \geq 0.574E. \qquad (13.6)$$

Thus the power constraints are no longer linear. Linear approximations are obtained by the same procedure that was applied to the nonlinear objective function. First the nonlinear functions on the left-hand sides of the constraints are evaluated at selected values of Y, $Y = 0, 1, 2, 3, 4$. The results are shown in Table 13.3. Then the nonlinear functions in Y

Table 13.3. Selected values of nonlinear power constraints in the two-season example.

Y (active capacity of reservoir B)	Left-hand side of constraint for —	
	Wet season	Dry season
0	1320	560
1	936	977
2	538	1407
3	126	1855
4	−299	2314

are supplanted by linear functions in $\alpha_0, \alpha_1, \ldots, \alpha_4$ as defined by Eq. (13.4). The resulting approximate constraints are

$$1320\alpha_0 + 936\alpha_1 + 538\alpha_2 + 126\alpha_3 - 299\alpha_4 \geq 0.574E,$$

and

$$560\alpha_0 + 977\alpha_1 + 1407\alpha_2 + 1855\alpha_3 + 2314\alpha_4 \geq 0.574E.$$

This pair of constraints can be inserted in place of the power constraints (W) and (D) of Table 13.2 and the resulting problem solved by the methods already discussed.

As this calculation illustrates, the device of chordal approximation can be used to handle nonlinear objective functions and constraints without adding to the computational burden, as long as they are separable. Of course, each approximate function introduced decreases the accuracy of the over-all model; this loss can be compensated for by employing a finer grid for the straight-line approximations.

If the functions involved are not separable, the method of chordal approximation becomes more complicated, but it may still be manageable. Suppose, for instance, that in the previous example the size of the conservation pool were a decision variable rather than a preassigned quantity. Denote this variable by a, substitute a for 9 in Eq. (13.5), and substitute the result in the constraints as expressed in Exp. (13.6). The wet- and dry-season constraints become respectively

$$(3.3 - Y)(271 + 14.3a - 7.15Y) \geq 0.574E$$

and

$$(1.4 + Y)(271 + 14.3a + 7.15Y) \geq 0.574E.$$

These constraints are now nonlinear functions of two variables, a and Y. The linearizing procedure depends on the fact that just as a function of one variable is graphically a curve, a function of two variables is graphically a surface. Furthermore, the analog of a chord through two points on a curve is a plane through three points on a surface. Therefore these two-variable functions can be approximated linearly by constructing a grid as in Fig. 13.3 and evaluating the functions at each of the intersection points of the grid. Then the linear approximation is a weighted average of the values of the function at the intersection points, where the weights satisfy the following rules: they are nonnegative, their sum equals unity, no more than three are positive, and all the points with positive weight are corners of the same little square of the grid. If we denote the weights by $\omega_{ij}(i = 0, 1, 2, 3, 4; j = 0, 3, 6, 9, 12, 15)$ the approximation takes the form

$$f(Y, a) \approx \sum_i \sum_j \omega_{ij} f(i, j),$$

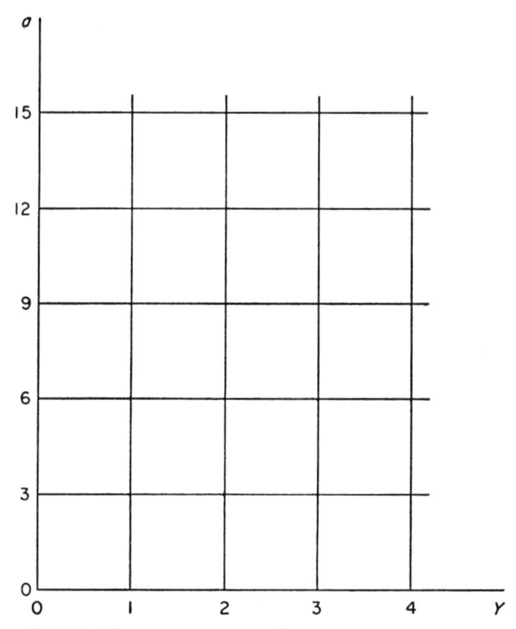

Fig. 13.3. Grid for linear approximation to a function of two variables.

where $f(i, j)$ is the value of the function being approximated at the point $Y = i$, $a = j$. By this device the nonlinear constraints in Y and a can be replaced by linear constraints in ω_{ij} and the linear-programming problem in terms of the weights can be solved by the methods already illustrated. Finally, when the optimal values of ω_{ij} satisfying these rules have been found, the corresponding values of the original variables can be recovered by means of the formulas:

$$Y = \sum_i \sum_j i\omega_{ij} - \sum_i i \sum_j \omega_{ij}$$

and

$$a = \sum_i \sum_j j\omega_{ij} = \sum_j j \sum_i \omega_{ij}.$$

Each of these sums will have at most three positive terms.

In principle, this approach can be used to linearize functions of more than two variables, but the number of auxiliary variables that must be introduced increases rapidly with the dimensionality of the function.[6]

[6] For other methods of solving mathematical-programming problems with nonlinear objective functions or constraints, see P. Wolfe, "The simplex method for quadratic programming," *Econometrica 27*, 382 (1959); J. B. Rosen, "The gradient projection method for nonlinear programming," *J. Soc. Indus. and Appl. Math. 8*, 181 (1960); and A. Charnes and C. E. Lemke, "Minimization of non-linear separable convex functionals," *Naval Research Logistics Quart. 1*, 301 (1954).

Now to summarize what this sample computation indicates about the usefulness of mathematical programming as an aid to designing water-resource development projects. The example was set up with two strong assumptions: first, that the problem could be analyzed by studying a small number of periods in isolation (two in the example), and second, that water inflow and all other factors could be predicted with certainty. Subject to these assumptions, mathematical programming determined simultaneously an optimal operating procedure and optimal values for the dimensions of installations and quantities of usable outputs, taking account of hydrologic, engineering, and economic considerations at one fell swoop. In general, mathematical programming can perform this task easily if the net-benefit formulas and the technologic constraints are linear functions of the decision variables, and with more trouble if they are nonlinear, provided that the relationships are not too complex. It appears that for relatively moderate computational expense this approach can provide useful first approximations to optimal designs.

But the two assumptions are certainly severe. The assumption of a small number of periods is appropriate only in humid regions or regions in which the annual runoff is highly regular. The assumption of certainty of prediction is almost never appropriate except as a crude first approximation. Thus it is important to relax these assumptions, and in the next two sections first one and then the other will be dropped. At the same time, all relationships will be taken to be linear in order to reduce the tedium of the exposition, since the present section has indicated how to take account of nonlinearity where necessary.

MORE PERIODS, PREDICTABLE HYDROLOGY

In the preceding section mathematical programming was shown to be helpful in finding the optimal designs and operating procedures for a fairly complicated system of water-resource installations provided that a short segment of time can be considered in isolation. This limitation arises from the fact that programming computations rapidly become more difficult as the number of equalities and inequalities to be handled increases. At the present time, readily available computing facilities can handle problems with approximately 250 constraints, and this can be considered the practicable limit on the size of problems to be undertaken.

This number of constraints is ample to describe the interrelationships among a quite elaborate system of reservoirs, irrigation districts, power plants, and so on within a single period of operation. But the complexity of the systems that can be analyzed within the computational limitation varies inversely with the number of periods that must be examined together. Consider first the situation in a well-watered region where each year's cycle of operations is self-contained. If the annual cycle can be regarded as consisting of two seasons, then 125 constraints are available to describe the relationships among the components of the system in each season. If the cycle must be divided into 12 monthly periods, only 20 constraints are available for each month, and that few constraints will suffice for only a very moderate number of components.

The situation is much more complicated in regions in which natural inflow is likely to be deficient in some years, so that overyear storage is required for efficient operation. Then the size of storage reservoirs is dictated in large part by the requirements for overyear storage, and these in turn must be estimated by considering the operation of the system over a substantial number of years. A plan of operation covering five years by months comprehends 600 elementary periods, or more than twice as many as the permissible number of constraints. Clearly if such analyses are to be undertaken, some method must be used for breaking up the total problem into a number of smaller problems of manageable size. This section is devoted to discussing such a method.

For technical reasons the overyear-storage problem will be divided into two parts. In real-life operation the managers of a system must decide on the amount of water to be left in storage at the end of a year while they still are ignorant of the amount of water that will become available in the next and subsequent years. As noted before, the unpredictability of water inflows presents highly intractable problems. Therefore, as a step toward solving the harder and more important problem, it seems advisable to assume in the present section that the pattern of flow for the next five or ten years can be predicted, and to consider here the problem of designing a system that makes efficient use of a sequence of unequal but predictable flows. The problems resulting from unpredictability will be treated in the next section.

It is well to remark, however, that the problem of design for a long sequence of predictable flows is more than an expository step. For one thing, this problem includes the analysis of an annual cycle of predictable flows with the year divided into months or even shorter periods. But, more important, it includes questions such as: What would have been the optimal design and operating procedures for the TVA network during the 1950's? Answers to this type of question are not entirely satisfactory for design purposes, since the hydrology of the 1950's is not likely to recur, but they are easy to obtain and they do provide an indication of the quantity of overyear storage capacity that is likely to be useful in the future. Since this type of problem can be solved much more easily and satisfactorily than a full-dress stochastic problem, the ability to solve it is very useful in its own right.

Our discussion will be conducted in terms of a very simple example. Remember that the method of analysis of the last section can, in principle, be extended to any problem, no matter how many periods are included, the only drawback being that when a sequence of many periods is involved, the computations become too onerous even for an electronic computer. The only illustrative problem that would do full justice to the method for reducing the burden of calculation to manageable size would be one so large that this reduction was necessary. But such an illustration would overwhelm both reader and writer with a mass of unilluminating detail. Hence the procedure will be illustrated by a small example that can be solved readily by standard methods, almost by hand, with the pretense that it is so large that devices for reducing the computational task are necessary.

The illustrative problem chosen is a modification of the one dealt with in the previous section. To give some of the flavor of analyzing a large number of periods, we shall increase to four the number of periods to be considered. On the other hand, many of the complicating features of last section's problem will be dropped in order to concentrate on the novel features entailed by having numerous periods.

The modified problem is shown schematically in Fig. 13.4. Note that reservoir C has been suppressed and that irrigation supply has been set at 3×10^6 acre ft per year. Four seasonal flows have been specified for each reach of the river, in place of two, and a new symbol, $a_t, t = 1, 2, 3, 4$, has been introduced to denote releases from reservoir B in the four periods. For the rest, the problem remains much as before. Other modifications and some new notation will be explained as the problem is described more specifically.

Since the quantity of irrigation water has been preassigned, it no longer enters the objective function, which consists now of only two terms: the capital cost of constructing reservoir B and the present value of the net hydroelectric benefits. Suppose that the construction cost of the reservoir is $32Y$, where Y denotes its capacity, in order to avoid repeating

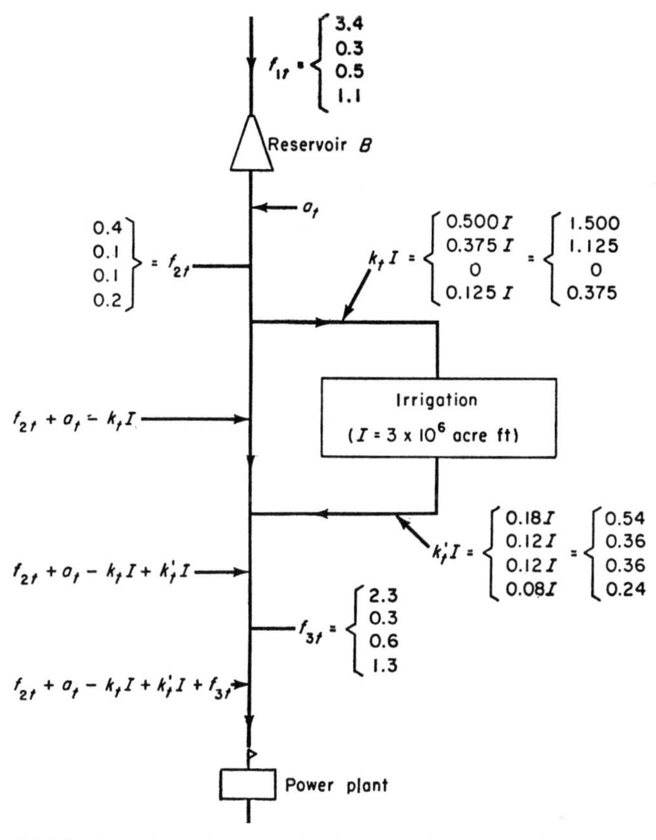

Fig. 13.4. Sketch of configuration used in four-period problem. Figures in each brace represent values for first through fourth periods of flows f_{1t}, f_{2t}, and f_{3t}, $t = 1, 2, 3, 4$. Releases from reservoir B in the four periods are denoted by a_t. Total irrigation water to be diverted, I, is preassigned at 3×10^6 acre ft, with the proportion of irrigation demand in each period denoted by k_t and irrigation-return-flow coefficients denoted by k_t'.

the tedious details of handling a nonlinear function. Similarly, take a linear approximation to the net hydroelectric benefits, present value of output minus power-plant construction costs. Then the objective function is simply

$$\pi = 221E - 32Y, \tag{13.7}$$

where π and E have the same meanings as before.

There are six constraints applicable to each period. The first is that the volume of water released from the reservoir must be sufficient to meet that period's irrigation demand, where the latter is a preassigned proportion, k_t, of the annual irrigation demand. The four values of k_t are given in Table 13.4. From the configuration in Fig. 13.4 this requirement

Table 13.4. Proportions of irrigation flow and energy demand for each period in the four-period problem.

Period t	Irrigation demand k_t	Irrigation return flow k_t'	Energy demand c_t
1	0.500	0.18	0.26
2	0.375	0.12	0.29
3	0	0.12	0.24
4	0.125	0.08	0.21
Total	1.000	0.50	1.00

can be written

$$a_t + f_{2t} \geq k_t I, \qquad t = 1, 2, 3, 4, \tag{C1}$$

where f_{2t} denotes the flow from the tributary that joins the main stem just above the irrigation-diversion canal and I is preassigned at 3×10^6 acre ft. The natural flows assumed in the computations are indicated in the figure.

The second constraint is that the volume of water released during any period cannot exceed the contents of the reservoir at the beginning of the period plus the flow into the reservoir during the period. To express this constraint, let S_t denote the contents of the reservoir at the beginning of period t. Then the constraint is

$$a_t \leq S_t + f_{1t}, \tag{C2}$$

f_{1t} being the preassigned natural flow into the reservoir during period t.

The third constraint is that the contents of the reservoir at the beginning of any period cannot exceed the amount left over from the previous period, or

$$S_t \leq S_{t-1} + f_{1,t-1} - a_{t-1}. \tag{C3}$$

The fourth constraint states that the contents of the reservoir at the end of any period cannot exceed the capacity of the reservoir, or

$$S_t + f_{1t} - a_t \leq Y. \tag{C4}$$

The third and fourth constraints together automatically ensure that the contents of the reservoir at the beginning of any period do not exceed its capacity, so that this requirement need not be listed separately.

The final two constraints relate to power generation. The fifth constraint is that the flow of water past the power plant must be sufficient to meet the requirement for power generation. To express this requirement assume, as in the previous example, that 6.95×10^6 acre ft of water are required to generate 1×10^9 kw hr of electric energy. The flow available at the power plant is the sum of the flow past the irrigated area, the return flow from the irrigated area, $k_t'I$ (where the return-flow coefficients, k_t', are as given in Table 13.4), and the natural flow from the eastern tributary, f_{3t}. This requirement is therefore

$$a_t + f_{2t} - (k_t - k_t')I + f_{3t} \geq 6.95E_t, \tag{C5}$$

where E_t denotes the quantity of electric energy generated in the tth period.

The sixth and last requirement is that the amount of electric energy generated in each period must be at least a specified proportion, c_t (shown in Table 13.4), of the annual energy output, or

$$E_t \geq c_tE. \tag{C6}$$

Subject to these requirements for each period, values of the decision variables a_t, S_t, Y, E_t, and E are to be found so as to make the objective function as large as possible. In all, 24 constraints connect the 14 decision variables. Although a problem of this scale is well within the capacity of even a moderate-size computer, it will be solved by a method that drastically reduces the number of constraints that have to be handled at any one time just as if it were a very large problem. The method to be used is the decomposition principle developed by George B. Dantzig and Philip Wolfe.[7]

The clue to the decomposition principle is contained in the constraints just listed. ($C1$), ($C2$), and ($C5$) each contain decision variables relating to period t and no others. In contrast, ($C3$) contains decision variables relating to two periods, and ($C4$) and ($C6$) contain variables, Y and E, that relate to all the periods. Now it is possible to replace by a single constraint all the constraints relating to the first period alone, and similarly for the other three periods. This is done by solving a linear-programming subproblem for each period, the subproblem including all the constraints that relate exclusively to that period — in this example, three. When this has been done, a master problem remains that includes

[7] George B. Dantzig and Philip Wolfe, "Decomposition principle for linear programs," *Operations Research 8*, 101 (1960).

all the constraints that relate to more than one period (in this example, ($C3$), ($C4$), and ($C6$) for each period) plus one supplementary constraint for each period. This device replaces a single problem in 24 constraints by a set of problems some of which have three constraints and one of which has 16 constraints. The advantage in this instance is not very great. But in general, the reduction in the size of the largest problem to be solved as a whole is equal to the number of periods multiplied by one less than the number of constraints in each period. This can be considerable, although it is only eight in our case.

The data of the problem are summarized in Table 13.5. This table was derived from Eqs. (13.7) and ($C1$) through ($C6$) by (1) replacing all the preassigned constants by their numerical values, (2) placing all decision

Table 13.5. Summary of constraints and objective function of the four-period problem, in detached-coefficient form.

Period	Constraint	S_1	a_1	E_1	S_2	a_2	E_2	S_3	a_3	E_3	S_4	a_4	E_4	Y	E	$\leq$ constant
	($C1$)		−1													−1.1
1	($C2$)	−1	1													3.4
	($C5$)		−1	6.95												1.74
	($C1$)					−1										−1.025
2	($C2$)				−1	1										0.3
	($C5$)					−1	6.95									−0.365
	($C1$)								−1							0.1
3	($C2$)							−1	1							0.5
	($C5$)								−1	6.95						1.06
	($C1$)											−1				−0.175
4	($C2$)										−1	1				1.1
	($C5$)											−1	6.95			1.365
	($C3$)	−1	1		1											3.4
					−1	1		1								0.3
								−1	1		1					0.5
		1									−1	1				1.1
	($C4$)	1	−1											−1		−3.4
					1	−1								−1		−0.3
								1	−1					−1		−0.5
											1	−1		−1		−1.1
	($C6$)			−1											0.26	0
							−1								0.29	0
										−1					0.24	0
													−1		0.21	0
Net-benefit function														−32	221	

variables on the left side of relationship signs and all constant terms on the right, (3) grouping together the relationships that pertain to a single period, (4) converting all "greater than" inequalities to "less than" inequalities, and (5) writing the result in detached-coefficient form. This arrangement highlights the insulation of the block of constraints relating to each period from variables pertaining to other periods.

Now suppose that a number of different solutions to the constraints for any period are known. For example, Table 13.6 contains three solutions to the fourth-period constraints (how these solutions were selected will emerge presently). If any one of these three solutions is substituted into the constraints for the fourth period, it will be found, of course, that all the inequalities are satisfied. But, what is more important, certain linear combinations of these solutions also satisfy the constraints. For example, the constraints are also satisfied by the synthetic solution consisting of one-half of the first solution plus one-quarter of the second solution and one-quarter of the third solution. This combination is shown in the last column of Table 13.6. In fact, any weighted sum of the three solutions, provided that the weights are nonnegative and total unity, will do so. This suggests that the three variables S_4, a_4, and E_4 be replaced by the weights to be applied to these three solutions and all the fourth-period constraints be replaced by the single constraint that the weights should total unity. If this substitution is performed for the fourth period and similar replacements for the other periods, the only constraints remaining will be those of the master problem plus the four constraints that force the weights in each period to total unity. The optimal scheme of weights that satisfies this shortened list of constraints can then be determined, in the knowledge that the levels of the original variables computed from those weights will satisfy all the period constraints.

Table 13.6. Three possible solutions to the fourth-period constraints and a combination solution.

Variable	Solution			
	1	2	3	Combination[a]
S_4	0	0	0	0
a_4	0.175	0.175	1.100	0.406
E_4	0.222	0	0.355	0.199

[a] This combination is ½ (solution 1) + ¼ (solution 2) + ¼ (solution 3).

When the optimal weights have been found and converted back to the original variables, these values will satisfy all the constraints but they will not necessarily be an optimal solution to the original problem. This process will produce such a solution if and only if the solutions to the individual problems to which the weights were applied (for instance, the three solutions for the fourth period in Table 13.6) included optimal solutions to all the individual problems.

Hence the decomposition method consists of two main parts: first, a procedure for finding solutions to the individual-period problems that collectively also satisfy the constraints of the master problem, and second, a procedure for generating additional solutions to the period problems until optimal solutions for all the individual periods have been discovered. When this has been done, an optimal solution to the over-all problem will also be at hand. The first of these procedures is a matter of straightforward algebra, but the second presupposes some familiarity with linear programming. Therefore this discussion will be limited to the strategic ideas of the method, referring the reader to the Dantzig-Wolfe paper for details.

The decomposition method starts with some set of values of the decision variables, whether economical or not, that satisfies all the constraints — all 24 in this case. Accordingly a power output $E = 1$, a level that obviously can be met, is chosen arbitrarily and E_t is set equal to c_t for all periods. In addition, S_1, the quantity of water in storage at the beginning of the wettest period, is set equal to zero. From these values, values of all the variables consistent with the constraints can be determined from Table 13.5 by going through the constraints seriatim, assigning at each stage the smallest permissible value to a_t and the largest permissible value to S_t. This starting set of values is shown in Table 13.7.

If the values of the fourth-period variables in this table are compared

Table 13.7. Starting values of variables in the four-period problem.

Variable	Value in all periods	Value in period — 1	2	3	4
S_t		0	2.300	0.219	0
a_t		1.10	2.381	0.719	0.175
E_t		0.26	0.290	0.240	0.210
Y	2.3				
E	1.0				

with the illustrative solutions for that period in Table 13.6, it will be seen that this solution is not one of those given in Table 13.6. The reason is that all of the solutions given in Table 13.6 are "basic solutions" to the fourth-period constraints, that is, solutions in which the number of variables with positive values does not exceed the number of constraints satisfied with strict equality. The solution to the fourth-period constraints given in Table 13.7 is not a basic solution in this sense, since substitution will show that, although two variables are positive, only the first constraint is satisfied with strict equality.

Basic solutions are of peculiar importance, because it is true of them, and of a certain other class to be discussed below, that all solutions can be expressed as weighted sums of them with weights totaling unity. The starting solution for the fourth period accordingly must be analyzed into the basic solutions that comprise it. This is done in Table 13.8, which is

Table 13.8. Analysis of starting solution for fourth period of the four-period problem. Zeros are omitted except in constant column.

				Slack[a] in constraint —			
Line	S_4	a_4	E_4	1	2	3	Constant
1		0.175	0.210		0.925	0.081	1.000
2			1.000			−6.950	0
3		0.175			0.925	1.541	1.000
4		0.166	0.210		0.876		0.947
5		0.175	0.222		0.925		1.000

[a] The slack in any constraint is the amount by which the constant column exceeds the result of substituting the starting solution in the constraint.

constructed as follows. The first three columns of line 1 contain the values of S_4, a_4, and E_4 in the starting solution to be analyzed. The next three columns display the "slack" in the fourth-period constraints, that is, the amount by which the right-hand side exceeds the left-hand side if the starting solution is used. The figure "1" in the final column indicates that if the values shown are substituted into the fourth-period constraints and the slack values are added, the result will be one times the constant column of those constraints. Line 2 records one solution to the following system of equations, in which x_2 and x_3 correspond to the slacks in the second and third constraints:

$$-1a_4 + 0E_4 + 0x_2 + 0x_3 = 0,$$

$$1a_4 + 0E_4 + 1x_2 + 0x_3 = 0,$$

$$-\ 1a_4 + 6.95E_4 + 0x_2 + 1x_3 = 0.$$

These equations are constructed by giving a_4 and E_4 the same coefficients as in the fourth-period constraints, and by giving x_2 and x_3 coefficients of unity in the equations in which positive slack was noted in line 1. If the values shown in line 2 are substituted into the constraints, the result will be zero times the constant column, as shown in the last column of line 2. Line 3 is derived by subtracting 0.21 times line 2 from line 1. The result is one of the basic solutions given in Table 13.6 and one of the basic solutions of which the starting solution is composed. Line 4 is obtained by subtracting 0.053 times line 3 from line 1, the factor being chosen so as to cancel out the slack in the third constraint of the fourth period. If the values in line 4 are substituted into the fourth-period constraints, the result will be 0.947 times the constant column. Hence this is not a solution to the constraints, but a solution can be obtained by dividing through by 0.947. This solution is shown on line 5. It is basic, since two variables have positive values and two equations have zero slack. The solutions on lines 3 and 5 are the basic solutions of which the starting solution is a weighted average, with weights 0.053 and 0.947 respectively. The weights, however, are of no interest; the computation is performed to reveal the underlying basic solutions.

When this same analysis is carried out for the other three periods, a starting list of basic solutions is generated. In addition, the list will include some "homogeneous solutions," that is, solutions to the period constraints with the right-hand side replaced by a column of zeros. For example, the starting solution for the third period, shown in Table 13.7, is composed of (1) the basic solution $a_3 = 0.5$ with weight 0.071, (2) the basic solution $a_3 = 0.5$, $E_3 = 0.224$ with weight 0.929, and (3) the homogeneous solution $S_3 = 0.219$, $a_3 = 0.219$, and $E_3 = 0.032$. This can be seen by direct substitution. For example, $0.5 \times 0.071 + 0.5 \times 0.929 + 0.219 = 0.719 = a_3$, as shown in Table 13.7.

A weighted sum of the basic solutions and homogeneous solutions in the starting list will satisfy all the individual-period constraints provided that the sum of the weights applied to the basic solutions for each period is unity. The standard linear-programming problem of finding the optimal weights that satisfy this restriction for each period and also the constraints of the master problem, disregarding the constraints pertaining to individual periods, will therefore yield the optimal weighting of the individual-period solutions included in the starting list.

The remainder of the procedure consists of improving the list of individual-period solutions by adding advantageous solutions that were not included originally. This improvement is accomplished by using the fact that the solution to a linear-programming problem contains the information necessary to estimate the amount by which the net benefits could be increased if any of the constraints were relaxed a slight amount. For example, recall that one of the stipulations of this starting solution was that the reservoir would be empty at the start of the first period. If, instead, the initial storage for this period had been set at 1×10^6 acre ft, and the other first-period values were unchanged, the capacity of the reservoir would have to be increased by 1×10^6 acre ft, at a cost of $32,000,000. At the same time, the solution to the problem of finding optimal weights for the starting list, which is not presented in detail, indicated that an additional 1×10^6 acre ft available at the beginning of the second period would permit an increase of $60,000,000 in net benefits. Clearly, then, the homogeneous solution $S_1 = 1$ should be added to the list.

When this addition is made, a new master problem results which, when solved, indicates some new desirable individual-period solutions. This process of adding desirable solutions to the individual-period problems and then searching for more is continued until a list of individual-period solutions is obtained for which the solution to the problem of finding optimal weights does not indicate any new individual-period solutiosg to be added. This even signalizes that the lst is complete, and the optimal scheme of weights can be used to determine the optimal levels of the various decision variablei by simple substitution.

Table 13.9 presents the final solution for this problem. The detailed computations are omitted since they cannot be interpreted in reasonable space without invoking a good many of the technical concepts of linear programming. The lower half of the table shows the amount by which maximum net benefit could be increased by relaxing the constraints of the master problem or, in other words, it shows the marginal or dual values of the resources associated with each of those constraints. For example, the flow f_{11} is worth $52.10 per acre ft, because if f_{11} could be increased by 1×10^6 acre ft to 4.4, net benefits could be increased by $52,100,000, as shown by the entry for ($C3$) in the first period.

Table 13.9. Solution to the four-period problem.

Optimal levels of decision variables					
		Value in period —			
Variable	Value in all periods	1	2	3	4
S_t		0.688	2.988	0.471	0
a_t		1.100	2.817	0.971	0.412
E_t		0.408	0.353	0.292	0.256
Y	2.988				
E	1.217				

Dual values associated with master constraints				
	Value in period —			
Constraint	1	2	3	4
($C3$)	52.1	52.1	20.1	20.1
($C4$)	32.0	0	0	0
($C6$)	0	361.7	361.7	139.5

Maximum net benefits = $173,400,000

Although this brief exposition could not make clear the details of this technique, the upshot should be evident — namely, that the mathematical-programming methods of the simple two-season case can be extended readily to problems that require the ascertainment of an optimal operating procedure for a time span composed of a substantial number of elementary periods, be they weeks, months, or years.

ALLOWANCE FOR UNCERTAINTY

The point of departure of all the techniques discussed in this chapter is mathematical programming, which has the salient virtue that it can be used to discover optimal decisions in quite complicated circumstances by means of highly standardized and readily mechanized computations. But one of the major limitations of mathematical programming in the present state of the art is that it cannot very readily encompass uncertainty. The problem of uncertainty was discussed in Chapter 3, and a procedure for dealing with it by means of linear programming will be developed in the next chapter. It will be noted that the methods suggested in both of these chapters apply only to very simple decisions. For example, the method of Chapter 14 determines optimal releases from a single reservoir, and assumes that the allocation of water among competing uses is a separate suboptimization problem.

This limitation of mathematical programming appears to be inherent in its logical structure. The method begins by expressing the constraints on decisions in the form of a series of linear inequalities, for example, inequalities stating that the contents of a reservoir cannot exceed its capacity or that the flow of water past a power plant must be sufficient to generate its firm-energy commitment. When water availability and other conditions are uncertain, these inequalities falsify the actual problem. A firm-energy commitment does not really mean that we must be prepared to supply the energy in the most adverse conceivable circumstances; we generally admit some risk of nonfulfillment. Thus, in reality, we are not required to honor all the inequalities in all contingencies, and the mathematical form of linear programming is more rigid and demanding than actuality. For this reason the constraints of a mathematical-programming model cannot be taken literally when uncertainty enters, and since mathematics is a very literal subject, if the advantages of programming are to reaped, the method must be applied to approximative models that do permit occasional violations of the constraints. Since the models are approximative, they cannot yield complete solutions to the problem of designing water-resource systems. Instead, they should be regarded as providing starting points for more flexible but less economical and convenient methods of finding optimal designs, such as simulation methods.

To convey the general strategy and concepts of our approximative approach, we shall expound a family of models. Our first example, far too simple to be of any practical interest, will deal with a single reservoir, a single use of water, and a single season per year. With this background we shall go on to a more realistic example, but still simple, designed to show how this conceptual framework can be extended to design problems of more practical interest.

THE VERY SIMPLEST CASE

Suppose that an irrigation project consisting of a single reservoir and associated structures is being contemplated. Guaranteed irrigation water is estimated to be worth \$15 per acre foot annually. In any year in which releases are less than the guaranteed amount, a loss of \$30 per acre foot of shortage is sustained, but for simplicity we assume that water in excess of the guaranteed amount is of no value. The annual amortization cost of the reservoir and irrigation works is \$1.50 per acre foot of capacity; there are no other costs. Annual inflow into the reservoir is a random variable with the probability distribution shown in Table 13.10. On the

Table 13.10. Assumed probability distribution of inflow into reservoir for the simplest problem with uncertainty.

Inflow, f_i (10^6 acre ft)	Probability, p_i
1	0.1
2	0.1
3	0.2
4	0.4
5	0.2
Average inflow = 3.5×10^6 acre ft	

basis of these data, how large a system should be constructed, and how large a commitment for water supply should be undertaken?

We shall assume that at the start of each year a reliable prediction can be made of the inflow that will be received during the year but that all we know about subsequent inflows is their probability distribution. The problem can then be expressed as follows. Let I denote the guaranteed annual quantity of irrigation water, K the capacity of the system, and y_i the quantity of water that will be supplied if inflow f_i is realized, all measured in 10^6 acre feet. From Table 13.10 we see that the possible values of f_i are 1, 2, 3, 4, 5. The variables y_i are not of direct interest, but they constitute a specification of the operating procedure to be followed.

In any year in which an inflow f_i is experienced, the net benefits will be

$$15I - 1.5K - 30(I - y_i)^+,$$

where $(I - y_i)^+$ equals $I - y_i$ if that number is zero or positive; otherwise it equals zero. (In more general terms, $u^+ = u$ if $u \geq 0$, and $u^+ = 0$ if $u < 0$.) The expected value of annual net benefits B then is

$$B = 15I - 1.5K - 30 \Sigma p_i(I - y_i)^+, \tag{13.8}$$

where p_i is the probability of inflow f_i as given in Table 13.10. We can take, at least as a first approximation to the optimal design, the design for which B as defined is as great as possible. The variables in Eq. (13.8) are far from independent. To express the restrictions to which they must conform, let S denote the quantity of water in storage at the beginning of the year. Then, since during a single year no more water can be released than the sum of the initial storage and the inflow, we have

$$y_i \leq S + f_i. \qquad f_i = i = 1, 2, 3, 4, 5$$

In addition, since the contents of the reservoir at the end of the year cannot exceed its capacity, we have

$$S + f_i - y_i \leq K. \qquad i = 1, 2, 3, 4, 5$$

At this point we introduce the simplifying device that renders this approach approximative rather than exact. We shall regard S, the initial storage, as a decision variable along with I, K, and y_i, although it is not one, in fact. Thus S will not stand for the actual storage at the start of any year but, in effect, for our aspiration level for initial storage. And we shall regard ourselves as being free to select any value of S we choose, provided that it does not exceed the mathematical expectation of the quantity of water in storage at the end of the year. This limitation is imposed to assure that the aspiration level is realistic in the sense that once it has been attained, it can realistically be expected to be reattained at the end of the year, so that the same outputs can be produced year after year. Algebraically, the constraint on S is

$$\Sigma p_i(S + f_i - y_i) \geq S,$$

which simplifies to

$$\Sigma p_i y_i \leq \Sigma p_i f_i = 3.5.$$

All the constraints we have listed are set forth in detached-coefficient form in Table 13.11. They imply among themselves that the initial storage, S, does not exceed the capacity of the reservoir, K, so that we do not have to list this constraint separately.

These are the constraints of a linear-programming problem, but the objective function, Eq. (13.8), is unfortunately not linear because of the corners in the expressions $(I - y_i)^+$. We can convert Eq. (13.8) to a linear function by the following device. We introduce some auxiliary variables, u_i, $i = 1, 2, \ldots, 5$, (where u stands for undersupply) that satisfy

$$u_i \geq I - y_i, \qquad u_i \geq 0, \tag{13.9}$$

Table 13.11. Constraints for the simplest problem with uncertainty, listed in detached-coefficient form.

Constraint	K	S	y_1	y_2	y_3	y_4	y_5	$\leq$ constant
1			0.1	0.1	0.2	0.4	0.2	3.5
2		−1	1					1
3		−1		1				2
4		−1			1			3
5		−1				1		4
6		−1					1	5
7	−1	1	−1					−1
8	−1	1		−1				−2
9	−1	1			−1			−3
10	−1	1				−1		−4
11	−1	1					−1	−5

and insert them in the objective function in place of the piecewise linear functions $(I - y_i)^+$, thus:

$$B = 15I - 1.5K - 30 \Sigma p_i u_i. \tag{13.10}$$

The constraints on the auxiliary variables of Exp. (13.9) are given in detached-coefficient form in Table 13.12. The constraints displayed in

Table 13.12. Auxiliary constraints for the simplest problem with uncertainty, listed in detached-coefficient form.

Constraint	I	y_1	y_2	y_3	y_4	y_5	u_1	u_2	u_3	u_4	u_5	$\leq$ constant
1	1	−1					−1					0
2	1		−1					−1				0
3	1			−1					−1			0
4	1				−1					−1		0
5	1					−1					−1	0

Tables 13.11 and 13.12, together with the objective function of Eq. (13.10), constitute a standard linear-programming problem. Solution is facilitated, moreover, because it is already in the form for applying the Dantzig-Wolfe decomposition principle; the constraints in Table 13.11 are treated as the master problem and those in Table 13.12 as the single subproblem. The solution is found straightforwardly either by applying the Dantzig-Wolfe algorithm or, since this is a very simple case, by combining the two tables into an ordinary linear-programming problem with 16 constraints. The following values of the variables are determined:

System capacity	$K = 3.0 \times 10^6$ acre ft
Guaranteed irrigation supply	$I = 3.6 \times 10^6$ acre ft
Expected annual net benefits	$B =$ \$46,500,000
Planned initial storage	$S = 1.6 \times 10^6$ acre ft

The operation of this system was simulated for a period of 50 years.[8] Eight shortages were experienced, the greatest amounting to 22 percent of the guaranteed supply, and the average annual net benefits were \$46,700,000. One interesting aspect of this solution is that the guaranteed supply is slightly greater than the average annual inflow, which after a moment's reflection is seen to be a very plausible policy.

This artificial example was presented only to introduce our general strategy and concepts. We are now prepared to go on to a more interesting example, one with three seasons and two uses of water that are not entirely complementary.

A MORE REALISTIC EXAMPLE

The key to this approach is the fiction that the quantity of water in storage at the beginning of each year is equal to its mathematical expectation, with expectation chosen at a level that balances the advantages of protection against possible drought, on the one hand, against the cost of impounding larger quantities of water, on the other. This device enables us to analyze the operation of the system in a single isolated year; in the language of engineering, it effectuates a "cut" in the chain of consequences of decision.

The same strategy applies when the year must be divided into dissimilar seasons, as is the case in every practical application. We then postulate that each season begins with the expected quantity of water in reserve. The following example illustrates the approach in a multiseason case.

Suppose that an irrigation and hydroelectric development is being considered for the region portrayed in Fig. 13.5. There are four essential decisions to be made: the capacity of the reservoir, to be denoted by K; the normal or target amount of water to be provided for irrigation,

[8] The simulation was performed by assuming 1.6×10^6 acre ft in storage at the beginning of the first year, drawing inflows for the 50 years from a table of random numbers in accordance with the probabilities of Table 13.10, in each year releasing the planned volume of water up to the limit of availability, and following the performance of the system year by year.

denoted by I; the generating capacity of the power plant, denoted by G; and the quantity of firm energy to be provided, denoted by E.

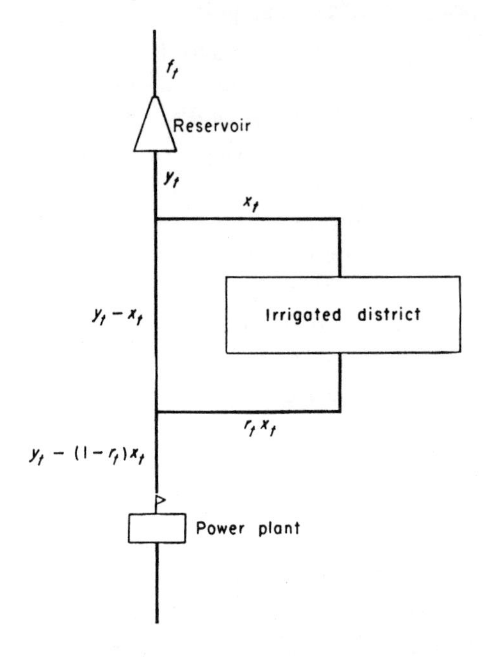

Fig. 13.5. Sketch of configuration used in three-season example with uncertainty. Inflow to the reservoir in each season is denoted by f_t and release in each season by y_t. Water diverted for irrigation is denoted by x_t, and irrigation return flow by $r_t x_t$.

These decisions depend on both hydrologic and economic considerations. The hydrologic data are given in Table 13.13, which shows the

Table 13.13. Assumed probability distributions of inflows for the three-season example with uncertainty.

Inflow, f_t (10^6 acre ft)	Probability distribution in season —		
	1	2	3
0		0.3	
1		0.4	0.3
2		0.3	0.4
3	0.3		0.3
4	0.4		
5	0.3		

probability distribution of inflow, f_t, in each of the three seasons into which the year is divided. It is assumed that the inflows in the separate seasons are statistically independent. This assumption facilitates the solution by making it possible to consider the individual seasons separately, as we shall see. The other data of the problem are set forth in Table 13.14.

Table 13.14. Assumed economic and other data for the three-season example with uncertainty.

Item	Value	Unit
Annual amortization costs		
Reservoir of capacity $K \times 10^6$ acre ft	$1.50K$	10^6 dollars
Power plant of installed capacity $G \times 10^6$ kw	$5.28G$	10^6 dollars
Irrigation works of capacity $I \times 10^6$ acre ft per year	$1.30I$	10^6 dollars
Annual operating costs		
Power plant of size G	$2.10G$	10^6 dollars
Irrigation works of size I	$0.50I$	10^6 dollars
Unit values for power		
Firm energy	0.007	Dollars per kw hr
Dump energy	0.0005	Dollars per kw hr
Supplementary thermal power	0.009	Dollars per kw hr
Unit values for irrigation water		
Normal supply	5	Dollars per acre ft
Shortage cost, season 1	8	Dollars per acre ft
Shortage cost, season 2	10	Dollars per acre ft
Seasonal pattern for irrigation		
Proportion in season 1	0.40	
Proportion in season 2	0.60	
Proportion in season 3	—	
Seasonal pattern for power supply		
Proportion in season 1	0.40	
Proportion in season 2	0.30	
Proportion in season 3	0.30	
Load factors for power supply		
Season 1	0.65	
Season 2	0.60	
Season 3	0.60	
Return-flow factors for irrigation		
Season 1	0.60	Acre ft per acre ft supplied
Season 2	0.40	
Conversion factor for power plant	144	Kw hr per acre ft through turbines
Length of each season	2920	Hours

To construct the net-benefit function, note that once the four main variables have been decided, the net benefits can be computed except for the shortage costs and the value of dump energy. Accordingly we derive from Table 13.14:

$$\text{Annual net benefits (in } 10^6 \text{ dollars)} = 3.2I + 0.007E - 1.5K - 7.38G + \text{value of dump energy} - \text{shortage costs} \quad (13.11)$$

To determine these last two terms, let x_1 and x_2 denote the quantities of irrigation water supplied in the first and second seasons, respectively. From the data on the seasonal pattern for irrigation, we see that the normal supply in the two seasons is $0.4I$ and $0.6I$, respectively, so that the cost of irrigation shortage is

$$8(0.4I - x_1) + 10(0.6I - x_2) = 9.2I - 8x_1 - 10x_2 \quad (13.12)$$

where it is assumed that the actual supply of irrigation water never exceeds the normal supply.[9]

Similarly, let q_1, q_2, and q_3 denote the flow through the turbines in the three seasons, which we assume never exceeds their capacity. Then, again from Table 13.14, the cost of the thermal power required to meet the firm-energy commitment is

$$0.009[(0.4E - 144q_1)^+ + (0.3E - 144q_2)^+ + (0.3E - 144q_3)^+],$$

where, as before, the notation u^+ is defined by $u^+ = u$ if $u \geq 0$ and $u^+ = 0$ if $u < 0$. The value of dump energy is

$$0.0005[(144q_1 - 0.4E)^+ + (144q_2 - 0.3E)^+ + (144q_3 - 0.3E)^+].$$

These two values can be combined by using the general rule that $(-u)^+ = u^+ - u$. Hence the excess of thermal-energy cost over the value of dump energy produced is

$$0.0085[(0.4E - 144q_1)^+ + (0.3E - 144q_2)^+ + (0.3E - 144q_3)^+] + 0.0005E - 0.072(q_1 + q_2 + q_3). \quad (13.13)$$

Finally, the formula for net benefits is obtained by substituting Exps. (13.12) and (13.13) into Eq. (13.11):

$$\text{Annual net benefits} = -6I + 0.0065E - 1.5K - 7.38G + 8x_1 + 10x_2 + 0.072(q_1 + q_2 + q_3) - 0.0085[(0.4E - 144q_1)^+ + (0.3E - 144q_2)^+ + (0.3E - 144q_3)^+].$$

This formula shows annual net benefits as a linear function of all the variables, except for the last bracket which is only piecewise linear. To

[9] This assumption is not restrictive. It means merely that all releases not needed for irrigation will be routed down the main stem rather than into the irrigation-diversion canal. It facilitates solution of the problem by allowing the use of linear expressions like $(0.4I - x_1)$ in place of nonlinear ones like $(0.4I - x_1)^+$. A similar nonrestrictive assumption will be introduced for the same reason in formulating the hydroelectric constraints.

complete the linearization, replace the three functions in the bracket by three new variables, e_1, e_2, and e_3 defined by

$$e_1, e_2, e_3 \geq 0,$$
$$e_1 \geq 0.4E - 144q_1,$$
$$e_2 \geq 0.3E - 144q_2,$$
and
$$e_3 \geq 0.3E - 144q_3.$$

The result is

$$\text{Annual net benefits} = -6I + 0.0065E - 1.5K - 7.38G + 8x_1 + 10x_2 + 0.072(q_1 + q_2 + q_3) - 0.0085(e_1 + e_2 + e_3). \tag{13.14}$$

To see why this replacement is legitimate, consider e_1. It is not defined to be equal to $(0.4E - 144q_1)^+$, but since it occurs in the net-benefit function with a negative coefficient, when we try to maximize net benefits it will assume its minimal permissible value, which is $(0.4E - 144q_1)^+$. In physical terms e_t is simply the amount of thermal energy supplied in period t.

We cannot maximize Eq. (13.14) directly since it includes some variables (those denoted by lower-case letters) that depend on some random variables, the inflows into the reservoir in each season. Instead, we shall maximize the expected value of annual net benefits. This maximization must, of course, respect certain constraints in each period. We shall deduce the constraints that pertain to the first season; those that apply to the other two seasons are derived similarly.

To deduce these constraints, we require a bit more notation. Let S_1 denote the assumed contents of the reservoir at the start of the first season; let y_{1f} be the releases to be made from the reservoir in the first season when the inflow is f, $f = 3, 4, 5$; and analogously define x_{1f}, q_{1f}, and e_{1f} to mean respectively irrigation supply, flow through turbines, and thermal energy supply when natural inflow is q. Then if the reservoir is to be neither drawn down below zero nor overtopped, we must have

$$S_1 + f \geq y_{1f} \tag{13.15}$$

and

$$S_1 + f - y_{1f} \leq K. \qquad f = 3, 4, 5 \tag{13.16}$$

We insist also that the expected contents of the reservoir at the end of the season be at least as great as S_2, the assumed contents at the start of the second season. Formally:

$$\sum_{f=3}^{5} p_f(S_1 + f - y_{1f}) \geq S_2,$$

where p_f is the probability of inflow equal to f, found in the first column of Table 13.13. This constraint simplifies to

$$S_1 + \sum_f p_f f - \sum p_f y_{1f} \geq S_2. \tag{13.17}$$

Next come a pair of irrigation constraints. The irrigation draft must not exceed the releases from the reservoir and also must not exceed the irrigation requirement. Therefore we impose

$$x_{1f} \leq y_{1f} \tag{13.18}$$

and

$$x_{1f} \leq 0.4I. \tag{13.19}$$

The power constraints assert that the flow through the turbines cannot exceed either the flow in the stream above the power plant or the capacity of the turbines. The first of these requirements gives rise to

$$q_{1f} \leq y_{1f} - 0.4x_{1f}. \tag{13.20}$$

To derive the second, we must compute the intake capacity of the turbines. A power plant of rated capacity $G \times 10^6$ kw can utilize a flow of $(G/144) \times 10^6$ acre ft per hour. When we apply the load factor for the first season, there are the equivalent of $2920 \times 0.65 = 1898$ hours in the season. Hence we have the turbine-intake constraint

$$q_{1f} \leq 1898(G/144)$$

or

$$144q_{1f} \leq 1898G. \tag{13.21}$$

Finally, the amount of thermal power provided must be at least equal to the excess of the firm-power commitment ($0.4E$) over the amount of hydropower ($144q_{1f}$), or

$$e_{1f} \geq 0.4E - 144q_{1f}. \tag{13.22}$$

Each of these inequalities, except for (13.17), must hold for each of the three possible inflows, $f = 3, 4, 5$. Therefore Exps. (13.15)–(13.22) comprise 22 constraints in all. There are 22 similar constraints for the second season; for the third, when there is no irrigation, 16 more. We now have the problem of determining values of all the variables (there are 40 of them) so as to maximize the expected value of Eq. (13.14) while satisfying all of these constraints. The expected value of Eq. (13.14) is obtained by replacing each of the variables that occurs in it by its expected value. Since the first four variables are not random, no replacement is necessary for them. To see how the expected values of the other variables are computed, consider x_1. It can take on three possible values, x_{13}, x_{14}, or x_{15}, depending on the amount of the inflow f. Since $x_1 = x_{13}$ whenever $f = 3$, the probability of this value is the same as the probability that $f = 3$, or $p_3 = 0.3$. In short, the probability that $x_1 = x_{1f}$ is p_f, and

$$\text{Expected value of } x_1 = \sum_{f=3}^{5} p_f x_{1f}.$$

This sum replaces x_1 in Eq. (13.14), and the replacement of the other random variables is similar.

Now we have arrived at a standard linear-programming problem with 60 constraints. A problem of this size is within the capacity of a moderate size computer (by 1961 standards), although if there had been many more constraints it would have been advisable to decompose the problem according to the methods described earlier in this chapter.

Table 13.15. Solution to the three-season example with uncertainty.

Variable	Value for whole year	Value in season — 1	2	3	Unit
Reservoir capacity, K	4.32				10^6 acre ft
Power-plant capacity, G	0.132				10^6 kw
Generating capability		250.84	231.55	231.55	10^6 kw hr
Firm energy, E	771.84	308.74	231.55	231.55	10^6 kw hr
Normal irrigation, I	6.70	2.68	4.02	0	10^6 acre ft
Expected net benefits	16.66				10^6 dollars
Planned initial storage		3.00	4.02	1.00	10^6 acre ft
Releases from reservoir					
Low inflow		2.68	4.02	0	10^6 acre ft
Medium inflow		2.68	4.02	0	10^6 acre ft
High inflow		3.68	4.02	0	10^6 acre ft
Irrigation supply					
Low inflow		2.68	4.02	0	10^6 acre ft
Medium inflow		2.68	4.02	0	10^6 acre ft
High inflow		2.68	4.02	0	10^6 acre ft
Irrigation shortage					
All inflows		0	0	0	10^6 acre ft
Hydropower output					
Low inflow		231.55	231.55	0	10^6 kw hr
Medium inflow		231.55	231.55	0	10^6 kw hr
High inflow		250.84	231.55	0	10^6 kw hr
Thermal power supply					
Low inflow		77.19	0	231.55	10^6 kw hr
Medium inflow		77.19	0	231.55	10^6 kw hr
High inflow		57.90	0	231.55	10^6 kw hr

The main features of the solution are shown in Table 13.15. Irrigation turns out to be the pre-emptive use of water. Indeed, no water is released in any season solely for power production, and the installed capacity of the power plant is just adequate to absorb the return flow from irrigation in the peak irrigation season. The penalties for irrigation shortages are so severe that the design seeks to avoid all such shortages and relies entirely on thermal power in the nonirrigation season. Of course, this would not have been the case if we had imposed either a low-flow constraint or an upper limit on thermal power supply.

As contrasted with the simpler example, the normal irrigation commitment is slightly less than the average annual inflow of water. As contrasted with all the nonprobabilistic cases, this plan calls for some water in storage at the outset of every season, even the third season when the reservoir begins to fill.

Table 13.16. Results of 50-yr simulation of the three-season example with uncertainty.

Variable	Planned value	Sample average	Unit
Annual net benefits	16.66	13.94	10^6 dollars
Cost of thermal power	2.73	2.81	10^6 dollars
Cost of irrigation shortage	0	2.64	10^6 dollars
Irrigation shortage			
Season 1	0	0.042	10^6 acre ft
Season 2	0	0.230	10^6 acre ft
Total	0	0.272	10^6 acre ft
Probability of irrigation shortage	0	0.32	
Irrigation supply			
Season 1	2.68	2.64	10^6 acre ft
Season 2	4.02	3.79	10^6 acre ft
Total	6.70	6.43	10^6 acre ft
Initial contents of reservoir			
Season 1	3.00	2.55	10^6 acre ft
Season 2	4.02	3.64	10^6 acre ft
Season 3	1.00	0.72	10^6 acre ft

The operation of this system was simulated for a period of 50 years.[10] The results of this simulation are summarized in Table 13.16. This table makes it clear that an optimistic bias was introduced by our assumption

[10] In the simulation an initial storage of 3×10^6 acre ft was assumed. Inflows for the three seasons for 50 years were drawn from a table of random numbers, in accordance with the probabilities given in Table 13.13. The performance of the system was then computed step by step on the assumptions that water would be released to meet current commitments in each season to the extent that water was available and that all excess water would be stored to the extent of available storage space.

that the initial contents of the reservoir equal its expected value in each season. This bias was to be expected on the grounds of a general theorem of Albert Madansky,[11] which asserts that the apparent maximum value of the objective function in problems of this sort is increased whenever a random uncontrolled variable is replaced by its expected value. The final entries in the table show where the trouble lies: the reservoir failed to refill to the planned level by a substantial margin in every season. Nevertheless, the optimistic bias was not very great. The sample cost of thermal power was only 3 percent more than the planned cost. Although irrigation shortages were frequent, they were mild. The average size of shortage was 13 percent of planned supply; the greatest experienced was 25 percent of planned supply (8 percent of the time).

On balance, this solution seems to pass muster as a first approximation to a sound design, although one suspects that the ultimate design will call for a less ambitious level of irrigation supply.

We can now see that the essence of this method of coping with uncertainty lies in the device of postulating that some variables that are really random always assume their expected values. The variables to be selected for this simplifying assumption are those that connect different segments of the problem, which are thereby decoupled and can be studied in isolation. The simplification attained by this device is enormous. In the three-season example there are actually 27 patterns of possible inflow to the system, but we dealt with only nine contingencies, three for each season. And this remark does not take account of the complications resulting from the influence of each year's operations on the initial state of its successor.

This simplification is bought at a cost. We have already noted the bias inherent in the method. A second drawback lies in the need to approximate the probability distributions of the basic random elements, the inflows in our examples, by discrete distributions that can take on only a finite and fairly small number of values since the number of constraints to be handled is roughly proportional to the number of possible values of the basic variables.

Still a third drawback can be perceived by contemplating how this approach would apply to a configuration like that of Fig. 13.1. In that figure there are two tributaries in addition to the main stem and therefore

[11] Albert Madansky, "Inequalities for stochastic linear programming problems," *Management Sci. 6*, 197 (1960).

three natural inflows in each season. If each of them can assume four possible values, there will be $4^3 = 64$ possible contingencies to be analyzed in each season. This problem could be rendered more manageable by making a "cut" somewhere, say by assuming that the flow into reservoir C always equals its expected value. Then the number of contingencies would be reduced to 16 above the cut plus 4 below, but reservoir C and the power plant would not be designed with any allowance for lower-than-average flow into the reservoir. It appears likely that the bias resulting from this simplification would be severe.[12] In general, when a system includes a number of independent tributaries in the random flows, a heavy computational burden appears inevitable.

CONCLUSION

This chapter has consisted of a sequence of examples of increasing complexity, in which mathematical programming was used to find preliminary designs for water-resource projects. First we introduced a method that can be used when one or two time periods can be considered in isolation from the rest of a project's life. Then we saw how this method can be extended to problems comprehending many more time periods. Finally we illustrated a procedure that can take account of the unpredictability of water flows and other random factors.

These examples and the investigations that led to them suggest that mathematical programming is a method of great flexibility and power, its field of application being limited chiefly by the expense of the computations it entails. But when even moderately large resource developments are at issue, the costs of mathematical-programming computations are trivial in comparison with the other expenditures and the benefits at issue. In actual practice one would gladly undertake computations that would be prohibitive for a pilot study such as this.

These remarks are not meant to assert that mathematical programming is the ultimate weapon for attacking the problems of planning resource developments. On the contrary, a mathematical-programming model will always be a severely schematized simplification of a real problem. The probability distributions of random elements, the linkages between segments of the system and successive time periods, the cost functions and operating characteristics of components must all be sim-

[12] Other, and less extreme, simplifications are also possible in this case, but all appear to entail risks of substantial bias.

plified in such a model if the method is to retain its prime virtue, manageability. Nevertheless, what these experiments have demonstrated is that the crucial characteristics of a fairly complicated system can be retained in a mathematical-programming description of it without rendering the model unduly difficult. The solution of this simplified model can then serve as a very useful starting point for more elaborate and expensive methods of analysis — in particular, simulation.

14 Mathematical Models: A Stochastic Sequential Approach

HAROLD A. THOMAS, JR.,
AND PETER WATERMEYER

THE mathematical model discussed in this chapter was designed to guide planners in their choice of optimal operating policies, optimal levels of development (as measured by target outputs), and optimal reservoir capacities. As before, optimal conditions are defined in economic terms as maximum expected net benefits. In common with other techniques of river-basin design discussed in this book, the model combines the technologic aspects of water-resource development with its economic aspects. In comparison with the other techniques, however, the model is unique because streamflows are treated as stochastic variates. In comparison with stochastic models suggested by other investigators, moreover, our model is unique because it takes into account reservoir inflow in a given period as well as storage at the beginning of that period.

The techniques employed may be said to include in particular the theory of reservoirs (which is analogous to the theory of inventories), linear programming and its extension to parametric programming, and benefit functions that consist of three separable components: benefits for meeting target outputs, losses for failing to meet them, and added benefits for surpassing them.

PREVIOUS RESEARCH

The research of others on which our model builds is fundamentally of two kinds: studies of the probabilistic characteristics of storage systems, and studies of dynamic decision-making.

PROBABILITY THEORY OF STORAGE SYSTEMS

Summarizing research on the probability theory of storage systems, J. Gani has pointed out that the theories of reservoirs and inventories are analogous.[1] Among the questions discussed by him and also by D. G. Kendall and P. A. P. Moran,[2] one was of particular interest to us, namely, how to obtain stationary distributions of reservoir content for various probability distributions of inflow when outputs are defined. In our terminology this involves a stochastic statement of the storage–yield relationship.

DYNAMIC DECISION-MAKING

Richard Bellman's theory of dynamic programming[3] was first applied to water-resource development by J. D. C. Little,[4] who constructed a mathematical model of a simple hydroelectric system. The model comprised a single storage reservoir with hydroelectric plant, a source of supplemental power generation, a known power demand, and an inflow characterized by its probability distribution. At the beginning of each specified time interval a decision was reached on draft that took into account current reservoir contents and preceding inflow. By this decision the expected cost of meeting the power demand for the remainder of the year was to be minimized. Later, to develop an operating policy aimed at minimizing operating costs of thermal power generation during the planning period while meeting a power demand, T. C. Koopmans[5] constructed a model of a power system that combined one storage reservoir and power plant with a thermal generating plant of unlimited capacity operating at increasing incremental cost. Like Little, Koopmans was concerned with dynamic decision-making, but his model differed from Little's in two respects that are significant to us: time was treated as a continuous rather than a discrete variable, and complete certainty about future inflows was assumed.

[1] J. Gani, "Problems in the probability theory of storage systems," *J. Roy. Stat. Soc., B, 19,* 181 (1957). The articles summarized in this paper which are of special interest to us include:

(*a*) K. J. Arrow, Theodore Harris, and Jacob Marschak, "Optimal inventory policy," *Econometrica 19,* 250 (1951).

(*b*) A. Dvoretzky, J. Kiefer, and J. Wolfowitz, "The inventory problem," *Econometrica 20,* 187 and 450 (1952).

(*c*) P. A. P. Moran, "A probability theory of a dam with a continuous release," *Quart. J. Math.* (Oxford) 7, 130 (1956).

[2] D. G. Kendall, "Some problems in the theory of dams," *J. Roy. Stat. Soc., B, 19,* 207 (1957), and P. A. P. Moran, *The theory of storage* (John Wiley, New York, 1960).

[3] Richard Bellman, "Some applications of the theory of dynamic programming," *J. Operations Research Soc. Am. 2,* 275 (1954).

[4] J. D. C. Little, "The use of storage water in a hydroelectric system," *J. Operations Research Soc. Am. 3,* 187 (1955).

[5] T. C. Koopmans, "Water storage in a simplified hydroelectric system," *Proc. 1st Intern. Conf. on Operations Research* (1957).

Building on the work of Little and Koopmans, John Gessford and Samuel Karlin[6] retained essentially the same objective function and assumed stochastic inflows as did Little, but they introduced as one refinement the development of optimal operating policies for both linear and convex cost functions. All of these water-resource models dealt with hydroelectric systems only. This does not imply, however, that they could not be adapted to other purposes of water development.

That the relationship between dynamic and linear programming is close was demonstrated by A. S. Manne[7] in his use of linear programming to make sequential decisions. In an illustration based on inventory control, he demonstrated how a typical sequential probabilistic model could be formulated and subsequently optimized by means of linear programming. Our own model, as we shall see, is an expansion of Manne's work in which we apply his theory to reservoirs. In our model, however, the variable defining the initial state of the system consists of both inflow and storage, rather than storage alone.[8]

THE MODEL: TECHNOLOGIC AND ECONOMIC ANALYSIS

STORAGE-YIELD RELATIONSHIP: A STOCHASTIC STATEMENT

To develop the storage–yield or storage–draft relationships for our mathematical model stochastically, let us assume that a reservoir is capable of holding M volume units of water in active storage and V volume units in dead storage, and that the amount of water in active storage at a given time may be any number of volume units from zero to $M(0, 1, \ldots, M)$. As the volume of water in storage increases, so does the head of water on the turbines in the powerhouse at the dam. A typical head–capacity function is sketched in Fig. 14.1.

Next let the subscript $i(i = 1, 2, \ldots, n)$ denote a given characteristic runoff period of the year, there being in each year n such periods, possibly

[6] John Gessford and Samuel Karlin, "Optimal policy for hydroelectric operations," in K. J. Arrow, Samuel Karlin, Herbert Scarf, *Studies in the mathematical theory of inventory and production* (Stanford University Press, Stanford, 1958), chap. 2. Gessford and Karlin also acknowledge their debt to the earlier work of Pierre Massé, *Les Réserves et la régulation de l'avenir dans la vie économique* (Hermann, Paris, 1946), vols. I and II.

[7] A. S. Manne, "Linear programming and sequential decisions," *Management Sci. 6*, 259 (1960).

[8] Manne has subsequently developed a single-reservoir, three-period model in which he assumes that both current inflow and initial storage are known. See A. S. Manne, "Product-mix alternatives: flood control, electric power, and irrigation," *Cowles Foundation Discussion Paper No. 95*, October 14, 1960.

of different lengths. Furthermore, let $t(t = 1, 2, \ldots, T)$ represent the cumulated runoff periods up to a given time. In deciding upon the number of periods to be used in a year, we must keep in mind the pattern of the demand for water during different periods as well as the available computer capacity. Choice of a large number of periods will place a heavy load on the computer, while too few periods will not permit adequate representation of the different uses of water at the various times during a year. For our purposes, therefore, the periods are selected to correspond with the several demands of our river-basin system.

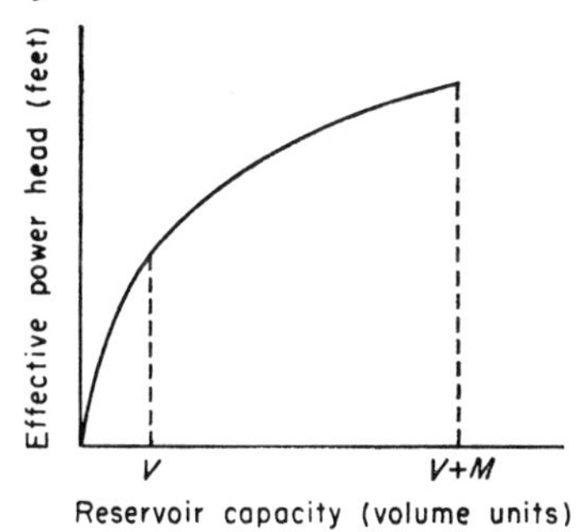

Fig. 14.1. Reservoir head–capacity function. V denotes dead storage, M active storage, and V + M total storage.

We assume also that the total inflow in a given period is a stochastic variate. The smaller the inflow volume unit that is chosen, the more precise the representation of actual flows. However, the volume unit should not be made so small and the resulting probability distribution so complex that the resulting computations become too cumbersome and time-consuming. The probability relationships for storage, inflow, and draft are developed below.

Let ${}_iX_t$ be the reservoir inflow during period i of a year and period t of the total time series, and let ${}_iP_j$ $(j = 0, 1, \ldots, a)$ equal the probability of inflow ${}_iX_t$ being j volume units where the sum of the probabilities $\sum_j {}_iP_j$ is unity. It will simplify our calculations if we assume that the inflows accumulate at a uniform rate from the start of the period, and that the concurrent rates of draft or release of water from the reservoir, ${}_iY_t$, also are uniform throughout a given period. Furthermore, let the active storage at the start of the period identified by i and t equal ${}_iR_t$ volume units, and ${}_iK_r$ $(r = 0, 1, \ldots, M)$ be the probability of ${}_iR_t$ equaling r, where $\sum_r {}_iK_r$ is unity. In combination, the probability of the reservoir inflow ${}_iX_t$ being j and the starting volume of water stored ${}_iR_t$ being

r is the product $({}_iP_j)({}_iK_r)$,[9] which we shall denote by ${}_ig_{j,r}$ where $\sum_{j,r} {}_ig_{j,r}$ is unity. Denoting the sum of the inflow j and the initial storage r by k, or $j + r = k$, we may write the probability of this occurrence as $\sum_j {}_ig_{j,k-j}$; $k = 0, 1, \ldots, (a + M)$. Since continuity requires that the probability of the reservoir being in state r at the end of the ith period must equal the probability of its being in the same state at the beginning of the $(i + 1)$st period, $\text{Prob}\{{}_iS_t = r\} = \text{Prob}\{{}_{i+1}R_{t+1} = r\} = {}_{i+1}K_r$, where ${}_iS_t$ is the volume units in active storage at the end of the ith period of a year and the tth period of the total time series.

Finally, let ${}_ih_{c,s}$ denote the probability of active reservoir storage ${}_iS_t$ being s after a draft ${}_iY_t$ of c during the period,[10] where $\sum_{c,s} {}_ih_{c,s} = 1$. It should be noted that the sum of the draft c and the final storage s in a given period must not exceed the sum of the maximum inflows a and the maximum active storage M in that period, or $(c + s) \leq (a + M)$. To be in balance, the amount of water available for use at the beginning of a given period (inflow plus storage) must equal the draft during the period plus storage at the end of the period, or $j + r = c + s = k$. The probabilities of this occurring must likewise be identical, or $\sum_j {}_ig_{j,k-j} = \sum_c {}_ih_{c,k-c}$.

Having defined the probability relationships for storage, inflow, and draft, we can now proceed to the development of an optimal operating policy. In general, an operating policy answers the question: what draft ${}_iY_t$ shall we allow when we know the sum of the inflow ${}_iX_t$ and the initial storage ${}_iR_t$ during the period? If we relate the values of ${}_iX_t + {}_iR_t$ to those of ${}_ih_{c,s}$ (that is, ${}_ih_{c,s} = {}_ih_{c,k-c}$, where $k = {}_iX_t + {}_iR_t$), then any set of ${}_ih_{c,s}$ values that sums to unity defines an operating procedure.

In our terminology we write an operating procedure in the following form:

$$\text{Prob}\{{}_iY_t = c \mid ({}_iX_t + {}_iR_t) = k\} = \frac{\text{Prob}\{{}_iY_t = c, ({}_iX_t + {}_iR_t = k)\}}{\text{Prob}\{({}_iX_t + {}_iR_t) = k\}}$$

$$= \frac{\text{Prob}\{{}_iY_t = c, {}_iS_t = (k - c)\}}{\text{Prob}\{({}_iX_t + {}_iR_t) = k\}} = \frac{{}_ih_{c,k-c}}{\sum_c {}_ih_{c,k-c}}.$$

[9] This is permissible since we assume no serial correlation between the inflows in succeeding periods.

[10] Draft is defined here as total outflow and is the sum of the useful flow and spill.

The result may represent a "mixed strategy" when, for a single value of ${}_iX_t + {}_iR_t = k$, there is more than one permissible value of the draft ${}_iY_t$, or a "pure strategy" when the draft has only a single value for a given sum of the inflow and storage ${}_iX_t + {}_iR_t = k$. A typical point ${}_ih_{c,s} = {}_ih_{c,k-c}$ relating inflow plus storage to draft is shown in Fig. 14.2.

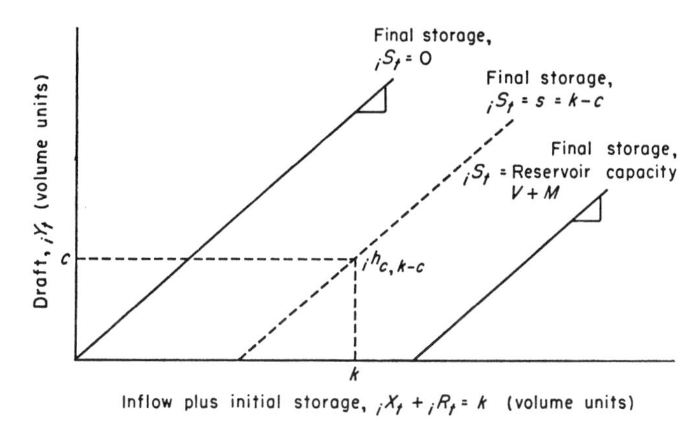

Fig. 14.2. Stochastic relationship between inflow plus initial active storage and draft. The term ${}_ih_{c,k-c}$ denotes the probability of the reservoir having a storage of $s = k - c$ volume units after a draft of c volume units in period i of a given year.

To construct a diagram of an operating policy, we would have to plot the remaining ${}_ih_{c,s}$ (or ${}_ih_{c,k-c}$) values associated with that policy as exemplified later in Fig. 14.5.

We are now in a position to derive an optimal operating policy by linear programming. At the outset we shall assume that the reservoir capacity, $V + M$, and level of development of the purposes to be served are given; later we shall optimize them. Before proceeding, however, we must develop constraint equations from the statement of the storage–draft relationship and define the objective function that is to be maximized. For the storage–draft relationship the two sets of linear equations suitable for solution in a linear program are the n equations (one for each value of i):

$$\sum_{c,s} {}_ih_{c,s} = 1 \tag{14.1}$$

and the $(a + M + 1)n$ equations:

$$\sum_j {}_ig_{j,k-j} = \sum_c {}_ih_{c,k-c} \tag{14.2}$$

(of which n, or one for each value of i, are redundant in Eq. (14.2) only).

However, Eq. (14.2) must be in terms of ${}_ih_{c,s}$ only. The ${}_ig_{j,r}$ values are transformed into ${}_ih_{c,s}$ values by making use of the continuity requirement that the probability of a reservoir being in state r at the end of the period $i - 1$ must equal the probability of its being in state r also at the beginning of the period i. Algebraically, therefore,

$$\sum_c {}_{i-1}h_{c,r} = {}_iK_r \tag{14.3}$$

and

$$ {}_ig_{j,r} = ({}_iP_j)({}_iK_r) = ({}_iP_j)(\sum_c {}_{i-1}h_{c,r}).$$

Equation (14.2) then takes the form

$$\sum_j {}_iP_j(\sum_c {}_{i-1}h_{c,r}) = \sum_c {}_ih_{c,k-c}, \tag{14.4}$$

there being $(a + M)n$ nonredundant equations.

OPTIMAL OPERATING POLICIES

Since, for the present, reservoir capacity and level of development are being held constant, we can consider capital costs and costs of operation, maintenance, and replacement (OMR) as fixed. Accordingly, maximizing expected gross benefits will give the optimal solution without computing net benefits.

If we designate the target output for a given purpose simply by ${}_iY_l$, benefits (U) are a single-valued function of draft or actual output ${}_iY_t$ and of target output ${}_iY_l$.[11] The basis for evaluation of benefits is ${}_iY_l$. However, when ${}_iY_t$ is less than ${}_iY_l$, benefits may be lost because water is not available for the purposes it is to serve; and when ${}_iY_t$ is more than ${}_iY_l$, additional benefits may accrue. These relationships are denoted by the L term in the equation

$$U({}_iY_t, {}_iY_l, V) = U'({}_iY_l, V) - L[{}_iY_l, ({}_iY_l - {}_iY_t)], \tag{14.5}$$

where $L({}_iY_l, 0) = 0$ and $L[{}_iY_l, ({}_iY_l - {}_iY_t)]$ is a nondecreasing function of $({}_iY_l - {}_iY_t)$. In general, the value of the last term of Eq. (14.5) is positive when ${}_iY_t$ is less than ${}_iY_l$ and negative when ${}_iY_t$ is more than ${}_iY_l$.

[11] If we are to consider variable-head power, we must define our benefit function somewhat differently. It then is necessary to specify for each period a target power output ${}_iE_l$ which depends in turn on a target draft ${}_iY_l$ and a target active storage ${}_iS_l$. In this case our loss/gain functions will be dependent not only on the difference between ${}_iY_t$ and ${}_iY_l$ but also on the difference between ${}_iS_t$ and ${}_iS_l$.

If we drop the dead-storage term, Eq. (14.5) becomes

$$U({}_iY_t, {}_iY_l) = U'({}_iY_l) - L[{}_iY_l, ({}_iY_l - {}_iY_t)].$$

Our problem is further simplified if we assume that both $U'({}_iY_l)$ and $L[{}_iY_l, ({}_iY_l - {}_iY_t)]$ are straight-line functions, as in Fig. 14.3. There, the

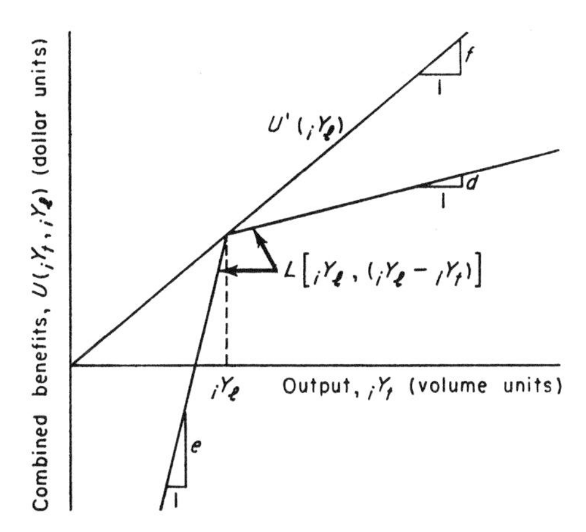

Fig. 14.3. Function relating combined benefits U to target and actual outputs, ${}_iY_l$ *and* ${}_iY_t$.

line with slope f represents the term $U'({}_iY_l)$, and the lines with slopes d and e the two components of $L[{}_iY_l, ({}_iY_l - {}_iY_t)]$, respectively. Accordingly, our equation becomes

$$U({}_iY_t, {}_iY_l) = f({}_iY_l) + d({}_iY_t - {}_iY_l)^+ - e({}_iY_l - {}_iY_t)^+,$$

where the superscript "+" for bracketed terms indicates that they are operative only when they are greater than, or equal to, zero.

The model itself, because it permits only one decision on the amount of draft ${}_iY_t$ in a given period, will handle only a single benefit function. However, this function need not result from a single purpose, as we shall see later when we discuss compound benefit functions arising from multipurpose developments.

We can now restate the objective function in a form appropriate for linear programming as

$$U({}_iY_t, {}_iY_l) = U(c, {}_iY_l) = {}_iu_c,$$

where $l = 1, 2, \ldots, m$ purposes. Since we wish to maximize the expected value of benefits, $E(U)$, the objective function becomes

$$E(U) = \sum_{i,c,s} ({}_iu_c)({}_ih_{c,s}) \qquad (14.6)$$

in terms of the notation of constraint equations (14.1) and (14.4). These can now be solved by selecting that set of ${}_ih_{c,s}$ values which maximizes the objective $[E(U)_{\max}]$. This solution establishes the optimal operating policy and fulfills the first purpose of our mathematical model. Of interest in this connection is H. M. Wagner's demonstration that optimal policies resulting from linear-programming solutions will always be pure rather than mixed strategies.[12]

OPTIMAL LEVELS OF DEVELOPMENT

To fulfill the second purpose of our mathematical model — that of identifying the optimal level of development — we recognize that we must find the set of target outputs ${}_iY_l$, $l = 1, 2, \ldots, m$ with the corresponding optimal operating policy that maximizes the objective function for any given reservoir capacity $V + M$. If the number of sets of target outputs to be considered is large, this becomes a tedious task of computation, but it can be eased through parametric programming. By this we mean an extension of linear programming that allows the coefficients of the objective function to vary. In our case the objective function is the benefit function, and the coefficients will change with target outputs. For parametric programming, the benefit function itself must be linear, whereas in our case it is composed of different linear sections. Hence we must modify the standard linear program by dealing with the linear portions in succession. Varying the target outputs implies that capital costs and OMR costs must be expected to change and that the objective function must be revised accordingly.

OPTIMAL RESERVOIR CAPACITIES

The third purpose of our mathematical model, it will be remembered, is to find the reservoir capacity $V + M$, and the associated optimal set of target outputs and operating policy, that maximize our objective. Since no technique presently known to us will perform this optimization internally, we have resorted to sampling in the range of relevant reservoir sizes to find the optimal value. As before, in these circumstances the capital costs and OMR costs of the reservoirs will vary, and the objective function will have to be further revised.

[12] H. M. Wagner, "On the optimality of pure strategies," *Management Sci. 6*, 268 (1960).

EXAMPLE

To illustrate the operation of our mathematical model, let us consider the development of a river-basin system that is characterized hydrologically by two periods — a wet and a dry season. During the dry season, the system is to provide power and irrigation; during the wet season, power and flood control.

The inflows during the dry season are of three magnitudes, ${}_1X_t = 0, 1, 2$ units of water, with probabilities of occurrence ${}_1P_j = 0.1, 0.7, 0.2$, respectively, and have a mean value of 1.1 volume units. During the wet season four magnitudes, ${}_2X_t = 0, 1, 2, 3$ units of water, occur with probabilities ${}_2P_j = 0.1, 0.3, 0.4, 0.2$, respectively, and have a mean value of 1.7 volume units. The active reservoir capacity M is assumed to be 2 volume units.

To organize these data for evaluation of the constraint equations (14.1) and (14.4), we need to find the probabilities ${}_ig_{j,r}$ and ${}_ih_{c,s}$. The technique is shown schematically in Table 14.1.

In accordance with Eq. (14.3), the K values are obtained from the following summations based on probabilities ${}_ih_{c,s}$ listed in Table 14.1:

Table 14.1. Determination of ${}_ig_{j,r}$ (probability of reservoir inflow ${}_iX_t$ being j and initial reservoir storage ${}_iR_t$ being r) and ${}_ih_{c,s}$ (probability of final active reservoir storage ${}_iS_t$ being s after a draft ${}_iY_t$ of c) for the two-season, single-reservoir example. For dry season, $i = 1$; for wet season, $i = 2$.

Dry season

(a) Values of ${}_1g_{j,r}$

${}_1R_t$	${}_1X_t$ = 0	1	2
0	$0.1({}_1K_0)$	$0.7({}_1K_0)$	$0.2({}_1K_0)$
1	$0.1({}_1K_1)$	$0.7({}_1K_1)$	$0.2({}_1K_1)$
2	$0.1({}_1K_2)$	$0.7({}_1K_2)$	$0.2({}_1K_2)$

(c) Values of ${}_1h_{c,s}$

${}_1S_t$	${}_1Y_t$ = 0	1	2	3	4
0	${}_1h_{00}$	${}_1h_{10}$	${}_1h_{20}$	${}_1h_{30}$	${}_1h_{40}$
1	${}_1h_{01}$	${}_1h_{11}$	${}_1h_{21}$	${}_1h_{31}$	
2	${}_1h_{02}$	${}_1h_{12}$	${}_1h_{22}$		

Wet season

(b) Values of ${}_2g_{j,r}$

${}_2R_t$	${}_2X_t$ = 0	1	2	3
0	$0.1({}_2K_0)$	$0.3({}_2K_0)$	$0.4({}_2K_0)$	$0.2({}_2K_0)$
1	$0.1({}_2K_1)$	$0.3({}_2K_1)$	$0.4({}_2K_1)$	$0.2({}_2K_1)$
2	$0.1({}_2K_2)$	$0.3({}_2K_2)$	$0.4({}_2K_2)$	$0.2({}_2K_2)$

(d) Values of ${}_2h_{c,s}$

${}_2S_t$	${}_2Y_t$ = 0	1	2	3	4	5
0	${}_2h_{00}$	${}_2h_{10}$	${}_2h_{20}$	${}_2h_{30}$	${}_2h_{40}$	${}_2h_{50}$
1	${}_2h_{01}$	${}_2h_{11}$	${}_2h_{21}$	${}_2h_{31}$	${}_2h_{41}$	
2	${}_2h_{02}$	${}_2h_{12}$	${}_2h_{22}$	${}_2h_{32}$		

$$
\begin{aligned}
{}_1K_0 &= {}_2h_{00} + {}_2h_{10} + {}_2h_{20} + {}_2h_{30} + {}_2h_{40} + {}_2h_{50},\\
{}_1K_1 &= {}_2h_{01} + {}_2h_{11} + {}_2h_{21} + {}_2h_{31} + {}_2h_{41},\\
{}_1K_2 &= {}_2h_{02} + {}_2h_{12} + {}_2h_{22} + {}_2h_{32},\\
{}_2K_0 &= {}_1h_{00} + {}_1h_{10} + {}_1h_{20} + {}_1h_{30} + {}_1h_{40},\\
{}_2K_1 &= {}_1h_{01} + {}_1h_{11} + {}_1h_{21} + {}_1h_{31},
\end{aligned}
$$

and

$$
{}_2K_2 = {}_1h_{02} + {}_1h_{12} + {}_1h_{22}.
$$

For the dry season, when the inflow plus the initial storage is zero (that is, $k = 0$), constraint equation (14.4) becomes:

$$
{}_1h_{00} = 0.1({}_1K_0) = 0.1[{}_2h_{00} + {}_2h_{10} + {}_2h_{20} + {}_2h_{30} + {}_2h_{40} + {}_2h_{50}],
$$

or

$$
{}_1h_{00} - 0.1[{}_2h_{00} + {}_2h_{10} + {}_2h_{20} + {}_2h_{30} + {}_2h_{40} + {}_2h_{50}] = 0.
$$

The coefficients of this and the remaining constraint equations are presented in Table 14.2.

To evaluate the objective function, let us assign the following unit magnitudes to benefits and losses:

(1) *Target benefit function, f:* $f_{\text{irrigation}} = 1$ dollar unit per volume unit; and $f_{\text{power}} = 2$ dollar units per volume unit, assuming a fixed power head.

(2) *Loss function, e:* $e_{\text{irrigation}} = 2$ dollar units per volume unit; and $e_{\text{power}} = 3$ dollar units per volume unit, assuming a fixed power head.

(3) *Additional benefits for exceeding target-output function, d:* $d_{\text{irrigation}} = 0.2$ dollar unit per volume unit; and $d_{\text{power}} = 0.3$ dollar unit per volume unit, assuming a fixed power head.

(4) *Damages resulting from floods:* 4 dollar units if the average draft is 3 volume units; 5 dollar units if the average draft is 4 volume units; and 7 dollar units if the average draft is 5 volume units.

To create a single function, we shall assume that these purposes are fully complementary and that the benefit and loss functions are, therefore, simply additive. In this way the example conforms to the limitation noted in our discussion of the objective function.

Let us now examine the nine combinations of target outputs for irrigation, power, and flood control shown in Table 14.3. Here we combine the benefit data with the target outputs to construct the sets of coefficients of the objective function. To illustrate, let us perform the necessary calculations for combination 4 with the visual assistance of Fig. 14.4, which shows graphically for this combination the benefits for irrigation, power, and flood control as functions of target and actual outputs. Table 14.4 can then be constructed on the assumption of simple additivity of

Table 14.2. Linear-programming matrix for the two-season, single-reservoir example. The values given constitute the coefficients of the constraint equations. The term $_ih_{c,s}$ denotes the probability that final active reservoir storage for season i will be s volume units after a draft of c volume units.

See footnotes	Values of $_ih_{c,s}$																											constant
	$_1h_{00}$	$_1h_{10}$	$_1h_{20}$	$_1h_{30}$	$_1h_{40}$	$_1h_{01}$	$_1h_{11}$	$_1h_{21}$	$_1h_{31}$	$_1h_{02}$	$_1h_{12}$	$_1h_{22}$	$_2h_{00}$	$_2h_{10}$	$_2h_{20}$	$_2h_{30}$	$_2h_{40}$	$_2h_{50}$	$_2h_{01}$	$_2h_{11}$	$_2h_{21}$	$_2h_{31}$	$_2h_{41}$	$_2h_{02}$	$_2h_{12}$	$_2h_{22}$	$_2h_{32}$	
a																												
$k = 0$	1												−0.1	−0.1	−0.1	−0.1	−0.1	−0.1										0
$k = 1$		1				1							−0.7	−0.7	−0.7	−0.7	−0.7	−0.7	−0.1	−0.1	−0.1	−0.1	−0.1					0
$k = 3$				1				1			1								−0.2	−0.2	−0.2	−0.2	−0.2	−0.7	−0.7	−0.7	−0.7	0
$k = 4$					1				1			1												−0.2	−0.2	−0.2	−0.2	0
b	1	1	1	1	1	1	1	1	1	1	1	1																1
c																												
$k = 0$	−0.1	−0.1	−0.1	−0.1	−0.1								1															0
$k = 1$	−0.3	−0.3	−0.3	−0.3	−0.3	−0.1	−0.1	−0.1	−0.1					1					1									0
$k = 2$	−0.4	−0.4	−0.4	−0.4	−0.4	−0.3	−0.3	−0.3	−0.3	−0.1	−0.1	−0.1			1					1				1				0
$k = 4$						−0.2	−0.2	−0.2	−0.2	−0.4	−0.4	−0.4					1					1				1		0
$k = 5$										−0.2	−0.2	−0.2						1					1				1	0
d													1	1	1	1	1	1	1	1	1	1	1	1	1	1	1	1

[a] Eq. (14.4) with $i = 1$ and $j + r = c + s = k$.
[b] Eq. (14.1) with $i = 1$.
[c] Eq. (14.4) with $i = 2$ and $j + r = c + s = k$.
[d] Eq. (14.1) with $i = 2$.

Table 14.3. Combinations of target outputs to be examined and the benefit coefficients of their objective functions for the two-season, single-reservoir example.

	Target outputs (volume units) for —				Draft (volume units)										
		Power (fixed-head power plant)			Dry season $_1Y_t$					Wet season $_2Y_t$					
Combination no.	Irrigation (dry season) $_1Y_1$	Dry season $_1Y_2$	Wet season $_2Y_2$	Flood control (wet season)	0	1	2	3	4	0	1	2	3	4	5
					Benefits[a] (dollar units)										
1	1.0	1.0	1.0	*b*	−2.00	3.00	3.50	4.00	4.50	−1.00	2.00	2.30	−1.40	−2.10	−3.80
2	1.0	1.0	2.0		−2.00	3.00	3.50	4.00	4.50	−2.00	1.00	4.00	0.30	−0.40	−2.10
3	1.0	2.0	2.0		−3.00	2.00	5.20	5.70	6.20	−2.00	1.00	4.00	0.30	−0.40	−2.10
4	2.0	1.0	2.0		−3.00	2.00	4.30	4.80	5.30	−2.00	1.00	4.00	0.30	−0.40	−2.10
5	1.1	1.1	1.7		−2.20	2.80	3.75	4.25	4.75	−1.70	1.30	3.49	−0.21	−0.91	−2.60
6	2.0	1.0	1.0		−3.00	2.00	4.30	4.80	5.30	−1.00	2.00	2.30	−1.40	−2.10	−3.80
7	1.0	2.0	1.0		−3.00	2.00	5.20	5.70	6.20	−1.00	2.00	2.30	−1.40	−2.10	−3.80
8	2.0	2.0	1.0		−4.00	1.00	6.00	6.50	7.00	−1.00	2.00	2.30	−1.40	−2.10	−3.80
9	2.0	2.0	2.0		−4.00	1.00	6.00	6.50	7.00	−2.00	1.00	4.00	0.30	−0.40	−2.10

[a] Flood damages are included as negative benefits.
[b] For all combinations with actual drafts $_2Y_t$ of 3, 4, and 5 volume units, the flood damages are 4, 5, and 7 dollar units, respectively.

benefit functions. With these values substituted in the objective function, the solution of the constraint-equation matrix by linear programming gives the following probabilities: ${}_1h_{00} + {}_1h_{10} + {}_1h_{11} + {}_1h_{21} + {}_1h_{22} = 0.060 +$

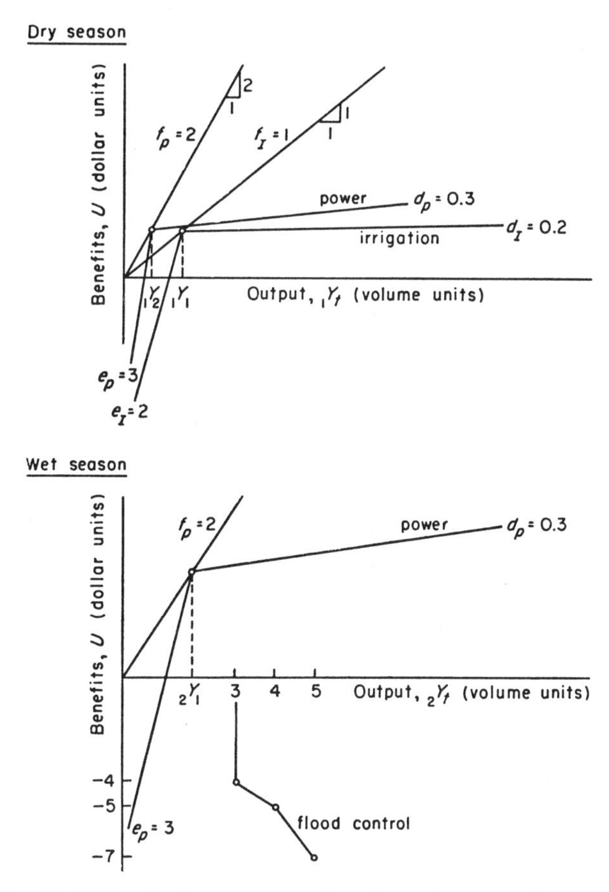

Fig. 14.4. Benefits U for single purposes as functions of target and actual outputs ${}_iY_l$ and ${}_iY_t$ for the two-season, single-reservoir example. The symbols d_I and d_P are the slopes of the L terms of the benefit function for irrigation and power purposes respectively when actual outputs exceed target outputs; e_I and e_P are the corresponding slopes when actual outputs are less than target outputs; and f_I and f_P are the slopes of the functions expressing benefits vs target outputs.

$0.447 + 0.338 + 0.134 + 0.021 = 1.000$ for the dry season and ${}_2h_{00} + {}_2h_{10} + {}_2h_{20} + {}_2h_{21} + {}_2h_{22} + {}_2h_{32} = 0.051 + 0.199 + 0.346 + 0.297 + 0.103 + 0.004 = 1.000$ for the wet season. These probabilities define the

Table 14.4. Combined benefits *U* for different drafts for combination 4 of Table 14.3.

(a) Dry season

	Draft ${}_1Y_t$				
	0	1	2	3	4
$U_{\text{irrig.}}$ =	−2.0	0.0	2.0	2.2	2.4
U_{power} =	−1.0	2.0	2.3	2.6	2.9
U_{total} =	−3.0	2.0	4.3	4.8	5.3

(b) Wet season

	Draft ${}_2Y_t$					
	0	1	2	3	4	5
U_{power} =	−2.0	1.0	4.0	4.3	4.6	4.9
U flood control =	0.0	0.0	0.0	−4.0	−5.0	−7.0
U_{total} =	−2.0	1.0	4.0	0.3	−0.4	−2.1

operating policies represented graphically in Fig. 14.5, the maximum net benefits $E(U)$, derived from Eq. (14.6), being 5.143 dollar units.

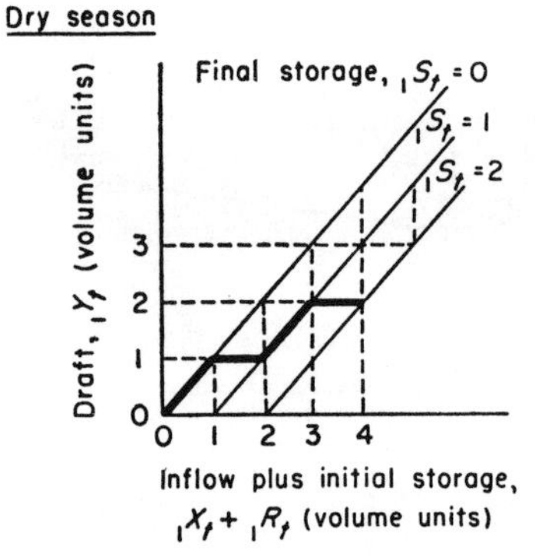

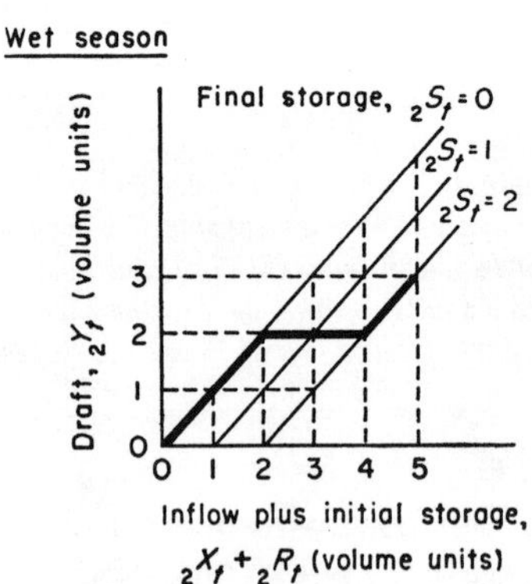

Fig. 14.5. Operating policies for the two-season, single-reservoir example.

Table 14.5. Maximum expected values of net benefits for combinations of target outputs in Table 14.3.

Combination no.	Expected value of net benefits, $E(U)_{\max}$ (dollar units)
1	5.264
2	5.863
3	5.542
4	5.143
5	5.425
6	5.435
7	6.131
8	6.751
9	6.109

When we examine the remaining combinations in the same way, the maximum benefits for each combination, presented in Table 14.5, are obtained. Fig. 14.6 then displays the corresponding optimal operating policies.

In this example, with a reservoir of fixed size, we have derived optimal operating policies for nine combinations of target outputs. However, we have found neither the optimal level of development, with its associated optimal operating policy, for this size of reservoir, nor the optimal size of the reservoir itself. Neither parametric programming for investigating the influence on net benefits of varying the levels of development nor sampling techniques for studying the effects of reservoir size upon net benefits were employed.

A CHECK ON THE CONSISTENCY OF THE LINEAR-PROGRAMMING SOLUTION: QUEUING THEORY

We propose to verify the consistency of our linear-programming solution by introducing a different technique based on the theory of queues. Given data will be the reservoir capacity and level of development of combination 4 in Table 14.3 and the optimal operating policy derived by the linear-programming solution. By queuing theory we shall then derive the associated probability distributions of draft and show that they are equivalent to the probabilities ${}_ih_{c,s}$ indicated in the previous section. Although we could have determined the optimal operating policy

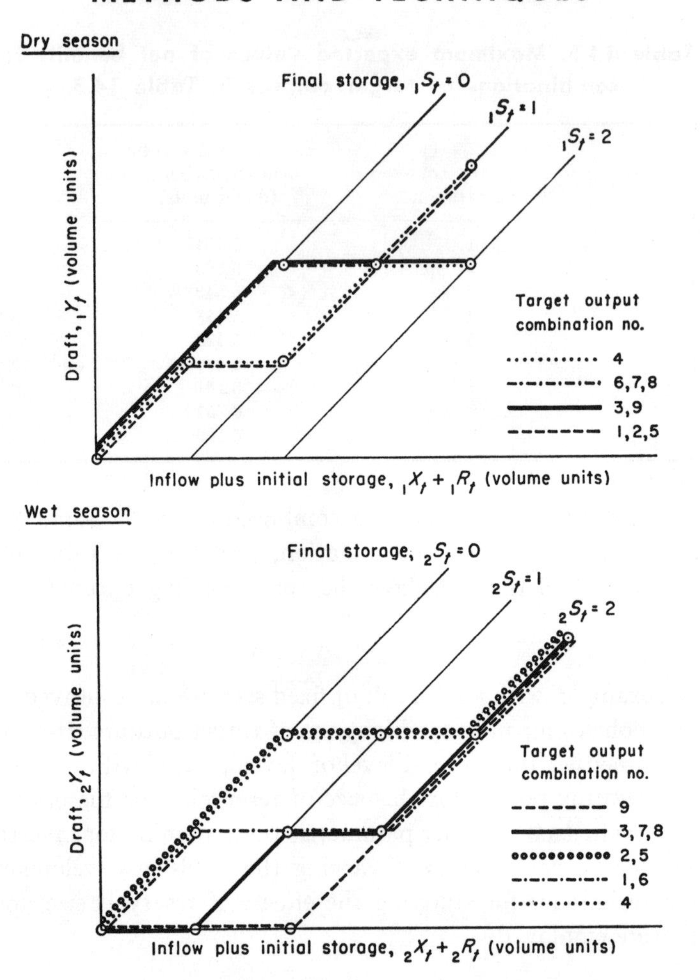

Fig. 14.6. Operating policies for nine combinations of target outputs for the two-season, single-reservoir example.

by queuing theory from the outset, the necessary calculations would have been cumbersome because, unlike linear programming, queuing theory does not include an internal optimizing procedure. Unless otherwise stated, the assumptions made for the linear-programming model hold also for the queuing model.

If we know (1) the active reservoir capacity in volume units M and (2) for each period i the probability distribution of the inflow ${_iX_t}$ and the operating policy ${_iY_t}$, given ${_iR_t} + {_iX_t}$, we can construct the pertinent transition-probability matrix by using as entries the probability ${_im_{rs}}$ that the reservoir starting period i in state ${_iR_t} = r$ will end it in state ${_iS_t} = s$.

Table 14.6. Transition-probability matrix for period i, showing values of ${}_jm_{rs}$, the probability that the reservoir starting period i in state ${}_iR_t$ will end it in state ${}_iS_t$.

Initial storage, period i, state ${}_iR_t$	Final storage, period i, state ${}_iS_t$				
	0	. . .	s	. . .	M
0	${}_im_{00}$	. . .	${}_im_{0s}$	. . .	${}_im_{0M}$
.	.		.		.
.	.		.		.
.	.		.		.
r	${}_im_{r0}$	. . .	${}_im_{rs}$	. . .	${}_im_{rM}$
.	.		.		.
.	.		.		.
.	.		.		.
M	${}_im_{M0}$	. . .	${}_im_{Ms}$	. . .	${}_im_{MM}$

A transition-probability matrix of this kind is shown in Table 14.6. The probabilities ${}_im_{rs}$ are found from the probability distribution of the inflow with the aid of the equations:

$${}_iS_t = {}_iR_t + {}_iX_t - {}_iY_t$$

and

$${}_iY_t = f({}_iR_t + {}_iX_t).$$

The ${}_im_{rs}$ values are then used to obtain the probabilities ${}_ip_{rs}$, where ${}_ip_{rs}$ denotes the probability of starting period i in state ${}_iR_t = r$ and ending period $i - 1$, of the following year in state ${}_{i-1}S_{t+n-1} = s$. If there are n periods in a year,

$${}_ip_{rs} = [({}_im_{r0})({}_{i+1}m_{00}) \ldots ({}_{n-1}m_{00})({}_nm_{00})][({}_1m_{00})({}_2m_{00}) \ldots ({}_{i-2}m_{00})({}_{i-1}m_{0s})] \quad \text{path 1}$$

$$+ \ldots$$

$$+ [({}_im_{rM})({}_{i+1}m_{MM}) \ldots ({}_{n-1}m_{MM})({}_nm_{MM})][({}_1m_{MM})({}_2m_{MM}) \ldots ({}_{i-2}m_{MM})({}_{i-1}m_{Ms})] \quad \text{path } (m+1)^{n-1}$$

$$= \sum_{s_2, \ldots, s_n} ({}_im_{rs_2})({}_{i+1}m_{s_2s_3})({}_{i+2}m_{s_3s_4}) \ldots ({}_{i-1}m_{s_ns}).$$

The example shown in Table 14.7 for the ith period is representative of nth-order transition matrices. In these matrices ${}_iK_r$ denotes the stationary probability that the reservoir will be in state ${}_iR_t = r$ at the beginning of period i. It has the same meaning as ${}_iK_r$ in the linear-programming model.

Table 14.7. Transition-probability matrix showing values of $_ip_{rs}$, the probability that the reservoir starting period i, in state $_iR_t$ will end in period $i - 1$ of the following year in state $_{i-1}S_{t+n-1}$.

Initial storage, period i, state $_iR_t$	Final storage, period $i - 1$ of following year, state $_{i-1}S_{t+n-1}$					Stationary probability that reservoir will be in state $_iR_t$ at beginning of period i
	0	$\cdots$	s	$\cdots$	M	
0	$_ip_{00}$	$\cdots$	$_ip_{0s}$	$\cdots$	$_ip_{0M}$	$_iK_0$
$\vdots$	$\vdots$		$\vdots$		$\vdots$	$\vdots$
r	$_ip_{r0}$	$\cdots$	$_ip_{rs}$	$\cdots$	$_ip_{rM}$	$_iK_r$
$\vdots$	$\vdots$		$\vdots$		$\vdots$	$\vdots$
M	$_ip_{M0}$	$\cdots$	$_ip_{Ms}$	$\cdots$	$_ip_{MM}$	$_iK_M$
						$\sum_r {_iK_r} = 1$

Applying the probability distributions of inflows given in the example of the preceding section and the parameters for the operating policies of the linear-programming model shown in Fig. 14.5, we can form the first-order transition-probability matrices shown in Table 14.8, the entries being the $_im_{rs}$ values. The second-order matrices with $_ip_{rs}$ values are constructed and inverted to give the $_iK_r$ values.

We now calculate the probability that the sum of the inflow j and the initial storage r is k, or Prob $(_iX_t + {_iR_t} = k) = \sum_j (_ip_j)(_iK_{k-j})$. For the dry season the probabilities of $_1X_t + {_1R_t}$ being 0, 1, 2, 3, 4 are respectively 0.0596, 0.4469, 0.3378, 0.1343, 0.0214 or a total of 1.0. For the wet season the probabilities of $_2X_t + {_2R_t}$ being 0, 1, 2, 3, 4, 5 are respectively 0.0507, 0.1993, 0.3465, 0.2965, 0.1028, 0.0042 or a total of 1.0. Finally, using the relationships between the draft $_iY_t$ and the sum of the inflow and initial storage $_iX_t + {_iR_t}$ given by the operating policies shown in Fig. 14.5, together with the probabilities of this sum, we calculate the probabilities of drafts $_1Y_t$ equaling 0, 1, 2 for the dry season respectively as 0.0596, 0.7847, 0.1557 or a total of 1.0. For the wet season the probabilities of drafts $_2Y_t$ equaling 0, 1, 2, 3, respectively are found to be 0.0507, 0.1993, 0.7458, 0.0042 or a total of 1.0. The corresponding linear-program probabilities are respectively 0.060, 0.785, 0.155 for $_1Y_t$ and 0.051, 0.199, 0.746, 0.004 for $_2Y_t$.

Table 14.8. Transition-probability matrices for periods 1 and 2.

(a) First-order matrices with entries ${}_im_{rs}$, the probability that the reservoir starting period i in state ${}_iR_t$ will end it in state ${}_iS_t$. For dry season, $i = 1$; for wet season, $i = 2$.

Dry season				*Wet season*			
	${}_1S_t$				${}_2S_t$		
${}_1R_t$	0	1	2	${}_2R_t$	0	1	2
0	0.8	0.2	0.0	0	0.8	0.2	0.0
1	0.1	0.9	0.0	1	0.4	0.4	0.2
2	0.0	0.8	0.2	2	0.1	0.3	0.6

(b) Second-order matrices with entries ${}_ip_{rs}$, the probability that the reservoir starting period i in state ${}_iR_t$ will end in period $i - 1$ of the following year in state ${}_{i-1}S_{t+n-1}$. For dry season, $i = 1$; for wet season, $i = 2$.

Dry season					*Wet season*				
	${}_2S_t$			Solution:		${}_1S_t$			Solution:
${}_1R_t$	0	1	2	${}_iK_r$	${}_2R_t$	0	1	2	${}_iK_r$
0	0.72	0.24	0.04	${}_1K_0 = 0.596$	0	0.66	0.34	0.00	${}_2K_0 = 0.507$
1	0.44	0.38	0.18	${}_1K_1 = 0.297$	1	0.36	0.60	0.04	${}_2K_1 = 0.472$
2	0.34	0.38	0.28	${}_1K_2 = 0.107$	2	0.11	0.77	0.12	${}_2K_2 = 0.021$
				1.000					1.000

APPLICABILITY OF THE MODEL

Advantages of our model are that it optimizes the operating policy, the levels of development, and, by sampling, the reservoir capacity. It has the further advantage that the input is stochastic. On the other hand, the following limitations should be recognized. The model is normally limited to operation of a single reservoir. Multiple purposes can be explored only if they can be combined into a single benefit function. The number of flow periods into which the year and record can be divided will be determined by the capacity of the computer. There must be no serial correlation between flow periods.

Operation of more than one reservoir can be studied when the reservoirs are in series and all inflows enter the uppermost reservoir while all drafts are taken from the lowermost reservoir, or when the reservoirs are in parallel and supply water for the same purpose, provided that cross-correlation between inflows is either 0 or 1. The analysis of reservoirs in series is limited further by problems of serial correlation between reservoir inflows that are analogous to serial correlation of inflows between periods.

The model will handle consumptive uses, such as irrigation and water supply; channel operational uses, such as navigation and waste-water disposal; fixed- and variable-head power; and flood control.

When the power head is variable, we must assume that energy output is a function of the active storage ${}_iS_t$ at the end of the period, as well as of the dead storage V and of ${}_iY_t$, the water released during the period, because the variables ${}_ih_{c,s}$ in the linear constraints are dependent on both ${}_iS_t$ and ${}_iY_t$. This implies that the head associated with the storage ${}_iS_t$ is an approximation of the average head during the period. This approximation is justified because the average of the end-of-period storages in any year, assuming that the year is divided into several periods, approximates the average storage in the reservoir in that year; and because it is advantageous to keep the reservoirs fairly full in hydroelectric developments, we find ourselves on that portion of the head–capacity curve where large withdrawals do not affect the head substantially.

If we wish to use the model for the analysis of flood control, we must assume that there is good correlation between average flows and flood damages during the period under study, the manner in which the flood period is chosen being of key importance. Incidentally, the correlation between average flows and flood damages is often fair even when the correlation between average and peak flows in a given period is poor. Because of the relatively long periods used in the model, the details of flood routing were not considered.

If multipurpose use of the reservoir is to be studied, it must be possible to combine the individual target outputs within a single benefit function. For each combination of target outputs for different purposes during each period (${}_iY_l$, where the subscript $l = 1, 2, \ldots, m$ denotes the target output for purposes $1, 2, \ldots, m$) we must find, through suboptimization techniques that are not discussed in this chapter, that allocation of draft ${}_iY_t$ to each purpose which maximizes benefits.

The IBM 700 series computers are capacious enough to program the operation of the model for about 12 periods in a year and a single reservoir. Inclusion of more than one reservoir would multiply the number of constraint equations, and this would have to be offset by decreasing the number of periods. This might not be objectionable, however, because in most cases no more than four to six periods per year are needed to take account of major variations in inflow and rate of draft required to provide outputs.

Serial correlation of inflows between periods is of importance in the theory of reservoirs, and failure to provide for it is, we have noted, a limitation on our model. It could be introduced into the model by redefining the decision variable in terms of ${}_iX_t$, ${}_iR_t$, and ${}_iY_t$, given the serial correlation between periods ${}_{i-1}P_i$, and the probability distributions of ${}_{i-1}X_{t-1}$ and ${}_iX_t$.

Part III GOVERNMENTAL FACTORS

15 System Design and the Political Process: A General Statement

ARTHUR MAASS

RELATING governmental institutions to the design of water developments is the task of this final chapter. In the past water-planners and engineers, in search of constraints to simplify their task of system design, have found it convenient to read inflexibility into governmental institutions and to treat them as immutable, if irrational, restrictions.[1] On the contrary, there is and should be considerable flexibility in legal and administrative forms, which are quite adaptable in the face of demonstrated economic, technologic, social, or political need.[2]

Our purpose is to identify those institutional factors which rightly are pervasive requirements and to show just how they limit what is otherwise an area of free choice among alternative governmental arrangements. These requirements are essentially of two types: those deriving from characteristics of the technologic and economic functions for water-resource development, and those deriving from the requirements of a constitutional democratic system of government. With respect to the former, it should be adequate to identify the requirements and indicate their consequences. Since, however, the requirements of a constitutional democratic government are introduced for the first time in this chapter, a more thorough discussion is called for. To this end we shall construct a model for the modern democratic state. Elaboration of the model will be based broadly on the pattern in the United States, but presented in such a way that the patterns of other constitutional democratic systems can be substituted without basic alterations.

[1] For example, the history of planning the development of the Colorado River from the Compact of 1922 to date. For a good summary of the lower river, see Vincent Ostrom, *Water and politics* (Haynes Foundation, Los Angeles, 1953); for the upper river, see the history of legislation authorizing the Colorado River Storage Project, U. S. Public Law 485, 84th Cong., 2nd sess.

[2] The history of water law in the United States provides an example. See Arthur Maass and Hiller Zobel, "Anglo-American water law," *Public Policy 10*, 109 (1960).

A MODEL FOR THE MODERN DEMOCRATIC STATE

Almost all theoretical discussions of the constitutional democratic state discriminate between the community or society on the one hand and the government or state on the other, each possessing unique functions.[3] This is not the place for a comprehensive treatment of this matter, but the dualism should be expounded in at least one generally recognized version, so that on its foundation we can build a model for analysis which defines precisely the several processes of government and the institutionalization of these processes.[4]

FUNCTION OF THE COMMUNITY

The community's unique political function is to reach agreement on the standards of the common life — the objectives. A constitutional democratic system is based on man's capacity to debate and determine the standards by which he wishes to live in political community with others. With two important exceptions these standards are not fixed. They are continually being resolved; they are ever emergent. And it is probably true that the more free and effective the democratic society, the more active and pregnant the discussion of what should be the governing objectives.

The exceptions are, of course, equality and liberty, which must be accepted as standards in any democratic community. These terms have specialized, rather than abstract, meanings in this context. Equality as a standard demands that each individual shall count and shall be enabled to make his own contribution. It does not mean equal conditions for all, or that all should be considered physically or intellectually equal. Similarly, liberty does not mean an absence of compulsion or law, but rather

[3] Often a three-part distinction is made between community, state, and government. This refinement is not necessary for our purposes.

[4] The version presented here is derived largely from A. D. Lindsay, *The modern democratic state* (Oxford University Press, London, 1943); Ernest Barker, *Reflections on government* (Oxford University Press, London, 1942); and Joseph Cooper, *The legislative veto: its promise and its perils* (Senior honors thesis, Harvard University, 1955), and *Legislative committees* (Ph.D. thesis, Harvard University, 1960).

A considerable group of political scientists denigrate model-building. Their interest and influence can be seen in the recent preoccupation of American political science and public administration with case studies, studies which systematically eschew any concern for principles and focus largely on the unique. We suggest that in this century the discipline has been too much concerned with nonconformities and not enough with normal structure and general principles; that twentieth-century American political science has probably moved too far away from the model-building focus of its distinguished late-eighteenth-century counterpart.

that each individual shall be enabled to control, to a meaningful extent, his own fate. While it is true that equality and liberty are requirements of democracy — that, in distinction to other standards, there must be acceptance of these — their meaning and application in specific situations are obviously open to interpretation. To a significant degree the community discussion of standards will involve the manner in which equality and liberty are to be realized.

The community in which discussion and resolution of standards takes place is one of voluntary association outside the law. The religious, the social, the economic, in short the common life of society, is lived by individuals in all manner of social relationships — churches, trade unions, institutions of all kinds — and these operate in "a sphere of initiative, spontaneity, and liberty," a sphere which therefore cannot be dominated by the state with its instrument of compulsion.[5]

Thus, the community's political function is to foster a process of discussion which results in agreement on objectives and in a propensity to reexamine them. This is quite different from considering the function of the community, as do some commentators, in terms of resolving conflicting interest demands on the basis of power.[6] Interest groups there are, to be sure; but we view them in a general context which emphasizes community consensus, and in an organizational context which seeks to elicit from the groups action based on their opinions of what is good for the community, rather than action deriving from their opinions of what is good for themselves abstracted from the community, or for themselves in naked struggles for power with other groups. In other words, the interest groups contribute to the discussion of, and agreement on, the standards; their demands provide an essential part of the data of politics. Furthermore, conflicting interest demands are resolved or adjusted by reference to common standards, not to pure power or even to power constrained by "rules of the game."[7] As a matter of fact, interests, according to the theory developed here, cannot be defined without reference to the common standards. They do not exist in a vacuum, but in a context of common values and purposes which inevitably enter into the very definition of the nature of interest.[8]

[5] Lindsay, reference 4, p. 245.

[6] Arthur F. Bentley, *The process of government* (Principia Press, Bloomington, 1908); John Dickinson, "Democratic realities and democratic dogmas," *Am. Pol. Sci. Rev. 24*, 25 (1930); David B. Truman, *The governmental process* (Alfred A. Knopf, New York, 1951).

[7] Truman, reference 6.

[8] John Plamenatz, "Interests," *Political Studies 2*, 1 (1954).

FUNCTIONS OF THE STATE

The state is to serve the community in order to make more effective the community's function relating to common standards. In doing so, the state must understand and in some ways interpret the principles which govern the common life, *but never seek to prescribe them.* More specifically, the state has three unique functions.

The first is to guarantee those political and economic conditions which are essential to enable all to participate in citizenship. Thus, the state must maintain for its members a minimum standard of political and legal rights — freedom of speech, press, association, voting, fair trial. Equally, it must maintain economic rights for all its members by guaranteeing minimum economic conditions — minimum wages, unemployment compensation, social services. The theory of the democratic state offered here recognizes, as did Aristotle's, that citizens possessed of political and legal rights may nonetheless be unable to participate in any real sense in the community because of their poverty; and it justifies the government's responsibility for preventing this.

To provide the opportunity for effective participation in citizenship, the state has a further responsibility to acquire and put at the disposal of all citizens accurate information about the community, so that the community can decide intelligently on objectives. This point is discussed in greater detail later in the chapter.

Finally, included in its first function, the state must defend the community from external attack and internal violence.

The second unique function of the state is to provide institutional means for focusing on issues and areas of community agreement and for translating these into rules and criteria sufficiently specific to serve as guides to governmental action, where such action is desired by the community.

The state's third function is to carry out activities in accordance with standards set in the process of discussion.

One can view the operation of the democratic community and democratic government as four concentric circles. The outer circle is the discussion in the community where the broad standards are agreed on. Here the state plays no part other than that described above as its first unique function. But the inner circles — the electoral, legislative, and administrative processes — are organized by the state. They are the means through which discussion and interpretation of standards are carried successively to greater and greater specificity, from general issue to concrete decision. Each circle is part of a continuous process of discussion and debate; each takes over from its outer circle, discharging a selective function, and hands on to its inner circle, performing an instructive function. At the same time, each circle or process has a duty and set of institutions of its own. While each receives instructions from outside, it has a specific deliberative discretion. And one of the main problems of democratic government, conducted in this way by successive stages of discussion, is to achieve respect for the principle of division of labor. For if the institutions of any stage or circle are ambitious and seek to transcend their duties, they will interfere with institutions of other stages and may throw the whole system out of gear.[9] Let us then identify very briefly the duties of the three processes organized by government.

The *electoral process* takes over from the community and its groups. The process involves discussion; at the moment of choice, men are selected as representatives and instructed to carry discussion to a further and more precise point in the legislative process. It is not essential that the programs of electoral candidates be detailed and specific; the later stages of discussion will accomplish this. Instead, "the essence of the selective function of the electorate consists in the choice of men who, in their personal capacity, and in virtue of their character, are fitted to discharge the task of deliberation and discussion at the parliamentary stage."[10]

The *legislative process*, as we have seen, takes over from the electorate and translates into rules of law the general programs endorsed by that body. The "legislators" (to be defined in pages 577–581), by special techniques for discussion and decision, including the capacity to use effectively more information than the electorate can use, accomplish something which their constituents are incapable of accomplishing. Also, since it is virtually impossible to achieve through the electoral process an integration of views "so coherent and cohesive that all subsequent decisions at the next level will flow merely as necessary consequences of common goals,"[11] the legislators are always involved in mutual enlightenment and reconciliation of the complex views they represent as a group.

Finally, through the *administrative process*, the legislative rules are

[9] Barker, reference 4, p. 38.
[10] Barker, reference 4, pp. 41–42.
[11] Cooper (1960), reference 4, chap. 1.

translated into criteria immediately useful for governmental action, and the actions of government are conducted in accordance with these criteria. The translation of rules into criteria is in effect the last stage of discussion and a part of the state's second function of providing institutional means for focusing on issues and translating them into action guides. The government's actions based on these criteria constitute the state's third function, to carry out activities in accordance with standards set in the process of discussion. But there can be no rigid line separating the two functions. In the exercise of their discretion in the conduct of governmental activities, "administrators" (defined also in pages 577–581) are inevitably and properly concerned with refinement of the governing criteria. Also, just as legislators are always involved in reconciling the complex views of the electorate because the electoral process cannot result in a fully cohesive and self-executing integration, administrators are always involved in further refining the standards of the common life because integration achieved in the legislative process is not self-enforcing.

LEADERSHIP, ACCOUNTABILITY, AND THE PUBLIC INTEREST

The process of democratic government, through successive and distinct stages of discussion, provides for both leadership and accountability. The principle of division of labor, it will be remembered, assigns to each stage a deliberative function of its own — a specific electoral, legislative, or administrative discretion — and in this way provides multiple opportunities for leadership. Though multiple, these opportunities, as we shall see, need not be divisive. Furthermore, the process supplies its own system of accountability. Even as it calls for discretion at each level, it lays down a standard by which to judge the manner in which the discretion is exercised and provides institutional means for ensuring that the standard is respected.

Crudely, the standard is conformity with agreements reached in the outer circles, which we shall call breadth. When the legislators exercise their discretion and seek agreement among themselves on policies and rules, their standard of conduct should be the outer circles of the electorate and of the community, viewed as associations in which individuals and groups are ever discussing and seeking agreement on the objectives which should govern their political union. Periodically the electorate, using this standard, will pass formal judgment on the legislators; between these periods the electorate and the community will provide intelligence to the legislators which the latter will evaluate and use as they see fit in the exercise of their discretion. Where the legislators are unable to reach agreement on an issue, they will often refer it back to the community, or periodically to the electorate, for further discussion and agreement.

Similarly, the standard to be followed by administrators in the exercise of their discretion will relate to the outer circles of the legislative and electoral processes and to the community. Consider for a moment the bureau chief in charge, let us say, of the distribution of electric power from an integrated hydro and steam system. He has two types of responsibility: first, to distribute the power in accordance with the rules laid down by the legislation; second, to report on what he has done and recommend changes in the legislative policy. In carrying out the first responsibility, his discretion may not be very broad. Of the determinants of his action, four-fifths may be ministerial in the sense that the policies are stated clearly and specifically by the legislation, and one-fifth may be left to his discretion. In his ministerial duties he must follow the specific provisions of the law with honesty and energy. With respect to his discretion, however, his standard of conduct should be breadth. He should seek to refine the law's intent by the broadest possible view of its incidence and effect. As Professor Arthur Macmahon has said:

". . . the operating administrator's . . . prime duty in carrying out the law is charted in the law's intent, declared or clearly implicit. In addition, still pursuant to the law but beyond its unmistakable guidance, the operating administrator must make innumerable judgments. Here enters his residual duty to take the broadest possible view of the consequences of any action." [12]

And in applying this standard, the administrator should look beyond the specific provisions of the legislation, which may be quite specialized in nature, to the broader consensus attained in the electoral process and in the community. (Obviously the administrator, or bureau chief, is responsible to his superiors in the bureaucracy for the exercise of this discretion. The relation of the administrator to the chief executive is discussed on page 583.)

In reporting and recommending, especially the latter, the administrator's discretion is almost complete; it is here, therefore, that the standard of breadth will be most effective. Here, too, if the administrator believes that the community and the electoral and legislative processes have not

[12] Arthur Macmahon, "Specialization and the public interest," in O. B. Conaway, ed., *Democracy in federal administration* (Graduate School, U. S. Dept. of Agriculture, 1955), p. 49.

reached a broad consensus adequate to guide him, he can and should propose that the policies be referred back for further discussion.

It should be noted that leadership, although it results from the division of labor, is also a by-product of the system of accountability. Legislators and top administrators participate in the electoral level of decision-making, defending or criticizing the record of past accomplishments; and through this participation, they can become leaders in attracting attention to, and seeking consensus on, important issues. Similarly, with respect to issues which are discussed in the legislative process and then referred back to the community or electorate for further discussion, the legislative discussion can be considered educating and leading in the sense that it directs the community's attention to the issues; it prepares the agenda for discussion at the community level.

One of the objectives of the theory of democratic government presented here is to emphasize the search for consensus on community values, and to deemphasize power politics based on individual interest or interest-group demands. This we can achieve, for as the process of democratic government moves through successive and distinct stages of discussion from general issue to concrete decision, it should move from value- or objective-oriented debate to interest-oriented debate. The latter is deferred, as it should be, until the later stages.

Suppose that the issue of whether or not the government should improve inland waterways for recreational craft is discussed in the community and in the legislative process; that agreement is reached in legislation which lays down broad standards for such waterways; and that the executive agency perfects design criteria and, in accordance with these criteria, designs a waterway which would bisect the property of J. Q. Citizen. J. Q., who had previously joined the consensus on the general policy, now objects vigorously, and interest-oriented debate commences. The point is that this debate, coming late in the governmental process — at the innermost circle of the administrative process — is easier to resolve in the broad public interest than if it had come earlier. J. Q., of course, is caught in Rousseau's familiar dilemma, where his will for the community is in conflict with his will for his own personal interest. It is not unreasonable to assume that a solution in the public interest can be reached with him more easily than if the interest-oriented debate had not been preceded by that on objectives.[13]

[13] Alternatively, it can be considered that in the early stages, when broad objectives were the subject of discussion, J. Q. knew that a waterway might cut his property, but he attached

DIVISION OF GOVERNMENTAL POWER

So far, in constructing our model for the modern democratic state, we have been concerned with the manner in which the community's objectives can be refined through the electoral, legislative, and administrative processes. To complete the model, we need to identify the units of government which conduct the processes and the more important relationships among these units. First, however, we should introduce another basic concept of constitutional democratic government, namely, the division of governmental power.[14]

Reasons for dividing governmental power. Governmental power is divided to help realize the basic objectives or values of a political community. To illustrate, assume that the basic values of a modern democratic state are broadly liberty, equality, and welfare. To promote liberty, governmental power can be divided so as to protect the individual and groups against arbitrary governmental action and against great concentrations of political and economic power — a restraining or "constitutional" effect. To promote equality, governmental power can be divided so as to provide broad opportunity for individual participation in public policy — a "democratic" effect. And to promote welfare, governmental power can be divided so as to assure that governmental action will be effective in meeting the needs of society — a service or facilitating effect. As a general rule, no one value can be maximized if all are to be achieved in high degree. Dividing governmental power so as to achieve the several values in the particular combination desired by the community is a great challenge for political engineering.

It is obvious that the governmental division of power, and the community's values as well, are related to the organization and division of nongovernmental power in the community — in short, to the community's social structure. The relations between governmental and nongovernmental divisions of power are reciprocal; the governmental division both reflects the community's power structure and itself influences it. At the same time, there is a reciprocal relationship between the community's values and its social structure. These complexities will not be

a very low probability to this possibility and was willing to take the risk in order to achieve a broad objective to which he subscribed.

[14] Much of this material is taken from Arthur Maass, "Division of powers: an areal analysis," in Maass, ed., *Area and power* (Free Press, Glencoe, 1959). The material is used with the kind permission of the publisher.

developed further here,[15] but it should be noted that government by successive stages of discussion is based on two assumptions: (1) that the social structure is such that, with institutional arrangements which foster it, the community will search for consensus through discussion; and (2) that institutional arrangements, including governmental divisions of power, can be developed which will foster the process of discussion. Theories of democratic government which emphasize the struggle for power among competing interest groups are based on contrary assumptions. We find sufficient evidence in American and British societies and institutions, and in those of many other nations, to justify building a general theory on the assumptions we have made.

Methods of dividing governmental power. Governmental power can be divided among officials and bodies of officials (a legislative chamber, for example) at the capital city of a political community; this we call the capital division of power. It can also be divided among areas or regions which exist or can be created within the political community; this we call the areal division of power. Further, power can be divided among official bodies at the capital and among component areas in a number of different ways. The most important of these are indicated in the diagram below, which can be applied separately to capital and areal divisions of power.

(1) Process	(2) Function	(3) Constituency	
			(a) Exclusive
			(b) Shared

As indicated in column (1), power can be divided according to the process used in governing. Thus, to effect a capital division of power, the process of legislation could be assigned to one body, the process of execution and administration to another, and the judicial process to a

[15] See Stanley Hoffmann, "The areal division of powers in the writings of French political thinkers," and Samuel P. Huntington, "The Founding Fathers and the division of powers," both in Maass, ed., reference 14.

third; or any of these processes could be shared by two or more bodies. To effect an areal division of power, the legislative process could be assigned to the central government and administration to provincial governments; or the process of legislation could be assigned to units of the central government at the capital, and the process of administration could be shared by those units and effectively decentralized units of the same government in the field — the Tennessee Valley Authority at Knoxville, for example.

According to column (2), power can be divided also by functions or activities of government. Thus, to effect an areal division of power, certain functions (such as the coining of money or the development of rivers for navigation) could be assigned to the central government, others to state or provincial governments, and still others to municipal and local governments. Similarly, to effect a capital division of power, functions could be assigned to agencies and departments of government which possess a significant degree of independence from one another. In most instances, of course, the division of functions among departments and agencies will be largely an administrative problem in specialization and division of labor, and will remain an administrative problem as long as functions can be rearranged when necessary for administrative reasons. Where, however, departments or agencies become legally or politically autonomous, division of functions at the capital assumes much greater importance.

Governmental power can be divided further according to constituency, column (3). To effect a capital division of power, one body — the upper chamber of a legislature — could be made to represent one constituency — certain groups in society; and the presidency, to represent another — society as the majority of qualified voters. Thus governmental power is divided by assigning to different units of government responsibility for representing different constituencies, in contrast to the Hobbesian scheme where all constituencies are perfectly represented in the single ruler. In a general way, it is axiomatic that an areal division of power assigns to the several levels of government representation of diverse constituencies.

Process, function, and constituency are interrelated as methods for dividing power. Functions can be divided among units of government which at the same time represent unique constituencies; and there is, as demonstrated by American experience, an almost infinite variety of combinations. Thus presentation of the three methods in separate columns of the diagram is not intended to demonstrate independence of any one from the others.

If we turn now to lines (a) and (b) of the diagram, the assignment of processes, functions, or constituencies to governmental units at the capital and to component areas can be either exclusive or shared. The function of coining money, for example, could be given exclusively to the central government, whereas the function of controlling pollution in streams could be shared by both central and provincial governments. Similarly, the process of legislation could be given exclusively to the legislature or could be shared between legislature and chief executive. And as we have seen, a constituency can be represented exclusively by one unit of government, as would be the case if electoral districts for the United States Senate were redrawn without regard to state or other political boundaries; or it can be interrelated in more than one unit, as was clearly the case when United States senators were selected by state legislatures.

Many elaborations of this simple two-part distinction are possible. With respect to shared powers, there are alternatives running from cooperation and careful coordination on the one hand to open competition on the other. That is, the assignment of power to the units or areas sharing a function, process, or constituency can be spelled out carefully to allow a minimum of overlap and discord, or they can be made to allow or promote competition and conflict.

In the further development of our model of a constitutional democratic system, we shall build explicitly on the principles presented earlier in this chapter rather than on the division of governmental power. The latter principle will never be out of sight, however. It can be considered as an ever-effective constraint, for as the author of a leading text on constitutional government and democracy has stated: "Division of power is the basis of civilized government. It is what is meant by constitutionalism."[16]

INSTITUTIONALIZATION OF THE ELECTORAL, LEGISLATIVE, AND ADMINISTRATIVE PROCESSES

On institutionalization of the electoral process, we observe only that our model provides operational criteria for evaluating and choosing among alternative electoral systems. There are three criteria: (1) that the system select delegates who as a group are broadly representative of

[16] C. J. Friedrich, *Constitutional government and democracy* (Ginn, Boston, 1950), p. 5.

the community (or outer circle), where the latter is viewed as a pluralist political association seeking agreement on objectives; (2) that it select men who, through given institutions of government, can carry on effectively the legislative and administrative processes (the inner circles)—that is, that the election result in the effective manning of institutions capable of exercising legislative and administrative discretion; and (3) that the resulting division of governmental power based on constituency, when considered along with divisions based on process and function, be adequate to satisfy community values.

To illustrate, if proportional representation and other techniques of direct democracy were judged by these criteria, they would be rejected in a great many instances; for legislative bodies selected by these means would likely be so splintered that they could not operate effectively. Further, in terms of the first criterion, such means would result more nearly in representation of the community as separate and conflicting groups than of the community in search of agreement on basic objectives.[17]

Let us consider at once institutionalization of the legislative and administrative processes. It should be understood from the outset that there is no exact correspondence between the legislative process and the functions of the legislature, or between the administrative process and the functions of the executive. The chief executive and the legislature both participate in both the legislative and administrative processes. Our purpose here is to distinguish the proper role for each in the two processes.

REASONS FOR A LEGISLATURE

The reasons for the existence of a legislature are not necessarily self-evident. They relate to exigencies of popular control over the administrative and legislative processes and to the unique or qualitative contribution that a legislature can make to the success of democratic government.[18]

Oversight of the administrative process. With respect to popular control over the administrative process, it has always been a premise of constitu-

[17] Unfortunately, the "principles of representation" found in much of the American literature offer no adequate theoretical justification for these important distinctions. See, for example, the article on "Representation" in *Encyclopedia of the social sciences* (Macmillan, New York, 1934), vol. 13, p. 309.

[18] For a similar development of these points, see Cooper (1955) and Cooper (1960), reference 4.

tional democratic government that bureaucracy suffers from inherent tendencies toward parochialism and toward aggrandizement of power by officials that destroys responsibility. Thus bureaucracy must be constantly subjected to informed criticism. The chief executive, the courts, professional standards, and direct relations between the community and the bureaus of government, while necessary, are not sufficient for this purpose. The chief executive, as an elected representative of the people, can subject the bureaucracy to popular control, but the supervision he provides has limitations. It can be exercised only by his top political aides in the executive, and "they are continually being pushed into becoming mere captives of their department by their own lack of information and expertise and by the department's quiet persistence, quiet obstruction, and command of facts."[19] The protection provided by the courts is likely to be largely negative and after the fact. Professional standards, although important in ensuring responsible conduct of public affairs, are no substitute for popular oversight. And since direct relations between a government bureau and the public develop almost inevitably into those between the bureau and a special public (the agency's clientele), it is difficult to ensure popular control by this means. In short, legislative oversight, as an essential supplement to the forms mentioned, guarantees the capacity of the people to call the bureaucracy to account.

Oversight of the legislative process. A legislature is needed also to ensure popular control over the legislative process. As in the administrative process, the legislature normally exercises oversight in the legislative process; the executive takes leadership. Because this point is often misunderstood, it would be well to state the reasons for executive initiative in the formulation and enactment of law.

The early stages of the legislative process are significantly a matter of reducing alternatives and concentrating on the more promising possibilities, and of ensuring that policy proposals are coordinated and consistent. The former requires, among other things, access to great stores of information and expertise; the latter requires central direction of the policy-formulation process. The executive is better able to provide both of these than is the legislature.

If the legislature were to attempt regular leadership in the legislative process, its efforts would in the long run be self-defeating. To obtain the necessary expertise and information, it would be forced to rely on a system of standing committees with large professional staffs. Yet to give

[19] Cooper (1955), reference 4, p. 9.

such permanent power and function to these committees would be to make more difficult, and probably impossible, subsequent coordination and central direction of the process by the legislature as a whole.

When in the United States presidential leadership is not forthcoming, when there is a leadership vacuum in the White House, there is no question but that the Congress can and often does take the initiative in the legislative process. On these occasions there is a leadership reserve in the Congress, and it is one of the main strengths of the American system of government. But this reserve is located largely in the committees and their leaders. It can be used with success episodically, but if exercised regularly, it could not be effective.

One might ask if there is not a similar problem in the executive. To get the expertise necessary for leadership, the chief executive must rely on the bureaus, and this very reliance makes it more difficult for him to exercise central direction. The United States President has developed with some success tools to make coordination and direction possible under these conditions — tools of budgeting, legislative clearance, administrative management, and the like. Why cannot the legislature do likewise? The primary reason is that the legislature, as we shall see below, should be organized to articulate the views of the nonexpert, rather than those of the expert, and therefore it cannot be organized or controlled as hierarchically as can the bureaucracy. Intrinsically and practically, the American legislature cannot provide a consistently effective and central direction of the legislative process, as the executive can.

Finally, the executive should normally be leader in the legislative process because of the importance in a democratic system of pinpointing responsibility so that the electorate can hold its agents accountable for this activity. Responsibility cannot be pinned down unless there is an agency of central direction; as we have seen, over the long run only the executive can provide this.[20]

Having justified executive leadership as the normal requirement, we can now delineate the legislature's role in the legislative process, which is to criticize and control on behalf of the nation. True, the American chief executive is himself elected and may consult many representatives of the public, and of special publics, in developing his proposals for

[20] Again, one might ask why cannot the legislature, or parties in the legislature, organize so that they can be identified and reached by the electorate? It is difficult to hold a party accountable unless its program is spelled out by acknowledged leaders; and for one party in the U. S. government, the leader is the chief executive. The President speaks for his party.

legislation. Still, there is probably no better index of the degree to which his proposals are actively or passively acceptable to the community, as a means of moving the community consensus to a further state of specificity, than the voice of the legislature. As Joseph Cooper has said, ". . . no other institution, public or private, is as well adapted to serve as the political barometer of the community . . . it is the most sensitive instrument of public opinion in the state. . . . Without a legislature there would be no guarantee that executive proposals would be adequately subject to modification by public opinion."[21] This results from the composition of the legislature and the methods by which its members are selected, but also from the fact that the American Congress serves effectively as a focal point for the expression and organization of opinion.

Where, in the judgment of the legislature, there is not sufficient community consensus to justify approving (or rejecting) an executive proposal — where, in other words, the legislature does not feel competent to exercise its discretion — it can refer the issue back to the community or electorate for further discussion. In the process of taking this decision, which is likely to involve public hearings and debate on the floor of the chamber, the legislature can focus public attention on the more general aspects of the issue and thereby perform a vital educating function. It is true, of course, that the chief executive also performs this function, but even a casual observer of national government in the United States or Great Britain, as examples, is aware of the powerful influence which legislative consideration of a major issue has on public attention.

Qualitative contribution of the legislature. We have shown, within the context of our model, that the exigencies of popular control over the administrative and legislative processes call for a legislature. Further justification for a legislature is found in the unique qualitative contribution that such a body can make in the modern democratic state. The elements of this contribution are basically two. First, the legislature can bring to both the administrative and legislative processes certain qualities of the nonexpert. "The collective nontechnical mind of the legislature contains insights and sensitivities beyond the perception and ken of any expert."[22] Second, the legislature institutionalizes the "open mind," the capacity for change, which is in some ways the ultimate strength of democracy. In both the administrative and legislative processes it can make the executive see the obvious and do something about it. "By

[21] Cooper (1955), reference 4, p. 15.
[22] Cooper (1955), reference 4, pp. 16–17.

providing an institutionalized framework in which change can occur, the legislature performs a function which is vital in an age when bureaucracy has become the prominent element in constitutional democratic states";[23] for the bureaucracy, more than the legislature, evidences a persistent tendency to carry on as it has in the past, to resist change.

To these contributions of a legislature can be added a third which is less uniquely associated with the collective quality of the organization; namely, that the legislature's constituency is a different one from the President's.[24] This fact provides an enlightening complement to the views of the President and makes the resultant of their interaction in both the administrative and legislative processes a more valid refinement of the community consensus than would otherwise be the case.

THE ROLES OF EXECUTIVE AND LEGISLATURE DEFINED AND RELATED

Having observed that the chief executive and the legislature both participate in both the administrative and legislative processes (that, in other words, we have a division of governmental power by processes shared), we have sought to distinguish the proper role for each institution in each process. This can now be reduced to the following simple form, with the reservation that the legislature can in the short run assume leadership in the legislative, though not in the administrative, process.[25]

Institution	Process	Role
Chief executive (President)	Legislative and administrative	Initiation
Legislature (Congress)	Legislative and administrative	Oversight

The roles of the two institutions can be further defined and related in terms of (1) the type of policy and of administrative performance which should concern each; and (2) the relations within each institution, and

[23] Cooper (1955), reference 4, p. 18.

[24] In the U. S., the legislature's constituency for the lower house is the sum of popular majorities in 435 single-member districts, and the President's is roughly the sum of popular majorities in the 50 states.

[25] It is interesting to note that U. S. political-science literature has accepted the President's role in the legislative process, but has consistently deplored Congress' role in the administrative process. A systematic exploration of the reasons for this error would make a fascinating study of the development of the science of politics and of political scientists in America.

between them, of the whole and its parts — that is, among the chief executive and his executive bureaus and the whole chamber and its legislative committees.

Type of policy and administrative performance. The legislature, in its exercise of oversight, should concentrate on *broad* policy and *general* administrative performance. The several justifications for a popularly elected assembly all support this principle. The unique qualities of the legislature — its institutionalization of the unspecialized mind and of the capacity for change — are most effective when applied to broad, integrated, and general issues rather than narrow, specialized, and detailed ones. Also, unless the legislature concentrates on broad issues, it will not be able to perform its educating function. "The people of the United States cannot be interested in whether or not Mill Creek, Virginia, is improved. nor even in whether Arizona or California should be allotted the greater share of the water of the Colorado River. But they can be aroused on national policy issues such as the prevention of speculation and monopoly in benefits derived from federal improvements."[26]

With regard to oversight of administration, legislative concern with detailed performance does not ensure efficiency and responsibility. It is largely through reliance on broad standards, applied to broad categories of an agency's performance, that responsible and efficient administration can be assured.

This is not to say that the legislature should ignore details. Oversight of administration, and of the legislative process too, requires a concrete foundation in facts. Detailed statistics on expenditures for pencils by the Corps of Engineers in a given fiscal year would not be very useful for evaluating administrative performance, to be sure; but statistics, equally detailed, on commerce of the nation's inland waterways in the same year would be. Although the legislature normally should not concern itself with detailed administrative performance, its capacity to command and use details should not be circumscribed.

The executive must deal with narrow issues and detailed administration, for it has responsibility, it will be remembered, for the state's third function — to carry on activities in accordance with the standards set in the process of discussion. But in initiating the legislative process, the executive too should be concerned with broad policy; and in reporting to the legislature and to the people on its activities, with general administrative performance. By taking the initiative in both legislative and ad-

[26] Arthur Maass, "Congress and water resources," *Am. Pol. Sci. Rev. 44,* 591 (1950).

ministrative processes, the Chief Executive sets the agenda of Congress; yet too often this fact is disregarded. When Congress concerns itself with picayune details, it is quite generally condemned by scholars and newsmen; but not infrequently, and perhaps generally, it is these details which have been served up to the legislature by the executive. An executive agency, for example, may report to Congress on specific projects for small-boat harbors without having previously initiated legislative proposals to declare policy for the government's participation in improving waterways for recreational craft and without having developed criteria based on this policy for the design of small-boat harbors. Where this is the case, responsibility for the resulting "pork barrel" should lie more heavily on the initiator than on Congress, the overseer.[27] "If an agency can be controlled effectively only when Congress focuses on major issues of integrated policy, a major criterion of its responsibility is the success with which that agency, in reporting to Congress through the President, points up the broad policy questions which require legislative determination and plays down administrative details."[28]

The whole and its parts. The executive bureaus should be responsible directly and primarily to the President for initiative in the legislative and administrative processes, and they should be responsible to the Congress only through the Chief Executive. The reasons for this principle are generally the same as those used to justify executive leadership in the legislative process, and they are a logical extension of the principle which emphasizes broad policy and general administrative performance. To be broad, the policy emanating from any one government bureau should be integrated with related policies of others; and to be general, administrative performance of related agencies should be coordinated. Since, as we have already seen, integration and coordination require effective hierarchical organization, and since the executive intrinsically is better able to provide this than is the legislature, the bureaus should report directly to the executive.

In a like manner, the committees of the legislature should be responsible directly and primarily to the whole chamber and should oversee

[27] It should be observed that "broad" and "general" are not synonymous with "federal" and "national." Thus a legislature which expresses a particular concern for the local aspects of national issues may be expressing a broad and general popular consensus on "localism." On this point, see Ernest Griffith, *Congress: its contemporary role* (New York University Press, New York, 1956), pp. 154 and 158.

[28] Arthur Maass and Laurence Radway, "Gauging administrative responsibility," *Pub. Admin. Rev. 6*, 188 (1949).

the executive establishment through the legislature as a single body. The unique contribution of the legislature to democratic government was derived largely from virtues of the body as a whole, rather than those of standing committees, each with a specialized competence and jurisdiction. At the same time, the standing committees of the American Congress are essential to the legislature's oversight of both the legislative and administrative processes in the normal case, and to its assumption of leadership in the legislative process when there is insufficient leadership in the executive.

Techniques for organizing the legislature to protect the committee system on the one hand and ensure that legislative action is effectively action of the whole on the other are complex and difficult. It may be more difficult for the legislature to coordinate its parts than for the executive. But this is not as disqualifying as it may seem. Because coordination is less essential in overseeing than in initiating the legislative and administrative processes, it is less compelling a requirement for organization of the legislature than for organization of the executive establishment.

Obviously it is direct relations between bureaus and committees that are the most serious challenge to the roles we have defined for the executive and legislature. Bureaus often maintain a network of informal contacts with committees and individual legislators. In some form such relationships are inevitable, and in fact indispensable. But in their pathological form they represent an institutional challenge to the theory of democratic government presented here. A single legislative committee, for example, may come to occupy a position of influence so commanding in matters affecting an agency that oversight of the agency's activities is restricted to the committee and denied to the remainder of the legislature and to the Chief Executive as well. The pathological form of these relations is a function of the organization and conduct of both the legislature and the bureaucracy. Action, therefore, to prevent or mitigate such extremes is equally a responsibility of Congress and of the Chief Executive. A bureau chief has a responsibility, in other words, to conduct his relations with the legislature in a manner that will prevent minority control over his agency's affairs. As in the case of congressional concern with narrow rather than broad issues, American political scientists and newsmen too often have singled out the Congress for blame.

IMPLICATIONS FOR WATER PLANNING OF GOVERNMENT'S RESPONSIBILITY FOR ORGANIZING THE LEGISLATIVE AND ADMINISTRATIVE PROCESSES

We have now arrived at a point where we can specify some of the institutional factors in planning for water development which derive from requirements of a democratic system of government. We have constructed a model whose foundation is a distinction between the functions of the community and those of the state, whose basic structure is a definition of the several governmental processes, and whose finish is the institutionalization of these processes. What remains is to apply the model to the relevant facts of river-system design, as these have been developed in this book.

FROM OBJECTIVES TO CONSTRUCTION

Economic factors in water-resource development, as presented in Chapters 2 and 4, call for three major stages in the process of system design and, associated with these, the presumption (if not requirement) of three levels of governmental organization. Accordingly, objectives would be set by legislative action; these would be translated into design criteria in a budget or planning agency at the center of government; and the actual design of river systems would be accomplished by planners in the field. These stages can be made congruent with the model for a constitutional democratic system. They represent part of an orderly movement from general principle to specific action.

First, the objectives of water development, in terms of efficiency, income redistribution, regional economic growth, control of speculation in benefits provided, and the like, should be determined in the legislative process of discussion. Here the consensus on objectives and values achieved in the community is interpreted and converted to a form usable as a guide for subsequent governmental activity. Here, in the normal case, the chief executive should propose a program of consistent objectives, and the legislature approve, revise, or reject it.

Second, values for benefits and costs, interest rates, and budgetary constraints should be fixed to reflect the objectives, and these should be incorporated in design criteria; and third, river systems should be designed in accordance with these criteria. These activities, representing translation of legislative rules into criteria immediately useful for governmental action, are part of what we have called the administrative process, the last stage of discussion. Since they require considerable specialized talent, they should be conducted largely by the executive branch — the preparation of design criteria at the center of government, and of system designs in area offices. The legislature should oversee the fidelity with which the executive has translated objectives into criteria and criteria into designs. In general, the former could be conducted through spot or periodic investigations by legislative committees; the latter, by such investigations or by additional legislation authorizing specific designs. Periodic investigations should suffice for designs of small projects and those which conform readily to the design criteria. Authorizing legislation would be justified for major undertakings and those which do not conform. In both cases further legislative oversight would be exercised by the appropriations committees when money is requested to commence construction.

As the executive seeks to translate the objectives into design criteria and the criteria into system designs, it may be confronted with conflicting legislative objectives. In some instances the legislature purposely may have transferred to the executive responsibility for reconciling conflicting goals; in others the conflict may be revealed only in the process of moving toward greater specificity. In either situation the legislature will want to review the reconciliation proposed or effected by the executive; and in cases where conflicts have been revealed initially in the later stages of the planning process, the executive may choose to delay adopting criteria until further action is taken in the legislative process.

Finally, projects should be constructed in accordance with the designs. This is a subsequent part of the administrative process and involves government's responsibility to carry out activities in accordance with standards set in the process of discussion. The major activity is executive, with the legislature overseeing for efficiency, honesty, and conformity to standards.

TWO ILLUSTRATIONS

All of this may appear self-evident in light of the model. A survey of certain aspects of water planning in the United States, however, should serve to demonstrate why it is desirable to be abundantly clear on these matters.

We might look first at the process for authorizing plans for navigation, flood control, and multipurpose projects in which navigation and flood control are significant elements. Here the most meaningful stages of the process of democratic discussion organized by government are omitted. The legislative process is by-passed, for there is little discussion of general objectives and little action based explicitly on them. Nor is there translation of legislative objectives into design criteria. Rather, the planning process begins with the design of systems or projects, and this with few useful policy guides to the planner.[29] Project designs developed in the field are reviewed at higher echelons in Washington, including the Executive Office of the President (Bureau of the Budget), and transmitted to the Congress. Only at this point does Congress become involved;[30] and the agenda for action which the policy initiators, the executive, have produced does not play to the legislature's strengths. The Congress is asked to give its approval to, and thereby authorize, plans proposed for the improvement of Twitch Cove, Maryland, to protect crabbers; of Cheesequake Creek, New Jersey, for pleasure craft; of the Columbia River, to provide hydroelectricity, irrigation, and flood protection over a wide area. Biennially Congress considers an omnibus Rivers and Harbors and Flood-control Bill which consists mainly of approval, in whole or part, of the project reports completed and submitted to the legislature in the interval since the last omnibus bill. The hearings and debate do not turn on objectives or even on design criteria, but on whether the specific proposals for Twitch Cove and Cheesequake Creek should be approved. Since on these questions the unspecialized and open mind of the legislature as a whole can have no informed view, the body must defer to the judgment of its representatives from Maryland's Eastern Shore and New Jersey's Sixth Congressional District. The cumulative impact of this procedure is called "pork barrel" by unkind critics.

The point is neither that the legislature should fail to oversee the manner in which the executive prepares project designs, nor that it should retreat from a procedure for authorizing the more important ones.[31] The point is rather that this should not be the legislature's main activity.

[29] For details on this point, see Arthur Maass, *Muddy waters: the Army Engineers and the nation's rivers* (Harvard University Press, Cambridge, 1951); and Maass, reference 26.

[30] In a great many cases the project surveys are authorized by legislation. While this procedure may once have had some significance in indicating to executive planners the objectives of Congress, this has not been the case for a great many years. The procedure is now almost entirely perfunctory.

[31] It should be noted, nonetheless, that Congress does not require project authorization for public works involving immense expenditures in housing, urban renewal, and roads.

Similarly, it is not that the executive should retreat from initiating project designs, but that this should not be its first initiative in the planning process. In other words the process should not begin, as it does, with project formulation. By beginning in this way, it encourages interest-oriented debate and repudiates, or at best passes over, the more important objective-oriented discussion, which is directed at community consensus.

Another interesting illustration relates to certain efforts between 1948 and 1952 to define and declare objectives for water-resource development. In 1948 the Task Force on Natural Resources of the Commission on Organization of the Executive Branch of the Government (First Hoover Commission), having confined its study of water resources to matters of organization, recommended that the Executive Office of the President conduct a major study of national water policy.[32] Largely in response to this recommendation, the President in 1950 established an ad hoc policy commission (President's Water Resources Policy Commission) in the President's office, and at the end of the year this commission made a far-reaching report, with proposals for legislation to establish objectives for water development.[33] These proposals were then subjected to an intensive and elaborate review by the Bureau of the Budget. In the end, however, they were not submitted to Congress with the recommendation that they be considered and adopted as basic water policy. Instead, the Bureau of the Budget incorporated in a budget circular, binding on all executive agencies, those policy proposals which it approved and which in its view could be proclaimed without additional legislative authority.[34] Thereafter the legislative committees objected repeatedly to this executive action; they sought "to reaffirm Congressional control over water policies."[35] The Budget Bureau's response to the legislative committees was, in our terminology, that the provisions of the circular were not so much objectives as design criteria, and that the Congress would have the opportunity to review them when it acted on the individual project-design reports of the agencies.

Measured against our model, establishing the ad hoc policy commission makes good sense. The President's failure to submit its proposals to

[32] U. S. Commission on Organization of the Executive Branch of the Government (First Hoover Commission), *Task force report on natural resources* (1949), p. 28.

[33] Report of the U. S. President's Water Resources Policy Commission, vol. 1, *A water policy for the American people* (U. S. Govt. Printing Office, Washington, 1950).

[34] Budget Circular No. A-47, December 31, 1952.

[35] See, for example, S. Res. 281, 84th Cong., 2nd sess., and legislative documents relating thereto.

Congress, coming as it did near the end of an administration, involves many issues that we cannot evaluate here. But it should be clear that in so doing the President failed to initiate the legislative process. In light of this, issuance of the budget circular raises serious questions. Even if one agrees with the Budget Bureau that the circular's provisions are more nearly design criteria than basic objectives, it is still true that the executive issued the criteria without meaningful prior action in the legislative process on the objectives from which the criteria must derive. To claim that the Congress can review these when it considers individual project reports is, again, to delay participation by the legislature until a time when it cannot be very useful or effective. In this instance one can say that the executive has contributed to poor water-resource planning and that the complaints of the congressional committees are prima facie justified.

Finally, we note that relations in the United States between the executive water-resource bureaus and the committees of Congress are so direct and intimate that clearly they violate that part of the model which relates the whole of each institution to its parts. Documentation on this is readily available, and the facts need not be repeated here.[36]

IMPLICATIONS FOR WATER PLANNING OF GOVERNMENT'S RESPONSIBILITY TO INFORM THE COMMUNITY

Government collects and analyzes intelligence for all four levels of discussion. For the administrative and legislative levels, this is part of its exercise of initiative. For the electoral and community discussion levels, it is government's responsibility to acquire and make readily available accurate information about the community. It will be remembered that government does not organize the process of community discussion; it only ensures the capacity of all citizens to participate actively in it.

Just as the legislature needs detailed facts if it is to direct its attention to broad policy and general administrative performance, so the community needs detailed facts to discuss objectives. As in the case of the legislative process, the question is not details or no details, but what kind of details. The answer is those which will encourage objective-oriented

[36] Maass, reference 29. These relations may account in part for the conduct of both the Bureau of the Budget and the legislative committees in the case of Budget Circular No. A-47.

instead of interest-oriented debate and will thereby lead to a broad consensus on the bases or objectives for specific programs.

Does this mean, then, that the intelligence which government provides the community on water resources should relate directly to the objectives of national economic efficiency, regional income redistribution, and the like, and have no relation to system designs? Not necessarily. System-planners can develop alternative designs from alternative objectives and, by illustrating the impact of the latter on the capabilities of water systems, use the designs to provide information on the objectives themselves. As a matter of fact, information of this type can be of special significance for the democratic process of discussion, providing an opportunity for it to recycle. More specifically, objectives can be set initially. Through processes leading to greater and greater specificity, these objectives can be converted into design criteria, system designs, and completed systems. Thereafter, these systems can be evaluated in terms of how they might have been designed or constructed under different objectives; such evaluation provides data for a continuing discussion of the objectives. Without this evaluation the process could degenerate, with the result that systems might be built in 1960 based on 1902 objectives. By providing adequate data to the community, government can help ensure that objectives derived initially are not so overemphasized or so blindly pursued for their own sake that they cannot be easily modified to keep in close touch with changing reality.

The data for community discussion come from the legislature, the executive, and from nongovernmental sources; here we are concerned only with the unique requirements imposed by this responsibility on executive agencies for water development.

PROBLEMS IN MAINTAINING OBJECTIVITY

Objectivity is the most important of these requirements, and it raises certain difficult problems concerning the organizational relationship between information and data activities on the one hand, and design and construction activities on the other.

Basic data. Consider the so-called basic data for design — data on population, national and regional income, projections of water requirements, hydrology, and the like. If a design and construction agency is responsible also for basic data, it may focus data collection and interpretation on the design objective of the moment. The result will be to stall the democratic process of discussion, because this process, in which objectives are continually being reevaluated and reestablished, demands a greater variety of data than that necessary to pursue a given objective. Similarly, many types of data must be collected and organized well in advance of their use for design purposes. One cannot commence gaging a stream at the time it is to be developed. Yet a program agency, in the toils of allocating a budget that always seems inadequate, may be led to neglect this all-important quality of basic data.

The task of basic-data collection requires, in other words, a uniquely broad view of the community's objectives; and this raises the question of whether or not an agency can operate effectively and objectively having simultaneously a very broad view for data collection and a narrower one for system design and construction.

Intelligence to evaluate objectives. We have already mentioned the desirability, for purposes of intelligence, of designing systems for alternative objectives. Where an agency has designed and possibly built a system for a given objective, it may be hesitant later to develop alternative designs based on alternative objectives, for fear that these will be used by unkind and perhaps unfair critics to censure the agency and slow down its action program. This raises the question of whether or not an agency can design or construct water systems in accordance with agreed-upon objectives and simultaneously provide the community with full and unbiased intelligence for the purpose of reevaluating these objectives.

Professional standards and public objectives. We have seen that the design objectives are deduced from broad values of the community, and that the expertise of the river-planner lies in his ability to translate these objectives into specific plans for physical improvement of the environment. There is a danger, however, that the river-planner, who is generally a physically oriented professional, will conceive of the principles and goals of his profession as the objectives he should implement.

A familiar example is the attitude toward risk in the design of flood-control structures. This attitude should be derived from the objectives of the community, and river-planners should provide the community with useful information so that the community can decide among alternative objectives. A fully informed community might, for example, prefer to accept a 25 percent risk of a damaging flood to a 5 percent risk, if with the money saved by building smaller flood structures, it could erect a municipal auditorium. Yet planning engineers are inclined to conceal these alternatives under so-called technical principles which may, for instance, establish 5 percent risk as an engineering standard for flood-control design. Thereby the planners substitute the standards of their profession for public objectives. They do this with neither intent nor hypocrisy. It is the product of bias rather than conscious action, the bias stemming from the planner's preoccupation with physical development.[37]

Professional bias affects several stages in the design process, but none more sensitively than providing intelligence to the public. The question is whether or not special institutional arrangements are necessary to protect against the danger at this stage.[38]

Interest-group views and intelligence. We have said repeatedly that the early stages of water planning should evoke objective-oriented rather than interest-oriented debate. As planning moves in the administrative process to greater specificity, however, decisions will be seen progressively in the light of their impact on particular interests, and the water-development agency will become involved in adjusting or accommodating these interests. In an agency's intelligence activities, however, interests should not be accommodated, and this raises the question of whether or not an agency that is involved in accommodating interests for one purpose can resist involvement for a related purpose.

Let us suppose that an agency undertakes to provide intelligence in the form of a general reconnaissance of possible agricultural developments in a particular area. It assigns the job to its investigators in the appropriate area office. The investigators find that the area is currently being dry-farmed in very large and relatively efficient corporate operating units; that if irrigation is introduced, the area can produce more efficiently, but with very much smaller operating units. The area office, looking ahead to future assignments in designing and building what would appear to be a good water-resource system, wishes to present its findings in a way that will promote agreement rather than discord, so it consults the local interests with this in mind. The local organization representing the vast majority of landowners objects strenuously to the finding that small ownerships would be more efficient under irrigation; they know that if the government develops the water, it will enforce a long-standing policy

[37] This point is developed with great insight by David Z. Farbman, *The master plan* (Senior honors thesis, Harvard University, 1960).

[38] See York Willbern, "Professionalization in public service: too little or too much?," *Pub. Admin. Rev. 14*, 13 (1954); Maass and Radway, reference 28, pp. 191–192 and references cited therein.

which would deny them any "unearned increment" on the land when it is divided into smaller units. To accommodate this powerful interest, the area office modifies its proposed report by omitting any reference to the size of operating unit which would be efficient under irrigation. The report goes forward to the regional office with enthusiastic support from all important local interests. A similar process takes place in the regional office with respect to regional interests, and in the national office of the agency with respect to the national interests with which it deals regularly. When, finally, the intelligence is made available to the public, and to the legislature, and even to the chief executive, it is watered-down intelligence. The most significant facts in terms of the broad interests of the public as a whole have been accommodated out.[39]

The point is not that the local interests should be denied an opportunity to comment. Their views should be sought and recorded; they are important intelligence. But they should not be accommodated from the bottom up. The intelligence ladder, in other words, should be purer in this respect than the operations ladder, which is concerned with the final stages of design and with constructing and operating water systems.

Remedies proposed. To guarantee objectivity in one or more of the situations presented above, two types of proposals have been put forward in this country in recent years. One calls for organizational separation of data and intelligence activities from action programs. Thus, with respect to basic data, it is argued that "the necessity for objectivity in the collection, organization, and presentation [to the public] of the facts of national life [requires that] the functions of fact finding be clearly distinguished from activities involving the setting of social goals or the promotion of . . . programs."[40] With respect to intelligence on alternative objectives and designs, an independent board of impartial analysis is recommended.[41]

The second type of proposal calls for independent review boards to check on design agencies and ensure that they avoid professional bias,

[39] Harry H. Ransom makes a similar point with respect to the watering down of intelligence estimates in his *Central intelligence and national security* (Harvard University Press, Cambridge, 1958).

[40] U. S. Commission on Organization of the Executive Branch of the Government (First Hoover Commission), *Task force report on statistical agencies* (1949), p. 13. This report was prepared by the National Bureau of Economic Research, under the direction of Frederick C. Mills and Clarence D. Long.

[41] See, for example, the recommendation that a Board of Impartial Analysis "report to the President and the Congress on the public and economic value of project proposals" in U. S. Commission on Organization of the Executive Branch of the Government (First Hoover Commission), *Report on Department of the Interior* (1949), pp. 2–6.

undue accommodation of special interests, or excessive concern with design objectives of the moment in collecting and presenting data and intelligence.[42]

Organizational independence and review boards are not the only techniques for promoting objectivity in the government's conduct of its activities for informing the public. Others include making explicit the dangers of bias, techniques of internal organization and control, and above all, realization of the system of accountability, with its standard of breadth, described earlier in this chapter. Furthermore, if design activities are organized for an orderly progress from broad objective to specific project, an institutional environment alert to the intelligence needs of the community should result.

There can and should be considerable flexibility in the manner in which these several techniques are combined to achieve the requirement of objectivity. No rigid and precise answer is known to us, nor has one been found in the U. S. Government.[43]

DEGREE OF INCLUSIVENESS REQUIRED

We have said that responsibility to inform the community imposes certain unique requirements on executive agencies for water development. The most important of these was objectivity, and we have discussed it at length. Another relates to inclusiveness, which we shall treat briefly.

Should the water agency try to present to the community all information, and all sides, and all alternatives of a policy issue? There would be relatively few situations, we think, in which it should present only one

[42] The proposals for review boards relate to the design activities of water-resource agencies as much as to their responsibility for providing information to the community. See First Hoover Commission, reference 41, pp. 2–6 and 75–76; U. S. Commission on Organization of the Executive Branch of the Government (Second Hoover Commission), *Task force report on water resources and power* (1955), vol. 1, pp. 24–27 and 90–95; U. S. Presidential Advisory Committee on Water Resources Policy, *Water resources policy* (1955), p. 19.

[43] For basic-data agencies, compare development of the organization of the U. S. Geological Survey (Department of the Interior), Bureau of Labor Statistics (Department of Labor), and Bureau of the Census (Department of Commerce). In this connection, see the fascinating case study of the Bureau of Labor Statistics, "The attack on the cost of living index," by Katheryn Arnow, in Harold Stein, ed., *Public administration and policy development* (Harcourt, Brace, New York, 1952).

With respect to proposals for independence of basic-data activities, we should mention the difficulty of developing an operative distinction between basic and nonbasic data.

For an excellent study of the similar problems faced by agencies that seek to combine action programs and research programs, where the latter may give results which challenge the very reasons for the former, see Ashley Schiff, *Fire and water: scientific heresy in the Forest Service* (Harvard University Press, Cambridge, 1962).

side and one alternative. But to pretend to present *all* in these categories — that is, for government to presume to have a monopoly of facts or of the ways in which they can be interpreted and put together into programs of action — would endanger the democratic system. Legislative committees should hear outside, as well as government, witnesses; and discussion in the community should be based on information provided by the community's plural associations as well as by government.

BALANCING DATA AND ACTION

Questions of organization for the collection of basic data are related also to the problem of reconciling the sometimes conflicting interests of the data expert and those of the system-designer and dam-builder. The former seeks, understandably, to perfect his product; the latter, to get ahead with his job. If a designer were to ask a hydrologist if the streamflow data on River *X* were adequate for design and improvement of the river, the latter would probably respond that he needed ten more years of observations to be able to interpret certain inconsistencies in the statistical analysis of the record. If after ten years the designer were to return to his colleague and put the same question, he would probably get the same answer. On the other hand, if the designer were to proceed without any assurances of adequate data, he would probably devise a most inefficient system, more likely from overdesign of spillways or floodways resulting from excessive caution than from underdesign.

Any overweighting of the interests of the hydrologist will usually result in too little water-resource development, and any overweighting of the interests of the designer, in inefficient development. What is needed is a definition of the optimal balance between the two, and some institutional means for enforcing it. The former we have sought throughout this book, particularly in Chapter 12. The latter will be devised from among the techniques discussed under the requirement of objectivity.

IMPLICATIONS FOR WATER PLANNING OF THE DIVISION OF GOVERNMENTAL POWER

The concept of governmental division of power, especially areal division of power, is important in determining the administrative organization for system design, including the geographical and functional extent of design responsibilities that should be assigned to a single unit of government.

Before we apply this concept, however, we should specify the facts of river-system design that are relevant to the problem.

Unified responsibility in a single agency for planning the development of the water and related resources of a river system has been given so often as the basic principle for administrative organization that it is now a commonplace.[44] This principle, however, conceals a number of complex relationships, and it is important to sort them out. The extent to which unified responsibility is an institutional requirement for water-planning depends on technologic, economic, and political factors and on the state of the art of design.

SIZE OF RIVER SYSTEM AND THE REQUIREMENT OF UNIFIED RESPONSIBILITY

With the use of shadow prices and related techniques, one could theoretically break down the planning of a river system to a point where each dam or structure could be designed by a different planner. Because of the nature of the technologic or production function for water resources, however, such a scheme would be most inefficient. As we have seen in Chapter 3, the production function includes (1) a storage–yield function, which becomes so complex when more than one reservoir is involved that trial-and-error methods simulating the operation of the entire system with various designs and operating procedures are used; and (2) an outflow hydrology, which is complicated by complementarity of uses, so that many outputs are typically joint products. For these reasons the local marginal productivity of each structure, which each planner would equate to the shadow price under a decentralized planning procedure, does not reflect the real contribution of that structure to total system benefits, and this makes it impractical to plan each structure separately.

All of this is well known to water-users, although they express it less elaborately. In the western United States, for example, it is a maxim that each river needs a watermaster — a central authority for decisions on allocation of water. The popular expression, "one river, one problem" goes to the same point.

Thus, technologic factors argue against any great decentralization of the planning process, and instead, for centralized planning and design of both small and large river systems as single units. But here we must

[44] See, for example, reference 32, p. 65.

stop to take account of the art of system design. Adequate planning techniques are not capable of dealing with very large systems as single units, so some reconciliation must be achieved between the optimal size of unit dictated by planning method and the size dictated by technology.

Take as an example the Mississippi River. As shown in Fig. 15.1, it

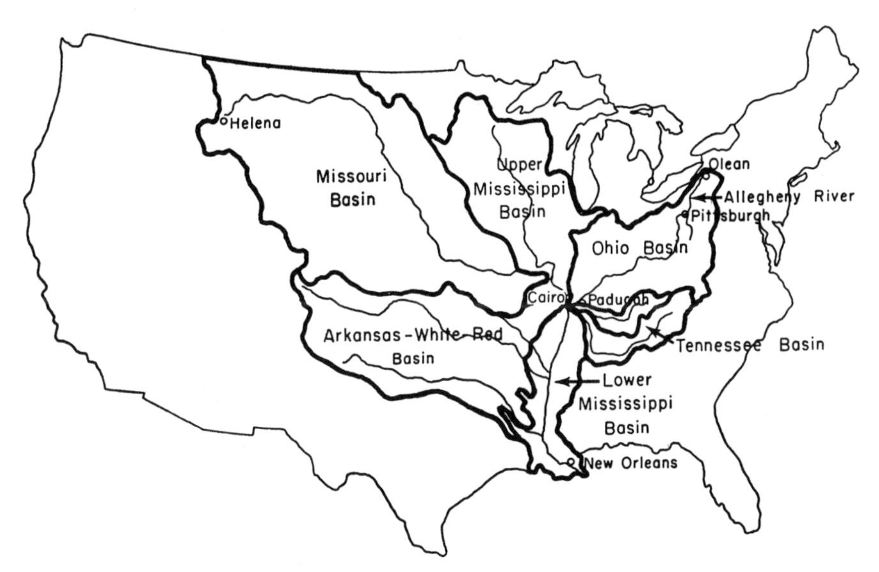

Fig. 15.1. Area drained by the Mississippi River and its tributaries.

drains about one-half of the continental United States, from the Rockies to the Appalachian Mountains, from Helena, Montana, to Pittsburgh, Pennsylvania, and Olean, New York. The Corps of Engineers' flood plan for the basin calls for more than 170 major reservoirs in the upper tributaries.[45] There is no question but that our design techniques — simulation or mathematical models — cannot be applied intensively to that many units for one purpose, to say nothing of the interrelationships among the several uses of storage at these and other reservoirs and river works on the Mississippi and its tributaries. But how important is it, from the viewpoint of technologic function, that the Mississippi be treated as a unit? Can it be segmented without significant loss in efficiency of development? What degree of segmenting is desirable to conform to the capabilities of planning method?

[45] Luna Leopold and Thomas Maddock, Jr. *The flood control controversy* (Ronald Press, New York, 1954), p. 52.

Fortunately, the technology of a very large basin like the Mississippi does not require that all units be studied simultaneously or equally intensively. For example, flood storage on the upper Missouri River in Montana, on the upper Allegheny in Pennsylvania and New York, or even on the upper Tennessee River will have little or no impact on damaging floods along the main stem of the lower Mississippi Valley. The effects of upstream dams decrease rapidly downstream, and large distances are involved here.[46] Similarly, water stored in the upper Missouri cannot create, or significantly improve the navigability of, a channel in the lower Mississippi. On the other hand, substantial flood storage at the mouth of the Tennessee River, as now provided in TVA's Kentucky Reservoir, can effect a significant reduction in flood stage at Paducah on the lower Ohio and at Cairo, where the Ohio joins the Mississippi, about 75 miles from the reservoir.[47] And a through 12-ft navigation channel could be provided from New Orleans to Pittsburgh by a combination of locks, storage, slack-water pools, and dredging and other forms of channel improvement.[48]

In planning water control on the main stem of the lower Mississippi, then, one would necessarily include in the system to be analyzed intensively Kentucky Reservoir and any other reservoirs that might be built near the mouths of the Ohio or Missouri Rivers. The effects on flood control and certain other uses in the lower Mississippi of reservoirs higher on the Tennessee, Ohio, and Missouri Rivers could with justification be approximated, or in some cases ignored. In planning the Tennessee River, on the other hand, a number of storage sites for flood control and other purposes (including Kentucky Reservoir) would be considered intensively, since they can contribute both to flood protection within the Tennessee Basin and, because of complementarity in the production function, to joint purposes as well.

The technologic unity of large systems like the Mississippi will therefore tolerate a significant amount of segmenting of the river for planning purposes, if provision is made for major overlapping units. Further, the technologic unity of such large systems will tolerate several degrees of

[46] Leopold and Maddock, reference 45, chap. 4, especially pp. 50–52.

[47] For TVA's most recent statement on the flood-control effects of its system, see U. S. Senate, 86th Cong., 1st sess., Select Committee on National Water Resources, *Flood problems and management in the Tennessee River Basin* (Committee Print No. 16, 1960), pp. 9–12. See also Water Resources Policy Commission, reference 33, vol. 2, pp. 722–724.

[48] A 12-ft channel actually has been built to Cairo and one to a controlling depth of 9 ft from there to Pittsburgh.

intensity in planning. Even if a combination of storages on each of the major tributaries of the Mississippi could in fact generate important effects on the main stem — and it has been suggested that this may be true for water quality in the lower Mississippi[49] — one could take this into account by skeletonizing each tributary in planning the lower river, that is, by conceptually aggregating the several related reservoirs on each tributary into one reservoir which approximates the same influence on downstream conditions.

If the technologic function will tolerate segmenting and skeletonizing in system planning, unified responsibility for the entire river system is not an institutional requirement of this function. There must be a plan for the larger unit, to be sure (a 9- or a 12-ft channel from New Orleans to Pittsburgh?), and it may be desirable for other reasons, including the governmental division of power, that a single agency have unified responsibility for preparing it, but the dictates of technologic function and of planning method do not make this essential. For these purposes it is sufficient that an agency with unified responsibility conduct the intensive integrated planning for each of the major segments, and that linking of the segments be ensured by institutional arrangements for coordination. We should warn the reader, however, that we do not have an adequate definition of major segment; the data necessary to define it have never been assembled.

It will be observed that not all large river systems are like the Mississippi. In some (the Nile, for example, or the Sacramento and San Joaquin considered as a single system), water is stored and transported to faraway points of consumption, or exchanged for other water which is consumed at those points. These rivers, which usually course arid areas and whose consumptive uses are of major importance, manifest technologic unity over great lengths. At the same time they have relatively fewer storage sites and relatively fewer important tributaries than rivers like the Mississippi. As are the Nile and the San Francisco in Brazil, these rivers are often long and narrow, and include large segments on which there are few if any important development possibilities.

In the planning of these rivers technologic unity is more impelling, and any significant amount of segmenting not justified. From the point of view of applying advanced design techniques, however, such segmenting might not be necessary. For the techniques could be applied intensively

[49] Edward Ackerman and George Löf, *Technology in American water development* (Johns Hopkins University Press, Baltimore, 1959), p. 503.

to the typically fewer tributaries and reservoir sites found on such systems.[50] Where there are important tributaries, or a considerable number of storage possibilities within a short stretch of river, skeletonizing might be justified.

UNITY OF RESOURCES AND THE REQUIREMENT OF UNIFIED RESPONSIBILITY

Many experts have carried the concept of unified responsibility or centralized planning beyond the river system to related resources in the valley.[51] This has been done for reasons derived from an interpretation of first, the economic and second, the technologic function of river-system planning.

Economically related activities. If a river is to be improved for navigation, river transportation facilities should be planned in the light of regional facilities for moving goods by rail and highway. If hydroelectric power is to be produced, it should be planned in the light of alternative and complementary sources of energy and of the regional system for distribution of electricity. If irrigation is to be introduced, it should be planned along with dry farming and ranching in the area.

From these relationships, some have concluded that unified responsibility should extend to the planning of regional activities which are closely related economically to the products of river development — that the agency which plans the river system should also plan the regional transportation, power, and agricultural systems. This conclusion does not necessarily follow. From the start river-system designers should have measures, which may be shadow prices, of the benefits of all possible purposes of development, and these should take account of alternatives, internal and external. The question is what agency should be responsible for preparing the benefit measures: the central budget agency, a regional budget agency, the central office of the water-planning agency, or the system-designers for the river? We shall discuss this matter below, but it is most unlikely that the answer will be the last alternative; yet it is the system-designers who exercise the unified responsibility which is related largely to the technology of water-resource development.

[50] For a good demonstration of the relative simplicity of the Nile in this respect, see H. A. W. Morrice and W. N. Allan, "Planning for the ultimate hydraulic development of the Nile Valley," *Proc. Instn. Civil Engrs. 14*, 101 (1959), especially Fig. 2.

[51] For a very readable statement of this view, see David E. Lilienthal, *TVA; democracy on the march* (Harper and Brothers, New York, 1944).

Physically related activities. Some believe that natural or ecologic — we should call them technologic — relationships between water and land resources are sufficiently potent to require that the agency which plans one plan the other. It is true under certain conditions that land-management practices can affect significantly the volume, timing, and quality of runoff which is available for use in the main channel of a river (for example, where siltation of reservoirs is a major problem, or where it is important to reduce runoff losses caused by vegetal-cover transpiration), but more often these practices are of minor consequence for the design of major river systems.[52] Also, the characteristics of the technologic function for land use and development are so different from those for water that it could be frustrating to combine the planning for both in a single agency. As Professor Charles McKinley has said:

"It requires a common master to organize and direct the intimate relationships . . . physically existing among [the] several uses [of 'water-in-the-channel']. Nothing like this complexity or intimacy of physical relationships ties one part of the land surface of a whole river watershed to all other parts. Timber on a local watershed can be liquidated without interfering with the practice of sustained yield on a neighboring tributary. It is usually possible to exercise good grazing control and maintain good ground cover in one pasture, even though across the fence ill management has destroyed the grass and produced a sagebrush waste. To be sure there are some problems of land management, such as fire control, pest or insect control, and some erosion situations that necessitate cooperation between the managers of adjacent land units. But even these situations requiring joint administrative effort are local and do not embrace whole river basin watersheds. The essential administrative unit for direct wild land management is the grazing district, the ranger district, or a minor tributary watershed. Each has its peculiarities of topography, soil composition, forest and ground cover, temperature and precipitation. Infinite variability in the land resource requires great variation in direct local management practices even while general social goals are pursued in common. . . . The immediate users of the land resources are the farmers for the 'tame lands,' the live-stock ranchers for the grazing and range lands and for the extensive forage resources in the timber of the National Forests, the loggers and saw-mill operators who buy and get out the timber, the hikers, campers, skiers, the hunters and fishermen who tramp the woods and mountains and whip the streams for recreation. While there is some overlap, on the whole the clientele and the

[52] "The concept of an organic unity of nature which compels a single administrative entity for both aspects of resources is only partly true and may be overworked," states Charles McKinley in "The valley authority and its alternatives," *Am. Pol. Sci. Rev. 44*, 618 (1950). See also Leopold and Maddock, reference 45. We do not refer here to irrigation practices and irrigation return flow, discussed on p. 602.

economic and social considerations of . . . land resource management differ markedly from those of the water-in-the-channel functions." [53]

Here again, although unified responsibility modified by the dictates of planning method is required by the technologic function for water-control planning, techniques of coordination are usually adequate for relating water planning to planning for land and other natural resources.[54]

What about planning the systems for distributing the products of water development — the main irrigation canals and water-supply conduits, and the backbone power-transmission lines? Does the principle of unified responsibility extend to these? Generally yes, although this will vary with local conditions. We have seen the importance of the return flow from irrigation canals for water-control planning. Similarly, the technical characteristics of an electric transmission system may have an integral functional relationship to generator capacity, and this in turn to water storage for power purposes. It should be noted that where the principle of unified responsibility is extended to the planning of distribution systems, the area for planning these systems may not correspond to what would otherwise be the area for water planning. Thus, the power-distribution area of the Tennessee Valley Authority extends beyond the Tennessee River drainage, and the distribution area for Delaware River water extends to New York City on the Hudson. In the case of major water diversions like the Delaware, it will often be advisable to consider the large tunnels and conduits as part of a technologically integrated river system. For power distribution we can introduce the planning concept of a river system with elastic boundaries — for any purpose for which it is necessary, the normal boundaries for planning water control can be stretched to comprehend distribution systems.

DIVISION OF GOVERNMENTAL POWER AND THE REQUIREMENT OF UNIFIED RESPONSIBILITY

The facts of river-system design relevant to administrative organization can be summarized as follows. Technologic and economic factors and the state of the art of system design require unified responsibility in a single agency for the intensive integrated planning of water in the channel

[53] McKinley, reference 52, pp. 617–618. Many observers do not agree with McKinley on this point. See, for example, Bernard Frank and Anthony Netboy, *Water, land, and people* (Alfred A. Knopf, New York, 1950).

[54] Are techniques of coordination adequate for planning fishery resources, especially the anadromous variety?

of river systems, or major segments of systems, which manifest considerable hydrologic unity. These factors, however, do not require unified responsibility for planning (1) water in the channels of large river systems with many tributaries, like the Mississippi, for which the technologic function will tolerate segmenting and skeletonizing; or (2) activities in a river basin economically and ecologically related to water control, provided values for benefits and costs which take into account both internal and external alternatives are furnished the system-designers. Furthermore, diversity of planning techniques may make it actually undesirable to assign to a single agency unified responsibility for planning land and water resources.

We have already said that the agency that plans a river system should have available from the beginning a set or alternative sets of design criteria, and that these should comprehend the manner in which benefits and costs, discount rates, and budgetary constraints are to be evaluated for system design. Because of the factors which should be taken into account in developing the criteria, many of which are outside of the interest and competence of the system-designers for any one river basin, and some of which are outside of the competence if not the interest of national water-planners, responsibility for translating objectives into design criteria should lie normally with a central budget or planning agency, working with the central office of the water-planning agency or agencies. Where, however, regional development is an objective, it may be desirable or necessary that a considerable part of this work be discharged by a regional budget or planning organization, working with regional water-planning agencies.

At the same time there should be close relations between those who develop the criteria and those who use them. A specific illustration is found in the procedure we have proposed in Chapter 4 for making budgetary constraints effective. A budget agency is to allocate the water-resource budget for all periods among all projects, by setting shadow prices to reflect the budgetary constraint. Because, however, the allocator's initial estimates may prove not to have effectuated the constraint when the designs are developed and their influences become known, there is an iterative process of estimate and revision involving both the allocator and the designer. On a more general level, the procedure we have proposed for a continuing evaluation of objectives and their design consequences involves an iterative process for relating the preparation of the criteria to their use.

To say that technologic and economic factors do not require unified responsibility in system design is not to say that related activities cannot be planned by the same unit (although we have raised a question concerning the desirability of combining all land and water planning). A government is largely free to combine or separate the planning of these resources, and its decision should be derived from analysis based on the concept of the division of governmental power. This analysis must be made independently for each institutional situation, for it will be remembered that the concept relates capital divisions of governmental power, areal divisions of governmental power, nongovernmental divisions of power, and community values. Based in part on such an analysis, some scholars have concluded that the TVA experience demonstrates the desirability of giving a single agency responsibility for planning a broad range of resources in an area.[55] Based on an analysis of institutional factors in the Columbia and Missouri River Basins, others contend that it would be unwise to assign to one planning agency responsibility for an extended spectrum of resource-development activities.[56] Having established that techniques of system design allow considerable scope for choice among alternative forms of administrative organization for design, and having indicated a technique of political analysis which can be used to choose among these, we shall carry the argument no further.

[55] See Lilienthal, reference 51. His view is based in part on an interpretation of technologic and economic factors which differs from ours. For a contrary conclusion on the TVA experience, see Philip Selznick, *TVA and the grass roots* (University of California Press, Berkeley, 1949).

[56] For the Columbia, see Charles McKinley, *Uncle Sam in the Pacific Northwest* (University of California Press, Berkeley, 1952); and for the Missouri, Henry Hart, *The dark Missouri* (University of Wisconsin Press, Madison, 1957), chap. 10.

INDEX

Index

INDEX

INDEX

INDEX

INDEX

INDEX

INDEX

INDEX

INDEX

INDEX

INDEX

INDEX

INDEX